Research Reports ESPRIT

Project 5341 · OSI95 · Volume 1

Edited in cooperation with the European Commission

A. Danthine   (Ed.)

The OSI 95 Transport Service with Multimedia Support

Springer-Verlag

Berlin Heidelberg New York
London Paris Tokyo
Hong Kong Barcelona
Budapest

Volume Editor

André Danthine
Université de Liège, Institut d'Electricité Montefiore, B28
B-4000 Liège, Belgium

ESPRIT Project 5341, OSI95 (High Performance OSI Protocols with Multimedia Support on HSLANs and B-ISDN) belongs to the Peripheral Systems, Business Systems and House Automation Sector of the second phase of the ESPRIT Programme (European Specific Programme for Research and Development in Information Technology) supported by the European Commission.

The goal of ESPRIT Project 5341, OSI95, was to design and formally specify a new transport service able to meet the requirements of the new applications and to offer the services available at network and sub-network levels. The project implied also the design and formal specification of a transport protocol.

Partners in the project are:
BULL (France), Alcatel Bell (Belgium), Alcatel ELIN (Austria), INRIA (France), Institut National des Télécommunications (France), Intracome (Greece), Olivetti Research Laboratories (UK), University of Lancaster (UK), University of Liège (Belgium), and University of Madrid (Spain).

CR Subject Classification (1991): C.2, D.2, H.5.1, F.4.3, B.4.3

ISBN-13:978-3-540-58316-5   e-ISBN-13:978-3-642-85165-0
DOI: 10.1007/978-3-642-85165-0

Publication No. EUR 15869 EN of the
European Commission,
Dissemination of Scientific and Technical Knowledge Unit,
Directorate-General Information Telecommunications, Information Market and
Exploitation of Research,
Luxembourg.

Typesetting: Camera-ready by the editor
SPIN: 10132184   45/3140-543210 – Printed on acid-free paper

Preface

The tremendous development of the communication environment with LANs, High Speed LANs and the ATM-based B-ISDN has opened the way for drastic changes in network performance and has introduced new services such as multicast service, synchronous service, and throughput guaranteed service, which are not available at the transport service level.

The new application environment, with its client-server paradigm and its multimedia orientation, has drastically changed the communication requirements when global networking is the ultimate goal.

In the communication architecture, the uniqueness of the transport service has been in the past an element of stability allowing the development of new applications on a well-defined service and, hidden from the application, the introduction of new technologies in the lower layers. However, the stability of the transport service is now becoming a drawback and it is felt that it is time to revisit this transport service and, consequently, the transport protocol. The main goal of ESPRIT Project 5341, OSI95 (High Performance OSI Protocols with Multimedia Support on HSLANs and B-ISDN) is to design and formally specify a new transport service able to meet the requirements of the new applications and to offer the services available at network and sub-network levels. A subsequent goal of the OSI95 project is to design and formally specify a transport protocol offering the newly defined transport service.

This report is structured in the following way. After an extended presentation of the new networking environment and the rationale for new standards, the various tasks and goals of the OSI95 project are presented. The sections of the book present the results of these various tasks in the following order:

- *The XTP study*
- *The requirements of the new applications*
- *The OSI95 transport service design*
- *ATM and the new communication environment*
- *Specific studies*
- *Time extension of LOTOS*
- *Standardisation.*

The following people served as reviewers:

P. Amer
T. Bolognesi
M. Diaz
S. Fdida
L. Ferreira Pires
R. Groz
P. Gunninberg
H. Hansson
R. Huis in't Veld
J.-P. Katoen
H. Kremer
T.F. La Porta
F. Y. Lai
D. Quartel
J. Quemada
J. Sifakis
K.J. Turner
G. Ventre
C. Vissers
I. Widya
C.L. Williamson
L. Wolf
L. Zhang
M. Zitterbart.

The editor and all the authors thank the reviewers for their valuable criticism and comments. They contributed to improve the quality and the presentation of the papers of this book.

The views and the judgments expressed in this book are solely those of the authors.

July 1994   *A. Danthine*

Table of Contents

The Networking Environment of the Nineties and the Need of New Standards

André Danthine
Professor at the University of Liège, Institut d'Electricité Montefiore, B28, B-4000 LIEGE, Belgium
Email: danthine@vm1.ulg.ac.be

The last ten years have seen a tremendous change in the communication environment. The changes in network performance have been at the origin of new protocols and of more efficient implementation methods. The changes in network service capabilities have raised the issue of offering, at the transport level, services available at the subnetwork or at the network level such as multicast service, synchronous service and broadband service. The changes in application requirements have introduced new requirements for communication. The Client/Server paradigm is looking for low latency, multimedia for low jitter and many applications are requesting bandwidth. Before describing the OSI95 project (ESPRIT II), this paper assesses the new networking environment of the nineties and introduces the need of new OSI standards for the transport service and for the transport protocol.

Keywords : High-speed networks, transport service, transport protocol, OSI, multimedia, broadband.

1 Introduction

The transport protocols TCP and TP4 have been designed in the late seventies at a time when the network environment was essentially based on point-to-point communications and when the packet switching was an emerging concept.

In ISO, the transport service was seen at the beginning as a unique service based on connection mode and, in this framework, one protocol class, TP4, was designed for the poor network service environment [Dan 90]. TP4, as well as TCP, were designed to be able to recover from the worst situation in terms of packet losses and packet disorders. It took several additional years to see an addendum to the transport service able to handle connectionless mode [ISO 8073 ADD2].

In the last ten years, we have seen a tremendous change in the communication environment. The LANs have drastically changed the scene and the high-speed LANs have accelerated the trend. More recently, the MANs have opened new possibilities not to mention the broadband ISDN which is expected to be deployed starting in the middle of this decade.

Not only the communication environment has drastically changed during the eighties, but we have also faced a drastic change in application requirements of which we will mention only the client-server paradigm and the multimedia-based applications.

It is this changing environment which is at the origin of ESPRIT II Project, OSI95[1], started in October 1990. In this paper we will explain why we believe that we need a new OSI standard for the transport service and for the transport protocol. Our opinion is based on the analysis of the consequences of the changes we have been facing
- in network performance
- in network services
- in application requirements.

2 Changes in Network Performance

Let us first point out the robustness of the OSI Reference Model which has been able to integrate the LANs, at least partially, in its layered architecture [Dan 89] and let us hope that the OSI RM will be able to integrate the B-ISDN.

The basic contributions of the LANs have been a drastic improvement of the network performance in terms of data rate as well as in terms of bit error rate (BER) and packet error rate. The error recovery was not any more a basic goal of the DL layer and this explained the tremendous success of the connectionless-mode.

The LANs have not only introduced an increase of the global communication capability due to the high value of their data transmission rate but, more important, they have offered the user an access data rate several orders of magnitude greater than what was available before.

If, as usual, the first implementations of the LAN controllers were not always able to offer the maximum access rate theoretically available at the MAC level [DHH 88a], the market, in about five years, was able to offer well-designed and well-implemented LAN interfaces.

It did not take long to realise that an access rate available at the MAC or LLC levels was far from being available above the transport layer [DHH 88a][Svo 89a][HeS 89][CCC 91]. This is not surprising, because opening a highway does not by itself change drastically the speed of the bicycles.

2.1 How to Improve ?

From the observations made in the second half of the eighties about the transport performance on top of LANs, detailed studies of mechanisms and implementations have been done [CJR 89], [GKW 89], [Mei 91], [JSB 90], [Zitt 91] and ended up in two different schools of thought.

The first school advocated the design of new transport protocols such as NETBLT [CLZ 87], VMTP [ChW 89], SNR NRS 90] and XTP [Ches 89], [Wha 89], [PE 91].

The second school is putting the burden of the poor transport performance to the bad implementation quality and is claiming the capability of existing or slightly

---

[1] OSI 95 is the acronym of ESPRIT II Project 5341 "High Performance OSI Protocols with Multimedia Support on HSLANs and B-ISDN".

modified transport protocols to better handle the performance available at the MAC level [CJR 89].

Both schools have arguments and we would like here to discuss them, keeping in mind that these LANs do not only offer a better access rate, they are also characterised by a very low probability of corrupted packets by transmission errors.

2.2 Mechanisms and Implementations

It is clear that protocol mechanisms must be adapted or modified when the network service is drastically changing. The following example will be sufficient to illustrate this fact.

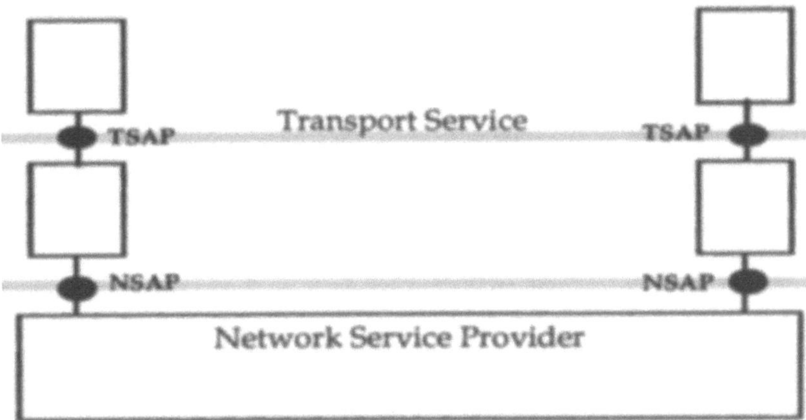

Fig.. 2.1 The basic OSI reference model

The figure 2.1 represents a simplified view of the basic OSI reference model and we assume, on the NSAP-to-NSAP association, a guaranteed throughput and a fixed round trip time. On top of such an well-defined N-service, we assume an ideal implementation of the transport entities which does not introduce any latency to process TSDUs into NSDUs. This does not mean however that the throughput and the round-trip delay on the TSAP-to-TSAP association, say a transport connection, will be equivalent to the values at the network level.

The product of "throughput x round-trip delay" - also called "bandwidth x delay" - is becoming, with high speed networks, an important characteristic to be dealt with. The value of such a product has evolved from 1 Kb (2 Kbps x 500 msec) to 6 Mb (100 Mbps x 60 msec), not to mention the very high speed environment. If the window associated with a transport connection does not allow the pipelining of enough megabits, the transport entity will be obliged to stop transmitting by lack of acknowledgements and of flow control credits. Therefore the performance at the transport level will be much lower than the performance at the network level even with an ideal implementation of the transport layer.

On the other hand, if protocol mechanisms may have such an influence on the performance even with an ideal implementation, it is also true that some mechanisms may induce very poor implementations. For instance a time-out per packet is not the best mechanism from an implementation point of view.

With a transport entity trying to send as many packets as possible, a lot of ACKs to handle represents a burden and the classical ARQ mechanism has therefore to be reviewed to reduce the penalties on performance.

Last but not least, there remain implementation problems which are independent of any protocol mechanism such as buffer handling and interaction with the operating system [CJR 89], [MaB 91].

There exist of course many choices of implementation from VLSI to pure software, more or less integrated in the operating system, using additional dedicated computing power or not [GKW 89], [JSB 90], [Zitt 91] .

The syntaxes of the PDU are not neutral with respect to the implementation results. Variable fields may save bandwidth but are more difficult to "siliconize". The place of the CRC field in the PDU is also important if hardware implementation with insertion and removal "on the fly" is envisaged.

With the availability of microprocessors of more than 10 Mips, with the possibility to use parallelism in the implementation of the transport entities, with the possible use of dedicated chips, we have a wide variety of possible implementations of the transport entities, all of them attempting to offer a transport service with a minimum drain on the host capabilities.

The conclusion to this section on the link between mechanisms and implementations is that even if it is true that the implementation of the protocol mechanisms is not always the most time consuming part of the transport activity, it will not be bad to have mechanisms which are well- aligned with the objectives and services of the protocol and which will ease the implementation problems. This does not prevent of course a careful choice of the syntaxes of the PDUs and an in-depth analysis of the memory management.

2.3 The Congestion Problem

During the eighties, we have not only seen the widespread usage of the LANs and recently of the high-speed LANs. We have also seen the building of very big internet-based networks following the efforts supported by the US DoD.

Such an internet network may give rise to very difficult problems of congestion which are almost impossible to tackle with the error recovery mechanism which has been introduced in TCP and TP4. The correcting action of the transport protocol in order to recover the lost packets due to the congestion is based on retransmission on time-out. These retransmissions contribute to an increase of the offered load and therefore to increase of the congestion problems.

This is why congestion avoidance mechanisms which try to operate the network at the knee of the response time curve [RaJ 88], [ChJ 89] have to be preferred to the congestion control approach. Time-out retransmission must be associated with a reduction of the load on the network [Jai 86].

In the congestion problems, we have also two possible viewpoints. Either the mechanisms are aiming at the protection of the provider of the service, i.e. the network itself, in order to avoid a complete collapse of communication capability. Either, they are aiming at offering, to a service user, the best service possible at any given time. We do not have today a clear trend between these two views.

Congestion control may be exercised by the network service user, and the slow-start method in TCP [Jac 88] is the best known way to do it, or it may be exercised by the network service itself through some kind of rate control.

In any case, we have to deal with some reduction of the rate exercised either by the network service user or by the network service provider. It is very difficult today to have a clear view of the best solution. There exist a lot of proposals based on different ways to collect the congestion information. The congestion control may indeed be exercised on the basis of the information collected by the user through the observations of the behaviour of the service provider. As measurement bases already proposed, we have the rate of packet loss, the increase of round-trip delay, the occurrence of a timeout on an unacknowledged packet. Another proposal intends to base congestion control on information provided by the network in the form of loss-load curves [WiC 91]. The provider of service may also be involved by setting a special bit in packets going through a congested area. This piece of information is used to report internal congestion to the transport recipient which is able to act on the flow control mechanism in order to ease the congestion problem [RaJ 88], [RaJ 90].

The congestion problem is much more difficult to handle with a connectionless network service and in [Zha 87b], it has been proposed to introduce, on a network association, the idea of a flow which will be used to handle a stream of packets. Setting up a flow, transmitting on a flow and terminating a flow appear similar to what we do for a connection but this flow is used only to handle resources on the way and is not associated with any service property.

The problem we have in the Internet today is not related to a high speed or high throughput environment. But the deployment of corporate communication networks will give rise to corporate internet networks where one will have to face the same problem due to congestion of bridges or routers. The high-speed environment will just make things a little more difficult. If the high throughput goal requires not to use time-out on data packets, it will be impossible to base congestion control on time-out occurrences.

Increasing the resources to avoid packet loss due to congestion may be done by increasing the buffering capability of the intermediate nodes, but this will drastically increase the round-trip delay and may not be considered as a good solution [Jai 90].

The QoS (Quality of Service) in OSI as well as the TOS in IP are specified in qualitative terms and are completely useless for the congestion control or the congestion avoidance.

For LANs, the probability that a frame be corrupted by a transmission error is very, very low, but the probability of packet loss due to congestion in the access device, to a host, is not to be neglected. With integrated LANs through bridges and routers, the problem may get more complex and the congestion may be responsible for packet losses which may jeopardise the quality of the service.

3 Changes in Network Service Capability

If it has been possible to integrate the basic functionalities of the LANs in the OSI Reference Model, there exist subnetwork services which are available neither at the network level nor at the transport level.

3.1 Multicast Service

One of these services is related to the capability of addressing not only a given service access point, but a group of SAPs through multicast or all SAPs through broadcast. As we will see in the next section, many applications have a direct interest to see this multicast service not limited to the subnetwork level but offered at the network level as well as at the transport level. Such an extension of service means clearly that it will be difficult to proceed by patching existing transport services and protocols.

3.2 Synchronous Service

In FDDI, we have two types of services : the asynchronous service and the synchronous service. The synchronous service offered by FDDI I is a service where the average throughput is guaranteed as soon as it has been reserved.

The stream of bits coming from a constant rate source or from a variable rate source may be packetised and sent through an association offering a jitter on latency which does not compromise the re-synchronisation at the acceptor [DRB 87]. This approach was applied in 1988 with the BWN, a 140-Mbps FDDI-like network developed in the framework of the ESPRIT project 73 [DHH 88b]. Videoconferencing with 2-Mbps codecs has been demonstrated on the connectionless BWN and this system was able to operate with a network loaded with more than 100 Mbps of background traffic. With FDDI and its synchronous service such an approach may be possible even with a higher rate and with a guaranteed average throughput. However such a synchronous service may not be offered today neither at the network service level nor at the transport service level.

3.3 Broadband Service

For the B-ISDN, even if the pace of the standardisation of the AAL (ATM Adaptation Layer) is slow, one may guess that, sooner or later, the AAL Type 4 will be extended to offer a service identical or comparable to SMDS or DQDB. It is not excluded that this connectionless service might be associated with a guaranteed average rate.

For AAL Type 1, the situation is not enough advanced to make any guess, but an isochronous service on SDU of more than 48 bytes may not be excluded and, if it happens, it is very likely to be associated with a guaranteed average or peak rate or to something in between.

To conclude this section, it seems necessary to be able to introduce in all the lower layers of the OSI Reference Model, better addressing capabilities such as multicast, as well as to find a way to link the throughput into the QoS not only in qualitative terms but in quantitative terms.

4 Changes in Application Requirements

Beside the classical applications such as file transfers and conversational remote access to databases, the eighties have seen new orientations in the applications and these new applications have introduced new requirements for communication.

4.1 Client-Server Paradigm

The Client-Server paradigm is one of the important changes which took place during the last decade. From a communication point of view, it implies a low latency and an extended addressing capability.

The search of a server may take place by broadcasting or multicasting a request and the blocking character of the request implies the requirement for a low latency. This low latency is difficult to achieve when using the complete OSI stack. A system such as ANSA [Her 88] has chosen a collapsed OSI architecture based on a convergence sublayer, located in the lower part of the layer 7, acting as a transport service and giving access to the SAP of the subnetwork at the layer 2 interface. Others systems have developed better focused transport service and protocol [ChW 89],[MaB 91].

Today, most of the client-servers are implemented on a local basis, but the clear trend to use this new paradigm at the corporate network level will require a re-evaluation of the architecture presently used. If we hope to see the distributed computing systems integrated in the OSI architecture, it is mandatory for the transport service to offer the needed addressing capabilities as well as a transaction-oriented service with low response time as requests are most often of the blocking type.

4.2 Graphics and Colours

We have seen a constant increase in the power of the microprocessors used for the workstations and PCs but an analysis of the evolution puts into evidence that quite an important part of the power has been invested in the user interface and in a more advanced form of presentation of the results to the users. Multiple windowing, graphics usage as well as use of colours have fundamentally changed the friendliness of the workstations and paved the way for the widespread usage of them.

Of course, the volume of data which has to be exchanged in interactions involving windows, graphics and colours, is much more important than for the classical ASCII pages. The user, however, is not ready to accept to have long delays associated with the dynamic of the data presentation. Therefore the higher volume of data to be exchanged through the communication system on one side and the need to achieve what may be called the "client expected latency" on the other side mean that we will have a request for a high data throughput with a still very important bursty character.

It is only if the global network service is offering the high value of the throughput associated with an acceptable latency that this kind of system will be usable on a wide area.

4.3 Multimedia

By multimedia we mean, as usual, a mixing of text, voice, graphics, image, audio and video. It is however important to make the distinction between two types of multimedia communications.

The videophone is a multimedia terminal as it is able to handle voice and video. But here the terminal is a source of a stream of bits which is very likely to be carried out on a physical or on a virtual circuit from one equipment to another equipment of the same type. The complete OSI architecture is very unlikely to be used as the information exchanged is not represented in machine-processable form, able to be processed, stored, retrieved and disseminated at any time.

The workstation where we will have the text in one window, a fixed image in another window and a video in a third one completed by some audio is an example of equipment requiring the second type of multimedia communication. Here all information which may be processed, stored, retrieved and disseminated are in machine-processable form. Here the video part of such an equipment will be seen as a sequence of data structures which have to be submitted to the presentation device at a rate compatible with the type of presentation we are looking for. As indicated earlier, such a video may be transmitted as a sequence of video frames, each frame being divided into packets at the transport level to be reconstructed at the remote end. The correct reconstruction requires from the transport service a sufficient throughput and a jitter on packet latency which is below some fixed value. Unfortunately, nothing today allows a user of TP4 or of TCP to request a quantitative value for the throughput on its connection, nor to express a limit to the jitter associated with the TSDU transport delay.

With video, the error recovery by retransmission does not work properly as it conflicts with the low jitter requirement. Such a data stream is not flow-controllable and if the transport service is not able to provide the required throughput, it will be reasonable to cancel the connection.

Most of today's multimedia applications are implemented on a stand-alone workstation. The trend to expand on network in the local area is already there and if some existing systems are working properly, it is very often because the transport service is able, through best effort, to offer a throughput sufficient for the application. This is true for an application implemented on a lightly loaded LAN, but may become less evident when the LAN will be more heavily loaded, when the LAN will be built with bridges and routers and, last but nor least, when we will try to have this application running on the wide area internet network. Here, we will face, as we go along these different environments, an increase of the latency and a possible reduction of the throughput.

All applications mentioned in this section require the possibility to associate, to a transport connection, quantitative compulsory QoS with regard to the throughput, transit delay and delay jitter, and the service provider may have to intervene on the connection if the performance is not fulfilled.

5 Need for a New Transport Service

In the previous sections, we tried to stress that the improvement of the transport service and protocol does not lie only on the performance, but that it will be necessary to extend the characteristics of the offered services. This opinion was already expressed in [ChW,89] to advocate a new generation of communication systems : *"By 'new generation' we mean that the design of protocols, networks and network interfaces should be rethought from the first principles in the context of the new environment. Refining existing designs and tuning conventional approaches in the area are inadequate, given the magnitude of changes that have taken place. At the same time, we need to be extremely careful to identify the key problems of the current standard protocols and come up with a convincing solution that is significantly better than these protocols. That is, we clearly recognize that an incremental improvement over existing protocols, such as the stream-oriented TCP or ISO TP4, would not warrant a new protocol or modifications to the existing protocols, given the enormous investment already in place."*

The design of a new transport service and a new transport protocol, is the basic goal of the ESPRIT Project OSI95 which started in October 1990. The consortium headed by BULL involved Alcatel Bell, Alcatel Austria ELIN, INRIA, Institut National des Télécommunications, Intracom, Olivetti Research, as well as the universities of Madrid, Lancaster and Liège and will be described in more details in the next paper.

6 Conclusion

We hope to have shown the very important need for specifying new transport services and new transport protocols based on new mechanisms. In the next paper we will describe the OSI95 project whose goal was to contribute to fulfil this need.

References

[Bie 90]   Biersack E.W., Feldmeier D.C., **Transport Protocol Issues for ATM-based Networks,** *EFOC/LAN 90,* Munich, June 27-29, 1990, pp. 104-113

[But 89]   Butscher B., **A Flexible Transport Service in the BERKOM Broadband Environment,** *Protocols for High-Speed Networks,* Zurich, May 9-11, 1989, Rudin H. & Williamson R. Ed., Elsevier (North Holland), pp. 289-301

[CCC 91]   Comer M.H, Condry M.W., Cattanach S., Campbell R., **Getting the Most for Your Megabit,** *Computer Communication Review,* 1991, Vol. 21, nr 3, pp. 5-12

[ChA 90]   Chow C-H, Adachi M., **Achieving Multimedia Communications on a Heterogeneous Network,** *IEEE Journal on Selected Areas in Communications,* 1990, Vol. 8, nr 3, pp. 348-359

[Ches 89]   Chesson G., **XTP/PE Design Considerations,** *Protocols for High-Speed Networks,* Zurich, May 9-11, 1989, Rudin H. & Williamson R. Ed., Elsevier (North Holland), pp. 27-32

[ChJ 89]   Chiu D-M, Jain R., **Analysis of the Increase and Decrease Algorithms for Congestion Avoidance in Computer Networks,** *Computer Networks and ISDN Systems,* 1989, Vol. 17, nr 1, pp. 1-14

[ChW 89]   Cheriton D.R., Williamson C.L., **VMTP as the Transport Layer for High-Performance Distributed Systems,** *IEEE Communications Magazine,* 1989, Vol. 27, n° 6, pp. 37-44

[CJR 89]   Clark D.D., Jacobson V., Romkey J., Salwen H., **An Analysis of TCP Processing Overhead,** *IEEE Communications Magazine,* 1989, Vol. 27, nr 6, pp. 23-29

[CIT 90]   Clark D.D., Tennenhouse D.L., **Architectural Considerations for a New Generation of Protocols,** *SIGCOMM 90,* Philadelphia, Sept. 24-27, 1990, *Computer Communications Review,* 1990, Vol. 20, nr 4, pp. 200-208

[CLZ 87]   Clark D.D., Lambert M.L., Zhank L., **NETBELT : A Bulk Data Transfer Protocol** Network Working Group RFC 998, March 1987

[Dab 89]   Dabbous W.S., **On High Speed Transport Protocols,** *Protocols for High-Speed Networks,* Zurich, May 9-11, 1989, Rudin H. & Williamson R. Ed., Elsevier (North Holland) pp. 135-141

[Dan 89]   Danthine A., **Communication Support for Distributed Systems - OSI versus Special Protocols,** *Proceedings of the 11th World Computer Congress on Information Processing,* San Francisco, 28 August-1 September 1989, pp. 181-190

[Dan 90]   DANTHINE A., **10 Years with OSI,** *INDC-90,* Lillehammer, Norway, 26-29 March 1990,(Ed. : D. Khakhar, F. Eliasen), Elsevier (North Holland), 1990, pp. 1-16

[DDK, 90]   Doeringer W.A., Dykeman D., Kaiserwerth M., Meister B.W., Rudin H., Williamson E., **A Survey of Light-Weight Transport Protocols,** *IEEE Transactions on Communications,* 1990, Vol. 38, nr 11, pp. 2025-2039

[DeC 90]   Deering S.E., Cheriton D.R., **Multicast Routing in Datagram Internetworks and Extended LANs,** *ACM Transactions on Computer Systems,* 1990, Vol. 8, nr 2, pp. 85-110

[DHH 88a]   Danthine A., Henquet P., Hauzeur B., Constantinidis C., Fagnoule D., Cornette V., **Access Rate Measurement in the BWN Environment,** Proceedings of High Speed Local Area Networks II, Liège, 14-15 April 1988, Elsevier Science Publishers BV(North Holland), 1990, pp. 89-115

[DHH 88b]   Danthine A., Hauzeur B., Henquet P., Constantinidis C., Fagnoule D., Cornette V., **Corporate Communication System by LAN Interconnection,** *EURINFO 88 - Information Technology for Organisational Systems,* Athènes, 16-20 May 1988 Bullinger H.J. & al., Ed., Elsevier (North Holland), 1988, pp. 315-326

[DRB 87]   De Prycker M., Ryckebusch M., Barry P., **Terminal Synchronization in Asynchronous Networks,** *IEEE Conference on Communications,* Seattle, 1987

[Fel 90b]   Feldmeier D.C., Biersack E.W., **Comparison of Error Control Protocols for High Bandwidth-Delay Product Networks,** *2nd IFIP Int. Workshop on Protocols for High-Speed Networks,* Palo Alto, Nov. 90, Johnson M., Ed., Eleesevier (N.Holland), pp. 1-20

[Fer 90a]   Ferrari D., Verma D.C., **A Scheme for Real-Time Channel Establishment in Wide-Area Networks,** *IEEE Journal on Selected Areas in Communications,* 1990, Vol. 8, nr 3, pp. 368-379

[Fer 90b]   Ferrari D., **Client Requirements for Real-Time Communication Services,** *IEEE Communications Magazine,* 1990, Vol. 28, nr 11, pp. 65-72

[GKW 89]   Giarrizzo D.,Kaiserwerth M., Wicki T., Williamson R.C., **High-Speed Parallel Protocol Implementation,** *Protocols for High-Speed Networks,* Zurich, May 9-11, 1989, Rudin H. & Williamson R. Ed., Elsevier (North Holland), pp. 165-180

[Her 88]   Herbert A., **Communications Aspects in ANSA**, *Computer Standards & Interfaces*, 8 (1988) pp 49-56

[Hes 89]   Heatley S., Stokesberry D., **Analysis of Transport Measurements Over a Local Area Network**, *IEEE Communications Magazine*, 1989, Vol. 27, nr 6, pp. 16-22, also in *Protocols for High-Speed Networks*, Zurich, May 9-11, 1989

[HSS 89]   Hehmann D.B., Salmony M.G., Stuttgen H.J., **High-Speed Transport Systems for Multi-Media Applications**, *Protocols for High-Speed Networks*, Zurich, May 9-11, 1989, Rudin H. & Williamson R. Ed., Elsevier (North Holland), pp. 303-321

[ISO 8073 ADD2]   ISO/IEC JTC1, **IPS-OSI - Connection-Oriented Transport Protocol Specification, Addendum 2: Operation of Class 4 over Connectionless Network Service**, ISO IS 8073 ADD2, 1989.

[ISO 8073bis, 91] ISO/IEC JTC1/SC6, Revised Text for ISO/IEC DIS 8073 —IT— Telecomm. and Inform. Exch. between Syst. — **Connection-Oriented Transport Protocol Specification** (4B42R), ISO/IEC JTC1/SC6 N 6776, October 7, 1991.

[Jac 88]   Jacobson V., **Congestion Avoidance and Control**, *SIGCOMM 88*, Stanford, Aug. 16-19, 1988, pp. 314-329

[Jai 86]   Jain R., **A Timeout-Based Congestion Control Scheme for Window Flow-Controlled Networks**, *IEEE Journal on Selected Areas in Communications*, Vol. SAC-4, nr 7, Oct. 1986, pp. 1162-1167

[Jai 90]   Jain R., **Congestion Control in Computer Networks : Issues and Trends**, *IEEE Network Magazine*, 1990, Vol. 4, nr 3, pp. 24-30

[JSB 90]   Jain N., Schwartz M., Bashkow T.R., **Transport Protocol Processing at GBPS Rates**, *SIGCOMM 90*, Philadelphia, Sep. 24-27, 1990, *Computer Communications Review*, 1990, Vol. 20, nr 4, pp. 188-199

[LaS 91]   La Porta T.F., Schwartz M., **Architectures, Features, and Implementation of High-Speed Transport Protocols**, *IEEE Network Magazine*, May 1991, pp. 14-22

[LeB 91]   Le Boudec J.Y., **About Maximum Transfer Rates for fast Packet Switching Networks**, *SIGCOMM 91*, Zurich, Sept. 1991, pp. 295-304

[Lid 90]   Lidinski W.P., **Data Communications Needs**, *IEEE Network Magazine*, 1990, Vol. 4, nr 2, pp. 28-33

[Mab 91]   Mafla E., Bhargava B., **Communication Facilities for Distributed Transaction-Processing Systems**, *Computer*, 1991, Vol. 24, nr 8, p. 61

[Mei 91]   Meister B.W., **A Performance Study of the ISO Transport Protocol**, *IEEE Transactions on Computers*, 1991, Vol. 40, nr 3, pp. 253-262

[NRS 90]   Netravali A.N., Roome W.D., Sabnani K., **Design and Implementation of a High-Speed Transport Protocol**, *IEEE Transactions on Communications*, 1990, Vol. 38, nr 11, pp. 2010-2024

[NoC 89]   Nordmark E., Cheriton D.R., **Experiences from VMTP : How to achieve low response time**, *Protocols for High-Speed Networks*, Zurich, May 9-11, 1989, Rudin H. & Williamson R. Ed., Elsevier (North Holland), pp. 43-54

[Par 90]   Partridge C., **How Slow Is One Gigabit Per Second ?**, *Computer Communications Review*, 1990, Vol. 20, nr 1, pp. 44-64

[PE 91]   Protocol Engines, **XTP Protocol Definition — Revision 3.6 — Editor's Second Draft**, November 4, 1991.

[RaJ 88]   Ramakrishnan K.K., Jain R., **A Binary Feedback Scheme for Congestion Avoidance in Computer Networks with a Connectionless Network Layer**, *SIGCOMM 88*, Stanford, Aug. 16-19, 1988, pp. 303-313

[RaJ 90]   Ramakrishnan K.K., Jain R., **A Binary Feedback Scheme for Congestion Avoidance in Computer Networks**, *ACM Transactions on Computer Systems*, 1990, Vol. 8, nr 2, pp. 158-181

[Rei 91]   Reisman S., **Developing Multimedia Applications**, *IEEE Computer Graphics and Applications*, 1991, Vol. 11, nr 4, pp. 52-56

[RPC 1077, 88] Gigabit Working Group, **Critical Issues in High Bandwidth Networking**, Leiner B., Ed., Nov. 1988, RPC 1077

[RSV 90)   Ramachandran U., Solomon M., Vernon M.K., **Hardware Support for Interprocess Communication,** *IEEE Transactions on Parallel and Distributed Systems,* 1990, Vol. 1, nr 3, pp. 318-329

[SJM 90]   Svobodova L., Janson Ph.A., Mumprecht E., **Heterogeneity and OSI,** *IEEE Journal on Selected Areas in Communications,* 1990, Vol. 8, nr 1, pp. 67-79

[Ste 90]   Steinmetz R., **Synchronization Properties in Multimedia Systems,** *IEEE Journal on Selected Areas in Communications,* 1990, Vol. 8, nr 3, pp. 401-412

[Svo 89a]   Svobodova L., **Measured Performance of Transport Service in LANs,** *Computer Networks and ISDN Systems,* 1989, Vol. 18, nr 1, pp. 31-45

[Svo 89b]   Svobodova L., **Implementing OSI Systems,** *IEEE Journal on Selected Areas in Communications,* 1989, Vol. 7, nr 7, pp. 1115-1130

[Wat 89]   Watson R.W., **What is Unique about Transport Protocol Design for High Performance Networks?,** *Protocols for High-Speed Networks,* Zurich, May 9-11, 1989

[Wha 89]   Whaley A.D., **The XPRESS Transfer Protocol,** *14th Conference on Local Computer Networks,* Minneapolis, Oct. 10-12, 1989, pp. 408-414

[WiC 91]   Williamson C.L., Cheriton D.R., **Loss-Load Curves : Support for Rate-Based Congestion Control in High-Speed Datagram Networks,** *SIGCOMM 91,* Sept. 1991, pp. 17-23

[WoK 90]   Woodruff G.M., Kositpaiboon R., **Multimedia Traffic Management Principles for Guaranteed ATM Network Performance,** *IEEE Journal on Selected Areas in Communications,* 1990, Vol. 8, nr 3, pp. 437-446

[Zha 86]   Zhang L., **Why TCP Timers Don't Work Well,** *SIGCOMM 86,* pp. 397-405

[Zha 87a]   Zhang L., **Some Thoughts on the Packet Network Architecture,** *ACM Computer Communication Review,* 1987, Vol. 17, nr 1-2, pp. 3-17

[Zha 87b]   Zhang L., **Designing a New Architecture for Packet Switching Communication Networks,** *IEEE Communications Magazine,* Sept. 1987, Vol. 25, nr 9, pp. 5-12

[Zitt 91]   Zitterbart M., **High-Speed Transport Components,** *IEEE Network Magazine,* 1991, Vol. 5, nr 1, pp. 5

[ZSC 91]   Zhang L., Shenker S., Clark D.D., **Observations on the Dynamics of a Congestion Control Algorithm : The Effects of Two-Way Traffic,** *SIGCOMM 91,* Zurich, Sept. 1991, pp. 133-147

The OSI95 Project

OSI 95 is the acronym of ESPRIT II Project 5341 "High Performance OSI Protocols with Multimedia Support on HSLANs and B-ISDN".

After the presentation of the consortium, the various tasks of this project are described in the order the contributions will be presented in this book.

Keywords : OSI95, ESPRIT

1 Introduction

It is the changing networking environment discussed in the previous paper [Dan 93] which was at the origin of the ESPRIT II Project, OSI95, started in October 1990 with the initial idea of 1995 as the deadline for its total achievement. The project ended up in February 1993 and it is this first phase that will be presented in detail here.

The consortium was headed by BULL(France) and involved, in alphabetical order, Alcatel Bell (Belgium), Alcatel ELIN (Austria), INRIA (France), Institut National des Télécommunications (France), Intracom (Greece), Olivetti Research Laboratories (UK), as well as the universities of Lancaster (UK), Liège (Belgium) and Madrid (Spain). The technical direction of the Project was assumed by the University of Liège.

The goal and the central task of this project was the design and the formal specification of a new transport service and of a new transport protocol.

This activity was supported by two other main tasks. The first one was the definition of the specific requirements coming from the distributed multimedia applications and ODP systems which may have an impact on the new transport service. The second one was related to the data link layer services provided by ATM networks, in order to identify new underlying network service facilities.

Three other tasks were also included in the project. One was related to an in-depth study of the XTP protocol introduced in 1989 and in development at the start of OSI95 with the ultimate goal of producing a chipset. The second task was related to the study of the congestion control, a important mechanism of the transport protocol, and to a preliminary study of the VLSI implementation of this protocol. The third one was related to an extension of the formal specification language LOTOS to handle the time related aspects of the protocol.

Last but not least, the project included a task aiming at the standardisation of its results by the creation of a new work item in ISO.

We will now describe in more detail the different tasks and goals, not in the order mentioned in this introduction but in the order the contributions will be presented in this book.

2. The XTP Study

Even if the intention was to work within the framework and spirit of OSI, which the partners feel will accommodate the new developments it was essential to take full account of all relevant research that has taken place outside of OSI, such as the work on new transport protocols. This involves analysing the features of recent transport protocols, particularly the currently proposed XTP (eXpress Transfer Protocol) protocol. The XTP protocol has been derived from the French Military real time Lan Standard GAMT 103. It was felt that it might become a de facto standard of high-speed transport protocol.

It is worth mentioning that XTP does not contain any definition of the service it is intended to provide, and does not seem, at the time, to be concerned with the B-ISDN environment. Furthermore, XTP has been neither validated for its logical coherence nor fully evaluated with respect to performance measures.

The XTP study included its formal specification in a standardised ISO language, its validation, an evaluation of its performance, and finally an implementation on UNIX.

In order to validate the logical correctness of the XTP protocol, its description had to be formally expressed. Due to BULL and INT[1] experience and tools availability, the Estelle specification language was chosen. Finally a prototype implementation of the XTP for demonstration purpose and for an initial performance evaluation was done by INRIA. The machine used for the porting was a UNIX-based commercially available workstation.

3. The Requirements of the New Applications

The characteristics of communications in today's environment differ from those for which the initial OSI protocols were designed. Originally the dominant factor was related to terminal/mainframe conversations and file transfer operations. In the case of terminal traffic, messages tended to be short and multiple transactions would be entered within a session that could be maintained for a long period. File transfer in addition implied that a connection between correspondents would persist for a relatively long time. Today's workstations, on the other hand, may operate for a significant length of time on a "large block" of data within their memory and only require communications service intermittently. However, there is a requirement for fast response and high transfer speed.

New distributed systems, as implemented in ESPRIT projects, such as COMANDOS or ISA, or as studied in the standardisation bodies under the ODP work item, imply a new pattern of communications between the different parts of integrated distributed applications. Also, support environments for ODP must support distribution transparencies which include, location, concurrency, replication and migration transparency. Other functions of SE-ODP which need support are traders,

---

[1]INT is the acronym of the "Institut National des Télécommunications "(France)

security and communications support. Investigation is needed to avoid duplication of functions in the SE-ODP and communications systems.

Many new applications will require the handling of multimedia objects composed of voice, text, image and video. Their multimedia nature implies that several datastreams must be handled in parallel; since these streams belong to a single application and are therefore not independent, ways of synchronising these streams must be found. Because of the requirement for multiple data streams, multi-connection facilities are necessary, for example to set up connections for voice and video simultaneously. Since the new applications will often involve more than two partners, protocols linking several entities in a network are required. None of these facilities is currently defined in the OSI standards.

The quality of service required by the applications, and the method of mapping this on to the lower-layer protocols, is extremely important. The interaction between these types of protocols and the session and presentation ATM layers is an area requiring investigation. Another protocol requirement is to support the dynamic creation of multiple parallel streams operating at different speeds, thereby giving the application the ability to shape the network to its needs. The automation of these facilities from an application and the optimisation of their use within, for example, a call-based charging regime also requires study.

The main aim was to look at these new requirements and discover how they affect the supporting protocols, and to propose what applications services and protocols will be necessary to meet these requirements. The work which involved the University of Lancaster, Alcatel Austria ELIN, Olivetti Research, DIT[2], INRIA and BULL consisted of two parts; one looked at the multimedia requirements, and the other dealt with the requirements of distributed applications, particularly the requirements of SE-ODP.

4. The OSI95 Transport Service Design

The basic goal of ULg was to design and specify a transport service and a transport protocol which would not jeopardise the potential bandwidth offered by the new environment created by the high speed LANs ($\geq$ 100 Mbps). The transport service was intended to be upward compatible with the standard connection-mode transport service, and the underlying service was the standard LLC type 1 service. The protocol had to be designed assuming a LLC access rate $\geq$ 50 Mbps. This implies that the layer 2 service is offered through a high performance controller with a chipset and/or ASIC or VLSI chips. The design of transport protocol had also to take account of the LLC service offered by the B-ISDN through its AALs (ATM Adaptation Layer).

One of the first step of the project of the new design was the detailed analysis by ULg of existing transport services and protocols (TP 4 and TCP/IP) to put into evidence all their drawbacks in the new environment, followed by a detailed analysis of XTP, to clearly identify the area where it contributes to the improvement of these drawbacks.

---

[2]DIT is the acronym of "Departamento de Ingenieria de Sistemas Telematicos" (Spain)

Protocols imply mechanisms, syntax and implementation. Some people claim that implementation is very often responsible for poor performance and it is true that a poor implementation gives rise to poor performance. But even good implementations are not able to correct the influence of poorly designed mechanisms and to deal with badly designed PDU syntaxes. Good implementations imply that the consequence of the scheduling, of buffers copying and other operating system related issues, be minimised for the performance of the station access rate. Furthermore, the design of a transport protocol must take account of the possibility of implementing it on silicon.

The goal of the project was not only to specify new transport service and protocol. It was also to apply LOTOS, a formal description technique, from the very beginning of the design process. The methodology for the design of the service and the protocol will be based on the constraint-oriented style in order to allow an incremental development of the protocol. This method, applicable to the design of large specifications, allows the addition of, the removal of or a modification in a mechanism without jeopardising the other part of the work. It was expected that this basic use of a formal description technique during the process of specification would end up in better protocol. The new methodology has been tested first on TP4 before being used for the OSI95 transport service.

The design of a complex LOTOS specifications required the use of adequate tools. The OSI95 project closely followed the work of ESPRIT/LOTOSPHERE and used their LOTOS toolset, LITE, as soon as they were made available to other ESPRIT projects.

5. ATM and the New Communication Environment

The ATM style of communication using short, fixed length cells, has gained widespread popularity. This has been enhanced by the CCITT I.121 recommendation of an ATM methodology and cell format for B-ISDN. It need hardly be repeated that ATM is suitable for both fixed and variable rate traffic and can support data and real-time traffic with equal ease.

One task, by Alcatel Bell, was to evaluate how these new communications techniques could be incorporated into the OSI data-link layer, particularly when real-time traffic was also considered. This task assessed the service offered by B-ISDN ATM networks; more particularly, it looked at the service of the different types of ATM Adaptation Layer and evaluated how these services would support OSI LLC types of service. The new features of ATM based networks (e.g. flexible bandwidth) could lead to the definition of new types of LLC in order to allow higher layers to exploit fully the capabilities of these ATM based networks.

Olivetti has extensive practical experience of slotted ring systems. Such short cell networks are compatible with ATM techniques and Olivetti developed network interface and device drivers to connect ATM networks into standard protocol stacks, including the Internet Protocol and Ultrix V4.2 socket interface. These hardware platforms could later be used for testing the transport protocol over ATM networks.

6. Specific Studies

The objective of the project was to define a high-speed transport protocol, TPX, optimised for high speed LANs. However, when LANs are interconnected, some users may decide to use this protocol over the interconnection. The purpose of the task was to study suitable congestion avoidance mechanisms that would be usable for high speed networks. The study included a survey of existing algorithms (slow-start, DEC-bit) and the simulation of rate-based control mechanism developed at INRIA.

Intracom identified the protocol VLSI implementation problems and provided guidelines for achieving a VLSI implementation of transport protocols. It was expected that the "siliconability" of a transport protocol, which was investigated in this workpackage, would provide some guidelines for the design of TPX.

7. Time Extension of LOTOS

It was the intent of the project to specify as much as possible the new transport service and protocol in the ISO language, LOTOS. It was however anticipated that the new communication and application environments would, in some specific cases, require some extensions to the language. In particular, formal specification of new transport services influenced by the availability of some ATM-based services could require extensions of LOTOS covering the modelling of isochronous as well as asynchronous systems.

An objective of this task was to define a model of an extended LOTOS language which could describe the time evolution of systems in a quantitative way, aside to modelling them in terms of events occurrence ordering only, as Standard LOTOS does. The timed LOTOS language model had to be able to describe both synchronous and asynchronous systems.

Another language extension was also developed to provide the facility of describing systems behaviour in a probabilistic way; thus allowing the reduction of the non-determinism of the Standard LOTOS Language when statistical information about the system being modelled is suitable and available.

The two extensions above, done by DIT, would make the LOTOS language suitable for analysing systems from the logical point of view, and from the performance point of view as well.

The DIT Timed LOTOS proposal was assessed on a well defined case study defined by ULg.

The DIT Extended LOTOS model (time + probabilities) was validated through modelling case studies selected from high speed networking technologies, services and protocols (XTP, FDDI, ATM,...).

8. Standardization

The interest of developing a new Transport service and a new transport protocol
would have been limited, if this effort has not been integrated in the standardization
activities. This was the reason why one of the goal of the project, strongly supported
by the CEC, was the creation of one or several New Work Items (NWI) on "High
Speed Protocols and Services" in various standardisation environments such as
ISO/IEC JTC1/SC6, ETSI/NA5, to avoid the repetition of the FDDI standardisation
which has been carried out entirely in the US and has given a competitive edge to US
companies for the development of FDDI-based products.

The creation of such a NWI was a prerequisite to the promotion of new transport
service and protocol in the Standardisation Organisations. The creation of such a NWI
was clearly a success criterion for the OSI95 project.

9. Conclusion

In this book, we will present, following the sections of this paper, the main results of
the first phase of the OSI95 project.

The XTP Study

The development of XTP started in 1989 and was still going on at the end of the first phase of the OSI95 project. The partners involved in the XTP study were faced with three successive modifications of the XTP specification. Furthermore, the XTP natural language specification describes only the protocol and there is no specification of the service offered to its users.

The ULg study of the XTP natural language protocol definition raised several questions that were clarified, at least partially, by the designers.

As expected, the formal specification, by Bull and INT, of the protocol raised several additional problems of undetermined behaviour. All these ambiguities had to be resolved either by contact with the designers or by selecting one of the possible solutions.

The task of formally specifying the XTP protocol in Estelle required as expected, an in-depth study of the methodology to be used taking especially account of the size of the problem.

The size and the complexity of the problem had been at the origin of a need to extend the Estelle toolset already available at the beginning of the project.

For the validation aspect, it has been necessary to define a missing service specification.

One of the results of this study is the confirmation of the absolute need of using formal methods for specification of any new protocol at a very early stage of the design.

For XTP itself, it is essential to stress that the goal was to design a new and more efficient protocol for the classical data communications. No new QoS has been introduced for covering the need of the new application environment such as the multimedia-based applications.

Even the completion of the XTP design and of its chipset will not change the need for the development of new transport service and protocol such as OSI95. However it is clear that some of the new mechanisms introduced in the pioneering work of XTP will find their way in many new protocols.

Modelling and Analysing the Xpress Transfer Protocol (XTP)

S. Budkowski, B. Alkhechi*, M.L. Benalycherif, P. Dembinski**, M. Gardie, E. Lallet, J.P. Mouchel La Fosse and Y. Souissi

Institut National des Telecommunications (INT), Systems and Networks Department, 91011 Evry, France

Email: stan@int-evry.fr

The paper describes our experience in developing a formal specification of the Xpress Transfer Protocol (XTP), studying its correctness and evaluating performance of some of its mechanisms. The validation issue is only outlined in this paper since it is a subject of a separate paper.

The XTP, a transfer layer (transport plus network layers) protocol for high performance real-time applications, is being designed by Protocol Engines Inc. (Mountain View, California, USA).

Estelle, a formal description technique standardised by ISO, has been used to formally specify XTP. The ESTELLE DEVELOPMENT TOOLSET (EDT) has been applied for supporting the development process.

The QNAP2 language and its associated tool has been used to model XTP (queueing network model) and to study some of its mechanisms with respect to their impact on protocol performances. This study has been realised by Bull S.A. and INT in the framework of the ESPRIT $N° 5341$ Project "OSI95".

Keywords: high-speed communication protocols, transfer protocol, transport protocol, Formal Description Technique, Estelle, validation, simulation, performance evaluation, queueing network model.

1 Introduction

The aim of this study was to formally specify and validate the XTP (Xpress Transfer Protocol) and to evaluate the performance of its main mechanisms. XTP is a transfer layer (transport plus network) protocol for high performance real-time applications. It is being designed by *Protocol Engines Inc. (Mountain View, California, USA)*.

Estelle [ISO 89, BUD 87], a formal description technique standardised by ISO, has been used to formally specify XTP starting from its English language definition published by *Protocol Engines Inc.*. It is worth noting that this definition changed three times during the period of the project starting from Revision 3.5 and ending with

---

* Bull S.A. until December 1992

** On leave of absence from the Institute of Computer Science, Polish Academy of Sciences, 00901 Warsaw, Poland

Revision 3.6 which incorporates the Addendum 1a proposed in the meantime. The Estelle specification of XTP presented in the paper corresponds to the Revision 3.6 defined in [PEI92].

To validate the correctness of the XTP protocol mechanisms and of its Estelle specification, the ESTELLE DEVELOPMENT TOOLSET (EDT) was used [BUD 92]. A new component UTDG (a Universal Test Drivers Generator) has been developed [LLMB 91] and integrated with EDT to allow automatic generation of simulation environments enabling separate validation of XTP parts.

The QNAP2 language (queuing network model) and associated tool has been chosen to model the performance issues. A preliminary study of a possibility of translating an Estelle specification to QNAP2 was also done [DEM 92].

This study has been realised by Bull S.A. and INT within the framework of the ESPRIT N° 5341 Project "OSI95".

2 Estelle Specification of XTP

2.1 An XTP Overview

XTP (Xpress Transfer Protocol) [PEI 92] is a high speed protocol which has been designed for software and VLSI execution environment. It takes into account transport and, to some extend, network layer functions. In XTP, resources of network and transport layers can be reserved in an only one connection set-up by means of a unique reference identifier. Consequently, XTP can be installed not only as an end system but also on intermediate nodes which are used to support network functions. In LANs environments, XTP may be on the top of either a MAC service (DARPA architecture) or a LLC type 1 service (OSI architecture). No service has been defined for XTP.

Packet Structure. In XTP, 9 types of packets (within 2 formats) can be exchanged. The RCNTL and CNTL (control) packets use the control format and the others (FIRST, DATA, PATH, DIAG, ROUTE, ...) the information format. Both formats use the same header structure (40-byte long) and the same trailer structure (4-byte long).

Connection Set-Up and Shutdown Management. XTP is a connection-oriented protocol. The state information related to an XTP connection (called *association*) in one endpoint is called a *context*. A context must be created (instanciated) before sending or receiving XTP packets. A created context, ready to take part of a connection, is called *listening*. Each context taking part in an established connection is called *active* and is identified by a 4-byte long *key*. Once a connection is established, XTP subsequent packets need not to carry a full network address but only the *key*. XTP admits an exchange key mechanism which allows a direct and quick allocation of state information.

Instead of the 3-way handshake connection set up mechanism used in TP4 and TCP, XTP uses a FAST CONNECT. An XTP entity A sets up an association with a remote entity B by sending to B a FIRST packet. The FIRST packet contains the receiver B address and the *key* of an *active* context of A. *It may also carry data.* At the A side a connection is regarded as opened once the FIRST packet is transmitted while at the B side, only after the FIRST packet has been received by a *listening* context in B (which then becomes active). In case there is no a listening context within B, a DIAG (diagnostic) packet is send back by the entity B.

During FAST CONNECT set-up, a negotiation of parameters is impossible. XTP offers, however, a way to select the different end-to-end operating modes. There are identified by 7 bitflags located in a header field *cmd*. They permit to enable/disable the checksum calculation of the information or control segment of the packet (NOCHECK), the error control (NOERR), the use of multicast mode (MULTI) and of reservation mode (RES), the sorting/prioritising packets (SORT), the flow control (NOFLOW), the aggressive error control (FASTNAK).

To shutdown an association both context taking part in the association have to be released which requires the closure of each data stream. Three following bitflags located within *cmd* header field are involved in the close procedure: WCLOSE- a data sender desire to close its output data stream, RCLOSE - a data receiver has closed its input data stream (no more data will be accepted), END - the context is being released (no more communication is possible). Closure of each data stream can be graceful or forced.

Acknowledgement and Error Recovery. The acknowledgement is a mean by which the receiver informs the sender that some data have been correctly received. XTP uses a cumulative piggy-backed acknowledgement by means of setting *dseq* header field in FIRST, DATA and CNTL packets (the *dseq* acknowledges data delivered to the remote application), and an explicit and selective acknowledgement by means of *rseq, spans, nspan* fields in CNTL packets. In XTP, CNTL packets are generated by the receiver only when the sender explicitly request it (by setting one of the 2 header bitflags SREQ or DREQ in outgoing DATA, FIRST or CNTL packets). By setting the DREQ bit the sender requests the receiver to sent him back a CNTL packet after the delivery of the data received before while by setting the SREQ the immediate response is requested. To detect that a CNTL packet requested is not received within an expected period of time, XTP uses a WTIMER timer. If WTIMER expires and no CNTL packet is received, XTP enters a *synchronising handshake*. During this procedure the sender is not allowed to transmit data. Another timer CTIMEOUT is used to limit the time XTP try to complete the synchronising handshake.

In XTP, retransmission is based on the reception and the parsing of CNTL packets sent by the receiver. There is no timers used for this purpose. Subject to bufferize out-of-sequence data, XTP provides not only a selective retransmission but also a go-back-n mechanism. Moreover, X TP allows an entity which implements selective retransmission to interoperate with an entity that only uses go-back-n.

End-to-end Flow Control. To prevent the receiving resources from overloading, XTP uses the sliding window mechanism. Every CNTL packet carries through the fields RSEQ and ALLOC, the sequence numbers limiting the window recommended by the receiver. The maximum size of the allocated window is 2^{32} bit long. Like in the traditional protocols, end-to-end flow control is optional, it is enabled by setting the NOFLOW header bit .

Rate Control. The rate control mechanism permits to protect the network resources against overloading. It limits the rate at which a sender is allowed to transmit data. The actors of the mechanism are not only XTP endpoints but also the intermediate nodes. At the sending entity, it takes the form of a bursty transmission of packets under the control of the timer RTIMER.

Multicast. In XTP, a multicast association is initialised by a FIRST packet with the MULTI bit turned on and a multicast address assigned to the address segment. In multicast mode, CNTL packets sent by multicast receivers in response to a received SREQ, are multicast to the group and not only unicast to the sender. All the mechanisms defined in unicast mode are provided in multicast mode.

2.2 XTP Estelle Structure

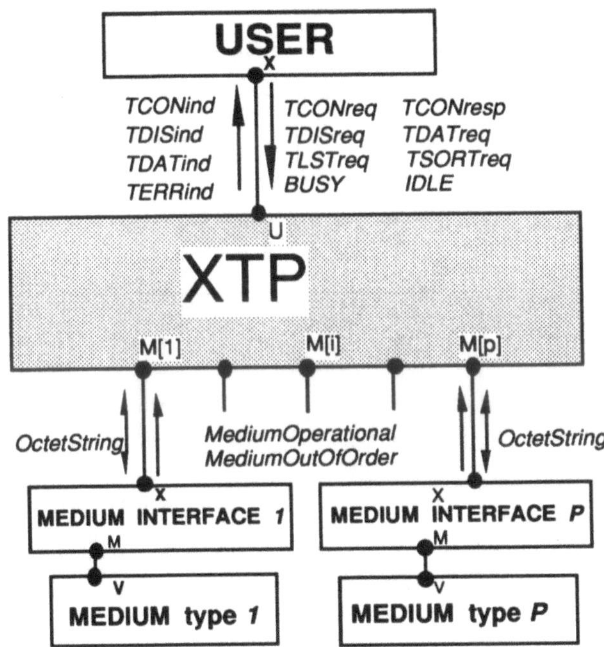

Fig. 2.1 Global structure of XTP and its environment

Figure 2.1 shows the global structure of the Estelle specification of XTP and of its environment. The USER module represents an application above XTP. The XTP module specifies the XTP protocol machine. An XTP entity may access, at the same time, several different media (e.g., Ethernet, Token Ring and FDDI); each MEDIUM module represents a given real medium. Each VIRTUAL MEDIUM module provides to the XTP module a unique medium-independent access interface (its low-interface is fully medium-dependent). The specification of behaviour of all modules but XTP is out of the scope of this paper (some specific behaviour of the USER module was, however, specified for validation purposes).

XTP Module Structure. The structure of the Estelle specification of the XTP module is shown in Figure 2.2. The XTP module is partitioned into five modules. The three (any fixed number in general) CONTEXT modules are dynamically created instances of the same generic module. The XTP module has two main functions:

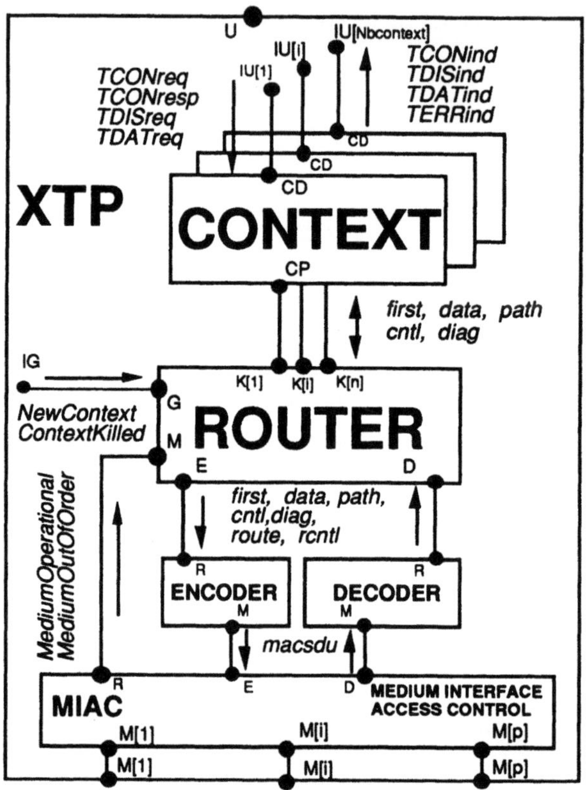

Fig. 2.2 XTP module structure

- *context management:* the XTP module creates (or initialises) and destroys context modules. Each context is identified by a local variable called a *key*. The KEY field of every packet is the image of the key variable value. The key has two parts: an index part which identifies the context, and an instance part which gives the "age" of the context to allow discarding old packets (its value is incremented each time the context is used for a new association).

- *sort management:* the XTP module serves all active contexts according to their priority (the priority, or sort value, is under the control of the user); it controls the delivery order (to the higher-layer clients) of packets (the MIAC module controls the delivery of packets to the network). For this reason, each existing context module is connected to an internal interaction point.

Each CONTEXT module, when created dynamically by its parent module XTP, is initialised to the LISTEN state upon receiving a request TLSTreq from the USER, or to the ACTIVE state upon receiving a request TCONreq from the user.

An association between distant contexts is created by transmitting a FIRST packet from one endpoint's ACTIVE context to another endpoint's LISTENING context. Then, upon reception of the FIRST packet, the LISTENING context moves to the ACTIVE state and creates all its children modules with the necessary links between their interaction points (see Figure 2.3).

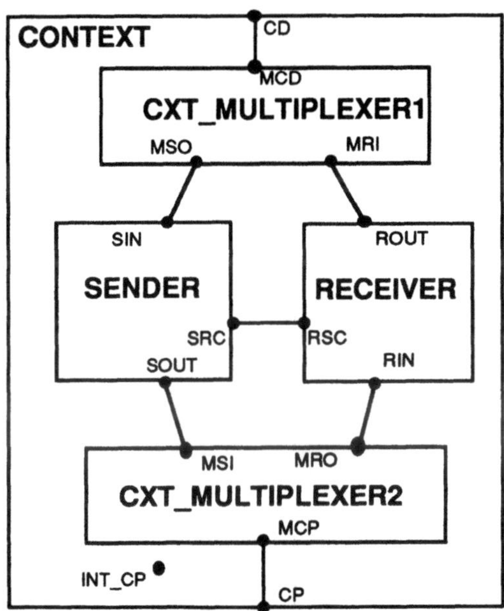

Fig. 2.3 Context module structure

To make the CONTEXT module able to transmit this FIRST packet to its future child CXT_MULTIPLEXER2, an internal interaction point "int_cp" is used, which is connected to the child CXT_MULTIPLEXER2 module only for sending the FIRST packet to the child, and disconnected immediately after.

The SENDER module controls the outgoing data stream. It receives requests (primitives) and data from a specific user (through the CXT_MULTIPLEXER1 module), and it transmits logical packets to the ROUTER module (through the CXT_MULTIPLEXER2 module).

The RECEIVER module controls the incoming data stream. It receives logical packets from the ROUTER module (through the CXT_MULTIPLEXER2 module), and it transmits indications (primitives) and data to the user (through the CXT_MULTIPLEXER1 module). It controls timers, and implements the flow control and error control procedures. The RECEIVER module acts on both received DATA and CNTL packets. It is also responsible for delivering data to the application. When a valid DATA packet is received, it is submitted to error control procedures and the receiver buffer variables are updated. If it contains an SREQ bit set, the receiver sends "send_control_packet(...)" message to the SENDER module to transmit a DATA_response control packet. When a valid CNTL packet is received, the receiver module sends "ind_received_cntl_packet(...)" message to the sender module to make it able to update its control variables. If the received CNTL packet contains an SREQ bit set, the receiver sends "send_control_packet(...)" message to the SENDER module to transmit a CNTL_response control packet.

Both the SENDER and the RECEIVER modules need internal buffers to store data received either from the local user or from the network. A ring buffer structure was chosen to manage the internal buffers. A user message may fill several XTP packets, therefore when this message arrives with the *btag* indicator set (this indicator means that the first 8 bytes of the data segment within the information segment contain tag information for the higher-layer application) the XTP sender must put this BTAG indicator only in the first packet (does not necessary means that the type of the packet is "FIRST") of the message because the tag information will exist only in this packet. A data structure is necessary to store the BTAG indicator each time it is set, and to associate it to the first packet of the message in the buffer. The same problem exists with respect to the EOM, DREQ, RCLOSE, WCLOSE and END flags which must be stored before they can be sent, if they have been set, in the last packet of the message.

Below we describe some concepts used within the SENDER module and the way they have been mapped in Estelle.

Sliding Window Management. Output flow control in XTP is based on a sliding window of sequence numbers. Alloc is the current do-not-exceed output sequence number. The sender is not allowed to transmit sequence numbers beyond nor including the value of alloc, which is updated from the alloc values contained in CNTL packets arriving from the network. When the sender reaches the limit of its output window (determined by alloc), or when it reaches the limit of its buffer (if the buffer is full of unacknowleged data), it sends a CNTL packet containing the sreq bit set to obtain a response from the distant XTP system increasing the alloc value, or acknowledging some data. The corresponding Estelle transition is summarised here:

```
FROM active TO active
 PROVIDED Flow_control(noflow,requested_cntl_packet,flow_variables)
      VAR cntl : xtp_control_packet;
      NAME sender_arrives_to_alloc_or_to_buffer_limit :
      BEGIN
            fill_cntl_sreq_packet(cntl);
            OUTPUT sout.control(cntl);
      END;
```

Flow_control is a boolean function which determines if a CNTL packet has to be sent (see section 4).

WTIMER Management. The sender enters the sync state when it executes the synchronising handshake after a WTIMER failure (WTIMER is the amount of time an XTP sender waits for a response to an sreq). This "synchronising handshake" refers to an exchange of control packets between the sending host and the receiving host after which the sending host is certain of the receiving host state. During this procedure the sender is not allowed to transmit data. The corresponding Estelle transition is summarised here:

```
FROM active TO sync
 PROVIDED (requested_cntl_packet)
 PRIORITY Medium
  DELAY (wtimer)
        VAR cntl : xtp_control_packet;
        NAME sender_enters_sync_state:
        (* There is no response to a requested cntl packet. *)
        (* Start the synchronising handshake.          *)
        BEGIN
                ctimeout_cond := TRUE;
                k_exp := 1; (* Initialisation of the K constant. *)
                retry_count := init_retry_count;
                IF (remote_context_is_alive) THEN
                        BEGIN
                        fill_cntl_sreq_packet(cntl);
                        OUTPUT sout.control(cntl);
                        END
                ELSE
                        BEGIN
                        first.header.cmd.sreq := TRUE;
                        first.header.sync := local_sync;
                        OUTPUT sout.first(first);
                        END;
        END;
```

ctimeout Management. ctimeout is a timer which is used to bound the amount of time XTP will try a message exchange during the synchronising handshake. This timer is enabled when the sender enters the sync or close state. When this timer expires the context must be aborted. The corresponding Estelle transition is summarised below (a similar transition has been defined to manage the retry_count parameter):

```
FROM sync, close TO error
 PROVIDED ctimeout_cond
  DELAY (ctimeout)
        NAME ctimeout_expires:
        BEGIN
                context_must_be_aborted := TRUE
        END;
```

The Router Module Specification. XTP definition [PEI92] gives little information about routing function since it is implementation-dependant. For completeness reason (and to have executable specification for simulation purposes) it was decided to describe this function within the Estelle specification. Some simplification has been, however, adopted e.g., there is no route modification (no processing of PATH packets) without affecting our results. The main functionalities of the ROUTER module are the following:

- management of the routing tables; the interaction point M receives messages concerning the availability of the different media;
- assignment of all packets that have reached their final destination to local contexts; the interaction point G receives messages concerning the availability of contexts;
- forwarding of all packets that have not reached their final destination;
- computation of the ROUTE field value for every outgoing packet;
- computation and control of the TTL field for each packet;
- release of routes (with the help of ROUTE packets);
- computation of the MAC address to be used and, if necessary, selection of the medium where to send the packet; these information are stored in routing tables.

In XTP, each route is identified by a value composed of three parts: a RETURN bit, an index part and an instance part. The ROUTE field in each XTP packet is the image of this value. The index part identifies the route, the instance part gives the "age" of the route. The sender of XTP packets can use either its local route identifier, or the remote route identifier (if it knows it). The RETURN bit is set to one in that case. The number of bits in each part are determined by the implementation. Each node must inform the opposite side of its local route identifier; this is done by a route exchange (see [PEI92]) which is controlled by three variables: the local route identifier which is always known, a boolean variable called *red* (red = Route Exchange Done) indicating if the route exchange has been done, and the remote route identifier.

Two kinds of addresses are used in XTP: the MAC addresses and the host addresses. The MAC address is a physical address. It is never conveyed in an XTP packet. The host address locates sender and destination machines. Host addresses are conveyed in the address segment of FIRST and PATH packets. XTP can use different host address syntaxes. For example, a host address can be an Internet Protocol address, an ISO standard address or an IEEE 802-style Source Route address (see [PEI92] for more information). Host addresses are supposed to be composed of a host identifier and a selector. The host identifier allows access to a specific machine, the selector allows access to a specific application on the selected machine.

Two tables are used to route packets. A table called "route_table" which identifies each route managed by the ROUTER module, and a table called "context_identifier" which identifies each CONTEXT module seen by the ROUTER module. The necessity of introducing these two tables is due to the fact that a station may play different roles (an intermediate system, an end system, or both of them).

The "route_table" is used:

1. to switch packets from one to another route (forwarding),
2. to determine that incoming packets are to be assigned to local contexts (this will be done by using the "context_identifier" table") and
3. to select routes, networks and MAC addresses for outgoing packets.

The "context_identifier" table is used:

1. to assign incoming packets to local contexts, depending on the *key* field of each packet and
2. to select an element in the "route_table" for outgoing packets coming from a local context (routes, networks and MAC addresses will be found in the selected element of the "route_table").

These two tables are composed of several elements. Each particular element in the "route_table" is selected by the index part of a route. The index part of a key is used to select an element in the "context_identifier" table. The number of elements in the "route_table" is the maximum number of routes that the ROUTER module can manage. This value (*Nbroute*) is implementation-dependent, and is local to a system. The number of elements in the "context_identifier" table is the maximum number of contexts that the XTP module can manage. This value (*Nbcontext*) is chosen at the implementation, and is local to a system.

Decoder and Encoder Module. The DECODER module transforms a real XTP packet (composed of bits) into a logical packet which is independent from the real structure (allowed to be independent from modifications introduced between the revision 3.5 and 3.6 of the XTP protocol definition)·.The ENCODER module realises a reverse transformation.

Miac Module. This module has several functionalities such as: common access to different media, separation of the bi-directional data flow into two separate data flows (one for the data coming from the lower layer and the other for the data going to the lower layer). It provides information to the ROUTER module about the availability of a medium and it sorts the outgoing packets according to their sort value.

Size and Limitations of the Specification. The whole Estelle specification including the simulation environment has about 7000 lines, is composed of 12 modules and more than 100 transitions. The following features of XTP were not taken into account in the Estelle specification: the multicast, path failure recovery, and path management features since they were not completely defined in revision 3.6, and the reservation mode and checksum function since they were outside the main XTP functionalities. The of sort management from the network side was not done since it was decided to demonstrate its feasibility realising it from the user side.

2.3 Some Anomalies and Obscurities Discovered in the XTP Definition

The following list makes references to the official XTP defining document[1] [PEI92].

1 In the table on page 17, the NOFLOW option has a potential change of "per context". This contradicts the sentence on page 19, which says that this option must be the

---

[1] This list represents the collective discoveries of researchers from Bull, the Institut National des Télécommunications (INT) and Bill Atwood (Concordia University, Montreal, Canada), who was visiting INT during June and July, 1992. It is a modified and extended version of the list initially edited by J.W. Atwood on July 28, 1992

same in all packets of an association. The "per context" interpretation is probably correct, in that Appendix D gives the user the ability to reject a proposed association if the NOFLOW option is not accepted. If so, however, the text on page 19 must be changed. *In the current version of Estelle specification the "per context" interpretation has been adopted.*

2 The second sync rule for a receiver on page 23 is invalid. It requires discarding packets (all packet types) with "old" sync values. However, if all DATA packets are sent with SREQ turned on (a permitted option), or if sync is increased for every DATA packet sent (another permitted option), then this rule will require discarding *all* out-of-order packets. This is clearly not intended. This rule should apply only to received CNTL packets. *In the current version of Estelle specification ALL packets with an old sync value are discarded, but sync value is not increased for every DATA packet, and thus not all out-of-order packets are discarded.*

3 The third sync rule for a receiver on page 23 is in conflict with the rules on page 27 concerning the interpretation of echo and techo. If techo conditions the interpretation of echo, then echo is valid only in CNTL-response packets. In this case, rule #3 must be changed to: "When a CNTL packet is received...". This choice makes the third paragraph of the implementation note on page 23 invalid, because the idea there is to use DATA -> DATA-response sequences to get new time estimates, but the echo field is not interpretable in DATA-response packets. It is only valid in CNTL-response packets, for which a valid techo is already defined. If techo does not condition the interpretation of echo, then rule #3 need not be changed, and the implementation note is valid, but the rule on page 27 must be changed to: "echo and techo are interpreted in all received CNTL packets." This interpretation is strengthened by the fact that page 28 says "The sync field received in any request packet [not just CNTL-request] must be copied to the echo field of the corresponding response packet." This implies that the originator of the sync field can expect returned echo values to mean something, in spite of the fact that techo will be zero if the response packet is a DATA-response packet. *The current version of Estelle specification follows the rule on page 27: echo is interpreted only when techo is non-zero*

4 Given that the dlen value of zero cannot be error checked, this value must be specifically disallowed, and a pointer given to the representation of zero-length segments (section 2.5.3.2). It seems necessary to add after the second sentence of section 2.3.9 the following: "The dlen value of zero is not permitted. See section 2.5.3.2 for the representation of zero-length data segments." It may also be desirable to add a DIAG code for this protocol error. *In the current version of Estelle specification dlen can never be zero*

5 On page 29 "Nspan is inspected in every CNTL packet" conflicts with page 27 "nspan is interpreted ... whenever the NOERR bit ... is zero.". *In the current version of Estelle specification nspan is not interpreted when the NOERR bit is set (page 27 is correct).*

6 On page 67, describing the processing steps for a received control packet, step 5 says that upon receiving a control packet, "local" alloc, rseq, dseq, burst, and rate control variables must be updated. This is confusing, because there are two "local"

variables: the ones to be put into outgoing packets, and the ones received from a peer context. A nomenclature should be found to better represent this. The confusion is further compounded by step 4, which gives rules for updating the "outgoing" rseq and dseq, under certain conditions. *In the current version of Estelle specification, only "local" alloc (which represents one past the maximum sequence number that the local sender can send) is updated at the context level. Burst and rate control variables are updated at the router level.*

7 On page 68, describing the processing steps for a received CNTL packet, step 8 seems to confuse actions to be taken for requesting retransmission (triggered by FASTNAK - see page 19) with actions to be taken to provide retransmission. *The current version of Estelle specification takes into account FASTNAK bit for requesting retransmission (according to page 19 definition).*

8 On page 70, the boundary condition "Output is permitted as long as credit remains greater than zero...' must be better specified. If credit = 1500, and a 1500-byte packet is to be sent, is this allowed or not? The above phrase can be interpreted to mean either that the check is after the decrementing and the sending, or after the decrementing but before the sending. *The current version of Estelle specification checks credit after decrementing AND sending .*

9 A particular case concerning FASTNAK is the following: A gap occurs because the last packet before a CNTL packet is lost. The gap is detected because the CNTL packet contains a seq value greater than the current local hseq. If FASTNAK is set, and if spans currently equals zero, it is permitted to report this gap. But there are no received data to report in spans[1]. The solution is to explicitly allow spans[1] = (hseq, hseq). It seems useful to add, after the first sentence at the top of page 72, the following: "If the gap is detected as a result of the arrival of a CNTL packet with a seq value greater than the local hseq, the local hseq should be set to seq, and the spans list should consist of the values (hseq, hseq)". *In the current version of Estelle specification hseq is set to seq and the spans list consist of the values (hseq, hseq).*

10 In the discussion on synchronisation handshake (page 77), steps 5 and 6 are written in such a way as to apply to all packets received. Based on the first sentence of section 3.6, it is believed that these steps do not apply to regular packets, but only to CNTL-response packets (i.e., those for which techo is non-zero). This should be clarified. This problem came to light when the DATA-response packet, whose delayed arrival had triggered the WTIMER failure and the subsequent synchronisation handshake, arrived (with the correct match for saved_sync) and caused the input pipe to stall, because it was not covered by the coded conditions. *In the current version of Estelle specification the echo is significative ONLY if techo is non-zero (page 27), thus the specification discards DATA-response packet as if they do not match saved_sync.*

11 In the discussion on synchronisation handshake (page 77), it should be made clear that all transmission of DATA packets ceases during the handshake, in order to prevent advances in the value of sync, thus voiding the intent of the handshake (which is to "clear the pipe"). *In the current version of Estelle specification no DATA is sent during synchronisation.*

3 Validation of XTP

The aim of validation was twofold. First the Estelle specification written on the basis of the English text definition had to be validated, and second the correctness of the XTP protocol itself had to be demonstrated. For the reason of the size of the Estelle XTP specification (about 7000 lines of Estelle code) the validation task as described above could not be based, in its dynamic part, on exhaustive verification approach. Only carefully designed testing by simulation was possible. The Estelle Development Toolset (EDT) [BUD92] was used for the purpose of XTP validation. A new component UTDG (a Universal Test Drivers Generator) has been, however, developed [LLMB 91] within the project and integrated with EDT to help the validation process. The main goal of UTDG is to automatically generate simulation environments enabling separate validation of the given specification parts.

In order to lower the complexity of the system to be validated, two major and quasi independent parts of the XTP protocol, namely the CONTEXT module and the ROUTER module, were validated separately. Once the separate validation of these modules was terminated, they were integrated and tested together. The focus was, however, on the validation of the CONTEXT module since it represents in fact the kernel of the XTP protocol.

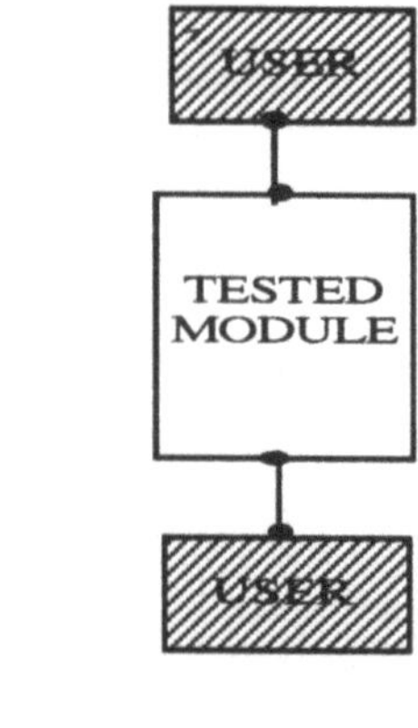

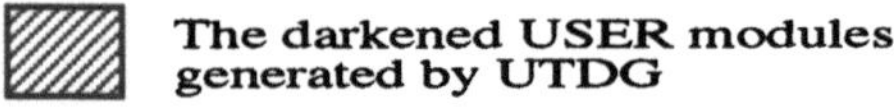

Fig. 3.1 Unit test

The step-by-step simulation was combined with an extensive simulation, the latter performing scenarios specifically adapted to a given functionality. During extensive simulations the macros and observers were used to describe the simulation scenarios and anomalies to be detected. Several especially written USER and/or MEDIUM modules were used for dialoguing with the tested modules. Each of these modules had a behaviour best suited to test the given functionality. For example, totally reliable

MEDIUM modules as well as those which can loose, duplicate or reorder packets (with a given probabilities) have been used.

The simulations have been conducted following the architectures depicted in figure 3.1 (unit test) and figure 3.2 (interoperability test). The tested modules are CONTEXT and ROUTER.

The Estelle specification of XTP was carefully inspected, debugged and rewritten in a large part. More than 40 run-time errors (subrange violation, non-initialised variables, ...) were, however, corrected and an important portion of the specification has been rewritten. Its final version may be really presented as a formal counterpart to its English reference with exception to functionalities (like multicasting) which have been announced as not handled in this version. Having in mind that Estelle is an implementation oriented language, the above means that prototype realisations of the protocol that "guarantee" conformance requirements seem to be quite easy to generate.

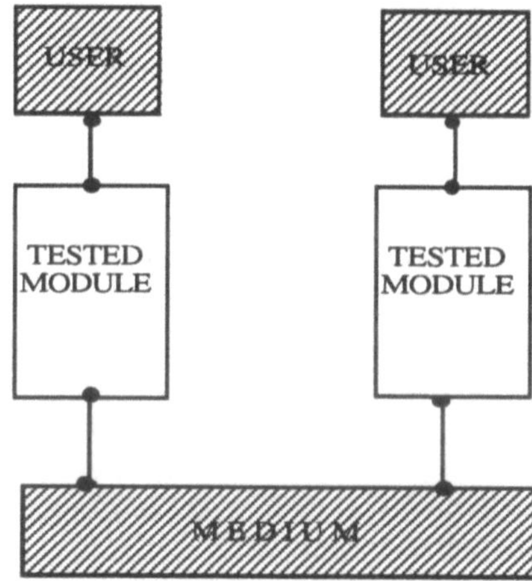

The darkened modules are either automatically generated by UTDG or especially written depending on the type of simulation (step-by-step or extensive simulations)

Fig. 3.2 Interoperability test

The XTP protocol itself occurred, in general, correctly specified, i.e., no major deviation from the intended behaviour has been noticed. The protocol may be only viewed sometimes as not **robust** enough, that is, not able to recover from persistently wrong behaviour at the other endpoint.

Several problems have been, however, detected during validation. They are listed in a separate paper [LDM93] especially devoted to XTP validation results.

The validation task has provided a lot of methodological experience in using simulation and simulation tools. One of its very concrete result was the design, implementation and integration into EDT of the UTDG tool. The tool occurred very useful in interactive simulation and helped in fast creation of different simulation environments. While it is always true that validation by simulation can only help in eliminating errors but cannot exclude them, it is also worth pointing out that systematic and well organised use of this method is highly convincing, especially for large and complex systems. This is certainly the case with XTP.

More information on used methods and results may be found in the companion paper [LDM93].

4 Performance Evaluation of XTP

The aim of this study is to give some first indications about the performance that can be achieved by XTP. The focus has been put on the throughput as a metric for evaluating XTP. The throughput is of course dependent on the offered bandwidth of the layer on which XTP directly relies. Here we suppose that XTP is used in the context of a LAN and relies directly on a MAC service.

It is very important to point out that XTP has been designed with a very different approach compared to TP4 [ISO 88] and TCP [RFC 81]. XTP is what we call a sender oriented protocol, in the sense that all the protocol procedures are triggered by the sender. This is not the case for TP4 and TCP, where the receiver manages many timers that allow it triggering control information exchange. In XTP, it is the charge of the sender to request the receiver status. The receiver status includes among others, acknowledgements, error report, window size and transmission speed. The XTP definition gives no rules, however, which indicate when and how often a sender has to ask for the receiver status (it is considered as implementation dependent). The lack of such rules leads (as shown in section 4.2) to different implementation policies of the protocol with different quantitative characteristics.

An XTP sender entity can request the receiver status by setting the SREQ or the DREQ bit in an outgoing DATA packet or in a CNTL packet. The reception of a packet with one of these two bits set, triggers the transmission of a CNTL packet back. In the following, we assume that XTP uses only the SREQ bit for control information exchange. (the use of the DREQ bit leads to some problems). When the sender sets SREQ it must also start a timer (WTIMER). When WTIMER expires the sender has to start a synchronisation handshake. This handshake consists of an exchange of CNTL packets. The sender sends a CNTL packet with the SREQ bit set, restarts WTIMER and waits for a CNTL packet in response. The handshake ends upon receiving a valid response.

The SREQ setting frequency that we call the "SREQ strategy" is actually conditioned by all the protocol procedures that require the receiver control information. In this paper, we focus on the impact of the SREQ strategy on the flow control and we assume that there is no transmission error and no packet loss. In the following, we consider a user A (on top of an XTP entity A) that sends data to a user B (on top of an XTP entity B). The data transmission through XTP can be split into three phases:

1 Data transmission : DATA packets are sent from A to B, SREQ is set from time to time in DATA packets, B respond by CNTL packet upon reception of an SREQ bit set
2 Synchronisation : during this phase, only CNTL packets are exchanged, no data transmission is allowed
3 Waiting for sliding the window : this phase is due to a closed transmission window.

The SREQ strategy plays an important role with regard to the time spent in each phase and in the number of processed packets of each type. To optimise the throughput, XTP must minimise the time spent in the two last phases. Moreover, in the first phase, XTP must also minimise the time spent in processing and transmitting CNTL packets (they do not carry any user data). It clearly appears that there will be a trade off between minimising control information exchange and minimising the time spent on phases 2 and 3. The overhead due to control information exchange, synchronisations and the time during which the transmission window is closed can be expressed through the following three ratios:

- **N_CNTL/ N_DATA**: the number of CNTL packets divided by the number of DATA packets.
- **T_SYNC / T**: the time spent in synchronisation handshake divided by the simulation time T.
- **T_CW / T**: the time during which the sender cannot send any data (because the window is closed) divided by the simulation time T

We propose four flow control strategies. These strategies differ in the control information exchange frequency and in the knowledge on which the decision of setting the SREQ bit is based.

Strategy 1: In this strategy, the SREQ bit is set when the upper bound of the window is reached. In other terms the SREQ bit is set in the last DATA packet of the window. The factor overhead due to CNTL packets is minimised as in the former one. The difference between strategies 1 and 2 is the moment at which the bit is set.

Strategy 2: The SREQ bit is set in each DATA packet. This strategy looks like that of TP4. But since no timer is reserved for acknowledgements, each data packet has to have a CNTL packet as a response. The main drawback of this strategy is the number of CNTL packets generated, which is at least as large as the number of transmitted DATA packets. The length of a data packet is then a sensitive parameter for this strategy.

Strategy 3: The SREQ bit is set in the first DATA packet sent. The arrival of a CNTL packet in response stops the WTIMER timer and triggers sending a new packet with the SREQ bit set. In other words, the strategy 3 sets the SREQ bit periodically after each round trip time (RTT).

Strategy 4: The SREQ bit is set in a DATA packet when the following conditions are fulfilled:

- for each previous DATA packet with the SREQ bit, a CNTL packet in response has been received by the sender, and
- the time needed to send the data bytes within the window (and not yet sent) is less than the round trip time.

4.1 XTP Model Using Queueing Networks

Queueing networks are a useful tool for the analysis of systems in which several entities compete to access common resources. In such a model, a queue is associated with a station (or service centre) in which customers arrive to receive a service. When the station is busy, serving other customers, incoming customers wait in the queue to be served. Upon completion of previous customer's service, waiting customers to be served next are selected according to some queueing policy.

The most challenging aspect is the process of tailoring the general methodology of modelling to a specific system analysis. We have chosen to give a detailed description of XTP without constraining hypothesis on its behaviour. This description yields a complex model. The XTP model has been written and simulated using the QNAP2 software [QNAP 84].

We present here the architecture of the QNAP2 model for XTP. This model is based on the structure suggested in the English specification given in [PEI 92]. Figure 4.1 represents an overall architecture of the QNAP2 model for XTP. In the following we only present the part of the model that is actually used for flow control analysis. This model is composed of two objects : HOST and NETWORK.

The Host Object: The HOST object represents an XTP entity with an internal user. We suppose that in the model there is an association between two hosts : host1 and host2 (they are two instances of the same object HOST) which communicate between them under a file transfer application. The HOST module is composed of two components : the user component and the XTP component.

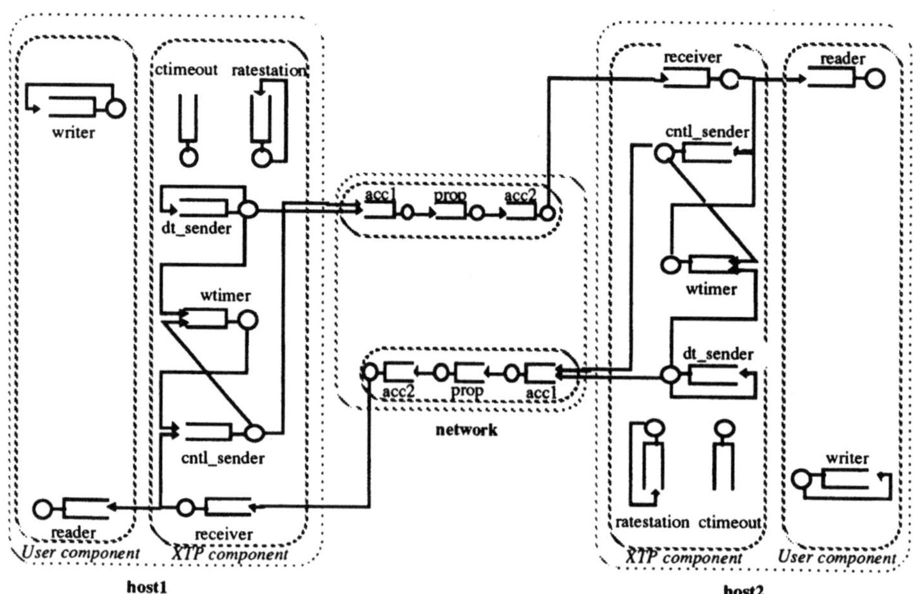

Fig. 4.1 The queueing network model of XTP

The user component: This component is the interface between the XTP entity and the application. It contains two stations (queues) : the writer station and the reader station. The writer station generates messages to be transmitted in packets for the remote host. The service of this station consists in creating customers (messages) to put them in the XTP buffer. It must also free the XTP sender if it is waiting for data from the application. The service time of the writer station is described by a distribution that depends on the kind of the application. The reader station delivers data to the application. Its service consists of taking the data to deliver and putting them out of the model. The time service of this station depends on the length of the data to be delivered.

The XTP component: The model supposes a sequential implementation of the protocol, which means that a protocol task will use the cpu in mutual exclusion with the others tasks. The XTP component manages both an outgoing information stream and an incoming information stream. For this reason, this component is divided into two modules : a sender that manages the outgoing information stream and a receiver that manages the incoming information stream. In addition, the sender module is divided into two stations : dt_sender for sending data, and cntl_sender for sending control information. Each XTP timer is modelled by a station having a constant service time.

The *dt_sender* station takes existing messages from the XTP buffer, transforms them into XTP data packets and transmits them to the network. This station is a 1-bounded queue with feedback, that means there is always a single customer which rounds in this queue. The only customer of this station can be blocked in four cases :

- when no user data is available for transmission ,
- when the upper bound of the transmission window is reached,
- when the specified data rate is reached,
- when a synchronisation handshake is in execution.

The *cntl_sender* station formats control packets, transmits them to the network and controls WTIMER. It is also responsible of all data retransmissions. The service of this station depends on the class of the current customer in its queue. If the current customer has the class *control* then the *cntl_sender* station creates a CNTL packet, assigns its attributes by reading the host variables, and transmits it to the network. If the current customer has the class *wtime*, which means that the *wtimer* station indicates to the *cntl_sender* the failure of the WTIMER, then the *cntl_sender* executes the procedure of the synchronising handshake.

The *receiver* station acts on received DATA and CNTL packets. If a valid DATA packet is received, then it is submitted to error control procedures and the receiver buffer variables are updated. If it contains an SREQ bit set, the receiver station sends a customer of class *control* to the *cntl_sender* station to transmit a DATA response control packet. Finally, if the received DATA packet permits to deliver data to the application, a customer is transmitted to the *reader* station containing all the potential data to be delivered. When the *receiver* station receives a valid control packet, it updates its control variables and tests if the received packet is a response to an SREQ bit set to stop the WTIMER by transiting out the current customer in the *wtimer* station.

The *receiver* station is responsible also of completing a started synchronising handshake.

THE NETWORK OBJECT: It represents the transport medium. It takes a transmitted customer from a host and sends it to the other host. The model assumes that the transport medium is an FDDI LAN. The time within this medium can be divided into three parts :

- access time T1 includes the transmission time and the token rotation time TRT,
- propagation time T2 represents the propagation time in the network,
- access time T3 represents the transmission time from the network to the XTP entity.

   The model parameters are the following:

- DATA packets have data field of 4096 bytes
- DATA packet are constructed (sender) and analysed (receiver) in 0.7 ms
- CNTL packets are constructed (sender) and analysed (receiver) in 0.25 ms

   The model factors are the following:

- TRT : the mean token rotation time : 0.48 ms and 0.32 ms
- The window_size; varying from 2*4096 to 50*4096 bytes

   With a data length of 4096 bytes per frame, the FDDI offered bandwidth is 40 Mbit/s for TRT=0.48 ms and 50 Mbit/s for TRT=0.32 ms.

4.2 XTP Flow Control Analysis

Figure 4.2 shows the throughput for strategies 1 to 4 as a function of the window size. For an FDDI offered bandwidth of 40 Mbit/s, Strategies 3 and 4 overlap. For an FDDI offered bandwidth of 50 Mbit/s the highest throughput, produced by strategy 4, is 45 Mbps (1383 packets/s). Hence XTP consumes for its procedures management 10 % of the offered bandwidth.

The curves representing strategies 3 and 4 have approximately the same shape. For increasing window size, throughput increases. Indeed, the higher the window size is, the less the N_CNTL/N_DATA is. Furthermore, one can observe that the increase of throughput is sharp for small values of window size (smaller than 6). From there, the increase is smooth if any. In fact, a window size value smaller than 6 is lower than the memory of the network (i.e. the product bandwidth*delay). It clearly appears that the strategy 4 leads to the highest throughputs. In fact, this strategy allows a minimum overhead generated by CNTL packets. Moreover, strategies 3 and 4 have more or less the same T_cw/T ratio. The advantage of the strategy 4 comes from a decreasing N_CNTL/N_DATA ratio for increasing values of window size. The more important is eventually the gap between the strategy 4 and the strategies 1 and 2.

Concerning the strategy 2, one can find again the breakpoint for window_size=6. Moreover, for a window_size bigger than 6, the throughput is constant. This is due to the fact that the N_CNTL/N_DATA ratio is equal to one and is independent of the window size. Likewise, the frequency of sliding the window and the amount with which the window is refreshed do not change too much with the increase of the window_size. In consequence, from window_size=6, strategy 2 is not sensitive to the window size. Having a N_CNTL/N_DATA ratio equal to 1, the strategy 2 spends, in processing CNTL packets, more or less equals 35 % of the time for processing DATA packets This overhead of course, increases with decreasing packet size.

By contrast to the other strategies, the curves representing the strategy 1 have no breakpoint for window_size=6. Throughput grows gradually for increasing window_size. This strategy is the worst for small values of T1 and small values of the window size. This can be easily explained by the fact that for those values, XTP sends a small number of data bytes and has to wait a long time before being allowed to send more data.

For strategy 1, the main drawback comes from the time wasted while the window is closed. Indeed, as the SREQ bit is set in the last DATA packet of a window, the sender have to wait for RTT before having a chance do send new packets. Nevertheless, this strategy has a small N_CNTL/N_DATA ratio (e.g. 4 % for window_size=30) and thus can offer a high throughput when using large sliding windows in order to get a big value of the window_size*round_trip_time product. Indeed, strategy 1 curve cuts across the one of strategy 2 (for window_size=11) and then the one of strategy 3 (for window_size=54).

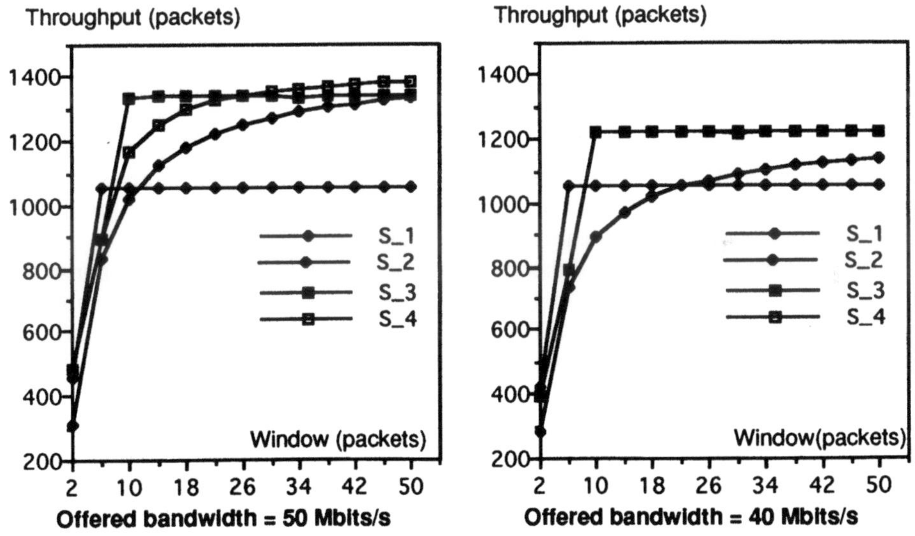

Fig. 4.2 XTP achievable throughput

XTP performance depends critically on the used "SREQ strategy". From our results, the strategy 4 is the more efficient when we consider the throughput as a metric. But the cost of making up this strategy puts a question mark with regard to the leading position of this strategy. In fact, non valid values of RTT and rate may lead to a degradation of the performance. That is to say that the strategy 4 depends on rate control and RTT estimation procedures which themselves depend on the control information exchange frequency and on the network parameter variations. Our simulations have not permitted to evaluate the effects of a varying network load (only constant offered loads have been considered). By contrast to the strategy 4, all the three other strategies can be easily realised but lead to a lower efficiency. Nevertheless, performances of the strategy 3 are close to that of the strategy 4.

5 Conclusions

The Estelle specification of XTP was carefully written, inspected, debugged and simulated. A significant part of it was rewritten due to either misunderstanding of the English text or simply due to the human errors. The use of the Estelle formal description technique allowed a relatively fast adaptation to the modifications of the XTP definition (recall that the definition was modified three times during the project). The EDT tool has proved its great power and flexibility. The Universal Test Drivers Generator (UTDG) tool was very useful in interactive simulation and helped in fast creation of different simulation environments. It supported the adopted validation method consisting in separate validation of parts of the whole specification before the integration and global validation. Some XTP features are still not specified mainly for the reason that they are still not well defined within the English text of protocol.

Concerning the performance aspects, XTP was carefully modelled and simulated (QNAP2) under the best conditions to see the highest throughputs it can produce. The SREQ mechanism has been studied and four different implementation strategies have been proposed and compared. It has been shown that XTP is able to provide a high throughput to its users ; up to 80 % of the raw bandwidth can be delivered to the user in a one-way file transfer application over an FDDI network.

In [AS 93] we have compared the SREQ strategies with respect to the response time and we have analysed XTP error control procedure. Further simulations should be made to study the capability of XTP to face a non constant offered load, the influence of the packet size, and XTP multicast procedures.

A part of information included in this paper was also presented in [BAB 93].

Acknowledgements

The authors wish to thank all anonymous reviewers for their helpful comments and suggestions.

References

[AS 93]   Alkhechi B., Souissi Y., **Performance Evaluation of Some Implementation Strategies of the Xpress Transfer Protocol**, *INT internal report 1993*, also submitted to International Conference on Network Protocols, 1993.

[BAB 93]   Budkowski S., Alkhechi B., Benalycherif M.L., Dembinski P., Gardie M., Lallet E., Mouchel La Fosse J.-P. , Souissi Y., **Formal specification, validation and performance evaluation of the XTP protocol**, *Proceedings IFIP 13h International Workshop on Protocol Specification, Testing and Verification, Liège 1993*

[BUD 87]   Budkowski S., Dembinski P., **An Introduction to Estelle: A Specification Language for Distributed Systems**, *Computer Networks and ISDN Systems Journal*, vol.14, N°.1, 1987

[BUD 92]   Budkowski S. , **Estelle Development Toolset**, *Computer Networks and ISDN Systems Journal*, Special Issue on FDT Concepts and Tools, vol.25, No.1, 1992

[CA90]   Cheung Y. T., Atwood J. W., **Specifying the Xpress Transfer Protocol using Estelle and Valira,** in *FORTE'90 Third International Conference on Formal Description Techniques*, November 1990.

[DEM 92]   Dembinski P., **Queueing Network Model for Estelle**, *Proceeding of the 5th International Conference on Formal Description Techniques (FORTE'92)*, Peros-Guirec, October 1992

[INT 93]   Alkhechi B., Benalycherif M.L., Budkowski S., Dembinski P., Gardie M., Lallet E., Mouchel La Fosse J.-P. , Souissi Y., **Formal specification, validation and performance evaluation of the XTP protocol**, *Report OSI95: OSI95/INT-BULL/Deliverable /P/V3*

[ISO 88]   ISO, Open System Interconnection - Connection-Oriented Transport Protocol Specification, ISO IS 8073, 1988.

[ISO 89]   ISO 9074: 1989(E) - **Estelle: A formal description technique based on an extended state transition model.** First edition 1989-07-15

[LLMB 91]   Lallet E., Lebrun Ch.-A., Martin J.-F., Budkowski S., **Un outil de génération automatique de l'environnement d'exécution de spécifications Estelle,** *Actes du Colloque Francophone sur l'Ingénierie des Protocoles (CFIP'91)*, (Ed. O. Rafiq), Hermes 1991

[LDM93]   Lallet E., Dembinski P., Mouchel La Fosse J .P., Alkhechi B., Budkowski S., Gardie M., **Validation of the Xpress Transfer Protocol,** *The OSI95 Transport Service with Multimedia Support on HSLAN's and B-ISDN*, A. Danthine, ed., Springer Verlag, 42-62

[PEI 92]   Protocol Engines Incorporated, "*XTP Protocol Definition - Revision 3.6*",1900 State Street, Santa Barbara, CA 93101, January 1992.

[RFC 81]   **Transmission Control Protocol** , *RFC 793*, 1981.

Validation of the Xpress Transfer Protocol (XTP)

E. Lallet, P. Dembinski[**], J.P. Mouchel La Fosse, B. Alkhechi[*], M. Gardie, O. Catrina and S. Budkowski
Institut National des Télécommunications (INT), Systems and Networks Department, 91011 Evry, France
Email: stan@int-evry.fr

The paper presents methods and results of a validation of the Xpress Transfer Protocol (XTP). The XTP is a transfer layer (combined transport and network layers) protocol for high performance real-time applications. Estelle, a formal description technique standardised by ISO, has been used in this work to formally specify XTP, enabling validation. The ESTELLE DEVELOPMENT TOOLSET (EDT) has been applied for supporting the specification and validation processes. This study has been realised by Bull S.A. and INT in the framework of the ESPRIT N° 5341 Project "OSI95".

Keywords: communication protocols, high-speed communication protocols, transfer protocol, transport protocol, Formal Description Technique (FDT), Estelle, validation, simulation.

1 Introduction

The aim of this paper is to describe our experience resulting from validation of the XTP (Xpress Transfer Protocol). The XTP is a transfer layer (transport plus network) protocol for high performance real-time applications. It was being designed by the *Protocol Engines Inc.*.

The demonstration of the correctness of a protocol becomes possible if the protocol is formally specified. Estelle [ISO 89, BUD 87], a formal description technique standardised by ISO, has been chosen to formally specify XTP based on its English definition published in [PEI 92]. It is worth noting that this definition has changed three times since the project started (from Revision 3.5 to Revision 3.6 with Addendum 1a proposed in the meantime). The validation presented in the paper concerns the Revision 3.6 of XTP protocol as defined in [PEI92].

The Estelle specification of XTP, whose structure is shown in Figure 1.1, has about 6000 lines; it is composed of 10 modules and more than 100 transitions.

[**] On leave of absence from the Institute of Computer Science, Polish Academy of Sciences, 00901 Warsaw, Poland
[*] Bull S.A. until December 1992

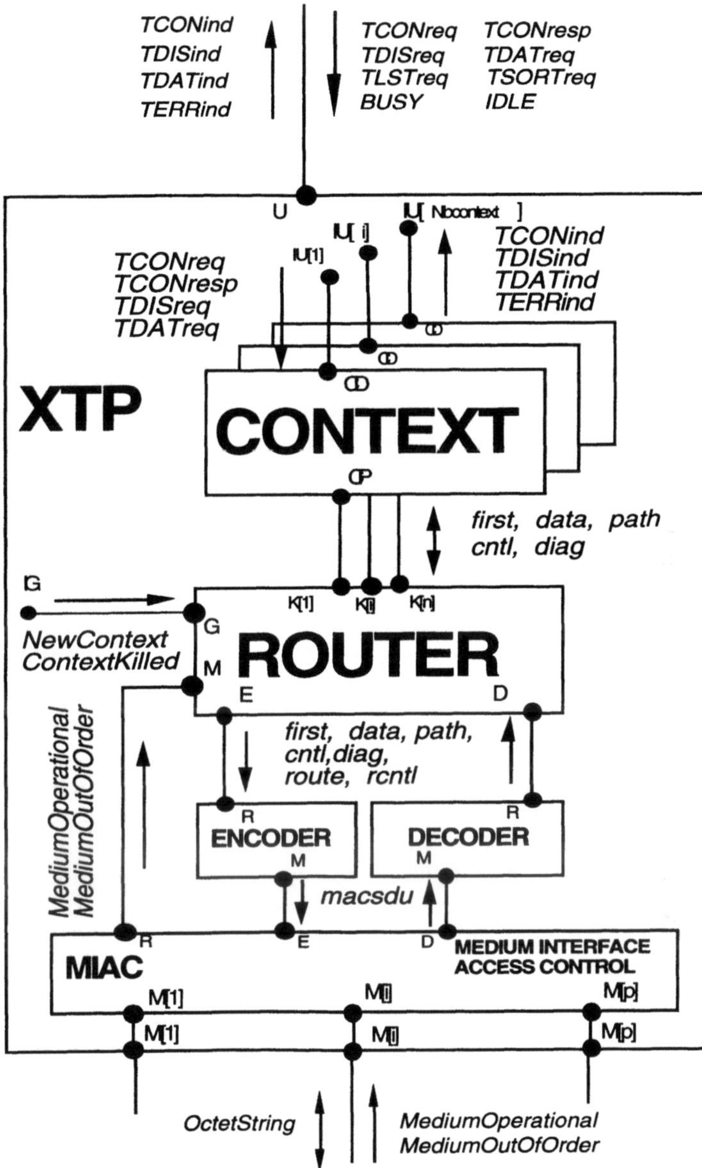

Fig. 1.1 XTP internal structure

Since [PEI92] does not define the service for XTP, we had to provide one in order to proceed with the formal specification and the validation of the protocol. At the same time we were free to conceive a service adequate for our purposes.

For the validation of the different protocol mechanisms, the service definition should offer a simple and flexible access of the "user" to XTP 's procedures, rather than being adapted from existing transport service standards. In fact, new services must be defined in order to exploit the original and powerful features of XTP (see ISO/ANSI work on HSTP).

Precisely, a set of Estelle interactions which carry in the parameters relevant information from/for the XTP procedures is defined at the XTP -Context interface and a similar one at the XTP -User interface. The interaction flow between the context and the user is just "filtered" by the XTP module for performing the context management, sorting, etc. For this purpose a set of internal interaction points is available in the XTP module. Consequently, a "low level" access to the XTP procedures is offered to the user.

This approach allows both for the validation of XTP and the study of an adequate service definition. A detailed verification by simulation of the different procedures is possible, without any restriction from the service, as well as the mapping of service primitives on the Estelle "lower level" interactions. Finally, a service may be defined and it can then be specified by conserving the interactions at the context level (and hence the context specification) and processing the abstract primitives in the XTP module's transitions.

A minimal set of interactions is defined at the XTP -User interface: listen request (TLSTreq), connect request/indication (TCONreq/TCONind), data request/indication (TDATreq/TDATind), disconnect request/indication (TDISreq/TDISind), error indication (TERRind) and sort request (TSORTreq). Two additional interactions are used for flow control on the interface XTP / User, namely, IDLE and BUSY (to indicate that the User is can or cannot accept new indication from XTP). It is worth noting that the disconnect pair of interactions is used only for rejecting an association (a connection). On the other hand, the connect and data request interactions carry a control parameter corresponding to the command *cmd* field of the XTP packets and an additional close parameter which allows them to control the closing of an association.

The following features of XTP were not taken into account in the Estelle specification: the multicast, path failure recovery, and path management features - since they were not completely defined in the Revision 3.6 and the reservation mode, checksum function and confirm mode - since they were outside the main XTP functions. The sort management was restricted to the application / XTP interface.

To validate the Estelle XTP specification and (indirectly) its informal definition the ESTELLE DEVELOPMENT TOOLSET (EDT) was used [BUD 92]. A new tool called UTDG (Universal Test Drivers Generator) has been developed [LLMB 91] and integrated with EDT. UTDG allows automatic generation of the simulation environments facilitating separate validation of XTP functions.

This study has been realised by Bull S.A. and INT within the framework of the ESPRIT N° 5341 Project "OSI95" [INT 93].

2 Brief Description of the Used Tools

ESTELLE DEVELOPMENT TOOLSET (EDT) [BUD92] was used for the purpose of the XTP specification and validation. The basic tool functions are presented below.

Estelle Compiler (Ec)
Estelle Compiler (Ec) translates an Estelle specification into C language source code.
The compiler consist of the following two tools :

- a translator, which given an Estelle specification performs a complete static analysis
  (lexical, syntactical and semantical) of the specification and, in case no errors has
  been found, generates specification representation in so called Intermediate Form
  (IF),
- a C code generator, which given the Intermediate Form representation of an Estelle
  specification returns C-code of the specification.

The Estelle Compiler (C code generator) may be invoked with an option with which
the generated C code is generated in such a way that run-time errors will be signalled
when the code is executed.

Simulator/debugger (Edb)
Edb is an Estelle symbolic and interactive simulator/debugger. The user of Edb may
conduct the simulation in different ways:
  • *Step-by-step simulation*, during which, in each step, the user chooses the transition
to be executed from among the set of the executable ones. Before making the choice
the user can interactively examine all the information that constitute the global state
of the specification under simulation. The user may also modify the values of many
objects such as local variables, the contents of FIFO queues associated to end points
of a channel. This last facility may serve, for example, to simulate loss or reordering
of packets by an unreliable transmission medium.
  • *Random simulation*, during which the simulation goes automatically for a given
number of computation steps, or for a predefined amount of time, and the executed
transition in each step, is randomly chosen from the executable ones.
  • *Extensive simulation*, during which a prewritten simulation scenario can be
automatically executed. For example, a scenario may consist in executing 100 times
1000 (randomly chosen) transitions each time restarting the simulation to try to
execute a different sequence of transitions (random simulation is therefore the
simplest case in this category).
  Each simulation session (of any type) can be traced to the required details,
memorised and then repeated. Access to the dynamic information, interactive and
automatic modifications, tracing, and development of different simulation scenarios
are done using the possibilities of the rich command language of Edb. This command
language offers the user, among others, the two following constructs:

- macros and,
- observers.

The Edb macro enables to group several Edb commands embedded or not in the
control structures and identify them with a given name. Macros are used to define
simulation scenarios.
  An observer describes a global situation that the user wants to analyse (for example
an anomaly to be detected), and an associated action, or group of actions, to be taken
(for example, to delete or alter the exchanged information, to trace some data, or,
simply to break the simulation). An observer can be enabled, disabled, deleted or

chained in a precise position within a list of several observers. The sequence of enabled observers is executed after each computation step.

Universal Test Driver Generator (UTDG)
It is a general purpose tool which facilitates the user validating separately a module or a part of a complex specification. Its main goal is to automatically generate the simulation environment for the given part of the specified system. To simulate a protocol specification it has to be usually complemented with some modules representing the system environment (typically the upper and lower layers of the protocol). These modules, called test drivers (their behaviour is generated by UTDG), are capable of stimulating the protocol by sending arbitrary sequences of interactions, either spontaneously or in response to the received requests. The sequences, which may be send, are not necessarily those allowed from the point of view of the service specified for the protocol. It means that a given protocol may be verified with the assumption that either the environments behaves correctly or erroneously. This latter may serve to assess the *robustness* of the designed protocol.

Estelle simulator/debugger (Edb) and the Universal Test Drivers Generator (UTDG) were extensively used during the validation task, while the Estelle compiler (Ec) helped to detect all static errors within the Estelle specification.

3 XTP Validation Process and Results

3.1 General

To validate a protocol with a help of software tools it is necessary to code the protocol natural language (English) definition into a formalism acceptable by a computer. For that reason XTP was written in Estelle. As a matter of fact, the validation process begins with this coding during which one is forced to carefully inspect the English text to precisely understand the protocol mechanisms and to find their adequate formal representation.

In this coding phase, in Estelle, the EDT compiler serves to verify the specification for syntax and static semantics errors. All such errors can be detected during compilation, since Estelle is a strongly typed language. Several corrections and improvements of the Estelle XTP specification are due to the results of its compilation.

Once an Estelle specification is syntactically correct, and it is completed with a specification of an appropriate environment (see Figures 3.1, 3.2 and 3.3.), it can be further debugged in search for usual runtime errors such as subrange violation, non-initialised variables, etc. As a technique of debugging the random simulation is often used (with environment behaving also in a random fashion) by which the run-time errors are quickly discovered. This form of debugging performed on subsequent Estelle versions of XTP led to correction of many of such errors.

Finally, the dynamic behaviour of debugged Estelle specification, is verified with respect to the logic and functions of the protocol as defined by the source English text. This is the most important part of the validation process. For a specification of XTP size, the verification cannot be based, in this dynamic part, on exhaustive approach. Only carefully designed testing by simulation can reach convincing results.

In such a simulation one tries to check first whether the separate execution of each protocol function (corresponding, for example, to a module in Estelle specification), gives the required results (as described in the English text). Then, if any part checked separately is correct, the protocol specification must be examined as a whole, within an appropriate environment, for presence of required (global) properties, and for absence of abnormal situations. By analogy to the testing terminology, these two validation (simulation) phases may be called unit and interoperability validation, respectively.

The whole validation process, from the specification-and-debugging to simulation, is obviously iterative because an error found during the process requires a change in the specification which may produce new errors, etc.

It is important to stress that validation, in the described above sense, serves to justify that the behaviour of the formal specification (Estelle specification in this case) under test "correctly represents" the behaviour of the protocol as defined in English. Indirectly, however, the protocol itself is verified in that errors found in the formal specification may point out erroneous, unclear or inconsistent definitions in the source English text. Such places have been found in the XTP Revision 3.6 [PEI92] during the presented validation task and are described below.

In order to lower the complexity of the system to be validated, two major and quasi independent parts of the XTP PROTOCOL, namely the Context (represented by the module CONTEXT in Estelle specification of XTP) and Router (represented by the module ROUTER), were validated separately. Once these modules were validated, they have been integrated and tested together. In this paper, the stress is on the validation of the CONTEXT module (Section 3.2 below), since it represent the core of the XTP PROTOCOL.

The simulation phase of XTP validation began with a carefully defined list of the functions of CONTEXT and ROUTER which could be tested separately, and another list of those functions which need a complete XTP entity to be tested. In case of CONTEXT, verified properties of selected functions of this list are described in some details (Section 3.2.1).

3.2 Simulation of the CONTEXT Part

The CONTEXT part of the XTP specification was simulated in the configuration shown in Figure 3.1, in which applications, represented by the USER_A and USER_B modules, use the two CONTEXT modules to establish a proper cooperation through a transmission medium represented by the module MEDIUM.

The internal structure of the CONTEXT modules is also visible in Fig. 3.1. The main components are the SENDER and RECEIVER modules controlling the outgoing and incoming data streams, respectively. The two other modules CMUX1 and CMUX2 serve as simple multiplexers for the streams.

UTDG was extensively used to quickly create different environments, i.e., to create USER and MEDIUM modules, especially when a step-by-step simulation with Edb was performed to examine a given CONTEXT function. A protocol mechanism under test frequently needed certain asymmetry in the configuration of Fig 3.1. Two examples below serve to illustrate this point. The first concerns the XTP flow control mechanism and needs different USERS, while the second deals with a specific exchange of information and needs an asymmetry in CONTEXT modules:

- One USER generates its outputs faster than the other may consume them, trying to saturate the remote RECEIVER. To this end, transitions of one USER are executed faster then the transitions of the other modules (it can be done by means of a special Edb command).
- Setting different values to some constants in the CONTEXT description one easily represents different internal buffer sizes of the two protocol entities. In that asymmetric case, a proper exchange of values of a special variable *alloc* may be checked. The exchange is required to limit the amount of data a SENDER may transmit (see also p.2 in Section 3.2.1).

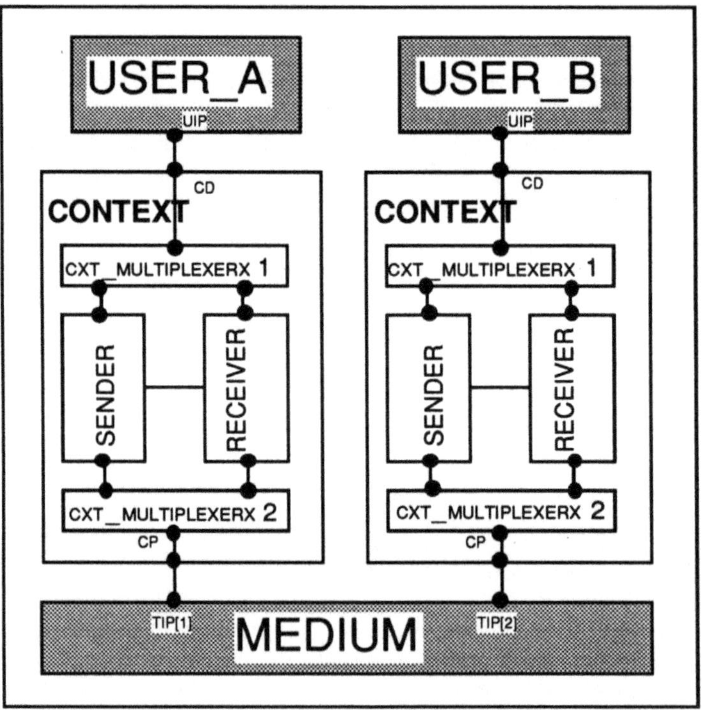

Fig. 3.1 Simulation configuration for CONTEXT

For extensive simulations, performing scenarios specifically defined to test chosen properties or functions in different situations, some environment modules were manually written with their behaviour best suited for the purpose. For example, MEDIUM modules with the following characteristics have been used :

manually written with their behaviour best suited for the purpose. For example, MEDIUM modules with the following characteristics have been used :

- a reliable medium,
- a medium that may loose packets with a given probability,
- a medium that may duplicate packets with a given probability,
- a medium that may reorder packets with a given probability,
- completely unreliable medium that may loose, duplicate and reorder packets with given probabilities.

A general rule while verifying a function of the protocol was to reduce the simulation to the most indicative cases (arguments) of the function, and to gradually make the simulation environment more complicated and less "specialised". To illustrate what is meant by that, let us take the following example:

Suppose one wants to verify the CONTEXT module with respect to different combinations of control and/or closing parameters. These control parameters (bits) (like : NOERR, SREQ, DREQ, FASTNAK, NOFLOW, NOCHECK) and closing parameters (WCLOSE, RCLOSE, END) may go with each packet.

They indicate, to the remote receiver that, e.g., the sender will not retransmit and the receiver should disable error correction procedure (NOERR bit set), or the sender requests immediate state information (SREQ bit set), or the sender initiates graceful or forced closure (WCLOSE or END bit set, respectively), etc. If one wants to check the receiver behaviour in response to a particular control parameter (parameters), he is interested in setting this bit (these bits) to true an he is not interested in values of other bits. Thus, for the purpose of such a simple verification, only a few combinations, out of all 2^n (for n parameters), are "interesting" to check. For example, from the protocol description, one can learn that only 4 of 8 closing combinations are really different, i.e.. "interesting" in the above sense.

A "specialised" simulation environment may consist, for instance, of a reliable transmission medium and one of external users playing only the role of a sender and the other of a receiver, etc. By loosing the constraints on the environment modules, e.g., making the transmission medium unreliable in a way, "less specialised" environments are obtained, multiplying the number of situations under test.

With the above in mind, the following steps might have been taken for the verification in the example :

1) Select a list of "interesting" parameter combinations

2) Check (trace), for each "interesting" combination from 1, the behaviour of the specification depicted in figure 3.1. where :

   - USER_A may only send messages and USER_B may only receive them,
   - USER_A sends only a connection request (not followed by any data message) that includes the interesting combination of parameters,
   - MEDIUM is reliable

3) Check (trace), for each type of unreliable MEDIUM module, the behaviour of the specification as in 2 but with :

   - USER_A which sends one sequence of messages, beginning with a connection request, followed by an arbitrary number of data messages, each of them containing one of the interesting parameter combinations of 1, and only the last message having its closing parameter set.

4) Check (trace) the behaviour of the specification as in 2 but with :

- both users which are completely general sending and receiving users, each of them able to send a sequence as in 3,
- MEDIUM which is completely unreliable.

The Edb scenario for the steps 2 and 3 may simply consists of performing a random simulation for a number of simulation steps that is big enough to allow the computation to stop (to deadlock) earlier (if it doesn't then it means that something goes wrong). A simulations "observer" is tracing fired transitions and interactions during the execution; The analysis of the trace should suffice either to confirm the correct behaviour or to detect the reason (s) of an error.

The Edb scenario for the step 4 may be given in such a way that sequences of messages described in 3 are continuously sent by both users and Edb is restarted each time an error occurred. The description of the error is stored on a file; That way the simulation way go for an unlimited amount of time, and the absence of errors after a sufficiently long period may be a convincing indication (enforced by steps 2 and 3) that this part of the protocol specification is correct;

In fact, very similar to the above scheme (but more elaborated, i.e., with some intermediate steps added) and scenarios have been used;

Many other configurations and scenarios have been defined depending on the tested function and the required quality of simulation.

<<Programmable>> users were also used for the validation of the whole XTP specification (see Section 3.4)

In the following two subsections, the description of the tested XTP functions uses specific XTP protocol terms and notation which cannot be explained in this paper. Therefore, it is assumed that the reader is familiar with them. Otherwise, it is only the authors' hope that the contents of this sections will give the reader an idea of the type of verification that has been carried on during this simulation phase.

3.2.1 Tested CONTEXT Functions

All the main functions of the XTP Context (represented by the module CONTEXT) have been tested (1 to 4 below). For each of these functions, several points have been first verified within the error free environment and with simulation scenarios that have not tried to check how the protocol specification deals with exceptional situations. These exception handling have been treated later (see the next section) when the protocol function was consider correctly specified with respect to this non-exceptional use.

Below, for each function, there is a brief description of several selected points that have been successfully checked.

1) Connection Set-Up:

a) *Reception of a TCONreq, containing data, by a CONTEXT module* (no internal submodule structure exists). It has been verified that:

• the CONTEXT creates all its children modules and moves to the ACTIVE state,

- the CONTEXT stores the data (verification of the initialisation of the variables of the internal buffer of the SENDER within the CONTEXT),
- if the data needs to be fragmented, the FIRST packet, containing the first part of the data, will be sent followed by DATA packets with the remaining data.

b) *Reception of a TLSTreq (carrying listen_confirm parameter) by a CONTEXT module* (no internal submodule structure exists). It has been verified that:

- the CONTEXT moves to the LISTEN state (waiting for the reception of the FIRST packet) and positions a flag (listen_confirm parameter value) to distinguish the case whether the CONTEXT shall ask, or not, the user if it may accept or must refuse a proposed connection).

c) *Reception of a FIRST packet by a CONTEXT module*. It has been verified that:

- if the CONTEXT is in the LISTEN state and the flag listen_confirm is not set, then it delivers a TCONind to the user and moves to the ACTIVE state,
- if the CONTEXT is in the LISTEN state and the flag listen_confirm is set, then it delivers a TCONind to the user and it remains in the LISTEN state waiting for TCONresp from the user to move to the ACTIVE state (the connection is accepted) or waiting for TDISreq (the connection is rejected) to send a DIAG packet and abort,
- if it is in the ACTIVE state and it recognises that the received FIRST packet is a replication of the previously received FIRST packet (*seq* value matches the *seq* value saved in the CONTEXT), then if the SREQ bit is not set, the packet is discarded by the CONTEXT, otherwise the CONTEXT sends back a CNTL packet representing its state,
- if it is in the ACTIVE state and it recognises that the received FIRST packet is not a replication of the previously received FIRST packet (*seq* value does not match the *seq* value saved in the CONTEXT), then the CONTEXT sends a DIAG packet and aborts.

2) Error Control Mechanisms:

a) *Detection of lost, duplicate and out-of-order packets.* It has been verified that:

- the data delivered to the user (application layer) have its sequence number strictly inferior to the sequence number of the first byte of a lost or out-of-order packet.
- when the NOERR bit is set in a FIRST packet sent by a CONTEXT, then:

  - CNTL packets received by the CONTEXT, in response to DREQ (or to SREQ), acknowledge the highest received sequence number,
  - the CONTEXT ignores NSPAN and SPAN received in CNTL packets, which means that retransmission is disabled,
  - the peer CONTEXT do not treat duplicated and reordered packets (the sequence of data delivered to the application may contain duplicates and may be not properly ordered) as well as lost packets (lost packets are replaced in the sequence of data delivered to the application by a space filled with blank characters).

b) *Recovery mechanism*: It has been verified that:

- lost, duplicate and out-of-order packets are retransmitted and inserted in the correct place within the sequence space of the already received data,

- there exists the interoperability between an entity which uses the go-back-end retransmission and another which uses the selective retransmission.

c) *Synchronisation handshake mechanism*: It has been verified that:

- each time an XTP entity waits too long for a response to a SREQ (Status request), or in other more precise words, each time the WTIMER expires, the data flow stream and even the retransmissions are interrupted and the synchronisation procedure is entered (see also Section 3.2.2)
- except for some special cases (see also Section 3.2.2), XTP entity is able to accomplish the synchronisation handshake in a finite delay
- data flow restart is correctly performed.

d) *DIAG packets management*: It has been verified that:

- XTP Context generates, when it is necessary, DIAG packets like, for example, in the case of the DIAG (code=2, val = 1) packet that follows rejection of a FIRST packet.
- DIAG packets that are received by the CONTEXT (from the ROUTER) are interpreted in such a way that the CONTEXT is aborted and the user gets a message with the reason indicated (TDISind with parameters).

3) End-to-End Flow Control:

a) *Sequence space management.* It has been verified that:

- initialisation of the two streams works properly, i.e., the values of SEQ and DSEQ fields in the FIRST packet actually initialise the input data stream and the output data stream, respectively.
- management of the SENDER and RECEIVER internal buffers preserves the sequencing of the two streams of bytes.
- the stream of data delivered to the remote user is identical with the stream of data generated by the sending user.

b) *Acknowledgement policy:* (based on the use of SREQ and DREQ bits). It has been verified that:

- whenever an XTP entity sends a packet (FIRST, DATA or CNTL packet) with its SREQ bit set, the CNTL packet received in response contains information that represents the exact state of the remote entity (see also p.4 in Section 3.2.1)
- when the DREQ bit is set, then the CNTL packet sent in response represents the acknowledgement of the data which have been effectively delivered to the user.

c) *Sliding window management:* It has been verified that:

- the lower bound of its window is communicated by the receiving entity to the peer sending entity in the DSEQ field of the header of all packets. This information is correctly managed in the SENDER part of the sending CONTEXT entity or, more precisely, the part of the SENDER's buffer which is liberated following the DSEQ information, is the correct one.

- sending entity sends data bytes with sequence numbers always strictly below than the upper bound of the remote window. This upper bound is transmitted in the *alloc* field of the control segment of a CNTL packet only (see also p.2 in Section 3.2.1),
- when the RES bit is not set, the *alloc* value represents the internal buffer space of the receiving XTP machine.

d) *NOFLOW option.* It has been verified that:

- The seq-based flow control is disabled when the NOFLOW bit is set.

e) *FASTNAK option.* It has been verified that:

- whenever the receiver detects an out-of-order packet in the input stream when the FASTNAK bit is set and NOERR is not, then the XTP entity immediately sends back a CNTL packet which represents its exact state.
- NOERR bit takes precedence over the FASTNAK bit, i.e., when the NOERR and FASTNAK bits are both set, then no CNTL packet is sent back on reception of an out-of-order packet.

f) *EOM mechanism.* It has been verified that:

- when a TCONreq or TDATreq is received from the application with the EOM parameter, then the respective FIRST or DATA packet is sent with its EOM bit set.
- when a FIRST or DATA packet contains EOM bit set, then the service primitive which delivers the *last* data byte of this packet is marked with EOM information.

g) *BTAG mechanism.* It has been verified that:

- when a TCONreq or TDATreq is received from the application with the BTAG parameter, then the respective FIRST or DATA packet is sent with the BTAG bit set.
- when a FIRST or DATA packet has the BTAG bit set, then the service primitive which delivers the *first* data byte of this packet is marked with the BTAG information.

4) Connection Release.

a) *Graceful close* (based on two WCLOSE/RCLOSE handshakes, one for each data stream). It has been verified that:

- all data have been delivered, including retransmitted data, before a CONTEXT is released (see, however, the case of an too early release described in Section 3.4).
- the CONTEXT which receives an END is released immediately, while the END initiating CONTEXT is released only after a delay long enough to detect the case in which the packet containing the END bit was lost.

b) *Forced termination.* It has been verified that:

- when a sending host receives a RCLOSE before it has sent WCLOSE itself, then the host releases its CONTEXT immediately after having delivered to its application all valid data, i.e., the stream of data items up to the last of them which is not out-of-sequence.
- when a host receives an abrupt END or a DIAG (code=invalid context), then its CONTEXT delivers all the valid data to the application, and then it is released.

Once the Estelle specification was corrected, the system depicted in Figure.4.3 was extensively simulated for a long time with different unreliable mediums. During that time of extensive simulation, an observer traced input and output streams of both applications (users) to check whether the end-to-end transmission was correct. The execution was automatically and periodically restarted to verify different computation paths. This experiment, together with earlier tests of particular functions, has been a convincing indication that the CONTEXT was correctly defined.

3.2.2 Main Problems

The list below is a collection of the most important problems that have been faced during the simulation of the CONTEXT. They indicate those parts of the CONTEXT the description of which are ambiguous or unclear and which may lead, in execution, to the situations considered erroneous or abnormal. The list below contains a brief description of each such situation together (in most cases) with a proposed change in appropriate definitions. Repeated simulation with the proposed changes inserted into the CONTEXT specification has not reproduced the situation.

1. **Retransmission of the FIRST Packet and Synchronisation.** When a FIRST packet has been sent with the SREQ bit set and the WTIMER expired before receiving the expected CNTL packet, the protocol definition is not precise about what is to be done. According to the WTIMER specification (Sections 3.5 and 3.6 of [PEI92]) the synchronising handshake procedure is to be entered, i.e., a CNTL(SREQ) packet should be sent. However, Section 3.2.1 of [PEI92] explains that in that case the original FIRST packet must be retransmitted. Combining the two slightly contradictory obligations, one may only conclude that the synchronising handshake procedure must start with the FIRST(SREQ) packet playing the role of the CNTL(SREQ) packet of the procedure. But then another problem arises. Some DATA(SREQ) packets might have followed the original FIRST packet before the WTIMER expired. Therefore (see [PEI92], Sec..2.3.5), the *sync* value of the retransmitted FIRST packet must be different (increased) which means that the CNTL packet responding to the original FIRST packet will not be recognised as such. Nevertheless, the above interpretation has been chosen and described in Estelle. A received CNTL packet that does not match for SYNC or SAVE_SYNC makes that the synchronisation procedure continues with CNTL(SREQ) packets sent instead of FIRST(SREQ) ones. The later choice is arbitrary but it is justified since the remote Context is known to be active and a CNTL packet is more informative than FIRST. This solution has been simulated and successfully tested. Interpreting the role of the retransmitted FIRST(SREQ) packet in the handshaking procedure in the above way may lead to a livelock situation described in **3** below.

2. **Data Exceeding the Capacity of the CONTEXT Internal Buffers.** It can happen that a packet carrying data (FIRST or DATA packet), satisfies the conditions of the MTU (Maximum Transmission Unit) of all the crossed media, but when it arrives to the end XTP Entity there is not enough place to store data of the packet in the XTP internal buffers. This case is possible as long as the first CNTL packet, carrying the first significant ALLOC value, has not been received from the peer XTP Entity. The definition document says nothing about the way to handle this transient phase. The

most efficient way seems to accept the data until they fill up the buffers, and simply to discard the remaining parts. With the next DATA, the receiving CONTEXT will discover the corresponding gap, and the normal retransmission mechanism will be triggered. A limit case is when there is no next DATA packet to be sent (after the FIRST or DATA packet under consideration). No solution exists and the data may be lost, for example, if the XTP entity will perform an abrupt Close procedure before receiving a first CNTL packet from the peer XTP. The simulation confirmed the existence of the problem.

3. Synchronisation Handshake.

• *XTP fails in its attempt to resynchronise:* An XTP CONTEXT which has not received a CNTL or DATA response within the WTIMER delay, enters in a synchronisation phase. During this phase, if the remote CONTEXT systematically sends back a CNTL response packets with a wrong SYNC value (e.g., because of an error in implementation), no mechanism will act to get out of the synchronisation phase. The CTIMER timer does not cover this case because there are always exchanged packets between the two CONTEXTs (CNTL requests and CNTL responses). A livelock situation occurs. This situation has been reproduced during simulation.

• *Out-of-order packets received during the synchronisation phase:* All the out-of-order packets received during this phase are discarded, due to the fact that their SYNC value does not match the Revision 3.6 SYNC rules. This is true for both types of the SYNC incrementation options for the DATA(SREQ) packets. This behaviour has been reproduced during simulation.

• *Wrong rules of the synchronisation handshake:* The synchronisation procedure can fail because of an unspecified reception, if its implementation follows exactly the rule 5 of the handshake procedure definition. The reason is that the rule's condition is not explicitly defined as the negation of that of the preceding rule 4. To properly understand the rule 5 one has to refer to Sec. 2.4 of [PEI92] where one of the point says that *echo* and *techo* in CNTL packets are interpreted only if $techo <> 0$. Implicitly, it means for the rule that the case in which *echo* matches for *sync* or *save_sync* and $techo = 0$ should be treated the same way as the one in which *echo* does not match for *sync* or *save_sync*

4. DREQ Management: The protocol definition indicates that the reception of packet with DREQ bitflag set causes *always* the sending a CNTL packet in response. No timer is, however, associated to the reception of this CNTL response packet. It means that in case this CNTL response is lost, no mechanism exists for its detection. Consequently, one can question the real usefulness of this command bit.

5. FASTNAK Option: Assume that DATA packets have been lost just before the reception of a CNTL packet with the property that *seq* value is higher than the local *hseq* value. In this situation, how to send back, without delay, a CNTL packet which will indicate the proper information about the span? One solution is to manage a new (*hseq*, *hseq*) span as soon as the CNTL packet with $seq > hseq$ is received. The simulation of the CONTEXT with such a solution showed that it works properly.

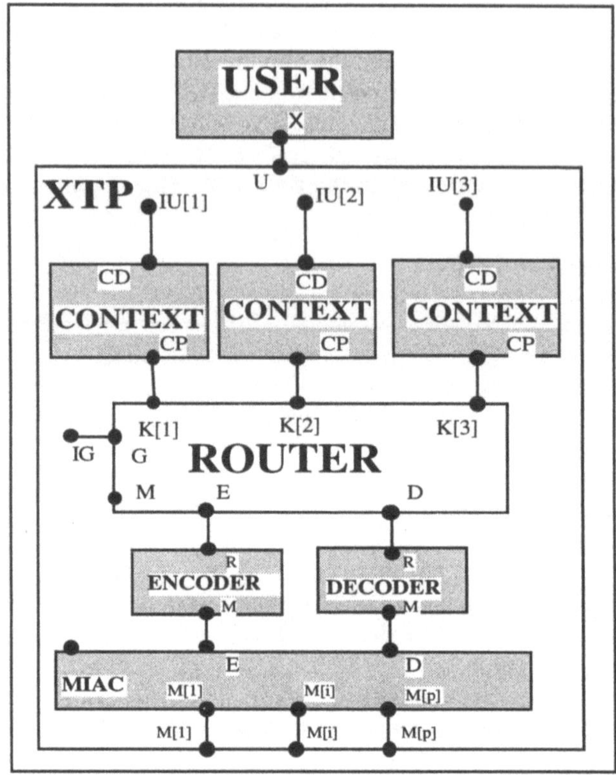

Fig. 3.2 Simulation environment for the ROUTER module. The darkened modules are automatically generated by the UTDG

3.3 Simulation of the ROUTER Part

The functions specific to the ROUTER have been less extensively checked than those of the CONTEXT. This is because the CONTEXT part is certainly the kernel of the protocol and the ROUTER part, though important, is much more implementation dependent and thus specified in less details in [PEI92]. The Estelle specification of the ROUTER has been debugged and simulated within the environment depicted in Fig. 3.2.

The following ROUTER functions have been verified:

- detection and filtering of the duplicated FIRST packet,
- path set-up mechanism,
- path release,
- route sharing,
- route exchange mechanism,
- switch management (path release),
- congestion control mechanism,

- TTL (time to live of a packet) management
- address family management
- context identification and key exchange mechanism;

It is worth mentioning that due to the imprecision of the descriptions of some of the network functions in [PEI92], efforts were made in order to obtain a ROUTER module operational - for the purpose of the simulation of the whole XTP entity, and at the same time functionally close to the English definition. Decisions to select appropriate actions had to be taken for the specification of some essential, insufficiently defined functions, such as path release. Also, several simplifications were made. Some functions, which were less interesting or were not enough defined in [PEI92] (e.g. the activity check done by intermediate nodes on a path) were not specified.

Concerning the path release, one may notice contradictions in the two parts of the description of the path failure recovery (in [PEI92] page 65/fig.3-20 and page 66/fig.3-21) and also the lack of details about the behaviour in a normal situation such as a release, which is simultaneously initiated by both endpoints of the path. As for the activity check, the timer indicated in [PEI92] for starting a path release from an intermediate node of a inactive path is a connection timer (CTIMER), whose value is known only at endpoints, thus, presumably, a new timer should be introduced.

The list of poorly defined or unfinished definitions includes also the less complex mechanism of route exchange for which two different methods are defined: by using the CNTL packets or the RCNTL packets, without explaining which and when to use them. Also, no recovery is defined if the packet which performs a key or route exchange is lost.

3.4 Simulation of the Whole XTP Part.

Finally, the XTP entity was simulated as a whole in the configuration shown in Fig.3.3, which models two interconnected networks. XTP_B was used both as a switch and as an endpoint. With this configuration a set of tests which covered both transport and network functionalities of XTP can be performed.

The tests focused on the integration of the modules separately validated in the previous phases, in order to prove the ability of XTP to offer intended services such as:

- connection (oriented) service with implicit/explicit connection establishment and
   bidirectional/unidirectional/foreshortened graceful close or forced termination;
- acknowledged/unacknowledged datagram;
- transaction.

To simplify the "production" of simulation scenarios corresponding to the services listed above, a "programmable" upper layer application (USER) was included in the specification. The USER is an automaton able to perform the basic operations defined for the XTP-USER interface and designed to be guided through its states by a "test scenario" consisting of Edb commands given from the Edb command line, a macro or a command file.

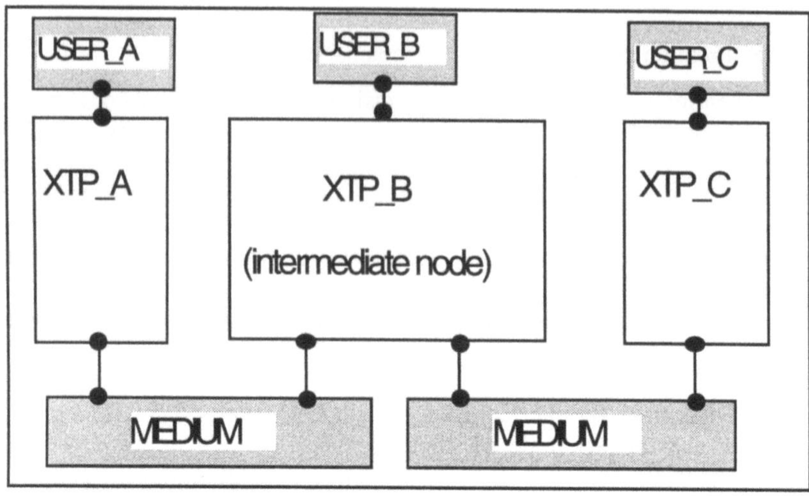

Fig. 3.3 Simulation environment for the whole XTP. The darkened modules are either manually written or automatically generated by the UTDG

In the user's body one can distinguish "request transitions" for sending service requests, enabled by the scenario and "indication transitions" for processing the XTP indications, enabled by the input interactions. Delay transitions are also used, e.g. for data rate control.

A USER "instruction" is a set of Edb commands which defines an operation (e.g. a service request) by assigning values to some of the USER's variables, including the values of the parameters of the interactions to be sent (some default values are also available) and enables a request transition by setting some flags. The USER may receive an "instruction" while the simulation is stopped, from the Edb command line or even when it is running, from an observer. The desired tracing of the execution may also be included in the "instruction" by defining and the necessary observer(s).

Any basic operation may be defined as an "instruction" and easily tested individually without having to make changes in the specification (and hence to recompile it). Several instructions may be necessary to bring the specification in an adequate state (such as a disconnect instruction that needs a connection to be "executed" previously).

The simplest scenario consists of one "instruction" and allows the study of different connection establishments, disconnections or data transfer. The USER stops after the execution of the transition (once or more times, as selected), waiting for an indication or the enabling of other request, according to the state that was reached. The XTP indications are processed (including for example the verification of the received data) and the user's state changes accordingly. Eventually, the whole test specification completes the service request and enters in a deadlock.

The description of more complex tests with sequences of instructions may be edited as an EDB command file and uses observers that identify the moments when different actions should be executed by the USER and hence the next instruction must be given. Extensive simulations were performed using such scenarios.

While the all items of the list of intended services, mentioned above, was finally verified (except for the explicit connection establishment, due to the lack of the confirm primitive in the specified service), several problems related to the graceful close have been detected.

For this mechanism, the Estelle specification has been written according to the description of the Input and Output Close Machines given in Appendix C of the XTP Revision 3.6 definition (note that the Input and Output Close Machines are associated respectively with the RECEIVER and SENDER of the XTP CONTEXT). These two Close Machines, combined with the rules on pages 54, 55 and 56 of the [PEI92] document, have created some interpretation problems which led to doubtful behaviour.

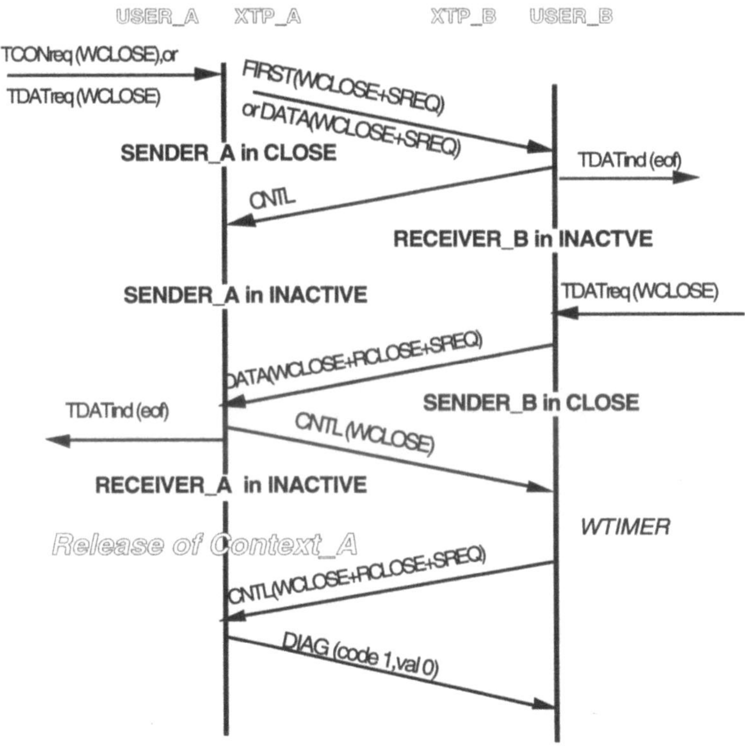

Fig. 3.4 Illustration of an too early release of the CONTEXT_A

• *Early release of CONTEXT* (to simplify, we suppose no retransmission is needed, but the same problem occurs with retransmissions): The XTP_A sends a FIRST or DATA packet containing data with the WCLOSE and SREQ bitflags set. The receiving

XTP_B delivers the data to its application (USER_B). If the CONTEXT_B does not use the CONFIRM mode, then as soon as all the data have been delivered to the USER_B, the RECEIVER_B module moves to the INACTIVE state. The SENDER_B sends all but last DATA packets with bitflag RCLOSE set and, after sending its last DATA packet with bitflags WCLOSE, RCLOSE and SREQ set, it moves to the CLOSE state. When the opposite CONTEXT_A receives this DATA packet with bitflags WCLOSE, RCLOSE and SREQ set, its SENDER_A module moves to the INACTIVE state while its RECEIVER_A module moves to the DELIVER state and starts to deliver data to the USER_A. At the end of the data delivery, the RECEIVER_A moves to the INACTIVE state. At this moment, the situation is the following:

- SENDER_A in the INACTIVE state,
- RECEIVER_A in the inactive state,

In such situation the XTP definition indicates that the CONTEXT_A has to be released. The CONTEXT_B will become aware of that when it will receive the DIAG packet (see Figure 3.4). The CONTEXT_A seems to be released prematurely since the CONTEXT_B will never receive the confirmation that all the data have been delivered to the USER_A (RCLOSE bitflag is used as a means for confirmation during the graceful close).

• *Unnecessary waiting for closing a connection*: This is not an error in a broad sense since it only suggests to follow an inefficient solution due to certain ambiguity of definitions. It may be, however, interpreted as an error in the particular context of the Xpress transfer protocol for which efficiency is of major concern.

When the local application (User) has closed its output data stream (LOCAL_WCLOSE event which corresponds to the reception of TDATreq(emptydata, WCLOSE) by the CONTEXT) it is possible that the SENDER has no data to transmit (XQEMPTY bit is true). The SENDER moves to the CLOSE state. According to the action definition assigned to the LOCAL_WCLOSE event when XQEMPTY is satisfied (p. 86, Table C-4), the following is to be performed:

  - SET_WCLOSE , and
  - START_TIMER.

The interpretation of the SET_WCLOSE (p. 84, Table C-1) is: "set WCLOSE bit on output packets". Since no more DATA packets are to be sent, an implementation may wait, being always in conformance with the XTP Revision 3.6 definition, until the WTIMER expires to send a CNTL packet with its WCLOSE bit set.

To avoid such inefficient solutions, it would be important to make either clear that the SENDER may move to the CLOSE state only if it earlier initiated closing (i.e., it has sent the last DATA packet with the WCLOSE bit set) or to redefine the SET_WCLOSE action to force sending a CNTL packet with WCLOSE bit set without any delay.

The validation of the XTP ended, similarly to the CONTEXT validation, with long extensive simulation sessions. They verified correctness of the cooperation between CONTEXT and ROUTER and confirmed the proper end-to-end transmission of data within unreliable environment.

4 Conclusions

The main results of the Estelle XTP validation are twofold.

The Estelle specification of XTP was carefully inspected, debugged and rewritten in a large part. More than 40 run-time errors (subrange violation, non-initialised variables, ...) were corrected and an important portion of the specification (about 40% of the CONTEXT module) was rewritten. Its final version may be viewed as a formal counterpart of its English reference with exception of functions (like multicasting) which have been announced as not handled in this version. Having in mind that Estelle is an implementation oriented language, the above means that prototype realisations of the protocol that "guarantee" conformance requirements seems to be quite easy to generate.

The XTP protocol itself occurred, in general, correctly specified, i.e., no major deviation from the intended behaviour has been noticed. The protocol may be only viewed sometimes as not robust enough, that is, not able to recover from persistently wrong behaviour at the other side endpoint.

The validation task has provided a lot of methodological experience in using simulation and simulation tools. One of its very concrete result was the design, implementation and integration into EDT of the UTDG tool. The tool occurred very useful in interactive simulation and helped in fast creation of different simulation environments. While it is always true that validation by simulation can only help in eliminating errors but cannot exclude them, it is also worth to point out that systematic and well organised use of this method is highly convincing, especially for large and complex systems. This is certainly the case of XTP.

In the companion paper [BAB 93a] the reader may find complementary information on the XTP specification in Estelle and on the study of performance evaluation of the control information exchange mechanisms .

A part of information included in this paper was also presented in [BAB 93b].

Acknowledgements

The authors wish to thank all anonymous reviewers for their helpful comments and suggestions.

References

[BAB 93a]   Budkowski S., Alkhechi B., Benalycherif M.L., Dembinski P., Gardie M., Lallet E., Mouchel La Fosse J.-P. , Souissi Y., **Modelling and analysing the Xpress Transfer Protocol (XTP)**, *The OSI95 Transport Service with Multimedia Support on HSLAN's and B-ISDN*, A. Danthine, ed., Springer Verlag, 20-41

[BAB 93b]   Budkowski S., Alkhechi B., Benalycherif M.L., Dembinski P., Gardie M., Lallet E., Mouchel La Fosse J.-P. , Souissi Y., **Formal specification, validation and performance** evaluation of the **XTP protocol**, *Proceedings IFIP 13h International Workshop on Protocol Specification, Testing and Verification, Liége 1993*

[BUD 87]   Budkowski S., Dembinski P., **An Introduction to Estelle: A Specification Language for Distributed Systems**, *Computer Networks and ISDN Systems Journal*, vol.14, N°.1, 1987

[BUD 92]   Budkowski S. **Estelle Development Toolset,** *Computer Networks and ISDN Systems Journal*, Special Issue on FDT Concepts and Tools, vol.25, No.1, 1992

[CA90]   Cheung Y. T., Atwood J. W., **Specifying the Xpress Transfer Protocol using Estelle and Valira,** in *FORTE'90 Third International Conference on Formal Description Techniques*, November 1990.

[INT 93]   Alkhechi B., Benalycherif M.L., Budkowski S., Dembinski P., Gardie M., Lallet E., Mouchel La Fosse J.-P. , Souissi Y., **Formal specification, validation and performance evaluation of the XTP protocol,** *Report OSI95: OSI95/INT-BULL/Deliverable /P/V3*

[ISO 89]   ISO 9074: 1989(E) - **Estelle: A formal description technique based on an extended state transition model.** First edition 1989-07-15

[LLMB 91] Lallet E., Lebrun Ch.-A., Martin J.-F., Budkowski S., **Un outils de génération automatique de l'environnement d'exécution de spécifications Estelle**, *Actes du Colloque Francophone sur l'Ingénierie des Protocoles (CFIP'91)*, (Ed. O. Rafiq), Hermes 1991

The Requirements of the New Applications

The multimedia-based applications have been developed first on stand alone systems, with, as communication support, the system bus with or without an additional dedicated bus.

The networking of multimedia-based applications requires the in-depth study of the communication requirements on the distributed transport service provided by local transport entities associated with LAN and WAN facilities integrated with bridges and routers.

The various streams involved, such as the video stream, the voice stream and the text stream, have different requirements of Quality of Service that have to be analysed and quantified.

The multimedia-based application implies not only the handling of various streams but also the support of the constraints between them. The solution proposed by OSI95 is the orchestration services for distributed multimedia synchronisation located, in the architecture, above the transport service.

The QoS expressed at the application level has to be mapped at the transport level into the semantics of the selected QoS performance parameters and into a set of consistent values for them. These QoS will have to be mapped at network and sub-network levels. This operation is particularly important when the network and the sub-network are able to offer more than the best effort QoS of the connectionless environment.

To support these views, OSI95 has developed a QoS architecture concept. This vertical distribution of QoS parameters at various levels of the architecture as well as the integration of all the QoS defined at every layer to provide the QoS required by the application may be considered as a part of the management of the open system.

The assessment of the OSI95 transport service regarding the requirements derived from the study of the multimedia-based application has been done in the final deliverable of Lancaster University.

Distributed Multimedia Communication System Requirements

Helmut Leopold[1], Andrew Campbell[2], David Hutchison[2] and Nikolaus Singer[1]
[1] Alcatel Austria Forschungszentrum, A-1210 Vienna, Austria
[2] Lancaster University, Lancaster, LA1 4YR, UK

The evolution of distributed computing is being influenced simultaneously by new multimedia applications requirements and by the increasing capability of high-speed multiservice networks. The performance of the communications system will be critical in meeting the needs of the emerging distributed multimedia applications; in particular new transport mechanisms should guarantee Quality of Service for a wide variety of end user systems. Specification and design of a high performance transport service and protocol should be based on a top down requirements approach. This paper describes the key requirements of distributed multimedia communication systems and their role in establishing a new Quality of Service Architecture (QOS-A).

Keywords: multimedia communications, open distributed systems, communications requirements, Quality of Service Architecture.

1 Introduction

Distributed computing environments are becoming more popular as can be evidenced by the widespread use of high performance workstations and interconnected LANs. The evolution of distributed computing is being influenced by two major factors: the new distributed multimedia applications requirements and by the increasing capability of high performance multiservice networks. Recently, multimedia computing has emerged as a major area of research motivated by the wide range of potential applications that are made feasible by combining information sources such as voice, graphics, hi-fi quality audio and video.

The combination of multimedia computing with distributed systems support offers great potential. A wide variety of application areas for distributed multimedia systems have been identified [WB 91] such as office information systems, scientific collaboration, conferencing systems and distance learning. The identification of these new application areas are evolving in parallel with the rapid advancement of multiservice communications technology.

Multiservice networks encompass high bandwidth MANs, such as DQDB and FDDI, and B-ISDN based on fast packet switching principles. Currently, broadband transport mechanisms such as ATM are capable of supporting a wide range of multimedia platforms with diverse quality of service requirements, from low speed voice transmission, to very high speed HDTV distribution. This technology push is influencing, and acting as a catalyst for, the specification of new distributed multimedia applications.

In parallel to these evolving areas of multimedia communications on the one hand and high performance multiservice networks on the other hand, a new standardisation area is becoming increasingly important - Open Distributed Processing (ODP). Besides other issues, ODP deals with the problem of information exchange within possibly heterogeneous distributed systems in general. In contrast to OSI, which is concerned with communication between end-systems only, ODP has a broader scope and considers activities both between and within end-systems. That means, instead of just dealing with the protocols between end-systems, the development of concepts and architectures for processing in a distributed environment is also treated.

The integration or harmonisation of these environments - B-ISDN, the OSI Reference Model (OSI-RM), the ODP architecture and the emerging multimedia end user service requirements - is identified as an essential issue. This has led to an important area of research. As an initial step, the relationship between the B-ISDN RM and the lower layers of the OSI-RM were investigated in [BOE 92]. The understanding of the basic differences and commonalities of ODP and OSI was the main objective of [SBC 92]. Finally, to identify the distributed multimedia requirements on the communication systems of tomorrow, [LBC 92] had the objective to investigate modern multimedia applications. The concern of this present paper is to summarise the 'distributed multimedia communication requirements' on a communication system. Thus we provide the first step in a top down approach for the ultimate goal of designing an enhanced transport service and protocol within the new communication environment.

This paper is structured as follows. The overall scope is presented in section 2 and influences on standardisation are briefly described in section 3. In section 4 a summary of the state of the art in distributed multimedia applications is presented. Then we introduce, in sections 5 through 9, the set of critical communication issues that should be supported by a new enhanced transport mechanism. Finally, related work and conclusions are presented in sections 10 and 11.

2 Distributed Multimedia Requirements Study

This paper results from work carried out by the ESPRIT OSI95 project. The aim of the Lancaster University/Alcatel Austria Forschungszentrum contribution to this project was to take a top down approach to determining requirements for OSI protocols by studying distributed multimedia applications [LBC 92]. This complements the work of other partners who were considering the impact of new communications systems, particularly ATM/B-ISDN. Both areas are the background for the main objective of OSI95, to design a new Transport service and protocol.

3 Standardisation Work

Standards have an important role to play in development of distributed multimedia systems. It is important for the commercialisation of multimedia that standards are agreed for the representation and interchange of multimedia. It is also important that existing standards are adopted where possible to allow multimedia to be integrated into the current generation of systems. However, currently OSI standards do not take into account the particular requirements of high performance distributed multimedia applications. For example, the OSI Reference Model (RM) does not support the kind of multimedia synchronisation, which we call orchestration, required by continuous media applications. It can therefore be anticipated that the OSI-RM will have to be extended or modified to support new multimedia profiles. A more detailed version of this paper is a UK contribution [CCG 92B] to ISO/IEC JTC1/SC6 New Work Item on Enhanced Communications Functions and Facilities (ECFF) for Lower Layers. The requirements identified in the contribution have been incorporated into the ISO Draft Guidelines [ISO 92].

Furthermore an important contribution to the new work item on a QOS Framework within ISO/IEC SC21 was done, to provide a first stage in the development of an integrated QOS Architecture which offers a framework to specify and implement the required performance properties of new multimedia applications over multiservice networks [CCG 92a], [CCG 93].

4 State of the Art in Distributed Multimedia Applications

There are a number of application areas where the introduction of multimedia can enhance the utility of existing systems. In addition, there are many new areas of application made possible with the emergence of multimedia. Also the topic of Computer Supported Cooperative Work (CSCW) is highlighted as an important development with implications for distributed multimedia technology.

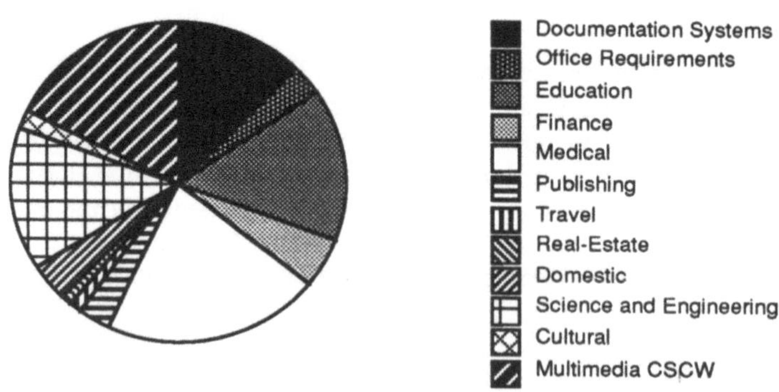

Fig. 4.1 Areas of application for distributed multimedia computing

In total, the application survey [WB 91], done within the OSI95 project, covers over 100 projects within Europe and the United States. The overall balance of projects is shown in figure 4.1. From this survey, the following general comments can be made about the state of the art in distributed multimedia applications:-

- *interest in distributed multimedia applications:* There is a great deal of interest in distributed multimedia applications, with a surprising level of activity uncovered in European programmes such as RACE, ESPRIT, DELTA and AIM.
- *diversity of application areas:* Distributed multimedia computing has applications in a wide range of areas such as the office environment, health, the financial sector, teaching, service industry, retail, science/engineering and in cultural activities.
- *importance of integration:* Integration is the biggest motivating factor in most multimedia applications. More specifically, the majority of applications are concerned with the integration of a variety of information systems into a common framework.
- *ad-hoc solutions:* Many of the applications have been developed on either analogue technology or on a hybrid analogue/digital solution. Few have experimented with completely digital solutions. This has allowed interesting prototypes to be developed but has deferred many of the technical problems associated with digital media.
- *importance of the end user:* A number of previous experiments with multimedia have failed (most notably videophones) because the technology has not been popular with end users. It is therefore important in future designs of distributed multimedia applications, to take into account the needs and wishes of end users and to involve them in the design process. This requirement is illustrated by the growth of the field of CSCW in recent years.

5 Key Issues in Distributed Multimedia Communications

The new high speed transport mechanism will provide enhanced protocols and services which will operate over very reliable, high bandwidth multiservice networks. The primary objective of such a transport service is to provide a high performance interface and guaranteed Quality of Service (QOS) to a wide variety of distributed multimedia applications. The transport protocol should provide QOS negotiation and renegotiation with the network provider on behalf of the Upper Layer Architecture (ULA), support for end-to-end connection management and providing different classes of services over a number of network environments. This section presents the major identified communication issues and categorises them as (i) general multimedia communication issues; (ii) group and multicast support; (iii) QOS; and (iv) synchronisation of multiple related streams, which we call orchestration.

It should be noted that the requirements can be divided into service and protocol related issues. Both issues are described to present a complete view of the problem. It should be clear that a requirement to a protocol service does not necessarily imply a specific protocol mechanism. Especially, since the underlying services have an essential impact on the used protocol concepts. It is very important to note, that within this paper the main objective was to follow a top-down approach. Thus no solutions for any communication requirements are discussed. This is a second step of our approach and outside of the scope of this paper.

6 General Multimedia Communication Issues

6.1 Continuous Media

The single most significant attribute of multimedia applications is the concept of continuous media, particularly audio and video, although the transmission of other data types such as image, text and graphics (i.e. still media) will normally be involved. The significance of continuous media communications lies in the real-time, isochronous nature of the traffic which is not adequately supported by existing communications sub-systems.

Continuous media consist of continuous sequences of samples that have strict timing dependencies [NIC 90], and thus are often referred to as streams [CBD 92]. Continuous media requirements are quite demanding, for example, minimum latency, guaranteed bandwidth, real-time transmission, bounded jitter, QOS guarantees, orchestration within single streams and between multiple streams, media and application specific protocols and error recovery strategies etc. These various requirements are discussed in greater detail below.

6.2 Error Sensitivity of Various Information Types

Any network will produce errors and there are two types of expected errors in high-speed networks: bit corruption due to noise and packet loss due to congestion. One function of a transport protocol is to provide the required end-to-end error rate desired by the application. Correct data delivery encompasses two issues: (i) correctness in the value domain; and (ii) correctness in the time domain. For the correctness in the value domain, we have two approaches to error control (It is important to note that error control encompasses both error detection and error correction, in the following text):-

- *Natural Redundancy in Multimedia:* High reliability is not always of pressing importance since media which includes natural redundancy such as voice and video are inherently tolerant of small amounts of error. Thus, instead of high reliability, error announcement (i.e. error detection and indication) at the receiver side gives the user the possibility to determine the grade of error rate and degradation in the requested service.
- *Error Sensitive Information:* In contrast to the requirement of information types with natural redundancy, information which is highly error sensitive, such as file transfer of control data, X-Ray pictures, etc., needs an error control mechanism for higher reliability.

For the correctness in the time domain, the communication system should support timeliness (i.e. bounded time limits) of media delivery. This can be achieved by providing a validity time to each message with an associated deadline. Thus, a global time, or a common reference time that provides system-wide time synchronisation is a basic issue in distributed real-time systems.

Distributed real-time application must be informed when the Communication System (CS) fails to transmit a message within its agreed time frame. Notification of such a failure allows the application to perform exception handling if necessary. We describe this as bounded invocation. Taking into account the timeliness requirement we get a third information type, which is sensitive to errors in both domains, the value and time domain (i.e. timeliness requirement).

- *Real-Time Error Sensitive Information:* Real-time applications which are sensitive to errors must transmit redundant information (i.e. Forward Error Correction (FEC)) in order to recover from loss without retransmission, which may cause unacceptable delays.

6.3 Error Recovery Mechanisms and Timeliness

Distributed multimedia applications often trade off reliability and timeliness, especially in high latency networks. A new communication system should support flexible error control selection to permit the intended trade off to be specified. Thus it is important that applications have means to select a suitable level of error control, ranging from full error recovery to only error indication, or no error treatment at all.

A possible set of error control procedures of a modern communication system could be (i) go-back-N; (ii) selective retransmission; (iii) Forward Error Correction (FEC); and (iv) no error correction. However, it is important to note, that the used error control mechanisms depends on the required service from the higher layer on the one hand and on the used underlying service on the other hand. That means, it is clear that such an internal protocol behaviour may not be visible to the protocol service user.

6.4 Error Recovery and Application Requirements

To allow distributed multimedia applications to make a trade off between reliability and timeliness on the one hand, and to give the application the possibility to perform error recovery at the application level on the other hand, the new transport mechanism must support flexible error control. For example, video can sustain a certain degradation in quality by simply redisplaying the previous frame, that is, continuity, which would be meaningless for speech. With regard to applications, some may prefer or tolerate slower delivery rates in order to improve quality, thus it is important that applications have a means of controlling and selecting error recovery strategies, as already stated above, ranging from full error recovery to just error indication, or even possibly no error treatment at all.

6.5 End-to-End Flow Control

The most critical requirement of a high speed transport is to sustain isochronous transmission and reception rates. To do this, the transport protocol [SDH 91] must be designed to support continuous transmission of data. This requires an open window flow control mechanism where the sender does not wait to send (rate-based flow control). End-to-end flow control techniques have a major impact on the effective throughput of the transport protocol. The receiver must process the received data stream in real-time and provide the buffering implicitly agreed upon during connection establishment.

6.6 Transaction Oriented Communication

Transactions occur in distributed systems such as distributed operating systems, databases or reservation systems. Although a Request/Response Communication (a client issues a request, a server responds to that request) is the basis for a transaction, it is important to note there are a lot of differences. The term transaction will imply the characteristics of failure atomicity, failure isolation, permanence and serialisability. The Transaction paradigm is an application oriented term and should not be misunderstood as a communication method for lower layers of the OSI model.

A similar communication issue is the Remote Procedure Call (RPC) at the OSI Presentation layer. The RPC Paradigm is a straight forward concept as described in [Nel 81], [BN 84]. It is based on the premise that distributed programming will be more readily adopted if inter-machine communications can be handled in a familiar and simple way. The procedure call is the primary means of invoking and transferring control and data between local procedures. In essence the RPC simply extends the procedure call concept to allow the control and transfer of data across communication networks. Rather than sending and receiving messages a program calls a remote operation and receives return values. This hides distribution complexities from the programmer and provides a familiar interface, that is, the procedure call. In an ideal world one would not be able to distinguish between remote and local procedure calls, distribution would be totally transparent. The following section describes some transactional oriented issues.

6.6.1 Reliable Request/Response Interaction Versus Unreliable Messaging Communication

In the case of a reliable request/response service both the request and the response have to be delivered reliably. Connection oriented mode, which supports reliable communications, requires a connection setup which adds additional overhead to the interaction. This overhead could be too high in relation to the duration of the whole interaction. If this is the case then connectionless services or fast connection setup and release could be more appropriate to real-time multimedia operations.

The various possibilities to implement distributed processing in general and RPC concepts on top of the OSI stack in particular, are outlined in [SBC 92].

6.6.2 Maintaining Connections with Long Lifetimes

If connections are used to support a reliable kind of communication, the duration requirements for connections may vary from milliseconds to long periods of time. The communication system should maintain connections that operate over long periods of time by polling the connection over its life-time. In addition, network resources required for long term connections should be managed efficiently; this is dependent on the underlying network provider but could include that dynamic allocation and deallocation of resources or the opening and closing of a connections when the communicating entities are active or not.

6.6.3 Multimedia Communication and RPCs

The basic RPC paradigm does not provide suitable conceptual basis for multimedia communications, primarily because multimedia requires a non blocking asynchronous structure that allows continuous data to be transmitted in a high performance environment. If the paradigm was modified to support the latter, it would still require additional extensions and modifications to support synchronisation, multiple stream transmission, timeliness, flexibility, QOS guarantees (see below), selective transparency etc., all of which, would produce a highly complex generalised paradigm with enormous overheads. RPC would lose its simplicity and familiarity, and thus the very features that have led to its wide acceptance and use.

Conceptually RPC is unsuitable because it cannot handle incremental results, remote display (i.e. I/O operations specify not only the data but also the source of or sink for the data) and continuous media within the context of the multimedia requirements.

It is proposed to offer the user a dual communications model by supporting synchronous blocking RPCs and non-blocking asynchronous remote pipes. A pipe is an independent service that may exist in its own right. It represents a transmission path, for streams, that will guarantee a particular set of characteristics, for example, an associated QOS, synchronisation strategy etc. These characteristics may be defined and updated via an appropriate interface. A pipe service should provide considerable flexibility in the selection of synchronisation and error recovery strategies, QOS, transport protocols etc..

Thus, the RPC mechanism may be used for control and the pipe abstraction for handling the actual transmission of continuous data, that is, streams.

6.7 Heterogeneity Requirements

Accommodating heterogeneity represents one of the biggest issues in distributed computing. Heterogeneity of hardware, software, models, architectures, etc., is a fundamental requirement of any distributed multimedia system as it provides the diversity that is necessary to solve specialised application problems. As well as being a requirement it is also unavoidable given industrial and commercial variability. Thus a distributed environment provides greater reliability, performance etc., but more significantly, allows users to access and utilise a much broader range of hardware and software resources. This issue was and is one of the main concerns of OSI and now also of ODP. The relationship between these two worlds, and their possible influence on the communication sub-systems is explained in more detail in OSI95 deliverable ELIN-3 [SBC 92].

6.8 Security

Security is a critical factor within emergent high speed networks. Security has been historically a difficult feature to provide in shared networks. It is envisioned that both unclassified and classified multimedia applications will utilise the new network environments, so protection mechanisms must be an integral part of the network access strategy. Features such as authentication, integrity, confidentiality, access control are essential to provide trusted and secure communication services for the new high performance applications.

However, although it is clear that security is an integral part of the protocol environment within modern high-speed networks, it is out of the scope of this paper.

7 Group and Multicast Support

The applications study [WB 91] highlights the importance of group and multicast services as characteristics of distributed multimedia communications. For example, many CSCW applications require sophisticated patterns of connectivity between different sites. Note that multicast is important for both transactions and streams. Consequently, the underlying transport mechanism and the network provider must support group connection management and multicast addressing respectively. Connection management must be flexible enough to support a wide range of data transfer, from a simplex connection between a camera and a remote display to more complex configurations that support multiple remote sources connected to a common local or remote sink.

Some communications media, such as satellite and Ethernet, inherently support broadcast and multicast communications without any additional communication system degradation. Also future broadband networks, such as ATM will provide multicast and broadcast support at the network interface, which will be achieved by replicating data as it is switched through the network. However, further research is required to determine the best way to exploit this facility at higher levels.

7.1 The Concept of Groups

A group is a concept to form an association between a set of users, peers or object entities. Grouping is motivated by the need to (i) encapsulate the internal state and hide interactions among the group members; (ii) provide fault tolerance by means of functional redundancy; (iii) reduce network and sender resources by delivering a single message to multiple destinations without the overhead of multiple discrete addressing; (iv) provide a high level communication interface that simplifies applications software complexity for simultaneous communications with multiple receivers.

7.2 Multicast Services

Multicast is the simultaneous transmission of a single message to multiple destinations (i.e. all destinations will receive the same message). A multicast service must provide the same QOS associated as with the more traditional point-to-point network access services, in terms of selection of end-to-end delay, jitter, throughput and reliability. A multicast service should be configurable to the communication need of distributed multimedia applications; that encompasses:-

- *communication topology:* That includes (i) multicast (one-to-many) communications; (ii) broadcast (one- to-all) communications; (iii) multi-collect (many-to- one) communications; and finally (iv) conference (typically many-to-many in video conferencing configuration) communications.
- *communication semantics:* That means either the sender will not have any knowledge about the number of recipients that actually received the message (un-acknowledged mode); or each message is acknowledged by each recipient (acknowledged mode); or the receivers retain the delivery of the message to the user, until a notification arrives from the sender (synchronised mode[1]).
- *collection and distribution functionality:* It is important to have a form of naming transparency; i.e. group names to address the receivers on the one side, and to process the replies properly on the other side (reply-handling transparency).
- *collection and distribution policy:* It should be possible to specify any set of receivers or senders and the timing relation of the communication. Examples are at least, at most, exactly; one, some, all; etc.

7.3 Multicast Error Control

Multicast error control poses a significant challenge for distributed multimedia applications, because the data transfer mode can be either (i) continuous media, (ii) transaction oriented communication, or (iii) bulk data transfer on the one hand, all the above stated multicast communication issues have to be supported on the other hand. Additionally, the dynamic nature of groups adds to the problem that members may join and leave groups with different QOS requirements. Thus the error control mechanism should also be flexible enough to meet new member's QOS requirements.

8 Quality of Service

Quality of Service parameters in current communications infrastructures permit the specification of user requirements which may or may not be supported by underlying

---

[1] This mode may be used to implement atomic multicast.

networks. Usually, these values are agreed between the carrier and the customer at the time the customer subscribes to a particular network service. Another function of these parameters is to form a basis for charging customers for pre-specified services.

With the emergence of digital video, audio, and other continuous media types, increased requirements are placed upon QOS support. For example, to support video connections high throughput is required and therefore high bandwidth guarantees will have to be made. Audio, on the other hand, will not require such a high bandwidth. End-to-end delay and delay jitter (i.e. variance in delay) are new factors which must be taken into account for continuous media transfers. In particular, interactive continuous media connections impose stringent delay constraints, derived from human perceptual thresholds, which must not be violated. Delay jitter must also be kept within rigorous bounds to preserve the intelligibility of audio and voice information.

It is therefore important to support a notion of a Quality of Service Architecture (QOS-A), whereby application requirements can be easily mapped down onto the communications sub-system, that is, high level requirements can be mapped onto communication channels and resources. This mapping should be clean, simple and efficient, protecting application programmers from communication details, that is, they should be stating what they require rather than how it is to be achieved.

Particular applications may require certain minimum standards, below which transmission may as well terminate, on the other hand other applications may require only an indication that the QOS has degraded, in this case the application is responsible for taking any action regarding the connection. It is up to the communications sub-system to guarantee that the required QOS is maintained throughout transmission. Users must have some means of defining the QOS they require at an appropriate level, these may include specific quality statements, don't care, best fit options or guaranteed.

8.1 Transport Level QOS Parameters

A set of suitable QOS parameters [LBC 92] which are meaningful to the transport level and the levels below, and which may be employed in characterising individual continuous media types, are (i) throughput, (ii) end-to-end delay, (iii) jitter, (iv) error rate and (v) priority (see table 8.1).

Table 8.1 QOS Parameters relevant for the Transport level

QOS parameter	specified range	example value	possible meaning
throughput	min, max, average	1M,2M,5M	bits/s
end-to-end delay	max	10	1 = 10 ms
jitter	max	10	1 = 1 ms
priority	value 1..n	3	1..256
error rate	max	9	$1 = 10^{-1}; 2 = 10^{-2}$

At connection establishment time it should be possible to quantify and express preferred, acceptable and unacceptable tolerance levels for each of these parameters. The requested parameters should then undergo full end-to-end negotiation. The finally agreed QOS should then be guaranteed for the duration of the connection (or at least a QOS degradation indication should be provided if the contracted values are violated). The background for these parameters is as follows:

- *Throughput and End-to-End Delay:* Multimedia applications require high bandwidth (from 12 Kbytes for a page of text to 150 Mbits/s for HDTV) and high performance delivery mechanisms with bounded end-to-end latency. These bandwidth requirements may be reduced with the use of appropriate compression techniques at the expense of latency. Based on these figures, new multiservice networks will be able to handle the transmission of raw data. This basic scenario seems to be acceptable until one considers real-time requirements, multiple transmissions, orchestration and the need for a guaranteed QOS, all of which impose significant demands on the communication stack.
  Also, late data in continuous media streams is as serious a problem as lost data, and therefore the new service must attempt to provide the required level of error control, taking into account the timeliness requirement.
- *Jitter:* There is never complete synchronisation between a sender and a receiver since variable delay (i.e. jitter) is introduced by a real network. Therefore, the sender and receiver may drift in and out of synchronisation. Continuous media streams require that jitter delay be kept within a small upper and lower bound. Jitter introduced by queuing in intermediate nodes may be reduced by the use of buffering at the receiver. The amount of jitter allowed must be traded off against the end-to-end delay requirements and the receiver buffer availability. It is important to note that the jitter detection is done within the network, and jitter correction at the receiver.
- *Priority:* The CS should support multiple levels of priority for multimedia traffic and the pre-emption of network resources for high priority applications. The transport service must be flexible enough to allow the transport service user to attach relative priorities across end system connections. It is feasible that this could be supported during connection set-up using an associated QOS parameter. High priority connections will get serviced before lower ones. Lower priority connection packets will be dropped first before high priority packets, should the network become congested, which is probable. Real-time systems require guaranteed delivery of high priority control messages such as a control message, even under peak-load condition[2].
- *Error rate:* Packet error rates are dependent upon the size of the Transport Protocol Data Unit (TPDU) at the transport layer and in some way the Transport Service

---

[2] Experiences at Olivetti Research Lab have shown that it would be advantageous if different priorities for the audio and video connections of, e.g., a video conference are used. Then one could drop video frames and keep audio going, for two reasons; selecting 'audio-only' still allows meaningful communication, whereas 'video-only' reduces the communication to miming at each other. In addition, video with 50\% dropout (recovered by playing old frames) is much more acceptable than audio only with 50\% drop-out (no recovery only silence or audible sound). This observation implies that some media are more robust when they are degraded than others.

Data Unit (TSDU) at the Transport Service Access Point (TSAP). Bit error rates are associated with the reliability of the network[3]. The user should just have to deal with packet error rates at the TSAP. Further research is necessary to work out an acceptable specification for these packet error rates.

The following section summarises additional QOS issues required to meet the needs of the new distributed multimedia applications.

8.2 Profile and Class of Service Selection

The transport service and protocol should be flexible enough, to support the full range of multimedia applications and the wide variety of communication characteristics, and work over different high performance networks. The idea of protocol selection (i.e. profile selection) is well established within the OSI-RM, and therefore will also be supported in our model.

Although we advocate the use of QOS parameterisation of connections, we do not envisage a single fully generic transport protocol that can cater for all types of traffic equally well. Instead, due to the diverse nature of distributed multimedia applications, different protocols will be required for different types of traffic and control data. We thus envisage that communication sub-systems of the future will be composed of horizontal and vertical subdivisions in a *protocol matrix*. The user will be required to select a protocol profile that will provide a suitable service for each traffic type without duplication of functionality at the various protocol layers.

Class of service selection will implicitly select between reliable and unreliable communications, and thus use various error control mechanism. In addition to the provision of profile selection, we see the need for an extension of the more traditional notion of class of service selection employed in current OSI standards. In the current OSI framework, classes of transport service are selected according to the network service over which connections are to operate; that is whether the network is connectionless, connection oriented, reliable, unreliable etc. For future multimedia applications the notion of class of service selection should be extended to encompass more user oriented functions.

8.3 Guaranteed QOS

Multimedia applications that can accurately forecast their resource requirements may prefer to reserve transport and network resources in advance via a connection oriented service. Also, transactions that require fast response often cannot wait for the connection time associated with resource allocation which is greater than twice the round trip delay. Applications that require any QOS higher than best-effort (offered by current networks) must request that resources be allocated to their connection in both the network and the transport (host-network) interface.

---

[3] In our text we described 'error rates' which cover both bit error rates and packet error rates. Clearly there is a distinction between these communications characteristics, but we adopt 'error rate' to encompass both.

An important requirement of continuous media QOS management at the network level is a suitable reservation scheme which is able to set up and guarantee network resources such as bandwidth and end-to-end delay for high performance continuous media communications between network subscribers. This could be achieved using resource reservation protocols based on such as SRP (Session Reservation Protocol) and ST-II (Stream Protocol) [Top 90] or directly provided by the network services, as proposed in ATM networks.

While peak bandwidth network resource reservation strategies lead to lower utilisation of network bandwidth, guaranteeing continuous media QOS is more important than maximising utilisation of the link bandwidth.

8.4 End-to-End Argument: End-System Responsibilities

In multimedia systems it is not sufficient to guarantee the transmission of information between two endpoints. It is also important that the operating system provides appropriate response. Hence, QOS issues such as throughput, latency and real-time responsiveness are end-system concerns as well as concerns of the communications infrastructure. This makes the relationship between, and integration of, the operating system and the communications infrastructure particularly important in distributed multimedia systems.

8.5 Dynamic QOS Control

In distributed multimedia systems it is not sufficient to specify a QOS level, protocol profile and service class at connection time which will statically remain in force for the life-time of the connection. A more flexible service is needed whereby users can dynamically alter the QOS of a connection while it is active. This is required because there are applications such as continuous media which have extremely diverse QOS demands which may vary during the course of a connection. Such dynamicity is illustrated in the following examples:

- If 'best effort' guarantees of QOS level are selected at connection time, the QOS level may degrade, and this will result in the transport user being informed of the degradation (via a QOS degradation indication). Rather than simply closing the connection down or accepting the degradation, the user may want to re-assess his priorities and, for example, close down other connections to save resources and then upgrade the first connection or retain the connection with the currently degraded QOS.
- The user may use the same connection successively for different purposes at different times. Examples are the transmission of full motion video interspersed with intervals of slow motion, or upgrading from monochrome to colour video, or telephone quality to CD quality audio. Another possibility is the in-service insertion of a compression module which may reduce the bandwidth requirement but increase the susceptibility to errors.

Of course, to permit QOS negotiation and renegotiation at the TSAP, the necessary support must be provided either within or below the transport protocol itself. Thus mechanisms are required to alter link-level bandwidths and/or processing and buffering resources at intermediate nodes. One possibility is the use of re-configuration at the network level in the context of protocols such as the real-time channel administration protocol [BM 91] where resource reservations are made at intermediate nodes. Even when an underlying connection must be torn down and re-established in order to achieve an alteration in QOS level, there are strong arguments for doing this transparently behind the transport service interface. For example, it allows the maintenance of buffers and protocol state over the successive connections which may minimise the delay before data flow may resume; it eases the management burden on the transport service user.

9 Orchestration

The development of communications systems to support distributed multimedia application which exploit continuous media introduces new synchronisation requirements [WB 91]. In particular, the upper layers require support for the following two varieties of synchronisation:

- *event-driven synchronisation:* This is the act of notifying that a relevant event or set of events has taken place, and then causing an associated real-time action or actions to take place. For example, a user clicking on the stop button relating to a video play-out could cause the play-out to stop instantaneously.
- *continuous synchronisation:* This is where two or more continuous streams need to be kept in step by an on going commitment to a repetitive pattern of event driven synchronisation relationships. An example scenario where it is required to form a continuous synchronisation relationship is a language laboratory where separate audio tracks in different languages are stored on a single server but are to be distributed to different workstations in a real-time interactive language lesson. Another example arises where it is required to associate captions from a text file with an on-going video play-out. In both of these scenarios there is a need to maintain a tight on-going linkage between real-time continuous media information streams.

We refer to these sorts of linkage as *orchestration*, and have developed an architecture to model [CCG 92] and realise such relationships. We employ the term 'orchestration' rather than 'synchronisation' for two reasons. Firstly, cross-stream relationships may encompass more than just temporal co-ordination, and secondly the term synchronisation is already overloaded and given a different emphasis in current OSI usage (the OSI concept pertains to checkpointing and state synchronisation between peer entities).

9.1 Orchestration Functionality

Focusing on the continuous synchronisation problem, the following functionality must be provided by the infrastructure to ensure that media flowing in separate (non-multiplexed) but related connections can be correctly co-ordinated in time [CCG 92]:-

- the ability to start related continuous media data flows precisely together. If the relationship is not correctly initiated, there is no possibility of maintaining a correct temporal relationship.
- the ability to start and stop the flow of continuous media information on sets of related connections together in real-time.
- the ability to create related connections with compatible QOS. This is so that the connections will maintain a compatible temporal transmission rate in the required ratio.
- the ability to monitor the on-going temporal relationship between related connections, and to regulate the connections to perform fine grained corrections if synchronisation is being lost. It is almost inevitable that related connections will eventually drift out of synchronisation due to such factors as the potentially long duration of continuous media connections in typical applications, the inevitable discrepancies between remote clock rates, and temporary 'glitches' occurring in individual connections.

10 Related Work

Note that a number of other projects have examined the requirements imposed by multimedia on transport services, most notably an ANSI X3S3 report [ANS 91]. We believe that the list presented above subsumes all the issues raised in the ANSI report. Within the EC RACE programme, project R.1071 [RAC 91] categorises the applications into a number of service domains. More recently, ANSI X3S3.3 has produced proposals for a new High Speed Transport Service (HSTS) and Protocol (HSTP), and these were studied and considered in OSI95.

11 Conclusion

The integration of distributed multimedia applications into a communications architecture that encompasses new multiservice networks poses significant challenges which we have identified in this paper. The new applications set particular requirements on the underlying support system, and ATM-based multiservice networks offer the possibility of very high bandwidths and flexible services. An important conclusion is that QOS provides a unifying theme around which most of the new requirements can be grouped. For applications relying on the transfer of multimedia, and in particular continuous media, it is important that QOS is guaranteed system-wide, including the distributed system platform, the transport protocol and the multiservice network. One novel component of this new infrastructure is the orchestration of continuous media streams that require an extended QOS provision including end-to-end QOS negotiation, renegotiation and degradation indication.

Following the work reported in this paper on identifying key communications requirements, our research has progressed to map these requirements onto a Distributed Multimedia Architecture shown in figure 11.1. The focus of this research is a QOS architecture for the new environment that would encompass all layers, providing an integrated framework for (i) orchestration; (ii) QOS negotiation, renegotiation and degradation indication; (iii) reservation of end-system (including the kernel/OS) and network resources; (iv) policing of these resources; (v) admission policy for connections; and (vi) resource control, monitoring and regulation strategies necessary for support of distributed multimedia applications.

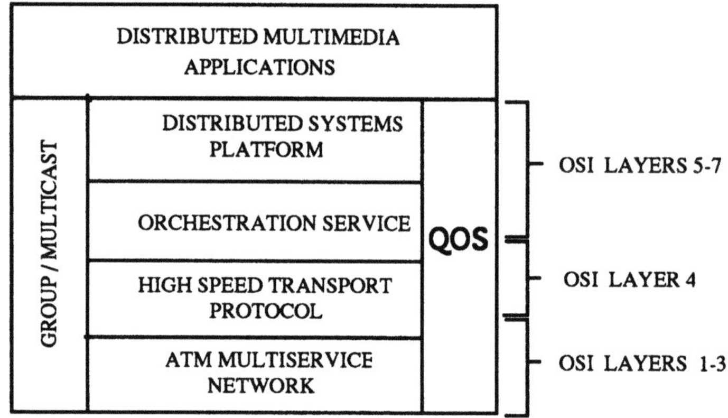

Fig. 11.1 Distributed multimedia architecture

This future work will detail the QOS functionality and interfaces at each discrete layer of the model and the mapping between these layers. The outcome of this phase of the research is presented elsewhere [CCG 93].

References

[ANS 91]   ANSI, **Results of the study on high performance protocols**, *Final Report to SPARC for Project 753 X3S3.3/91-58R*, ANSI X3S3, April 1991.

[BM 91]   Banerjea A., Mah B.A., **The real-time channel administration protocol**, *Proceedings of the Second International Workshop on Network and Operating System Support for Digital Audio and Video*, IBM ENC, Heidelberg, Germany, 1991.

[BN 84]   Birrell A., Nelson B., **Implementing remote procedure calls**, *ACM Transactions on Computer Systems*, Vol 2, No 1, February 1984.

[Boe 92]   Boerjan J., **ATM LLC assessment**, *OSI95 Deliverable Bell-1/P/V2*, Alcatel Bell Telephone, Francis Wellesplein 1, Antwerpen, B-2018, Belgium, October 1992.

[CBD 92]   Coulson, G., Blair, G.S., Davies, N., Williams, N., **Extensions to ANSA for Multimedia Computing**, *Computer Networks and ISDN Systems*, 25, 1992, pp 305-323.

[CCG 92a]   Campbell A., Coulson G., Garcia F., Hutchison D., Leopold H., **Integrated quality of service for multimedia communications**, *Contribution to ISO/IEC/JTC1/SC21/WG1*, August 1992.

[CCG 92b]   Campbell A., Coulson G., Garcia F., Hutchison D., Leopold H., Singer N., **Key Issues in Distributed Multimedia Communications: A Contribution to Discussion on the New ISO Work Item**, *UK expert paper contribution to ISO plenary meeting of ISO/IEC JTC1/SC6*, San Diego, USA, July 1992 BSI/IST 6/-/2/738, Primary input to the Draft Guidelines Document ISO/IEC JTC1/SC6/N7309 and submitted to the New Work Item on QOS in BSI/IST 21/-/1/5, Lancaster University and Alcatel Austria Forschungszentrum, April 1992.

[CCG 92]   Campbell A., Coulson G., Garcia F., Hutchison D., **A continuous media transport and orchestration service**, *In SIGCOMM '92*, Baltimore, Maryland, USA, August 1992.

[CCG 93]   Campbell A., Coulson G., Garcia F., Hutchison D., And Leopold H., **Integrated quality of service for multimedia communications**, *In IEEE INFOCOM '93*, San Francisco, USA, April 1993.

[ISO 92]   ISO, **Draft guidelines for enhanced communication functions and facilities for the lower layers**, *Draft ISO/IEC/JTC1/SC6/N7309*, ISO, May 1992.

[LBC 92]   Leopold H., Blair G.S., Campbell A., Coulson G., Dark P., Garcia F., Hutchison D., Singer N., Williams N., **Distributed Multimedia Communication System Requirements**, *OSI95/Deliverable ELIN-1/P/V4*, Alcatel Austria Forschungszentrum and Lancaster University, Ruthnergasse 1-7, A 1210 Vienna, Austria, April 1992.

[Nel 81]   Nelson B.J., **Remote Procedure Call**, *PhD thesis, Report No. CMU-CS-81-119*, Carnegie-Mellon University, 1981.

[Nic 90]   Nicolaou C., **An architecture for real-time multimedia communication systems**, *IEEE Journal on Selected Areas in Communications*, Vol 8, No 3, April 1990.

[RAC 91]   RACE, **IBC Application Analysis**, *Deliverable 71/I/IFC/DS/B/011/bi*, RACE project 1071, August 1991.

[SBC 92]   Singer N., Blair G., Campbell A., Coulson G., Frimpong-Ansah K., Garcia F., Hutchison D., Leopold H., **OSI/ULA - ODP Architecture Mapping**, *OSI95/Deliverable ELIN-3/P/V1 or 3BY 00097 0003 UPZZA ED.01*, Alcatel Austria Forschungszentrum and Lancaster University, Ruthnergasse 1-7, A 1210 Vienna, Austria, December 1992.

[SDH 91]   Shepherd W.D., Garcia F., Hutchison D., Coulson G., **Protocol support for distributed multimedia applications**, *In Second International Workshop on Network and Operating System Support for Digital Audio and Video*, IBM ENC, Heidelberg, Germany, November 1991.

[Top 90]   Topolcic C., **Experimental internet stream protocol, version 2 (ST-II)**, *Internet request for comments no. 1190*, October 1990.

[WB 91]   Williams N., Blair G.S., **Distributed multimedia application survey**, *Internal Report MPG-91 11*, Computing Department, Lancaster University, Bailrigg, Lancaster LA1 4YR, UK, April 1991.

Orchestration Services for Distributed Multimedia Synchronisation *

Andrew Campbell, Geoff Coulson, Francisco García and David Hutchison
Computing Department, Lancaster University, Lancaster LA1 4YR, UK
Email: mpg@comp.lancs.ac.uk

Rapid developments in networking technology over the past few years have led to the emergence of distributed applications which incorporate continuous media data types, particularly digital video and audio. Such applications have stringent real-time synchronisation requirements which have been documented in the literature. However, little research has been carried out into suitable mechanisms to support such synchronisation. This paper presents a multimedia synchronisation architecture and a detailed description of the lower layer services of the architecture. The paper also provides a rationale for the services by describing a real world application area and illustrates how the services can be exploited in this application area. Because the services described incorporate a variety of co-ordination functions over multiple transport connections a more general term, orchestration, is introduced to describe the low level synchronisation services.
Keywords: Computer-Communication Networks, Network Protocols, Distributed Systems.

1 Introduction

Rapid developments in networking technology over the past few years have led to the emergence of distributed applications which incorporate *multimedia* or *continuous media* information exchange [ATW 90] (especially digital video and audio). Such applications introduce new design challenges at all levels from network protocols and operating systems to application support platforms. This is because multimedia applications introduce novel requirements [WBH 92] such as the need to represent continuous media storage and transmission, quality of service (QOS) configurability and real-time synchronisation of continuous media streams.

This paper addresses the requirement for real-time synchronisation of continuous media streams. Our approach is based on the assumption that the underlying transport protocol supports a degree of QOS configurability and has responsive back pressure flow control. Previous work at Lancaster has addressed these issues and is reported in [BGH 92]. In this paper we describe layers on top of such a protocol which support

---

* This paper has been published in "High Performance Networking IV", (C-14), A.Danthine and O. Spaniol (Editors) and is reprinted here with the kind permission of Elsevier Science Publishers B.V. (North Holland).

application level synchronisation between multiple information streams [CCG 92]. These services also provide support for synchronisation and rate control within single streams. Because of the scope of the services, we introduce a new term, *orchestration*, which is defined as the dynamic management of information flow and QOS in a multimedia session involving a set of connections and end devices.

The paper first sets out, in section 2, requirements for synchronisation in distributed multimedia applications which are set in the context of a real world application scenario. Section 3 then introduces our orchestration services including an architecture which places the services within an experimental distributed multimedia application platform developed at Lancaster. Finally, section 4 presents an application example from the scenario of section 2 which exercises the orchestration services, and section 5 presents our conclusions.

2 Synchronisation Requirements

2.1 Application Scenario

Previous work in the field has identified the need for real-time synchronisation between related activities in multimedia applications (e.g. [Ste 90], [Lit 90], [HHS 91]). This section illustrates such requirements in the context of a real world application scenario which arose from a collaboration between Lancaster University and ICI, Runcorn, UK [Wil 91] .

The scenario is one of remote scientific co-operative working. ICI maintains a number of specialised microscopes at various sites throughout the country and employs scientists who use the various microscopes on a regular basis. Currently, scientists need to travel between sites to use microscopes and to collaborate with remotely sited colleagues on microscope output data. In the latter case, microscope output (usually slow scan video) is dumped to videotape and taken along by car.

To reduce travelling overheads and improve the efficiency of collaboration, we have designed a prototype system which allows remote collaborative working between scientists at the various sites. Presently the system runs over a local network but should be capable of running over a wide area network without fundamental design changes. Each scientist has a audio/ video multimedia workstation with the capability to control and display the slow scan video output from remote microscope devices. Such output can also be *multicast* to a number of sites which are linked by a multiparty video telephone component. Finally, scientists engaging in remote collaborative working can record microscope output to disc and create multimedia documents including co-ordinated video, text and voice annotations. They may also send such documents to remote sites as 'video mail'.

2.2 Synchronisation Scenarios

To begin to address the requirements of multimedia synchronisation, we have identified two categories of synchronisation as follows:-

- event-driven synchronisation: This is the act of notifying that a relevant event or set of events has taken place, and then causing an associated action or actions to take place. This must all be done in a *timely* manner due to the real-time nature of continuous media communication. For example, a user clicking on the stop button relating to a video play-out should cause the play-out to stop instantaneously.
- continuous synchronisation: This is an on-going commitment to a *repetitive* fine grained pattern of event driven synchronisation relationships such as the 'lip sync' relationship between the individual frames in an audio and video components of a play-out. Continuous synchronisation is ultimately based on event synchronisation but is a useful concept in itself as it permits potentially complex patterns of event synchronisation, perhaps involving various degrees of 'slack' and tolerance, to be encapsulated and handled as a whole.

To illustrate the applicability of the concepts of event driven and continuous synchronisation, we can extract a number of situations from the application scenario introduced above:-

- event synchronisation: A caption is to be shown at a particular point in a microscope video segment.
- event synchronisation with user interaction: The user hits a graphical user interface button to start/stop continuous information flow (perhaps over multiple flows simultaneously).
- lip synchronisation: This is the most commonly cited form of multimedia synchronisation. It appears in our scenario as the need to synchronise the video and audio components in the playback of a recorded videophone message (i.e. video mail). In such cases, video and audio are almost always stored in separate files and sometimes on separate storage servers which are optimised for different media [LOU 92].
- continuous synchronisation other than lip synchronisation: This is illustrated by the playback of two simultaneously recorded video segments which record the same experimental sample from two different perspectives. An example of this is where recordings of different magnification are made and then simultaneously replayed to convey an impression of the context of the higher magnification.
- continuous synchronisation requiring varying degrees of 'tightness': Simultaneously recorded video perspectives must be played in precise frame by frame synchrony so that relevant features may be simultaneously observed. On the other hand, lip synchronisation in multimedia documents does not need to be absolutely precise when the main information channel is auditory and video is only used to enhance the sense of presence. It is useful to permit degrees of tightness of continuous synchronisation as looser synchronisation is often sufficient and can be achieved with a relatively low overhead.
- continuous synchronisation of many streams: This occurs in multimedia documents where an audio annotation, perhaps with accompanying video, is associated with microscope output clips involving one or more video streams.
- continuous synchronisation from disparate sources and sinks: The need for continuous synchronisation arises in a number of different physical node configurations. For example, video and audio from separate remote sources often need to be synchronised at a common sink. Conversely, the playout of a single

segment of stored microscope may need to be displayed simultaneously at different remote sinks so that scientists discussing the output over a videophone can simultaneously refer to the same features. There are also situations where two or more streams need to be synchronised which all originate from different sources and are played out at separate remote sinks. For example, two remote scientists may each view different separately stored perspectives on the same experiment while discussing related events over the videophone. Finally, there is a need to synchronise separate multicast playouts where, for example, a number of scientists interactively collaborate over a continuously synchronised playout when the components are separately stored.

2.3 Infrastructure Requirements for Continuous Synchronisation

Although it is possible in some situations to support continuous synchronisation simply by multiplexing the different media onto a single connection in the correct ratios, there exist strong arguments against this as a general solution [TEN 90]:-

- the overhead and complexity of multiplexing/ demultiplexing is significant, especially when different encoding/ compression schemes are used for different media; this can lead to excessive real-time delays, especially where it would otherwise be possible to interface the transport protocol directly to hardware such as frame grabbers, codecs etc.;
- the opportunity to process separate connections in parallel is lost, thus reducing potential performance;
- multiplexing leads to a combined QOS which must be sufficient for the most demanding medium; this may be both expensive and unsuited to some component media types;
- multiplexing is not an option where media originate from different sources.

If multiplexing is rejected as a general purpose strategy for the support of continuous synchronisation, an analysis of the continuous synchronisation problem suggests that the following support should be provided by the infrastructure. Sections 3 and 4 illustrate how our design satisfies these requirements.

- the ability to start and stop and pause related continuous media data flows precisely together. If a temporal relationship is not correctly initiated, there is no possibility of maintaining correct synchronisation.
- the ability to monitor the on-going temporal relationship between related connections, and to regulate the connections to perform fine grained corrections if synchronisation is being lost. It is almost inevitable that related connections will eventually drift out of synchronisation due to factors such as the potentially long duration of continuous media connections in typical applications, and temporary 'glitches' occurring in individual connections and the scheduling of source and sink application threads.

Finally, note that the need for the comprehensive continuous synchronisation support detailed in this paper is only strictly necessary when all the CM sources to be orchestrated are *stored*. This is because with live media, there is no possibility of control over when the information flow starts (e.g. it depends when the camera is switched on!), and also no possibility of altering the speed of a live media flow. Whenever live sources are to be continuously synchronised (e.g. the output from a camera and a microphone), the major requirement is to ensure that the *latency* of the connections is the same. Other QOS parameters such as delay, jitter and error rates can be separately controlled over individual connections as desired.

3 Orchestration Architecture

This section presents an architecture which addresses the need for the temporal co-ordination of multiple related continuous media transport streams identified in section 2.

It can be seen from the architecture diagram in figure 3.1 that orchestration is a multi-layered activity. Each layer provides *policy* to its lower neighbour and *mechanism* to its upper neighbour. This design provides both flexibility and efficiency because the lower layers are simply provided with targets, and all exceptions, error handling and re-structuring are handled in the layers above.

3.1 Upper Architectural Layer

The top level of the synchronisation architecture forms part of the Lancaster multimedia application platform [CBD 92]. This is an object-based set of services based on the ANSA distributed systems architecture [APM 89]. At the application platform level all entities in the system are represented as abstract data type interfaces with named operations which can be invoked by RPC. Such entities include documents, the individual components of documents, 'devices' such as video windows and speakers, and even continuous media connections themselves. Abstract data type interfaces are referred to through *interface references* which are location independent 'handles' which can be freely passed around the system.

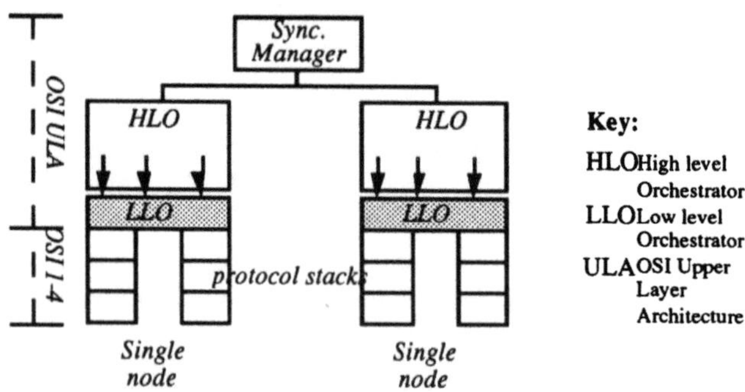

Fig. 3.1 Three level orchestration architecture

The *synchronisation manager* is the platform level view of the orchestration services and, like all the other platform services, appears to applications as an abstract data type interface. The synchronisation manager is responsible for finding the physical locations of the transport connections underlying the platform level communications abstractions, and thus choosing a single node from which the lower levels of orchestration will be co-ordinated. The node selected, known as the *orchestrating node*, is that common to the greatest number of connections (see figure 3.2). For example, if it was required to orchestrate separate video and audio tracks of a film stored on separate storage servers, the common sink would be designated as the orchestrating node by the synchronisation manager. The platform level of the architecture is not discussed further in this paper. See [CBD 92] for more details of this aspect of the architecture. A more detailed description of the synchronisation specific aspects of the platform can be found in [Bla 92].

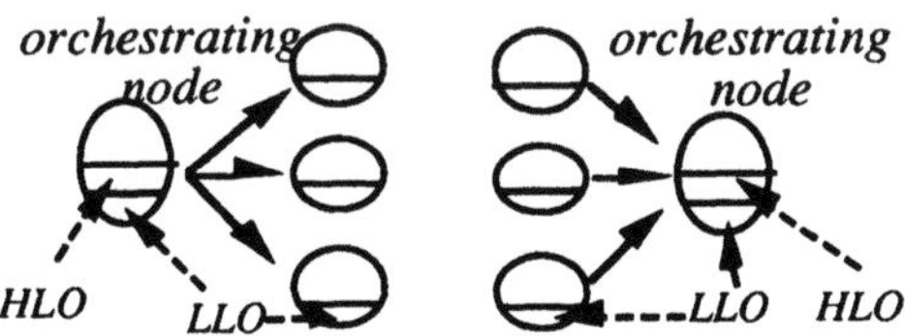

Fig. 3.2 Orchestrating at the common node

3.2 Lower Architectural Layers

Below the platform level, the remaining two orchestration components are responsible for realising the behaviour and policy required by the synchronisation manager. At this level, the orchestration process is realised as *high level orchestrator* (HLO) entities which monitor and regulate multiple transport connections via a *low level orchestrator* (LLO) interface. For each orchestrated group of connections, a single HLO runs on the orchestrating node, and an LLO instance runs on all source and sink nodes of all the orchestrated connections. The HLO only interacts with its local LLO instance, but the multiple LLO instances interact with each other via Orchestrator PDUs (OPDUs), on out of band connections.

3.2.1 Orchestration Control Framework

The out of band connections for OPDU transfers must have guaranteed bandwidth to support the necessary real-time communication of orchestration primitives and, in general, a separate connection is required between the orchestrating node and each source and sink node involved. However, depending on the topology the numbers of connections can often be reduced in practice: e.g. when a number of connections are sourced and sinked on the orchestrating node itself. In our current implementation we exploit duplex control connections associated with each simplex continuous media connection [SHG 91].

There are three sorts of interaction between the HLO and the LLO instance on the orchestrating node, each of which involves a separate set of primitives. The first group of primitives are used for management purposes to establish and modify orchestrated groups of connections. The second set operates over a grouping of transport connections and provides the ability to atomically *prime*, *start* and *stop* the flow of data in these connections both atomically and instantaneously. The third set allows the HLO to control the rate of information flow on individual orchestrated connections, and thus forms the basis for the implementation of continuous synchronisation across multiple connections.

Figure 3.3 illustrates the pattern of interaction for a single connection between the HLO and the local LLO where the third set of primitives is being applied. The HLO supplies the LLO with *rate targets* for each orchestrated connection over specified *intervals*. These targets require that each orchestrated connection runs at the necessary rate for the required synchronisation relationship between the orchestrated connections to be maintained. The LLO attempts to meet the required rate target over each interval for each connection, and reports back at the end of the interval on its actual success or failure. Then, on the basis of these reports, the HLO may set new targets for the next interval which compensate for any relative speed up or slow down among the orchestrated connections. If no new target is set for the forthcoming interval, the LLO uses the rate specified in the previous request until further notice. The LLO operates on a *best effort* principle; it is the responsibility of the HLO to take appropriate action (e.g. set new targets or re-negotiate the connection QOS) if the LLO consistently fails to meet targets. The length of interval chosen largely determines the granularity or 'tightness' of the synchronisation required (as specified by the application). As mentioned in section 2, loose synchronisation based on long intervals is relatively cheap in terms of message exchanges and synchronisation overhead.

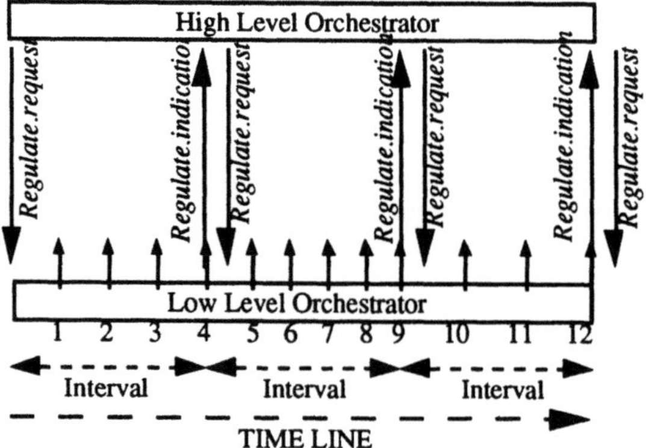

Fig. 3.3 Interaction between HLO and LLO

3.2.2 Data Transfer in Orchestrated Connections

The small numbered arrows in figure 3.3 represent the delivery of quanta of CM information which are released by the sink LLO instance to the application thread at times determined by the HLO initiated targets. These quanta are known as *Orchestrator Service Data Units* (OSDUs), and are the units of CM information meaningful to applications (e.g. video frame or text paragraph).

The orchestration services maintain a special *OSDU sequence number* field for each OSDU, which starts from zero when the connection is first used; a second such field, known as an *event* field, is employed for use by the Orch.Event primitive (see later). Both these fields form part of an OPDU which is sent along with each OSDU. OSDU and OPDU boundaries are maintained by the transport service. This is possible in our system because, at connection establishment time, the transport service is given the maximum size of an OSDU as a QOS parameter, and this (plus the size of the OPDU) is interpreted as a lower bound on buffer size allocation.

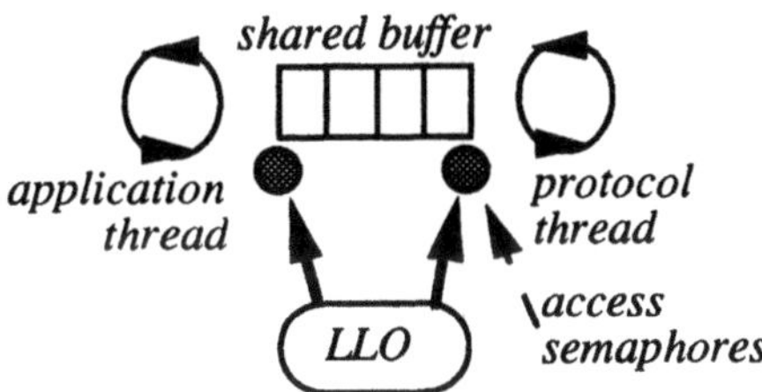

Fig. 3.4 Shared buffer interface between application and protocol threads

The application's *data.request* and *data.indication* interface to the orchestration service is implemented as a circular shared buffer (see figure 3.4) which maintains mutual exclusion and access control by means of semaphores. The protocol and application run as separate threads and do not need to explicitly synchronise via the access semaphores if they are running at compatible rates. When applications write/ read OSDUs into/ from the circular shared buffers, they write/ read from the beginning of a buffer, and may also write/ read the current OSDU size to/ from an auxiliary memory location. The LLO operates largely by controlling the flow of data by means of the shared buffer access semaphores. The interface and mechanism of the LLO is described in the following section.

4 Low Level Orchestrator

The LLO service interface consists of three sets of OSI-like primitives corresponding to the three types of HLO/LLO interaction described above. As indicated above, the LLO interface is not expected to be used directly by applications but is intended for use by an HLO instance and ultimately via a synchronisation manager interface by an application running in an object-based computational model.

We now present the LLO interface primitives in more detail together with their implementation in terms of protocol message exchanges.

4.1 Management Primitives

The management primitives and their parameters are illustrated in table 4.1 below.

4.1.1 Orch.request

We assume that before an HLO instance attempts to instantiate an LLO orchestrating service, the connections to be orchestrated have already been established.

The initiating HLO issues a Orch.request to the orchestrating LLO instance at its local node. This causes the orchestrating LLO to open transport associations between itself and the LLO instances at the source and sink of all connections in the orchestrated group. Subsequently, the orchestrating LLO instance uses these associations to pass an Orch.request to all the LLO instances involved. Each source and sink LLO instance passes an indication up to the application thread which owns the connection and waits for either an Orch.response or an Orch.release.request depending on whether or not the application wants the connection to be involved. Each LLO instance then replies on its private connection to the initiating LLO with either an Orch.confirm or an Orch.release.request packet. If accepted by all the remote LLO services, the HLO will eventually be passed an Orch.confirm, or if rejected a Orch.Release.indication giving a reason why orchestration was rejected. Apart from refusal by the application, rejection may also occur because some LLO instance has no table space available, or because one or more of the specified connections do not exist etc..

Having authenticated the set of connections to be orchestrated, the orchestrating LLO instance enters the connection identifiers and the corresponding source and sink addresses in a private table accessible via the orch-session-id.

Table 4.1 Orchestration management primitives together with their associated parameters

Primitives	Parameters
Orch.request	orch-session-id, list-of-vc-ids
Orch.indication	"
Orch.response	orch-session-id, vc-id, status
Orch.confirm	"
Orch.Release.request	orch-session-id, vc-id
Orch.Release.indication	"
Orch.Release.response	orch-session-id, vc-id, reason
Orch.Release.confirm	"
Orch.Add.request	orch-session-id, vc-id
Orch.Add.indication	"
Orch.Add.response	orch-session-id, vc-id, status
Orch.Add.confirm	"

Orch.Remove.request	orch-session-id, vc-id
Orch.Remove.indication	"
Orch.Remove.response	orch-session-id, vc-id, status
Orch.Remove.confirm	"

4.1.2 Orch.Add and Orch.Remove

The Orch.Add and the Orch.Remove primitives are employed to either add or remove a connection or connections from an orchestrated group. The message sequence is similar to that for Orch.request. Note that when connections are removed from an orchestrated group they are not disconnected and thus data may still be flowing. If a single connection is closed, the local LLO instance is informed by the transport service and issues a Orch.Remove.request to the orchestrating LLO instance.

4.1.3 Orch.Release

An entire orchestration session is released by issuing a Orch.Release.request. Again, the message sequence is similar to that for Orch.request. Orchestration will also be released implicitly if all the connections in an orchestrated session are closed.

4.2 Group Operation Primitives

The primitives to *prime* connections, and atomically *start* and *stop* the data flow on groups of orchestrated connections are illustrated in Table 4.2.

4.2.1 Orch.Prime

The prime mechanism has the effect of filling the end-to-end pipeline of buffers in a connection and is used to ensure that multiple streams of remotely stored CM data can be started together in a co-ordinated manner. It is also useful in ensuring that time critical data can be pre-fetched and made available when required. A third application of Orch.Prime arises when it is required to flush the buffers in an end to end connection. This need arises when a user stops a media play-out and then wishes to seek to another part of the media before resuming. If the buffers were not flushed in this situation, a short burst of media buffered from the previous play would be discernible.

Following the issue of an Orch.Prime.request by the initiating HLO, the orchestrating LLO forwards Orch.Prime.requests to all involved source and sink LLO instances. At each LLO instance, Orch.Prime.indication primitives are passed to the application threads associated with the connection. On receipt of the Orch.prime.indication, each application thread is expected to flush any internal buffers and start generating data or preparing to accept data as appropriate. If any application thread is not in a position to do this it can set the error-flag in the Orch.Prime.response primitive.

As data begins to arrive at the sinks, the sink LLOs in the primed state allow the receiver's communications buffers to fill, but prevent the data from being delivered to the receiving application threads. When the receive buffers are eventually full, each sink LLO notifies the orchestrating LLO, which eventually relays the received Orch.Prime.confirm packet to the originating HLO. At this point, the source application thread will also be blocked by the protocol's flow control mechanism, but the pipeline is filled and ready to go.

Table 4.2 Orchestration primitives for priming, starting and stopping

Primitives	Parameters
Orch.Prime.request	orch-session-id
Orch.Prime.indication	"
Orch.Prime.response	orch-session-id, error-flag
Orch.Prime.confirm	"
Orch.Start.request	orch-session-id, start-time, default-rate
Orch.Start.indication	"
Orch.Start.response	"
Orch.Start.confirm	"
Orch.Stop.request	orch-session-id
Orch.Stop.indication	"
Orch.Stop.response	"
Orch.Stop.confirm	"

4.2.2 Orch.Start

This is intended to be issued after the successful completion of an Orch.Prime. The primitive re-starts the transport protocol and also unblocks the previously filled receive buffers so that data may be consumed by the sink application thread. In terms of messages, the Orch.Start.request issued by the HLO is forwarded to each sink LLO instance concerned. To ensure simultaneity of action, the orchestrating LLO must keep information on the maximum delay of its out-of-band associations with the LLO instances at the sinks of each connection. This is possible in our experimental system as the transport service provides delay bound configurability as part of its QOS control interface. The orchestrating LLO must then time stamp each Orch.Start.request packet with the value $'now'$ + $max(delay_1, delay_2, ..., delay_n)$. On arrival of these packets at the LLO instance at each sink, the message is held back until the current time becomes equal to the timestamp value. Note that this requires a globally synchronised clock for correct operation. This can be supplied by mechanisms such as satellite time co-ordination or network time protocols such as NTP [Mil 89].

If an Orch.Prime has been issued before the present Orch.Start, data will already be waiting at all the sinks, and all the receiving application threads in the orchestrated group will start to receive data at (almost) the same instant. A Orch.Start.indication is sent to each sink application thread as a result of the Orch.Start.request in an analogous manner to that described for Orch.Prime. However, where the system is already in a primed state, these threads will not need to take any special action as they are already set up to produce/ consume data, but are blocked by the underlying transport protocol's back-pressure flow control mechanism.

After Orch.Start.request packets have been received at each sink LLO instance, the LLO instances reply to the orchestrating LLO by means of an Orch.Start.response packet, and the final response is relayed to the originating HLO when all expected packets have been received.

4.2.3 Orch.Stop

Orch.Stop 'instantaneously' freezes the flow of data in the specified connections. Internal messages exchanges are only necessary between the orchestrating LLO and the sinks: back pressure in the transport protocol is relied on to stop data flow at the source. Note, however, that the flow of data can not actually be stopped until the underlying protocol's flow control mechanism can take effect. As with the Orch.Prime primitive, the receive buffers are made unavailable to the application sink thread before they are drained so that data is available for a subsequent primed start. Simultaneity of action is attained in a similar manner to Orch.Start via a timestamping mechanism.

Note that a potential problem with both Orch.Start and Orch.Stop is that an orchestration protocol message may be lost or delayed beyond its expected latency thus causing some connections to be either left out or uncoordinated. This problem could be overcome by using a standard two phase commit algorithm [Ber 81] but the overhead here could be significant. We are investigating this pragmatically as our implementation develops.

4.3 Regulation Primitives

The third group of primitives (see table 4.3) operate on single transport connections within an orchestrated grouping. Thus each primitive is issued with both an orch-session-id and a vc-id. These primitives enable the controlling HLO to regulate and monitor the flow rate targets described in section 3.1. As stated above, LLO instances will attempt to meet these targets on a best effort basis. Primitives are also provided to report back to the HLO on the actual performance achieved at the end of each interval.

4.3.1 Orch.Regulate

4.3.1.1 Orch.Regulate.request
The Orch.Regulate.request primitive is issued by the HLO instance to set a flow rate target for the forthcoming interval. Note that there are no confirm and response packets associated with this primitive as communication is not passed up above the LLO layer at the remote end. For this reason, the indication variant of this primitive does not require a response. The same applies to the Orch.Event primitive described later.

Parameters to Orch.Regulate.request include the orchestration session ID, the ID of the connection to be controlled, a target OSDU# to be delivered at the end of the forthcoming interval, the length of the forthcoming interval, an interval# to identify the corresponding Orch.Regulate.indication, a default flow rate to be used for subsequent intervals, and a max-drop# parameter. The interval# is also used to ensure that the desired effect occurs within the desired interval: if the interval# refers to an interval which has already passed, the HLO will be returned an Orch.Regulate.indication with an appropriate error flag.

The target-OSDU# parameter denotes the OSDU sequence number which should ideally be delivered to the sink application thread at precisely the end of the interval. The required flow rate target is calculated as ((*target-OSDU# - current-OSDU#*) / *interval-length*). Absolute OSDU sequence numbers are used to avoid ambiguities due to OSDU loss: if targets are specified in OSDUs/interval then lost OSDUs may cause synchronisation to be lost even when relative rates appear correct. The default rate parameter is used for subsequent intervals if no further Orch.Regulate.requests are issued. Subsequent intervals also use the most recently requested interval length.

Table 4.3 Orchestration Primitives for Regulation and Monitoring

Primitives	Parameters
Orch.Regulate.request	orch-session-id, vc-id, target-OSDU#, interval-length, interval#, default-rate, max-drop#
Orch.Regulate.indication	orch-session-id, vc-id, interval#, OSDU#, dropped#, proto-block-times, app-block-times, error-flag
Orch.Regulate.Source.request	orch-session-id, vc-id, interval-length, interval#, interval-start-time, drop#
Orch.Regulate.Source.response	orch-session-id, vc-id, interval#, dropped#, proto-block-times, app-block-times
Orch.Event.request	orch-session-id, vc-id, event-pattern
Orch.Event.indication	"
Orch.Delayed.request	orch-session-id, vc-id, source-or-sink, interval-length, OSDUs-behind, max-drop#
Orch.Delayed.indication	"
Orch.Delayed.response	"
Orch.Delayed.confirm	"

When the LLO is attempting to meet the requested flow rate target for a connection, there are three possible cases: it may be on target, behind target or ahead of target. If the connection is on target, no action need be taken. If, however, either of the other cases is true, the following compensatory strategies are available to the LLO:-

- if a connection is behind, its sole compensatory strategy is to drop OSDUs. The max-drop# parameter to Orch.Regulate.request states the maximum number of OSDUs which the connection may discard in order to achieve its flow rate target. All such discards are performed at the source by incrementing the source shared buffer pointer. This permits the source application thread to immediately insert another OSDU and thus overwrite the previous one before it is sent. This strategy

may help a delayed connection to catch up and meet the link delivery target in case it is lack of transport bandwidth which is causing the delay.
- if, on the other hand, a connection is ahead of schedule, the compensatory action is simply to block. Note that, as with the Orch.Stop primitive, the flow control mechanism of the underlying transport protocol must be capable of rapid adaptation for this to be feasible. The rate based mechanism used in our protocol [SHG 91], is adequate for this purpose. Note also that, for both these compensatory actions, the LLO must take responsibility for attempting to spread compensatory actions over the length of the target interval to avoid unnecessary jitter.

If these compensatory actions are not available to the LLO (e.g. a max-drop# of zero will often be chosen where a no-loss medium such as voice is involved), then the necessary corrections must be taken by the HLO on the basis of the information gathered via the Orch.Regulate.indication primitive. Such actions will typically involve issuing Orch.Delayed primitives to participating application threads, and re-negotiating the QOS of individual connections.

4.3.1.2 Orch.Regulate.indication
The Orch.Regulate.indication primitive is used to report back to the HLO on the performance *actually achieved* by each orchestrated connection. The interval# parameter is used to match the indication to a prior request. If an interval expires and no new request has been received, the LLO will automatically start a new interval on the basis of the default parameters in the most recent Orch.Regulate.request, and will continuously generate asynchronous Orch.Regulate.indications at the end of each interval. The statistics reported include the OSDU# *actually* delivered at the end of the interval, the number of OSDUs *actually* dropped, and the times spent blocking by both the application and protocol threads at both the source and sink ends of the connection. This blocking time information is gathered by associating timers with the shared circular buffer semaphores described in section 3.

The blocking time information is used by the HLO instance to determine which part of the system was responsible for any failure to meet the flow rate target. Based on this information, the HLO can take compensatory action if required. For example, if the application threads spent an excessive amount of time blocked, the protocol throughput was presumably too low and the HLO may re-negotiate the QOS of the connection. Alternatively if the protocol threads were blocked, the application threads were presumably slow in producing/ consuming data. In this latter case the HLO will probably issue a Orch.Delayed primitive.

4.3.1.3 Orch.Regulate.Source
The message sequence required for the Orch.Regulate primitives uses the Orch.Regulate.Source primitive. This primitive is internal to the orchestration protocols and is not visible to the client HLO. Because of this there are no indication and confirm variants of this primitive.

When an Orch.Regulate.request is issued by the HLO, the orchestrating LLO forwards it to the sink LLO of the connection concerned. The sink LLO then attempts to impose the flow control strategy contained in the request. It also issues an Orch.Regulate.Source.request primitive to the source LLO instance. This primitive is used to co-ordinate state information between the LLO instances at the source and sink of each connection. Initially, Orch.Regulate.Source.request notifies the source of the interval length and the time of commencement of the next interval. The source is expected to asynchronously generate Orch.Regulate.Source.response primitives at the end of each interval. Orch.Regulate.Source.request is also used to request the source LLO instance to drop OSDUs as and when this strategy is ordered by the sink LLO which acts as 'master' for this purpose. The sink LLO instance receives an Orch.Regulate.Source.response at the end of each interval and combines the information in this packet with its own local information to build an Orch.Regulate.indication which is sent to the orchestrating LLO and thence to the originating HLO.

4.3.2 Orch.Delayed

This primitive is issued by the HLO in response to the situation where it can be deduced that an application thread is responsible a connection being behind the required schedule (see section 4.3.1.2). The effect of an Orch.Delayed.request is to cause an indication to be delivered to the application thread(s) causing the delay. The intended interpretation of a Orch.Delayed.indication is that the thread is not running sufficiently fast to produce/ consume data at a rate required by the client of the location independent orchestration service. Applications so informed may take any appropriate action such as requesting more processor resources or dropping OSDUs and should then reply with an Orch.Delayed.response.

4.3.3 Orch.Event

This primitive is used to register an interest in a particular application defined event associated with some OSDU; it thus provides support for event-driven synchronisation.

To register an interest in some application defined event, a Orch.Event.request is issued which is forwarded to the LLO instance at the sink end of the specified connection, together with a bit pattern representing the event. Subsequently, the sink LLO instance will match this bit pattern, which is not interpreted in any way by the LLO, against the bit patterns in the event fields of the OPDUs associated with incoming OSDUs (see section 5) on the set of orchestrated connections. If the event in the OSDU matches the registered bit pattern, a Orch.Event.indication is raised both locally to the sink application thread and also to the originating HLO via the orchestrating LLO instance. To cause an event to be initiated, the event fields of OSDUs may optionally be set by the source application thread when writing an OSDU.

An example of use of the event mechanism is when a change of encoding is being signalled in the data stream such as the introduction of a particular compression scheme. It would obviously be possible to implement such a scheme in an ad-hoc manner in the application layer, but this would require that application threads examine each incoming OSDU. The present scheme avoids complicating application code, permits system dependent optimisations to be made, and also permits OSDUs to be dumped directly into, say, a video frame buffer.

5 Mode of Use

We now attempt to place the orchestration services in perspective by illustrating their use in the microscope application scenario of section 2. Note, however, that the LLO services are self contained and could in principle be used by a number of alternative upper layer designs. In the forthcoming description much of the detail of the synchronisation manager and the HLO layers of the architecture is omitted. The main intention is to show how the LLO interface primitives are used in a realistic scenario.

Consider an application configuration in the scenario of section 2 where a multimedia document needs to be played back to two separately located scientists who are conferring over a video telephone. The document contains two video clips which view the same sample at different levels of magnification, together with occasional voice annotation and text captions. The various components of the document are separately stored but must nevertheless be synchronised on playout both together and across the two scientist's workstations. The details of the required synchronisation (i.e. timing of text captions, starting time of annotation, degree of permitted slack etc.) are contained in a 'script' encapsulated within the multimedia document structure. Figure 5.1 illustrates the logical topology associated with this configuration.

When it is required to play the document the following sequence of events takes place at the application platform level (see section 3.1). Firstly, the controlling application program obtains a location independent handle onto the document and passes it to a *playout service*. The playout service obtains handles on the various media sources in the document, obtains or creates (platform level) playback devices (i.e. video windows, speakers, caption windows) on the required workstations, and connects the sources and sinks together by means of the platform's connection (stream) abstraction. Once the playout service has performed the necessary platform operations to connect the configuration - which results in the underlying transport connections being established - it begins to parse the script associated with the document.

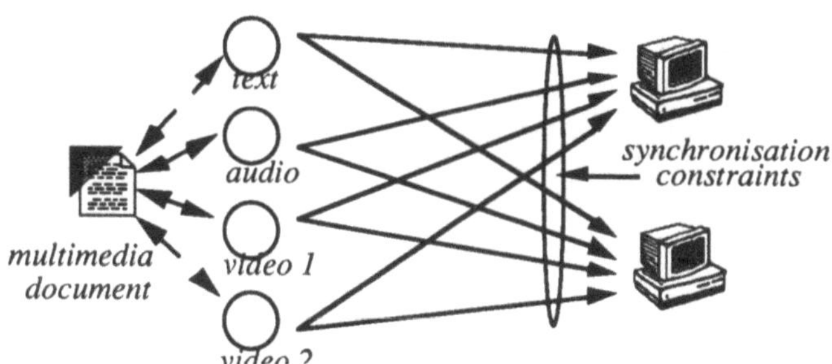

Fig. 5.1 Example topology

When parsing the script, the playout service determines the streams which are to be synchronised and passes their handles to the top level component of the orchestration architecture illustrated in figure 3.1: i.e. the synchronisation manager. At this time, the playout service also initialises and starts the source and sink devices in the session via RPC. This, however, will not cause data to flow as the flow of data is controlled by the orchestration services. Next the synchronisation manager determines the physical location of the sources and sinks of the streams involved and selects the orchestrating node. It then instantiates an HLO instance on the orchestrating node and passes it details derived from the document script. These details include the ratio of OSDU delivery for each stream to a reference time line, and 'on' and 'off' events for each stream. For example, the 'on' event for a caption or voice annotation could be the delivery of a particular OSDU in a video stream, and the 'on' event for the video streams would be the start of the session itself.

The first action of the HLO is to create a platform level control interface handle onto itself which it returns to the controlling application. This interface contains *start*, *stop*, *prime* operations which the application can invoke via RPC to control the set of streams as an atomic unit. After this the HLO instantiates the orchestration session by issuing the Orch.request primitive. Subsequently, on receiving *prime* and *start* invocations from the application, the HLO issues the Orch.Prime.request and Orch.Start.request primitives which start data flowing in the orchestrated set of connections. The data flow will then be controlled by the HLO by means of the Orch.Regulate primitive according to the document policy. At certain points in the playout, as determined by the script, the HLO will pass caption OSDUs through to the caption window. The event causing this action could be either an explicit OSDU number, or an Orch.Event.indication depending on the contents of the script. If the platform level application wants to momentarily pause the playout, it issues a stop invocation which results in the HLO issuing an Orch.Stop.request and thus freezing the flow of data. Note that as synchronisation is event driven rather than being associated with absolute time, pauses will not affect subsequent caption and voice annotation timings when the playout is restarted.

Finally, it is possible to dynamically add new streams to an ongoing session; for example, the scientists may want to run a library video clip along with the current sample. This is achieved by passing the handle of a new source/stream/sink configuration to the synchronisation manager which derives the physical locations of the sources and sinks and passes them to the HLO. The HLO then issues the Orch.Add.request primitive for the new stream. Subsequently, Orch.Prime, Orch.Start and Orch.Stop will take effect on the new stream in addition to the established session.

6 Conclusion

We have identified the need for comprehensive real-time synchronisation support in distributed multimedia systems, and have described the low level components of a design which performs synchronisation of continuous information streams in such a system. The low level orchestration services provide a mechanism to group together existing transport connections and control the flow of information in these connections according to given synchronisation constraints. Primitives are provided to prime, start and stop the information in the grouped connections as an atomic unit, and also to regulate the rate of flow in each connection with a flexible degree of granularity. Although the system acts mainly on transport connections, primitives are provided which supply hints to end systems when they are running too slowly to meet the required end-to-end synchronisation constraints.

The low level orchestration services are embedded in a larger design which presents the functionality of the lower layers to applications through a distributed object-based computational model. The orchestration services can be used as a component of a range of possible higher level services including playout services for complex multimedia components with encapsulated synchronisation specifications.

In terms of implementation, we currently have the low level orchestration services in place and have also completed the high level application platform. The implementation has been carried out in the context of a collection of standard workstations augmented with transputer based *multimedia network interface* (MNI) units [BCC 93] which handle both network interfacing and all continuous media source and sink mechanisms such as video and audio capture, playback and storage. The workstations are connected by means of a real-time transputer based emulation of an FDDI network. We use a specially designed rate based transport protocol which provides simplex connections with a high degree of QOS configurability [SHG 91]. More details of the implementation can be found in [DCW 91].

Finally, our future plans involve the development of the high level orchestrator component which acts as the link between the platform abstractions and the low level orchestration mechanisms. This component is expected to be fairly complex as it is required simultaneously to monitor and control a number of orchestrated connections in real time according to an arbitrarily application defined specification. However, because of the support provided by the lower layers, the HLO will not have to consider real-time communication and distribution issues as these are delegated to the LLO services.

Acknowledgement

Part of this work was carried out within the MNI project (funded under the UK SERC Specially Promoted Programme in Integrated Multiservice Communication Networks (grant number GR/F 03097) and co-sponsored by British Telecom Labs), and part within the OSI95 project (ESPRIT project 5341, funded by the European Commission).

References

[APM 89]   APM Ltd., **The ANSA Reference Manual Release 01.01**, Architecture Projects Management Ltd., Poseidon House, Castle Park, Cambridge, UK, July 1989.

[ATW 90]   Anderson, D.P., S.Y. Tzou, R. Wahbe, R. Govindan, Andrews M., **Support for Continuous Media in the DASH System**, *Proc. 10th International Conference on Distributed Computing Systems*, Paris, May 1990.

[BCC 93]   Blair, G.S., Campbell, A., Coulson, G., Garcia, F., Hutchison, D., Scott, A., Shepherd, W.D., **A Network Interface Unit to Support Continuous Media**, *To appear in IEEE Journal of Selected Areas in Communications (JSAC)* , 1993.

[Ber 81]   Bernstein, P.A., Goodman, N., **Concurrency Control in Distributed Database Systems**, *Computing Surveys Vol 13 No. 2*, pp 185-221, June 1981.

[BGH 92]   Blair, G.S., Garcia, F., Hutchison, D., Shepherd, W.D., **Towards New Transport Services to Support Distributed Multimedia Applications**, *Proc. Multimedia '92: 4th IEEE COMSOC International Workshop*, Monterey, USA, April 1-4, 1992.

[Bla 92]   Blair, G.S., Coulson, G., **Meeting the Real-time Synchronisation Requirements of Multimedia in Open Distributed Systems**, *Internal Report*, Computing Department, Lancaster University, Bailrigg, Lancaster LA1 4YR, UK, 1992.

[CBD 92]   Coulson, G., Blair, G.S., Davies, N., Williams, N., **Extensions to ANSA for Multimedia Computing**, *Computer Networks and ISDN Systems*, 25, 1992, pp 305-323.

[CCG 92]   Campbell, A., Coulson G., Garcia F., Hutchison, D., **A Continuous Media Transport and Orchestration Service**, *Proc. ACM SIGCOMM '92*, Baltimore, Maryland, USA, August 1992.

[DCW 91]   Davies, N., Coulson, G., Williams, Blair, G.S., **Experiences of Handling Multimedia in Distributed Open Systems**, *Proc. SEDMS '92, Newport Beach CA, April 1992*; also available from Computing Department, Lancaster University, Bailrigg, Lancaster LA1 4YR, UK. November 1991.

[HHS 91]   Hazard, L., Horn, F., Stefani, J.B., **Notes on Architectural Support for Distributed Multimedia Applications**, *CNET/RC.W01.LHFH.001*, Centre National d'Etudes des Telecommunications, Paris, France, March 91.

[Lit 90]   Little, T.D.C., Ghafoor, A., **Network Considerations for Distributed Multimedia Object Composition and Communication**, *IEEE Network Magazine*, November 1990, pp 32-49.

[LOU 92]   Lougher, P., **The Design of a Storage Server for Continuous Media**, *The Computer Journal* (Special Issue on Multimedia), Vol 36, No 1, January 1992.

[Mil 89]   Mills, D.L., **Internet Time Synchronisation: the Network Time Protocol**, *Internet Request for Comments No. 1129* RFC-1129, October 1989.

[SHG 91]   Shepherd, W.D., Hutchison, D., Garcia, F., Coulson, G., **Protocol Support for Distributed Multimedia Applications**, *Second International Workshop on Network and Operating System Support for Digital Audio and Video*, IBM ENC, Heidelberg, Germany, Nov 18-19 1991.

[Ste 90]   Steinmetz, R., **Synchronisation Properties in Multimedia Systems**, *IEEE Journal on Selected Areas in Communications IEEE JSAC Vol. 8 No. 3*, pp 401-412, April 1990.

[TEN 90]   Tennenhouse, D.L., **Layered Multiplexing Considered Harmful**, *Protocols for High-Speed Networks*, Elsevier Science Publishers B.V. (North-Holland), 1990.

[WBH 92]   Williams, N., Blair, G.S., Head R.A., **Multimedia Computing: An Assessment of the State of the Art**, *Journal of Information Services and Use*, October 1992.

[Wil 91]   Williams, N., Blair, G.S., **A Distributed Multimedia Application Study**, *To appear in Computer Communications*, and available as an Internal Report, Computing Department, Lancaster University, Bailrigg, Lancaster LA1 4YR, UK, June 1991.

Integrated Quality of Service for Multimedia Communications

Andrew Campbell, Geoff Coulson, Francisco García and David Hutchison
Computing Department, Lancaster University, Lancaster LA1 4YR, UK
Email: mpg@comp.lancs.ac.uk

The integration of distributed multimedia systems support into a communications architecture encompassing the new multiservice networks poses significant challenges. A key observation about the new environment is that Quality of Service (QOS) provides a unifying theme around which most of the new communications requirements can be grouped. For applications relying on the transfer of multimedia information, and in particular continuous media, it is essential that QOS is guaranteed system-wide, including the distributed system platform, the transport protocol and the multiservice network. Enhanced protocol support such as end-to-end QOS negotiation, renegotiation, indication of QOS degradations and co-ordination over multiple related connections are also required. Little attention, however, has so far been paid to the definition of a coherent framework that incorporates QOS interfaces, management and mechanisms across all the layers. This paper describes the first stage in the development of an integrated Quality of Service Architecture (QOS-A) which offers a framework to specify and implement the required performance properties of new multimedia applications over multiservice ATM-based networks.

Keywords : Quality of Service Architecture, Communication Protocols, Multimedia Services, High Performance Networks, Distributed Systems.

1 Introduction

The evolution of distributed computing is being influenced simultaneously by the emergence of high-speed multiservice networks and the requirements of new distributed multimedia applications [LBC 92]. Due to the requirements of distributed multimedia applications, future communications infrastructures should offer comprehensive QOS configurability for a wide variety of media types and specific user requirements [Wil 91]. An important first step in meeting such a requirement is the specification of a *Quality of Service Architecture (QOS-A)* which offers an integrated framework for QOS specification and resource control over all architectural layers from distributed application platforms to the network layer.

The approach of this paper is to present a set of key QOS requirements and map these requirements onto a provisional QOS-A which has emerged from an experimental system designed and implemented at Lancaster. Because of the likely complexity of a fully general QOS-A, we limit the scope of our discussion in this paper to aspects of a QOS-A for the support of continuous media communications.

We also concentrate on ATM at the network layer rather than consider the full range of multiservice networks. However, a generalised QOS-A should eventually be extensible to incorporate other areas of QOS provision such as real-time control systems, file transfer and real-time transaction processing.

The paper is structured as follows. Section two identifies the requirement for a QOS-A in the light of the emerging requirements of distributed multimedia applications. Section three reviews current notions of QOS in OSI and the current ATM proposals from CCITT. Section four then presents research at Lancaster which has defined a baseline QOS architecture. This section also looks at functions and mechanisms for QOS support and attempts to place them within the evolving QOS-A. Finally, section five briefly examines related work in the field and section six presents our conclusions.

2 The Need for a QOS-A

2.1 QOS Requirements for Continuous Media

With the emergence of multimedia information exchange, increased requirements are placed upon communications support. Multimedia is particularly characterised by *continuous media* such as voice, video, high quality audio, and graphical animation, which place much greater demands on communications than still media, such as, text, images and graphics. Different types of continuous media require different levels of latency, bandwidth and jitter, and they also require guarantees that levels of service can be maintained. For example, video connections require high throughput guarantees but telephone audio requires only modest bandwidth. Error control should also be configurable: e.g. unencoded video is highly tolerant of communications errors whereas compressed voice can tolerate almost no errors and file transfers should be 100% error free. Delay jitter (i.e. variance in delay) is an additional factor which must be taken into account for continuous media transfers and must be kept within particularly rigorous bounds to preserve the intelligibility of audio and voice information.

In distributed multimedia applications the concept of QOS becomes applicable on a full *end-to-end* basis. In addition to the communications sub-system, this has implications for operating system scheduling for threads which are producing/ consuming information for quality controlled connections. End-to-end QOS also involves *distributed application platforms* which are layered on top of the operating system to provide distribution transparencies and object based computational models for the benefit of programmers of distributed multimedia applications.

Finally, the concept of QOS is also applicable to areas other than the traditional arena of point-to-point connections. For example, the need to support multicast and group communications in distributed multimedia systems leads to considerations such as the ordering semantics of group message delivery [Bir 82] which can be treated as a QOS issue. Also, the requirements of multimedia synchronisation such as 'lip-sync' impose QOS constraints over multiple transport level connections. We refer to this latter requirement as *orchestration* (see later); QOS properties in this context are concerned with the 'tightness' of orchestration required and the strategies to adopt when QOS provision degrades.

2.2 QOS-A Requirements

Because of the increased range and complexity of QOS provision required by the emerging distributed applications, it becomes essential that the necessary extensions to QOS provision are not done on a piecemeal basis. Instead, we advocate the notion of a comprehensive QOS-A, whereby application requirements can be mapped through all the levels of the system. Thus, communications abstractions at the application platform level should provide QOS abstractions which can be mapped down through all intermediate layers to the multiservice network access point in a coherent and integrated way. This mapping should be clean, simple and efficient, protecting application programmers from communication details; that is, they should be stating what they require rather than how it is to be achieved.

In addition to mapping functions, the QOS-A should provide a framework for the support of QOS throughout the layers. In particular, there is a need for enhanced protocol support which is lacking in current protocol specifications. The framework will include *management functions* and selectable QOS support *mechanisms*. Examples of management functions are:

- end to end QOS negotiation including admission control for new connections;
- policing to ensure that users are not violating negotiated QOS parameters;
- monitoring to ensure that negotiated QOS levels are being maintained by the service provider.

QOS support mechanisms take the form of a pool of available procedures which can be configured and inserted into a protocol stack. Depending on the QOS contract established at negotiation time, the QOS-A would be responsible for building a suitable stack profile by selecting from the available set of mechanisms. Examples of such mechanisms are error control modules, parameterisable scheduling modules and jitter smoothing modules. Once again, many of these mechanisms will be applicable at multiple system layers. For example, scheduling appears in the context of end user threads and also in network switches.

3 Current Notions of QOS

Traditionally, the term 'quality of service' in the communications context referred to certain characteristics of network services as observed by transport users. These characteristics were not controllable by users, and described only those aspects of services attributable to the network provider. QOS parameters in current communications infrastructures, like OSI and CCITT do permit the specification of some user requirements but these are almost never supported by the underlying network. For example, the current OSI standards treat QOS in a layer specific way, and QOS definition has been looked at by separate committees (viz. the presentation, session, transport, network and data link committees) working in isolation. Thus the relationship between QOS layers is not clearly defined and there is no consistent, integrated notion of QOS which relates user requirements to the network provider services.

The following sub-sections review the degree of QOS provision in sample architectures at the session, transport and network layers. The session and transport layer specifications are taken from the ISO's Reference Model for Open Systems Interconnection (OSI-RM) [Hen 88] and the network specifications are taken from the CCITT's series I recommendations for ATM cell switching [CCI 90].

3.1 OSI Perspective

In the OSI-RM, QOS parameters associated with the application layer's P-CONNECT primitive are generally mapped directly down to the associated QOS parameters at the session layer. Thus the QOS parameters associated with a P-CONNECT service element exist solely to give the application process access to the corresponding parameter of the session service element, S-CONNECT. The functionality of the session layer QOS parameters is then mainly concerned with monitoring and maintaining session services to a level agreed by the negotiation between peers as part of connection establishment.

Table 3.1 OSI Performance-oriented QOS Parameters

Parameter	Description
Throughput	The maximum number of bytes, contained in SDUs, that may be successfully transferred in unit time by the service provider over the connection, on a sustained basis.
Transit Delay	The time delay between the issuing of a data.request and the corresponding data.indication. The parameter is usually specified as a pair of values, a statistical average and a maximum. Those data transfers where a receiving service-user exercises flow control are excluded. The computations are all based on a SDUs of a fixed size.
Residual Error Rate	The estimated probability that an SDU is transferred with error, or that it is lost, or that a duplicate copy is transferred.
Establishment Delay	The delay between the issuing connect.request and the corresponding connect.confirm.
Establishment Failure Probability	The estimated probability that a requested connection is not established within the specified maximum acceptable establishment delay as a consequence of actions that are solely attributable to the service-provider.
Transfer Failure Probability	The estimated probability that the observed performance in respect to transit delay, residual error rate or throughput will be worse than the specified level of performance. The failure probability is, as such, specified for each measure of performance of data transfer, discussed above.
Resilience	The estimated probability that a service-provider will, on its own, release the connection, or reset it, within a specified interval of time.
Release Delay	The maximum delay between the issuing of a disconnect.request primitive by the service-user and a corresponding disconnect.indication primitive issued by the service provider.
Release Failure Probability	The estimated probability that the service-provider is unable to release the connection within a specified maximum release delay.

There are, however, aspects of the S-CONNECT QOS parameter which relate directly to the reliable data transfer environment required to service the session. These aspects are included in the QOS parameter of the transport layer T-CONNECT service element. The session layer QOS parameter is in fact a list of parameters, each of which relates to a particular QOS performance parameter. There are parameters covering each of the phases of the session, i.e. connection establishment, data transfer and connection release. The parameters are also classified into two major groups: *performance oriented* and *non-performance-oriented* [Hen 88]. The non-performance-oriented parameters do not directly affect the performance of the communications but are concerned with protection, priority and cost aspects. The complete set of parameters together with their interpretations is given in tables 3.1 and 3.2.

Table 3.2 OSI Non-performance-oriented QOS Parameters

Parameter	Description
Protection	This is the extent to which a service provider attempts to prevent unauthorised monitoring or manipulation of user data. The level of protection is specified qualitatively by selecting either *(i)* no protection; *(ii)* protection against passive monitoring; (iii) protection against modification, addition or deletion, a combination of *(i)* and *(ii)*.
Priority	High priority connections are serviced before lower ones. Lower priority connection packets will be dropped first before high priority packets, should the network become congested.
Cost Determinants	A parameter to define the maximum acceptable cost for a network connection. It may be stated in relative or absolute terms. Final actions on this parameter are left to the specific network providers

3.2 CCITT I-Series Perspective

The CCITT, in their series-I recommendations, have recognised the need for QOS configurability in the emerging ATM standards for B-ISDN and a fairly comprehensive set of parameters has been defined. QOS of bearer services in ATM networks is applicable at three control levels:

- *call control level*: this is concerned with the establishment and release of the call. A call is rejected by the call acceptance control algorithm if the requested bandwidth is not available at the time of call set-up request;
- *connection level*: this is concerned with allocation of resources for the data transfer phase. A call is rejected if there is no available path (sequence of links) to its destination. At the connection level resources have to be allocated at each intermediate hop between the source and the destination;
- *cell control level*: this is concerned with the data transfer phase itself. Once a connection has been accepted, the cell stream must be policed to ensure that the user does not exceed the values contracted in the call-setup.

A user wishing to establish a connection signals his QOS requirements to the ATM network. The signalling message includes a declaration of the QOS characteristics of the user data which have somehow been mapped down from the higher layers, and enables the connection acceptance control function to allocate the required QOS resources if the connection is accepted. The connection is then assigned a source policing function which monitors the cell stream and causes cells exceeding the declared traffic rate to be either discarded immediately or marked to be discard later if necessary.

At the call control level, the available parameters are similar to those defined in OSI: i.e. establishment delay, establishment failure probability, release delay etc.. At the connection level the parameters outlined in table 3.3 are applicable:

Table 3.3 CCITT QOS Parameters

Parameter	Description
peak arrival rate of cells	The maximum resources required by the application at peak load.
peak duration	The average duration of the maximum load.
average cell arrival rate	The average amount of network resources requested by the source [Ana 91]. This is the number of cells measured during the duration of the connection divided by the duration.
burstiness	The ratio between the peak cell rate and the average cell rate. Examples: voice = 2, interactive = 10, Standard video = 1 to 10, HDTV = 1-2 and high quality video telephony = 2 [DeP 92].
cell loss ratio (CLR)	The ratio of number of lost cells to transmitted cells. This type of error usually occurs because of congestion in the switches.
cell insertion ratio (CIR)	This type of error occurs when the address field in the header is corrupted to another valid network address.
bit error rate (BER)	Defined as the number of bits which are delivered erroneously divided by the total number transmitted. These sorts of errors are mainly caused by transmission system.

The cell level employs the traffic characterisation supplied by the connection level and uses it to ensure that the application does not exceed the peak and average traffic levels agreed at connection time. As an example of traffic characterisation, the QOS parameters of variable bit rate encoded video could be: peak rate = 50Mbps, average cell arrival rate = 25Mpbs, burstiness = 2 and the peak duration = 10ms.

3.3 Evaluation

The clearest point to emerge is that QOS is currently looked on largely as a service provider issue whereas the requirements identified in section 2 imply that QOS should also be a user level issue. It is also clear that the OSI standards are currently incomplete and inconsistent for specifying QOS properties. In particular, the protocol specifications and service definitions do not include any notion of QOS management and the semantics of responsibilities and guarantees are not clear. Furthermore, those functions which are defined are almost never supported by protocols and networks.

Another important point is that the OSI upper layers have no notion of QOS: QOS parameters are simply mapped through to the transport layer. This is also true in alternative upper layer architectures such as the object-based ODP architecture [ISO 89]. If users want to specify QOS they are forced to drop below the level of abstraction provided by these architectures and interact with layers that are supposed to be hidden. Furthermore, there are very limited facilities for QOS negotiation at the user level. A user must simply specify the parameter values required and let the lower layers either accept or reject the proposal.

The CCITT's ATM recommendations are more comprehensive in scope with a fairly detailed traffic characterisation model. Here it is the mapping between the higher layers and the ATM adaptation layer, and also the mechanisms required to support particular QOS specifications which are lacking in substance. These are precisely the concern of a generalised QOS-A. An important step in our work will be resolving the present inconsistencies in the relationship traffic characterisation parameters of ATM and the OSI-RM. Other requirements are the development of protocol support for QOS in terms of the various QOS management functions and support mechanisms, examples of which were given in section two above.

Finally, a major limitation of all current notions of QOS is that the value of a QOS parameter, whether negotiated or not, remains the same through the lifetime of a connection: i.e. once negotiated a QOS parameter is never re-negotiated. Another implication of this is that the service-provider is committed to provide the QOS over the lifetime of the connection. There is, however, no guarantee that the service-provider will be able to maintain the originally specified values: in fact maintaining end-to-end service levels in the face of variable load is an unsolved problem involving resource scheduling at multiple levels. Even when the QOS of a connection does deteriorate the service provider is under no obligation to signal such a change in QOS to the users of the connection. The provider may, however, disconnect the connection unilaterally.

The essential characteristics of the current state of QOS provision may therefore be summarised as follows:

- *lack of overall framework*: the framework for QOS must extend from the distributed application platform through the transport subsystem and the network. It must also encompass QOS considerations in areas such as orchestration and groups;
- *inconsistency*: the framework must build on and reconcile the existing notions of QOS particularly in the OSI-RM and the ATM series I recommendations;
- *incompleteness*: the framework should include extensions to current QOS provision as detailed in the following section;
- *lack of mechanisms to support QOS guarantees*: research is needed in basic mechanisms such as scheduling so that contracted QOS levels can, in fact, be maintained.

4 An Extended View of QOS

4.1 Baseline Architecture

4.1.1 Service Model

The most fundamental aspect of the QOS-A is the interface at which desired levels of QOS can be requested, negotiated and contracted. In a layered architecture there are multiple instances of this interface; each instance has a customer above and a service provider below (see figure 4.1). The central function of the QOS-A is to permit end-to-end QOS negotiation from the top user level down to the network layer and up again at the remote site.

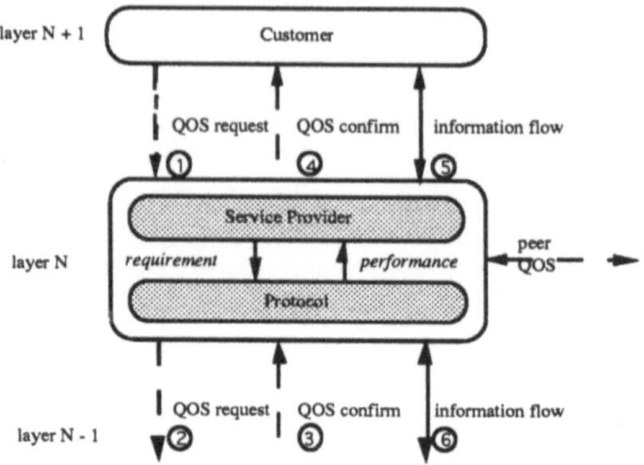

Fig. 4.1 QOS-A Service Model

A successful negotiation at each interface level results in a *service contract* with two major clauses:

- an agreed *level of service* which the service provider level must undertake to maintain, and
- an agreed *level of traffic* which the customer must undertake not to exceed.

Both the level of service and the level of traffic are expressed in terms of a common, layer specific, *traffic characterisation* language based on the fundamental QOS dimensions. Examples of layer specific traffic characterisations appear in section 4.3.

4.1.2 Architectural Layers

The baseline for the development of the QOS-A is the layered architecture depicted at the left hand side of figure 4.2. This has been derived from our experimentation to date with distributed multimedia applications. A detailed description of our current infrastructure and its implementation is given in [DCW 91]. The remainder of figure 4.2 illustrates the aspects of the QOS-A to be described in this section.

The upper layer in the layered architecture consists of distributed applications platform which is provided by an ODP compatible distributed systems platform augmented with services to provide multimedia communications, QOS configuration and synchronisation [CBD 90].

Below the platform level is a layer of services used to add value to the functionality provided by the lower transport layer. Specifically, these services control jitter and rate regulation for continuous media streams. They also provide these services, together with cross-stream synchronisation, across multiple application related connections. Because these services are concerned with co-ordinating multiple sources and sinks we refer to them as *orchestration* services; a full description of these services can be found in [CCG 92].

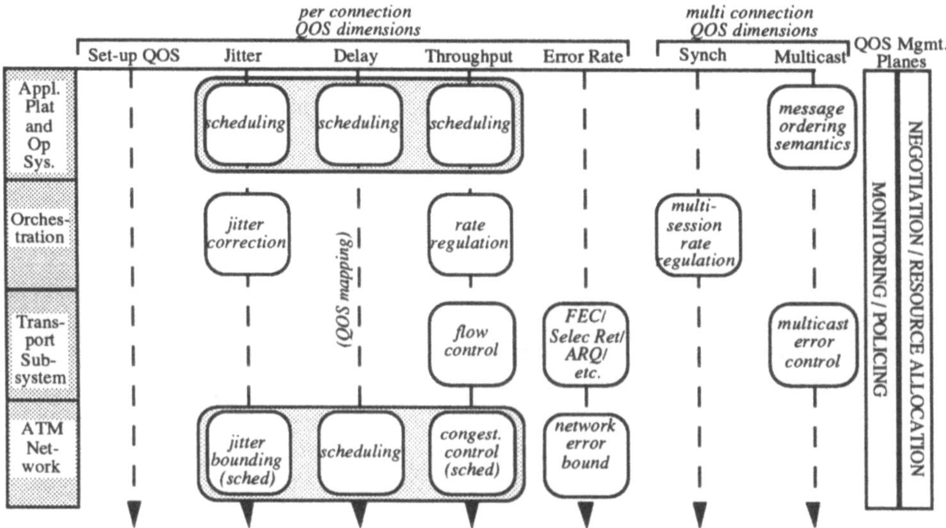

Fig. 4.2 QOS-A

Below the orchestration services is a transport service and protocol which is specifically designed for continuous media communications. It is highly configurable in terms of QOS and offers full end-to-end QOS negotiation and re-negotiation. Full details of the transport services are available in [SHG 91].

The communications infrastructure is provided by a multiservice network. Currently we are using a real-time FDDI emulation, but are in the process of upgrading the communications to use an ATM network. To achieve this aim, we require new hardware interfaces to our current multimedia workstations [Bla 93], and also an implementation of the ATM adaptation layer software.

4.1.3 QOS Dimensions

In figure 4.2 we have attempted to extract a canonical and orthogonal set of *dimensions* within which traffic can be characterised in our chosen domain of continuous media communications. The chosen set of dimensions are: *set-up QOS* (i.e. the OSI establishment and release parameters), *jitter* (i.e. variation in delay), *delay*, *throughput* and *error characteristics*. In addition to these fundamental dimensions, two additional dimensions are included: *synchronisation* between media streams, and aspects of *multicast* quality of service. The essence of these latter two dimensions is that they are applicable over multiple connections whereas the others apply to single connections. In fact, the multi-connection dimensions also subsume the fundamental dimensions but additional quality of service characteristics arise as emergent properties.

Later sections describe how traffic may be specifically characterised at the various layers, and how levels of service along the canonical QOS dimensions are maintained through profile selection at the different layers.

4.2 QOS Management Functions

4.2.1 QOS Negotiation

A further aspect to the above mentioned contract is the *degree of commitment* in the above clauses: e.g. is the provider committed to maintaining the level of service in all conditions or are there circumstances in which the level of service may be relaxed? A related question is what sanctions will be imposed by the provider if the contracted traffic level commitment is exceeded by the user? Both of these points can also be related to the *cost* of the service; presumably a higher commitment by the provider for the same nominal service will cost more, as will the option of a lower commitment from the user.

To express degrees of commitment either a relative measure such as priority levels can be used, or absolute measure such as a percentage. Absolute measures can also be expressed as a step function with values such as {*deterministic, probabilistic, best-effort*}. An absolute scale is also appropriate for cost measures. Even if only a step function were ultimately available at the bottom level, percentages measures of commitment at higher levels could be used to express trade-offs. For example, in a videophone connection consisting of separate video and audio channels, a slightly lower commitment probability for the video channel would be appropriate Even if the two probabilities chosen mapped to the same step function value at the lower layer (e.g. probabilistic), the higher commitment of the audio channel would ensure that if one channel had to be taken down due to lack of resources, the more important audio channel would be preserved.

As pointed out in section 2 the same QOS interface is not necessarily appropriate for all layers in a layered architecture. However, in abstract terms, the interface at each layer will consist of some subset of the following general components:

- a notation for quality characterisation along the QOS dimensions,
- a notation for commitment specification,
- a notation for cost specification, and
- a protocol for QOS negotiation.

There are other possibilities in the negotiation process which may be of use at different layers. For example, rather then simply proposing a level of service, upper and lower bounds on acceptability could be proposed by the user and the provider could return a contract as close to the upper bounds as possible. Also, renegotiation of the QOS on a live connection could be permitted. Finally, the user could specify alternative *degradation paths* to be taken when commitments are not met by the provider. Possible alternatives are: ignore the situation, simply inform the user, or inform the user and reconfigure according to a user specified degradation path. As an example of the latter a hifi audio channel could be degraded to a 64Kbps voice audio channel. At the same time, the system would inform the user who would adjust the source and sink codecs as appropriate.

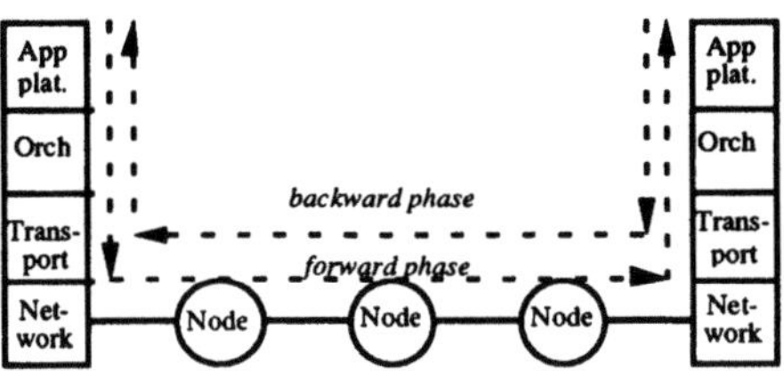

Fig. 4.3 End-to-end QOS Negotiation

The mechanism of negotiation is illustrated in figure 4.3. In broad terms, negotiation is a two phase process. On the forward phase, from the source towards the sink, each intermediate resource holding node or end-system layer attempts to allocate resources to the request. Each layer in the system contains an *admission control* module which determines whether or not the request should be considered. For example, the operating system will decide if it is able to create a new real-time thread and the network layer will determine whether or not it can allocate a new connection. If the request is accepted by admission control, as many resources as are available are dedicated to the request. Eventually, at the sink end, the allocated QOS is compared against the requested requirement along each QOS dimension and the amount of over-commitment, if any, is calculated. On the second phase, from sink to source, any over-commitment is divided among the intermediate nodes and the over committed resources are released. Examples of negotiation mechanisms from the literature include [And 91] and [Top 90] which address negotiation in the network. Full end-to-end QOS negotiation is discussed in [Nic 91].

4.2.2 Monitoring and Policing

QOS must be monitored at all layers of the QOS-A to ensure that the negotiated levels of QOS are being maintained. Each layer will collect statistical information associated with the on-going connection performance and make an assessment of the QOS measured against QOS requested. Monitors may then either attempt to take action to restore QOS levels themselves or they may choose simply to inform the upper layer that there is a problem.

Policing, on the other hand, is primarily seen as a network (and perhaps transport) function and may not be carried out at all layers. In particular, the application platform and orchestration layers are simply not able to violate QOS agreements by the nature of QOS support at those layers. Examples of monitoring and policing at the various levels in figure 4.2 are as follows:

Application platform An important consideration at the platform level is the monitoring of end-to-end communications and will thus take into account possible QOS degradations arising from the end-system in addition to those arising from the communications subsystem. Application platform level monitoring will be mainly concerned with monitoring the performance of the operating system thread scheduling mechanism to ensure that data which arrives correctly is also delivered correctly to its final destination.

Orchestration A fundamental part of orchestration is the ability to monitor the on-going temporal relationship between of connections, and to regulate the connections to perform fine grained corrections if synchronisation is being lost. It is almost inevitable that related connections will eventually drift out of synchronisation due to such factors as discrepancies between remote clock rates and network congestion caused by temporary blocking at intermediate switches. The degree of multimedia synchronisation accuracy required by the user is viewed as a monitoring issue in the QOS-A. The monitor may choose to attempt to correct the drift (perhaps by renegotiating transport QOS), or if the drift is too great it may simply inform the upper level.

Transport The transport level monitoring entity captures statistical information related to the connection performance between two TSAPs. In the case of guaranteed QOS, the QOS state for each parameter is periodically measured to determine the *(i)* typical response time; *(ii)* average throughput over an interval *(iii)* amount of time buffering/blocking in the network per packet; *(iv)* worst case responses time; and *(v)* best case response time. If any of the negotiated QOS parameters degrade below the specified minimum tolerance, then a *QOS.Indication* [CCG 92a] is raised by the monitoring mechanism detailing which QOS parameter has degraded and its current measured value, the indication also includes a statistical profile of all monitored parameters. On receiving a QOS.Indication, the application is free to make a judgement based on the current connections performance; for example the application may accept a poorer QOS in the light of network congestion, on the other hand it may initiate a full end to end renegotiation and also upgrade the level of commitment of the connections QOS.

Network Once a connection has been successfully negotiated between two end points, the source must be policed at the edge of the network to ensure that it does not exceed the traffic profile declared at ATM call-setup time. This is particularly important if a statistical multiplexing approach is assumed. Should the user exceed the agreed throughput levels, the policing function will intervene and discard cells. In addition a network level QOS.Indication will be generated indicating a *bandwidth violation* has been detected on the connection. On reception of the indication, the application can either regulate its traffic or renegotiate a higher throughput to meet its end to end throughput requirement.

4.3 Layer Specific QOS Considerations

4.3.1 Application Platform Layer

4.3.1.1 Functions
In our experimental configuration, this layer includes both the operating system and a distributed application platform based on the ANSA architecture [APM 89]. In

ANSA, programmers view an object-based computational model in which all interaction is specified in terms of RPC invocations of named operations in location independent abstract data type interfaces. Thus it appears that QOS requirements should be attached to interface descriptions so that constraints on the QOS of invocations on that interface can be specified. The language used to specify such constraints should be network independent and its expression should be in terms of user level concepts such as frames rather than packets or bits.

Another consideration is that application programmers may not wish to be concerned with the precise details of a possibly complex set of QOS parameters. There should be a shorthand way of specifying commonly used 'channel types' by name such as HifiAudio or StandardVideo. These standard channel types could also be parameterised with arbitrary semantics: e.g. Monochrome-Video(MPEG, 512, 256) would mean that the data stream will be compressed according to the MPEG standard and the throughput can be deduced from the required window size of 512 x 256 pixels. Such shorthand specifications could also implicitly include default QOS parameter ranges, probabilities of commitment and degradation paths as described above. As long as a common underlying QOS representation is used, new QOS channel types can be provided simply by adding new *QOS-mapper* services to the running system. These are run time services which resolve channel type names and parameters into the representation used at the next level down. A sample set of commonly used QOS channel types [LBC 92] is illustrated in table 4.1.

Table 4.1 Standard Channel Types

Channel Types	Bandwidth	Jitter	Delay	Traffic type	Error
StandVideo	25 Mbps	10 ms	250 ms	probabilistic	10^{-3}
SlowScanVideo	1 Mbps	10 ms	250 ms	probabilistic	10^{-2}
MPEGVideo	10 Mbps	1 ms	250 ms	deterministic	10^{-9}
VoiceAudio	64 Kbps	10 ms	250 ms	probabilistic	10^{-1}
HiFiAudio	2 Mbps	5 ms	500 ms	deterministic	10^{-5}

An alternative notation intended for the expression of arbitrary QOS specifications is reported in [Ste 92]. This is a detailed and comprehensive QOS language based on function composition, simple arithmetic operators and probability. A design for parameterised QOS channel types is presented in [Nic 91].

4.3.1.2 Mechanisms

The QOS support *mechanisms* at this level are almost entirely concerned with scheduling at the operating system level. As can be seen from figure 4.2, operating system scheduling effects the QOS dimensions of jitter, delay and throughput. However, this layer is not involved in QOS support mechanisms for error handling, synchronisation or multicast.

The way in which scheduling is parameterised to support the jitter delay and throughput dimensions is dependent on the particular scheduling policy used. In our system we are using a modification of the deadline/workahead scheduling implementation developed at the University of California at Berkeley and described in

[Gov 91]. This scheme applies earliest deadline first scheduling to packets generated by devices or arriving from network connections according to a deadline timestamp contained in each packet. If the scheduler has sufficient processing resources, it is able to schedule the threads expecting these packets such that the packet deadlines are not violated. The scheduler is also careful not to schedule a thread too early and thus is able to control jitter.

The platform level is also involved in the QOS of multicast connections. At this level, multicast appears in the computational model as the invocation of groups of interfaces. A range of qualities of service are possible here most of which pertain to message ordering guarantees at the multiple interfaces. For example, virtual synchrony [Bir 82] is a particularly strong QOS where it is guaranteed that if there are multiple clients of an interface group, all messages from the different clients are delivered to all the service interfaces in the same order. This also illustrates the general principle that a QOS specification determines the construction of a tailored 'profile' which includes modules such as virtual synchrony if required. The same principle is applicable to all the layers.

4.3.1.3 Example

Finally, as an example of platform level QOS we present the specification of a video channel in terms of an ANSA interface specification. This example will be taken up again as we examine the lower system layers in subsequent sections.

```
TYPE VideoSource = INTERFACE
BEGIN

GetVideo : OPERATION = [ ]
        RETURNS [ VideoFrame ]
        WITH QOS "StandardVideo";

END.
```

4.3.2 Orchestration Layer

4.3.2.1 Functions

Orchestration services provide value added communications services approximately at the session layer in OSI terms. However, the functionality of these services is not concerned with the traditional OSI telematics session functions. Instead, it offers services which are of use to continuous media applications: primarily rate control and jitter correction. The orchestration services also operate over multiple connections and provide synchronisation between continuous media streams flowing in separate connections. An example of such synchronisation is lip-sync between audio and video channels. To support such synchronisation relationships between channels, the orchestration services apply a compatible rate control to the flow of data in each connection and adaptively adjust these rates when brief disruptions occur. For example, if a number of packets are lost from an audio play-out, a corresponding number of packets may be dropped from an associated video channel.

The degree of multimedia synchronisation accuracy required by the user is viewed as a QOS issue in the architecture. This is expressed by the user at a platform level interface using the concept of per stream intervals which are interpreted as synchronisation points. A global clock service ensures that intervals can be interpreted similarly throughout the network. Other QOS concerns are the degradation paths to be taken when synchronisation is lost. These may include requests to *renegotiate* the QOS of transport connections or the adoption of longer intervals. A full description of the motivation for the orchestration services is beyond the scope of this paper. The same is true of the mechanisms used in orchestration. Full details may be found in [CCG 92].

4.3.2.2 Example

To return to the example above, our video QOS may be expressed in the following terms at the orchestration layer.

```
Delivery rate: 25 frames/sec
Permissible jitter: 10 ms
Sync interval: 1 second
```

4.3.3 Transport Layer

4.3.3.1 Functions

Traditional transports such as OSI Transport Class 4 have mainly serviced the communication needs of file transfer and low-bandwidth interactive applications. It is generally accepted that these protocols are not suitable for multimedia and real-time communications, and that dedicated transport services and supporting protocols are needed to exploit the performance of the new high speed network environment such as the ATM based multiservice networks [SDG 91]. Consequently, the following enhanced transfer services have been identified: *(i)* continuous media service; *(ii)* transaction oriented service; and *(iii)* bulk data transfer service. These transfer services should operate over connection-oriented and connectionless services and should also permit multicast topologies.

Continuous media transfer requires new transport mechanisms that support the following functionality:

- QOS negotiation is required across all the QOS dimensions in figure 4.2 except for the cross stream synchronisation dimension;
- indications of QOS degradation to the upper layer are required. This is an aspect of monitoring discussed in section 4.2.2.
- re-negotiation of QOS on live connections is required. For example, users may wish to re-use the same connection for different media to minimise connection establishment delays;
- a flexible approach to error recovery [Fel 92]; This is needed because error recovery needs vary considerably with the encoding used for continuous media data. For example, uncompressed video is highly tolerant of errors whereas MPEG compressed video is highly intolerant. A further consideration is that error recovery must not involve retransmission as packets will then arrive to late to be useful. For this reason forward error correction (FEC) is a promising technique here.

- multicast;
- simplex connections are useful as resources do not have to be allocated for full duplex operation when, say, a one way video transmission is required.

Lancaster's experimental *transport protocol* is a rate-based transport mechanism which supports the communications needs of continuous media. Currently it runs over the FDDI emulation mentioned above but is being extended to interface to the ATM adaptation layer (AAL type 5). The transport layer defines the QOS performance in terms of parameters comprising burst size, burst duration, delay, jitter, priority, packet error rates and error profile. See [SHG 91] for further details.

4.3.3.2 Mechanisms

As can be seen from figure 4.2, the QOS dimensions applicable at the transport layer are throughput, error correction and multicast issues. Jitter and delay are not treated within the transport layer and these parameters are simply mapped through to the network layer. Throughput is configurable in the transport layer by manipulation of buffer resources and flow control mechanisms. To avoid the overhead of window based flow control, our experimental protocol uses rate based flow control which is well suited to continuous media data. Throughput is affected by the burst size and burst duration in the rate control scheme.

For errors, a variety of error correction profiles such as FEC, selective retransmission or go-back-N can be chosen depending on the QOS required. In multicast mode the error control dimension is expanded to include considerations of differential packet losses visible at each destination. 'Leaky bucket' [ANS 91] error correction algorithms can be brought in to the profile to provide reliable multicast delivery if such a QOS is required.

4.3.3.3 Example

Returning again to our example, the QOS specification at the transport layer may be as follows.

```
Burst size: 100 Kbps
Burst rate: 100 per sec
Delay: 1 sec
Jitter: 20 ms
Priority: 10
Error profile: FEC
Error rate: 2%
```

Note that the jitter specified at this level is coarser than the top level requirement as the orchestration layer is taking some responsibility for jitter control.

4.3.4 Network Layer

4.3.4.1 Functions

The key requirement at the ATM layer is to achieve a high degree of resource utilisation by statistically multiplexing traffic while simultaneously meeting the users source traffic QOS requirements. Because of this requirement QOS at the ATM level is more concerned with traffic characterisation for efficient resource management than for user convenience. Note that this difference in perspective is also apparent in the emphasis on source traffic policing rather than service level monitoring. Because the ATM QOS parameters listed in section 3.3 have been formulated with this aim, and also allow QOS expression over the full set of QOS dimensions described in section 4.1.2, they are probably sufficient regardless of the traffic types involved.

One traffic characterisation issue still to be addressed at the ATM layer is the issue of level of service commitment. For ease of resource management, it is convenient to partition commitment into a small number of fixed levels rather than a continuous scale [CSZ 92]. A possible partition is as follows:

- *deterministic service*: the highest priority service the ATM network provider can support. Typically, this is a guaranteed QOS for hard real time performance applications;
- *probabilistic service*: this may suffer from QOS degradation from time to time because of the statistical nature of the network service. This service is particularly suitable for many continuous media applications that can compensate for loss of QOS because most media has built in redundancy. QOS monitoring, as described in section 4.2.2, is an important aspect of this service;
- *best effort service*: the lowest priority service, is synonymous with datagram services. No network resources are allocated or monitored because the network provider is not obliged to guarantee any level of service. This commitment level only receives whatever network resources are available after the other levels have been serviced.

4.3.4.2 Mechanisms

As is clear from figure 4.2, a network level traffic scheduling mechanism is an central component of the QOS-A as this directly supports bounded delay, jitter and guaranteed throughput. Scheduling is important because the dominant factor in delays and jitter is the network queuing delay introduced by intermediate switches rather than the propagation delay between switches. The network must provide strong performance guarantees for deterministic and probabilistic traffic (even during peak-load and over-load conditions) by correctly scheduling cell departure times at the end-systems and at each switch so that delays can be kept within the agreed bounds.

All three levels of QOS commitment could be supported in the network using a combination of deadline scheduling and priority queuing. In a priority based scheme, each traffic class (deterministic, probabilistic and best effort) is characterised by a given priority. This ensures that deterministic traffic has priority over probabilistic traffic which in turn has priority over best effort traffic. An alternative strategy is possible which allocates fixed network resources e.g. deterministic and probabilistic traffic receive 85% of the bandwidth and best effort only 15%. This guarantees the performance of deterministic and probabilistic communications and avoids starvation of datagram communications, but may lead to lower overall utilisation of bandwidth because statistical gain may not be realised.

Research on the network scheduling problem is reported in, for example, [Zha 91] and [CSZ 92]. A significant amount of work is currently being undertaken in this field.

Apart from active scheduling, ATM networks will require a suitable resource management mechanism which is able to allocate network resources such as bandwidth and end-to-end delay at set-up time. A resource reservation mechanism will allocate network resources at the end systems and at all intermediate switches between the source and destination. A policing mechanism is also required to monitor the user behaviour at the network edge. The network has to exercise flexible resource allocation and congestion control to exploit the potential increase in network efficiency resulting from the use of statistical multiplexing.

4.3.4.3 Example

The final stage in the QOS mapping example is the generation of the following ATM layer QOS specification.

```
QOS commitment level: deterministic
QOS parameters
        peak cell arrival rate: 10 Mbps
        average cell arrival rate: 8 Mbps
        burstiness: 5
        peak duration: 100 ms
Call control parameters
        establishment delay: 4 seconds
        establishment failure prob: 0.05
    (i.e. 95% chance to connect)
        release delay: 4 seconds
Reliability parameters
        BER: 10^-7 with FEC
        CLR: 10^-5 with FEC
        CIR: 10^-3
Delay parameters
        delay: 250 ms
        jitter: 20 ms
```

It is clear from this final stage of the example that the mapping process is non-trivial and further work is necessary to automate this process in practice.

5 Related Work

5.1 Research

There is currently very little literature on the integrated treatment of QOS. One contribution [Slu 91] examines the requirements for QOS support in Open System standards and proposed enhancements to the existing OSI RM to support a QOS framework. Several projects in the RACE programme are concerned with QOS, in particular QOSMIC (R.1082). However, RACE restricts itself to a network-level view of QOS and therefore there is little consideration of higher level QOS issues.

In the higher layers a number of research teams has looked at QOS as a transport layer issue. Work at the IBM European Networking Center [HHS 91] has also investigated the integration of transport QOS and resource management (scheduling) in the operating system. Work in the transport service area at Berkeley is reported in [Wol 91].

An important requirement for the generalised QOS-A in non ATM networks is a suitable reservation scheme which is able to set up and guarantee network resources such as bandwidth and end-to-end delay for high performance continuous media communications between network subscribers. This could be achieved using resource reservation protocols based on such as SRP [And 91] and ST-II [Top 90]. Related work on resource management in internets is reported in [Fer 90].

Clark, Shenker and Zhang [CSZ 92] describe an Integrated Service Packet Network which can support three levels of service commitment: *(i) guaranteed service* for real-time applications; *(ii) predicted service* which utilises the measured performance of delays and is targeted towards adaptive or continuous media applications which can compensate for momentary loss of QOS; and *(iii) best effort* datagram service where no QOS guarantees are provided. Also, a unified traffic scheduling mechanism is developed which is based on a combination of weighted fair queuing and static priority algorithms.

In [Zha 91] a comparison of network scheduling algorithms is presented with respect to their suitability for guaranteed bounding of throughput, delay and jitter. The algorithms reviewed include Virtual Clock, Fair Queuing, Delay-Earliest-Due-Date, Jitter-Earliest-Due-Date, Stop-and Go and Hierarchical Round Robin.

5.2 Standardisation

Standards have an important role to play in development of a QOS-A for high performance distributed computing. The current OSI and CCITT standards do not take into account the particular requirements of an integrated QOS model. It can therefore be anticipated that the OSI RM will have to be extended to support the new QOS requirements [Cam 92].

In ISO, a New Work Item on QOS has recently been initiated (ISO/IEC JTC1/SC21, and in the UK IST21/-/1/5) which aims to address QOS in a consistent way. This activity will cover QOS very broadly, but has begun by investigating user requirements for QOS [ISO 92a] and some architectural issues [ISO 92b]. Lancaster University has provided early input on our QOS-A [CCH 92] work into this activity, and the intention is to participate actively at both national and international level.

One potential difficulty in achieving a unified QOS-A is that the ISO and CCITT have contrasting views on QOS. The CCITT perspective is network oriented and the ISO approach is geared towards the user. Another important viewpoint, which is just being taken into account, is that of the Open Distributed Processing (ODP) QOS requirements. This has been recently initiated in the UK with a liaison between ODP and IST/21/-/1/5 panels.

6 Conclusion and Future Work

This paper has argued for a comprehensive architectural framework for QOS support in the light of new applications with varied QOS requirements and new networks able to support QOS configurability. We have presented a draft QOS-A based on our experimental work and have discussed requirements and possible mechanisms for QOS support.

However, much remains to be done. Firstly, the number of QOS dimensions in the architecture can be expanded. Candidates include security and cost dimensions and there are probably others from application domains other than the field of continuous media support. Other work remains to be done in the vertical aspects of the architecture. In particular, the specification of QOS at the different layers is a major aspect of further work, as is the process of mapping between the layers. Also required is the definition of a standard framework for QOS negotiation, resource management, monitoring and policing over all the layers. Finally, research must be carried out on QOS protocol support mechanisms. This applies particularly to scheduling which is fundamental at the network layer and is also important at the operating system level where continuous media is involved.

As the next phase of our research we intend to refine and implement the QOS-A in an experimental configuration consisting of workstations connected via an ATM switch. The workstations will be equipped with audio and video cards and will run our existing transport and orchestration software on a multimedia network interface (MNI) [Bla 93] which we have built in a previous project. The workstations will also run a real-time operating system and our multimedia enhanced ANSA application platform to provide a test bed for thread scheduling for continuous media streams. On the ATM side, we also intend to experiment with scheduling mechanisms and to provide a prototype adaptation layer with negotiation, policing and resource management facilities.

Acknowledgement

Part of this work was carried out within the MNI project (funded under the UK SERC Specially Promoted Programme in Integrated Multiservice Communication Networks (grant number GR/F 03097) and co-sponsored by British Telecom Labs), and part within the OSI95 project (ESPRIT project 5341, funded by the European Commission). Furthermore, this research is being continued within the QOS-A project (funded under the UK SERC Specially Promoted Programme in Integrated Multiservice Communication Networks (grant number GR/F 77194) in cooperation with Netcomm Ltd.

References

[Ana 91]   Anagnostou, M.E., **Quality of Service Requirements in ATM-Based B-ISDNs**, *Computer Networks and ISDN Systems*, Vol 14, No 4, May 1991, pages 197-204.

[And 91]   Anderson, D.P., Herrtwich, R.G, Schaefer, C., **SRP: A Resource Reservation Protocol for Guaranteed Performance Communication in the Internet**, *Internal Report* University of California at Berkeley, 1991.

[ANS 91]   ANSI, **High Speed Transport Protocol (HSTP) Specification**, *X3S3.3/91-264*, American National Standards Institute, September 1991.

[APM 89]   APM Ltd., **The ANSA Reference Manual Release 01.01**, Architecture Projects Management Ltd., Poseidon House, Castle Park, Cambridge, UK, July 1989.

[Bla 93]   Blair, G.S., Campbell, A., Coulson, G., Garcia, F., Hutchison, D., Scott, A., Shepherd, W.D., **A Network Interface Unit to Support Continuous Media**, *IEEE Journal of Selected Areas in Communications (JSAC)* , Vol. 11, No. 2, 1993.

[Bir 82]   Birman, K.P., Joseph, T.A., **Exploiting Virtual Synchrony in Distributed Systems**, *Operating System Review Op. Sys. Review Vol. 21 No. 5*, pp 123-138.

[CCG 92]   Campbell, A., Coulson G., Garcia F., Hutchison, D., **A Continuous Media Transport and Orchestration Service**, *Presented at ACM SIGCOMM '92, Baltimore, Maryland, USA*, August 1992.

[CCH 92]   Campbell, A., Coulson G., Hutchison, G., **A Suggested QOS Architecture for Multimedia Communications**, ISO/IEC JTC1/SC21/WG1 N1201, *International Standards Organisation*, UK, November, 1992.

[Cam 92]   Campbell, A., Hutchison, D., **Contribution to the new ISO Work Item: Key Issues in Distributed Multimedia Communications**, *Draft BSI/IST6/-/2/738*, British Standards Institute, UK, April 1992.

[CCI 91]   CCITT, "**Draft Recommendations I Series**, CCITT Geneva, 1991.

[CSZ 92]   Clark, D.D., Shenker S., Zhang, L., **Supporting Real-Time Applications in an Integrated Services Packet Network: Architecture and Mechanism**, *Proc ACM SIGCOMM' 92*, pp. 14-26, Baltimore, December,1992

[CBD 90]   Coulson, G., Blair, G.S., Davies N., Macartney, A., **Extensions to ANSA for Multimedia Computing**, *To appear in Computer Networks and ISDN Systems*, also available as MPG-90-11, Computing Department, Lancaster University, Bailrigg, Lancaster LA1 4YR, UK, October 1990.

[DCW 91]   Davies, N., Coulson, G., Williams, N., Blair, G.S., **Experiences of Handling Multimedia in Distributed Open Systems**, *Proc.SEDMS '92, Usenix Symposium on Distributed and Multiprocessor Systems*, Newport Beach, CA, USA, also available from Computing Department, Lancaster University, Bailrigg, Lancaster LA1 4YR, UK, November 1991.

[deP 91]   De Prycker, M., **Asynchronous Transfer Mode: Solution for Broadband ISDN**, ISBN 0-13-053513-3, Ellis Horwood, New York, 1991.

[Fel 92]   Feldmeier, D., **Architectural Concepts for High Speed Communications Systems**, *Internal Report*, Bellcore, Morristown, NJ, USA, February 1992.

[Fer 90]   Ferrari, D., Verma D., **A Scheme for Real-Time Channel Establishment in Wide-Area Networks**, *IEEE JSAC*, Vol 8, No 3, April 1990, pages 368-377.

[Gov 91]   Govindan, R., Anderson, D.P., **Scheduling and IPC Mechanisms for Continuous Media**, *Thirteenth ACM Symposium on Operating Systems Principles*, Asilomar Conference Center, Pacific Grove, California, USA, SIGOPS, Vol 25, Pages 68-80.

[HHS 91]   Hehmann, D.B., Herrtwich, R.G., Schulz, W., Schuett, T., Steinmetz, R., **Implementing HeiTS: Architecture and Implementation Strategy of the Heidelberg High Speed Transport System**, *Second International Workshop on Network and Operating System Support for Digital Audio and Video*, IBM ENC, Heidelberg, Germany, 1991.

[Hen 88]   Henshall, J., Shaw, S., **OSI Explained: End-to-end Computer Communication Standards**, ISBN 07458-0253-2, Ellis Horwood, New York, 1988.

[ISO 89]   ISO, **Basic Reference Model of Open Distributed Processing**, *Working document on structures and functions*, ISO/IEC JTC1/SC21 N4022, 10 November 1989.

[ISO 92a]   ISO, **User Requirements for Quality of Service**, ISO/IEC JTC1/SC21/WG1 N1146, *International Standards Organisation*, UK, May 1992.

[ISO 92b]   ISO, **Quality of Service Framework - Outline**, ISO/IEC JTC1/SC21/WG1 N1145, *International Standards Organisation*, UK, March 1992.

[LBC 92]   Leopold, H., Blair, G., Campbell, A., Coulson, G., Dark, P., Garcia, F., Hutchison, D., Singer, Williams N., **Distributed Multimedia Communications System Requirements**, *OSI95/Deliverable ELIN-1/C/V3*, Alcatel ELIN Research, A-1210 Vienna, Ruthnergasse 1-7, Austria, April 1992.

[Nic 91]   Nicolaou, C., **A Distributed Architecture for Multimedia Communication Systems**, *Ph.D Thesis*, University of Cambridge, Computer Laboratory, April 1991.

[SHG 91]   Shepherd, W.D., Hutchison, D., Garcia, F., Coulson G., **Protocol Support for Distributed Multimedia Applications**, *Second International Workshop on Network and Operating System Support for Digital Audio and Video*, IBM ENC, Heidelberg, Germany, Nov 18-19 1991.

[Slu 92]   Sluman, C., **Quality of Service in Distributed Systems**, BSI/IST21/-/1/5:33, *British Standards Institute*, UK, October 1991.

[Ste 92]   Stefani, J.B., **Sur les Proprietes Temporelles d'une Machine d'Execution ESTEREL**, *Internal Report*, Centre National d'Etudes Telecommunications, 38-40 rue du General Leclerc, 92131 Issy-les-Moulineaux, Paris, France, April 1992.

[Top 90]   Topolcic, C., **Experimental Internet Stream Protocol, Version 2 (ST-II)**, *Internet Request for Comments No. 1190* RFC-1190, October 1990.

[Wil 91]   Williams, N., Blair, G.S., **Distributed Multimedia Application Survey**, *Internal Report N. MPG-91-11*. Computing Department, Lancaster University, Bailrigg, Lancaster LA1 4YR, UK, April 1991.

[Wol 91]   Wolfinger, B., Moran M., **A Continuous Media Data Transport Service and Protocol for Real-time Communication in High Speed Networks**, *Second International Workshop on Network and Operating System Support for Digital Audio and Video*, IBM ENC, Heidelberg, Germany, 1991.

[Zha 91]   Zhang, H., Keshav S., **Comparison of Rate-Based Service Disciplines**, *Proc. ACM SIGCOMM'91*, October 1991, pages 113-121.

The OSI95 Transport Service Design

The communication requirements of multimedia-based applications will not be satisfied by the best effort QoS. One of the contributions of OSI95 has been to propose QoS Enhancements without requiring guaranteed QoS which are available only if enough resources may be firmly reserved for a given connection.

The new semantics for the QoS allows a service user to express more accurately its requirements. The new negotiation rules ends up with a clear contract between the service users and the service provider which has now a well-defined obligation of behaviour driven by the result of its monitoring.

The goal of the project was not only to specify a new transport service. It was also to apply LOTOS, a formal description technique, from the very beginning of the design process. But to do so the knowledge of LOTOS is not enough and we must apply a methodology of design in a consistent way. The method for applying LOTOS at an early design stage, which has been developed in OSI95, has been first applied to the ISO Transport Protocol Class 4 in order to prove its feasibility. This work has led us to the detection of some ambiguities or errors in the standard. These problems have been collected in a list of 10 defect reports (number 143 to 152) submitted to ISO/SC6/WG4. This result reinforces our will and faith in the method based on LOTOS for the design of OSI95 formal specification.

The proposed OSI connection-mode transport service is offering, besides the enhanced QoS for a limited number of performance parameters, more flexibility in the type of transport connection (uni-directional TC or bi-directional TC), more flexibility of the error control on a TC, a QoS re-negotiation facility and an out-of-band data transfer facility associated with a TC.

Last but not least, the connection release facilities have been widened and the rationale for doing it is discussed in-depth outside any political or radical attitude.

The connectionless-mode, which finds its way very slowly in the ISO world, has been greatly extended to offer facilities of interest from some classes of applications.

One of the contributions of XTP is the introduction of a fast connect facility in a connection-mode transport service. A detailed study of the formal definition of this facility shows that the models used in the past are questionable but that several models may be used if this facility is required.

The OSI95 transport service is obviously more complex than the existing ISO transport service. The OSI95 transport service LOTOS specification was written with the constraint-oriented style well adapted to an incremental design. It consists of 102 (sub)processes, 2200 LOTOS lines (non commented) for the processes and 2200 lines of abstract data types. The interest of a time extension of LOTOS was confirmed and the use of a LOTOS toolset has been essential for doing such a specification in one man.year, even if the simulation has to be limited to small groups of sub-processes due to the time and resource consuming aspects of the tools.

The QoS Enhancements in OSI95

André Danthine, Olivier Bonaventure[1] and Guy Leduc[2]
[1] Research Assistant of the University of Liege
[2] Research Associate of the National Fund for Scientific Research (Belgium)
Université de Liège, Institut d'Electricité Montefiore, B28, B-4000, LIEGE, Belgium
Email: danthine@vm1.ulg.ac.be

The new communication environment has brought new requirements on the qualities of service for which the present "best effort" semantics is inadequate. The response to this evolution goes through the definition of a new model of QoS for the lower layers. In the "best effort" model, when a service provider accepts a transmission with a given QoS, it does not commit itself to any duty about the way it will take account of this QoS. The "guaranteed" QoS requires resources reservation mechanisms which are not always available. We present in this paper a new semantics for the QoS. It allows a service user to express more accurately its requirements, and although it does not include yet a concept of guarantee of the result, the fact is that it assures the users that the provider will monitor the selected parameters and return some defined feedback about the way it succeeds in meeting their requirements. New negotiation rules have also been specified, which are consistent with the new semantics. An example illustrates the practical use of the notions we have introduced which are at the origin of the OSI95 transport service.
Key words: Quality of Service (QoS), New Communication Services, Multimedia

1 Introduction

The term Quality of Service (QoS) is the collective name referring to the service performance which determines the degree of satisfaction of the user of a specific service. The CCITT make a clear difference between *user-oriented* QoS which is a service performance measure from the user's point of view and the *network-oriented* QoS which is the quality of the bearer service that is necessary to provide a certain terminal with the requested user-oriented QoS [CCITT I.350]. In this paper we will concentrate on the user-oriented QoS, defined at the Service Access Point (SAP) of a layer of the architecture, and drop from now on the user-oriented qualifier.

The role of the QoS becomes more and more important with the present evolution of the applications. The client/server-based applications demand low-latency request/response-oriented services. The multimedia applications, with their particular needs on throughput and quality of transmission, tend to extent on local, or even wider, networks.

It is believed in this context that it is necessary to define for the lower layers a new model of QoS involving a new semantics of the QoS parameters and the definition of new parameters.

Before introducing the new semantics, let us review the present situation.

2 The QoS Paradigm

The goal of this paper is to study the quality of the service offered by a transport service provider to transport service users, in a peer-to-peer connection mode service.

The figure 2.1 represents the paradigm that will be used in this study. The only observable events are the occurrences of transport service primitives at the two transport service access points (T-SAP).

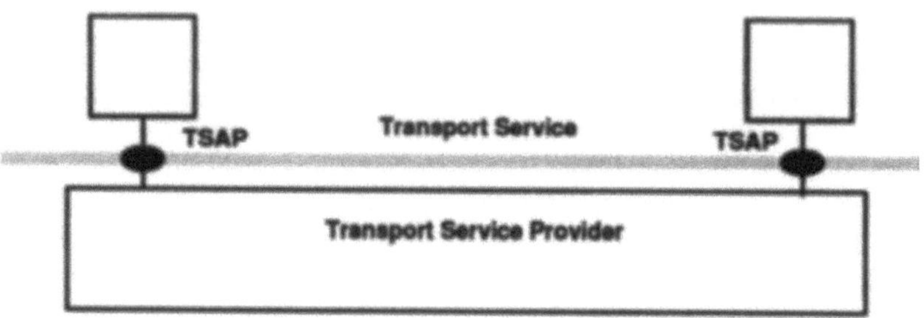

Fig. 2.1 The Transport Service Model

The term Quality of Service (QoS) is the collective name given to certain characteristics associated with the invocations of the service facilities as observed at the SAP. The QoS is specified in terms of a set of parameters. The performance QoS parameters, in particular, are those QoS parameters whose definition is based on a performance criterion to be assessed.

For the service user[1], the quality of service is determined by the values of some end-to-end parameters. The most important performance criteria for a connection are related to the throughput, the transit delay and the reliability.

It is very important to have a clear view about the way to define a performance QoS parameter. Of course the assessment of a performance criterion requires the introduction of timing considerations. Since the only events observable by a service user, when it is using a service facility, are the primitives that occur at its SAP, the only notion of time which can be relied on to introduce timing considerations is the notion of time of occurrence of a service primitive at a SAP. Such a time is an absolute time which is not usable in isolation.

Time has to appear in the definition of the performance QoS parameters in the form of time intervals between occurrences of service primitives. Thus, any performance QoS parameter has to be defined in relation with the occurrences of two or more related service primitives. These primitives may be related in very different ways. They may be related just because they pertain to a same connection, or because they are occurring successively at a same SAP, or because one or several parameters of one of the primitives occurring at a given SAP have been replicated in the other primitive(s) occurring at peer SAP(s), etc.

---

[1] In the following, we will use "service user" and "service provider" instead of "transport service user" and "transport service provider"

The figure 2.2 gives examples of time intervals usable in relation with the performance parameter definition. More complex relationships resulting from combinations of several basic relationships are also possible.

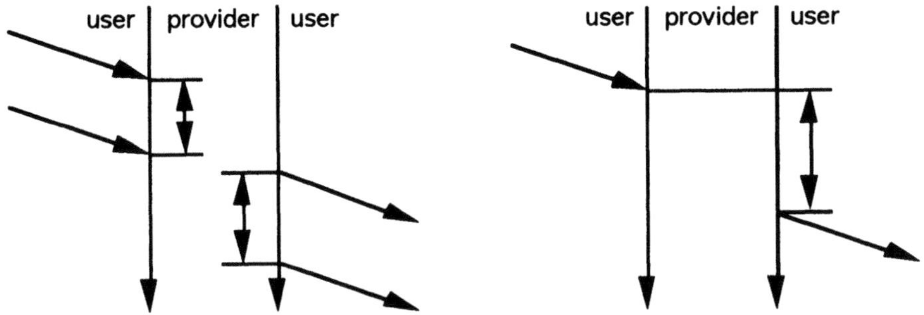

Fig. 2. 2 Time intervals usable in relation with the performance parameter definition

In the connection mode, the values of the QoS parameters are negotiated at the connection establishment time. In the connectionless-mode, the QoS are selected by the calling user and related to a single SDU.

The QoS paradigm of this section is consistent with the view of the user-oriented QoS in [CCITT I.350] which states that "from a user's point of view, quality of service may be expressed by parameters which

- focus on user-perceivable effects, rather than their causes within the network.
- should not depend on the internal design of the network.
- have to take into account all aspects of the service from the user's point of view, which can be objectively measured at the service access point.
- may be granted to the user at the service access point by the service provider.
- are described in network independent terms and create a common language understandable by both the user and the service provider."

This view of the QoS is clearly aligned to our view that QoS parameters must be based on observable events at the SAPs.

2.1 QoS Parameter Definition

A performance QoS parameter, by its definition, must be able to express the requirement of the service users. An instantaneous value of the throughput may be of more interest for a service user than an average value that may be more convenient to use for the service provider.

Furthermore, our interest in a service model must not hide the fact that the implementation of the service may require the monitoring of the QoS parameters and their definitions must be done with this characteristic in mind. Examples of possible definitions are given below. Other definitions or slightly different ones may be found in [FRV 93].

2.1.1 Transit Delay

The **transit delay** associated with an invocation of a peer-to-peer SDU transfer facility may be defined *as the time interval between the occurrence, at the SAP of the sending user, of the DATA request primitive that conveys the SDU and the occurrence, at the peer SAP, of the corresponding DATA indication primitive.* Here, the DATA request and indication primitives are related by the fact that the SDU parameter of the DATA indication has been replicated from the SDU parameter of the DATA request. This definition may be used in the connectionless-mode case as well as in the connection-mode case.

2.1.2 Transit Delay Jitter

In connection mode, it is conceivable that a performance parameter calculated at each invocation of a service facility intervenes in the definition of a more complex performance parameter whose calculation is based on a certain number of invocations of this service facility. A **transit delay jitter** may be defined for one direction of data transfer of a connection as *the difference between the longest and the shortest transit delays observed on this direction since the connection establishment.* Here, each pair formed by a DATA request and the corresponding DATA indication primitives is related to the other pairs of primitives by the fact that they pertain to a same direction of data transfer on a connection.

2.1.3 Throughput

An example of a performance parameter whose definition is based on the time of occurrence of service primitives at the same SAP is the throughput of a direction of data transfer on a connection. Unlike the definitions of the previous two performance parameters, the definition of the throughput does not only use time intervals between occurrences of service primitives, but uses an additional quantity, namely the length of the transferred SDUs.

In the specifications of the ISO services [ISO 8072], the throughput for one direction of transfer of a connection is defined in terms of a sequence of n (with $n \geq 2$) successfully transferred SDUs.

When particularised to the case $n = 2$, this definition becomes:

"the throughput for one direction of transfer is defined to be the smaller of:

   a) the number of SDU octets contained in the last transferred SDU divided by the time interval between the previous and the last T-DATA requests; and

   b) the number of SDU octets contained in the last transferred SDU divided by the time interval between the previous and the last T-DATA indications",

where the throughput measured in a) and b) may be referred to respectively as the sending user's throughput and the receiving user's throughput for the direction of transfer considered. With $n = 2$, this definition corresponds to an instantaneous throughput.

The figure 2.3 represents the ISO sending user's throughput when $n = 2$.

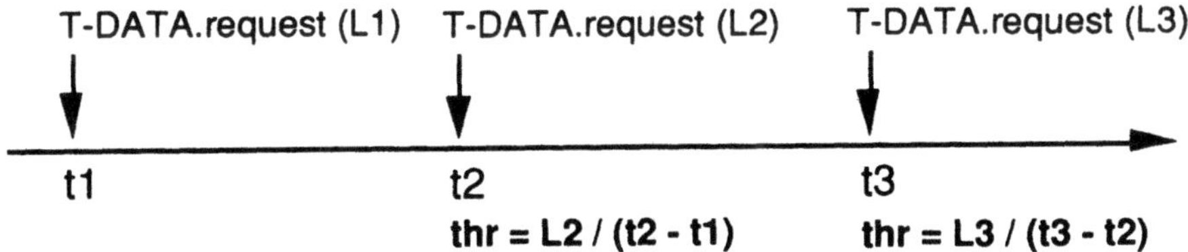

Fig. 2. 3. ISO definition of the sending user's throughput

An alternative way to define the throughput associated with an invocation of the SDU transfer facility on one direction of transfer of a connection is the following: the **global throughput** for one direction of transfer is still defined to be the smaller of the **sending user's throughput** and the **receiving user's throughput** but these two throughputs are now defined differently from the ISO's view:

- the **sending user's throughput** is now defined as the *number of SDU octets contained in the last transferred SDU divided by the time interval between the last and the next T-DATA requests.*

- the **receiving user's throughput** is now defined as the *number of SDU octets contained in the last transferred SDU divided by the time interval between the corresponding last and next T-DATA indications.*

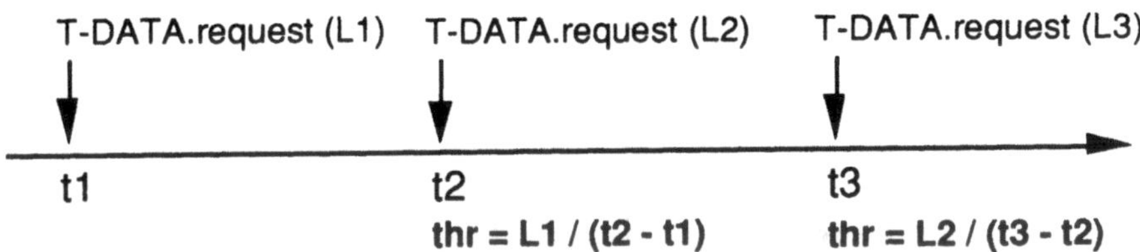

Fig. 2.4. OSI95 definition of the sending user's throughput

This definition corresponds also to an instantaneous throughput. It is this definition of the throughput that has been adopted in the framework of the OSI95 Enhanced Transport Service Definition [DBL 92b], [DBL 92c], [BLL 92b]. The figure 2.4 represent the OSI95 definition of the sending user's throughput [BLL 93]

With the latter definition, the behaviour of the service provider between two occurrences of DATA request primitives on the sending side is influenced only by the time that is elapsing. This is due to the fact that the length of the SDU that is used for the calculation of the current throughput value (i.e. the value associated with the last invocation of the SDU transfer facility) is known. Only the time interval between the last DATA request and the next expected DATA request is unknown until the occurrence of this next primitive. This means that the constraints that the service provider has to obey on the sending side will be expressed only in terms of the time already elapsed since the last DATA request, regardless of the possible lengths of SDU in the next DATA request. Of course a similar conclusion is true on the receiving side.

With the ISO's definition, the behaviour of the service provider between two occurrences of DATA request primitives on the sending side depends not only upon the time that is elapsing but also upon the possible lengths of SDU in the next expected DATA request. This results from the fact that the length of the SDU that is to be used for the calculation of the current throughput value (i.e. the value associated with the next invocation of the SDU transfer facility) is obviously unknown, as well as the time interval between the last DATA request and the next expected DATA request, until the occurrence of this next primitive. This means that the constraints that the service provider has to obey on the sending side will be expressed in terms of the time already elapsed since the last DATA request but also in terms of the possible lengths of SDU in the next DATA request. Of course a similar conclusion is true on the receiving side.

In relation with the introduction of § 2.1, it is clear that the OSI95 definition of the throughput may be easily monitored and may be linked to a protocol mechanism such as the access rate control.

2.2 Types of QoS Negotiations

In the peer-to-peer case, the three actors of the negotiation are the calling (service) user, the called (service) user and the service provider. All negotiations are based on the classical 4-primitive exchange; request, indication, response and confirmation (figure 2.5).

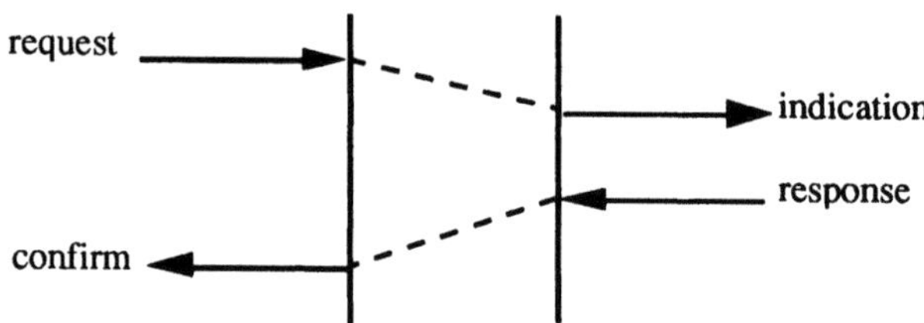

Fig 2.5 The classical 4-primitive exchange

For some performance parameters such as the throughput, the higher the better. For some other performance parameters such as the transit delay jitter, the smaller the better. Throughout this text, we will use the terms "weakening" and "strengthening" a performance parameter to indicate the trend of the modification. Weakening a throughput means reducing its value but weakening a transit delay jitter means increasing its value.

2.2.1 Triangular Negotiation for Information Exchange

In this type of negotiation, the calling service user introduces in the request primitive the value of a QoS parameter. This value may be considered as a suggested value because the service provider is free to weaken it as much as it wants before presenting the new value to the called user through an indication primitive. The called user may also weaken the value of the parameter before introducing it in the response primitive. This final value will be included without change by the service provider in the confirm primitive. At the end of the negotiation, the three actors have the same value of this QoS parameter (figure 2.6).

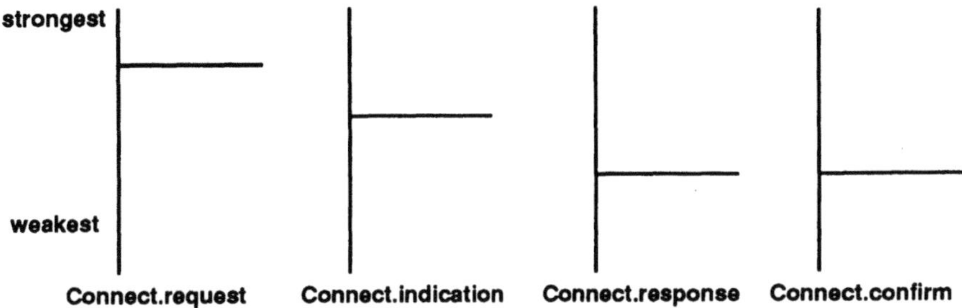

Fig 2.6 Triangular Negotiation for Information Exchange

Taking account of the freedom for the service provider to weaken the value suggested by the calling user, the service provider will reject directly the request only if it is unable to offer the service whatever the value of the QoS.

The calling user has always the possibility to request a disconnection if it is unsatisfied by the value resulting from the negotiation.

The goal of such triangular negotiation is essentially to exchange information among the three actors and to fix the weakest value agreeable to the three actors.

The ISO Transport Service uses this type of negotiation for the performance-oriented QoS [ISO 8072]. The classes 1 and 3 of the ISO Network Service [ISO 8348] are also based on the same scheme.

2.2.2 Triangular Negotiation for a Contractual Value

With the previous negotiation, it is clear that, in such a scheme, it is not possible for the service user to request a well-defined value for a QoS parameter. Coming back to our service oriented view and to the liaison between the QoS value and the service user requirements, it is necessary to introduce a negotiation scheme which allows the service users to express clearly their requirements.

In this type of negotiation, the goal is to obtain a contractual value of a QoS parameter which will bind both the service provider and the service users. Here the calling user introduces, in the request primitive, two values of a QoS parameter, the minimal requested value and the bound for strengthening it. If it accepts the request, the service provider is not allowed to change the value of the minimal requested value. However the service provider is free, as long as it does not weaken it below the minimal requested value, to weaken the bound for strengthening before presenting the new value of the bound for strengthening and the unchanged value of the minimal requested value to the called user, through an indication primitive (fig 2.7). It will be the privilege of the called user to take the final decision concerning the selected value. This selected value of the QoS will be returned by the called user in the response primitive and is acceptable for the service provider. This selected value will be included without change by the service provider in the confirm primitive. At the end of the negotiation, the three actors have agreed on the value of this QoS parameter (fig 2.7).

The service provider may have to reject the request if it does not agree to provide a QoS in the requested range.

The called user may also reject the connection attempt if it is not satisfied with the range of values proposed in the indication primitive.

With respect to the minimal requested value introduced by the calling user, the only possible modification introduced by the negotiation is the strengthening of the minimal requested value but limited by the bound for strengthening value. The service provider may weaken the bound for strengthening and the called user may strengthen the minimal requested value but up to the limit accepted by the service provider.

It is this scheme of negotiation that is used in OSI95 for two types of requested values.

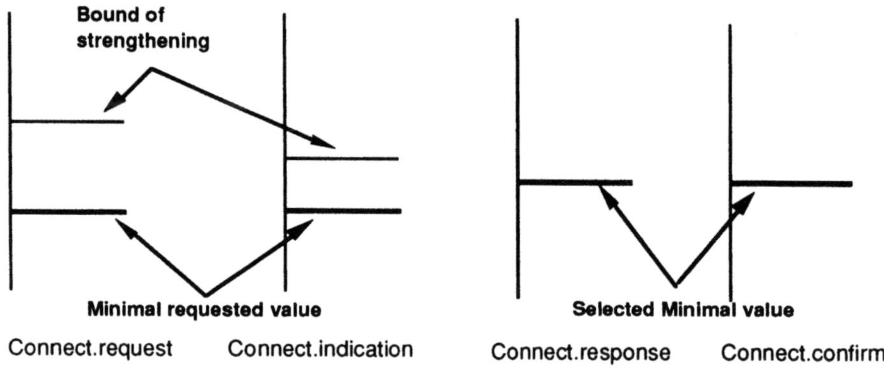

Fig. 2.7 Triangular Negotiation for a Contractual value

3 Best Effort QoS

3.1 The Semantics of the QoS in the ISO Transport Service

Coming back to the result of the negotiation for information exchange of § 2.2.1, it is clear that it is difficult to attach a strong semantics to the resulting value of the QoS parameter. It is only a value on which all three actors agreed.

The semantics of the QoS in the ISO Transport Service is that of the Best Effort, and *"users of the Transport Service should be aware that there is no guarantee that the originally negotiated QoS will be maintained throughout the Transport Connection lifetime, and that changes in QoS are not explicitly signalled by the Transport Service provider."* [ISO 8072]

Even when the calling TS user prohibits the negotiation of a particular QoS parameter and therefore, expresses its QoS as an "absolute" requirement in the CONNECT.request, the service provider will still be allowed to violate the QoS value without notice during the lifetime of the connection.

The QoS parameter does not require a permanent monitoring by the service provider because it is not possible to specify a particular behaviour of the service provider if the real value of the QoS parameter is weaker than the agreed value. The service users do not expect any particular behaviour in such case. They are just expecting the *"best effort"* from the service provider.

In such a loosely defined environment, if a service user introduces, in a request primitive, the value of a QoS parameter, it is not always clear whether this suggested value is related to a boundary or to an average value. In the latter case, the measurement sample or the number of SDUs to be considered is often far from obvious. However, this is not a great problem if no monitoring has to be done.

For a negotiation in which the service provider is not allowed to weaken the value suggested by the calling user, it is possible for the service provider to reject the request due to the requested QoS value but, as already mentioned, in case of acceptance, nothing will be done if the service provider does not reach the QoS value. In this case, the service user is not even informed about the situation by the service provider as no REPORT indication primitive has been defined.

It is therefore not surprising that in many cases, the QoS is expressed in qualitative terms without any specification of a given value. This confirms the lack of relationship between the QoS parameter and a real performance parameter. The only way for the service users to assess the value of a QoS is to monitor it.

The situation we just described is the today situation of transport service and transport protocols in ISO.

If, to operate in a correct way, an application requires a well-defined set of performance parameters, the present approach will not be suitable.

3.2 The Monitoring and the Protocol Functions

With such a poor semantics of the performance QoS parameters of the ISO TS, it may be surprising that the situation has been accepted for more than 10 years.

This results from the fact that some performance QoS values result directly from the protocol functions which are implemented by the service provider. The best example of this situation is the error control scheme in protocols such as TP4 [ISO 8073]. The goal of this protocol is to deliver the data in order and without corruption to the receiving user. To achieve this, TP4 uses two protocol functions : the detection of errors and the retransmission of the lost or corrupted data. The errors are detected with a checksum that covers each DATA TPDU. This checksum can only detect errors with a known probability which depends on the length of the DATA TPDUs covered by the checksum. When a DATA TPDU has been lost or corrupted, it will have to be retransmitted. With these protocol functions with known (i.e. mathematically provable) properties, TP4 achieves its goal with a very low value of the residual error rate. When the protocol detects (usually after a certain amount of retransmissions without acknowledgement) that it is no more able to transmit the data reliably, it releases the connection.

In such a situation, the monitoring of the residual error rate by the service provider is impossible and the associated QoS negotiation is without interest because requiring a residual error rate weaker than the value provided by the error control function is without interest and requiring a residual error rate stronger than the value provided by the error control functions may not even force the service provider to reject the connection request.

For the data communication, the ISO transport service has been considered as adequate because it was offering a reliable service which was the basic requirement. The throughput, the transit delay and the transit delay jitter were not considered as critical factors of performance and the "best effort" situation we just described was acceptable for most of the past applications.

3.3 The Semantics of the QoS in TCP/IP[2]

Due to the lack of service definition for TCP and for IP, it is difficult to speak of QoS parameters, which are parameters of the service primitives of the OSI Reference Model. To understand the QoS provided by TCP we have to analyse the protocol and the protocol data units and to evaluate the service provided by the protocol entities.

During the connection set-up in TCP, there is no selection or negotiation of any QoS parameter and so we cannot speak of the QoS of a TCP connection. Therefore, when the calling TCP user asks for a connection, it does not have to select the values of the QoS parameters which will apply during the connection.

IP does not allow a negotiation of the QoS parameters, but this is not surprising, as IP offers a connectionless service.

Instead of relying on QoS parameters associated with the connection, the calling TCP user is allowed to select dynamically the QoS of each TPDU sent through the IP layer. However this possibility, for the TCP user to select the QoS for each data sent on the connection, is not required by all implementations [Bra 89].

Assuming no influence of the transport layer entities, the QoS, associated with a TPDU and provided by TCP to its user, will be the QoS provided by the IP layer.

3.4 ToS and QoS

Each IP datagram contains a field named *Type of Service* (ToS) which can be considered as carrying the QoS parameters of IP. The ToS is expressing the wishes related to the priority, the delay, the throughput and the reliability of the datagram.

The ToS in IP can only be used as a qualitative indication of the quality of service required for the datagram. Only a few implementations effectively support the ToS field that should be used as an aid in routing. Therefore the semantics of the QoS in both TCP and IP can also be classified as Best Effort.

---

[2]We chose to analyse both IP [Pos 81a] and TCP [Pos 81b] in the same section because TCP is rarely used without IP and because of the link between the QoS offered at TCP and IP level.

Some attempt to define a more global goal, through the ToS values, has been recently introduced [Alm 92] but the ToS field is still only an advice given to the network. This limited support for the ToS parameters is inherent to the connectionless-mode operation of IP.

4 The Need for an Enhancement

In [Dan 93], the need for new standards or at least for an enhancement of existing ones, has been analysed, based on the consequences of the changes we are facing in network performance, in network services and in the application environment.

At the service level, the need for an enhancement means the need for a new semantics for the QoS. If, in the past, the basic requirement, at the transport service level, was a reliable transfer of data associated with a best effort for the other performance QoS parameters, the situation has to be improved.

A new application based on a client-server paradigm may have a specific requirement for the transit delay and the round-trip time.

Some multimedia-based applications will not be able to operate properly if the throughput offered by the transport service falls below some specific value.

The time dependence between successive video frames may not be preserved if the delay jitter goes above a given value.

Those three examples are mentioned to show that the best effort approach for the throughput, the transit delay and the transit delay jitter may not be an acceptable mode of operation for the service provider.

To enhance the present situation, it is possible to associate with the QoS the concept of a guarantee. By so doing we will associate a very strong semantics with the QoS. This has of course some implications which will be analysed in the next section.

5 The Guaranteed QoS

It would be nice to be able to have the concept of a guaranteed QoS, especially when the guaranteed values result from a negotiation for a contractual value presented in section 2.2.2.

The possibility for the service provider to give such a guarantee is related to the existence of enough resources associated with some protocol function for allocating the resources and managing them during the connection.

As already discussed, a residual error rate on a connection may be guaranteed by the service provider if enough storage resources can be allocated to the connection and if it operates an error control function with known properties.

A minimum throughput on a connection may be guaranteed by a transport service provider if each transport protocol entity can be allocated enough processing and storage resources and if the underlying network service provides the guarantee of a throughput compatible with the requested one, taking into account the layer overhead (figure 5.1).

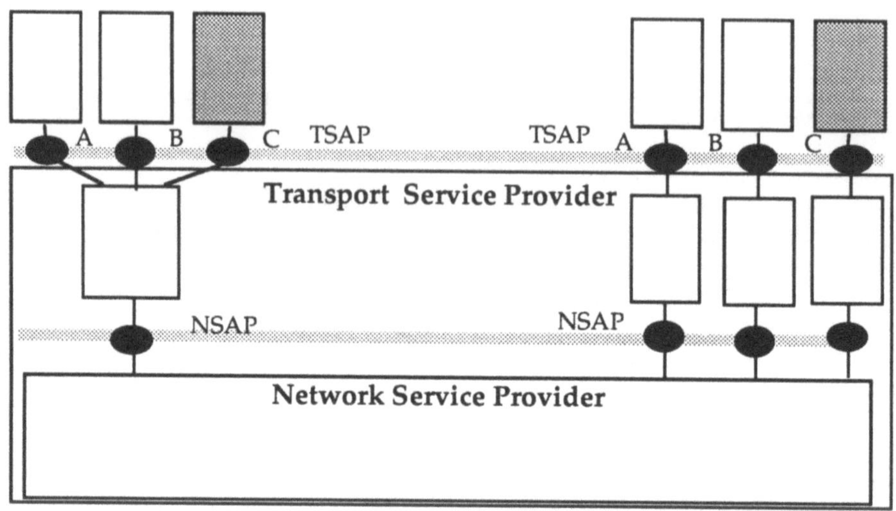

Fig. 5.1 The Transport Service and the Network Service

With the best effort, a request to establish the transport connection CC (figure 5.1), will usually be handled without taking account of the resources already used by the AA and BB connections and of the limited throughput of the network service provider. The result will be that the three connections will share the processing resources of the transport entities and will also share the throughput capacity of the network service provider. Any opening of a new connection will increase the level of multiplexing at the transport level and may affect the performance QoS parameters of the already established connections.

With a guaranteed QoS semantics, the situation will be different. Any attempt to establish a new connection with a guaranteed throughput (such as CC in figure 5.1) will imply the evaluation of the remaining processing and storage capacities of the transport entities and of the possibility for the network service provider to guarantee the needed throughput taking account of the layer overhead.

The basic point here is the permanent availability of resources allocated to the connection. It is only in this case that the service provider will be able to "guarantee" the QoS value requested by the service users. In this case, the monitoring of the QoS values is not necessary as the requested value will be achieved by the service provider, except in exceptional pathological situations.

With a guaranteed QoS semantics, the transport service, if not able to allocate the necessary resources, may be obliged to reject a request to establish the transport connection CC in order to protect the QoS of the already established connections. The same non pre-emptive approach has also been proposed in [FRV 93].

In practice, it is not obvious to obtain a guaranteed throughput from the network service even when the network service is provided by a single subnet. This is however possible with the synchronous service of FDDI and the ATM may also offer an equivalent service.

When the network service is provided by an internet, it is almost impossible to get today any guarantee on a throughput. This is the case with IP and with the ISO internet.

5.1 The Internet Stream Protocol, ST-II

As IP offers a pure connectionless service, it cannot provide any guarantee on the achieved QoS nor allow resources to be reserved. TCP only adds reliability above the IP layer, but it does not support other QoS parameters.

Proposals for an extended IP protocol [Dee 92, 93], [Tsu 93], [Fra 93] recognise the need to include Quality of Service parameters in an extended IP, but they do not completely support it now because there is no common understanding yet of what kind of QoS is needed in such an extended IP. If this extended IP remains entirely connectionless, there is little hope that it will offer any guarantee on the QoS. [Dee 93] states that a new reservation protocol is being developed for SIP. This protocol will be used to set-up a route and reserve resources on each node along the route.

This was already the philosophy of [AHS 90] and of the Internet Stream Protocol, Version 2 (ST-II) which is *"an IP-layer protocol that provides end-to-end guaranteed service across an internet."* [Top 90]. One of the main goals of ST-II was to provide a point-to-point simplex and a point-to-multipoint simplex data transfer for the applications with real-time requirements. In the ST-II specification, these simplex data paths are called "stream".

"An ST stream is :
- the set of paths that data generated by an application entity traverses on its way to its peer application entity(s) that receive it,
- the resources allocated to support that transmission of data, and
- the state information that is maintained describing that transmission of data."

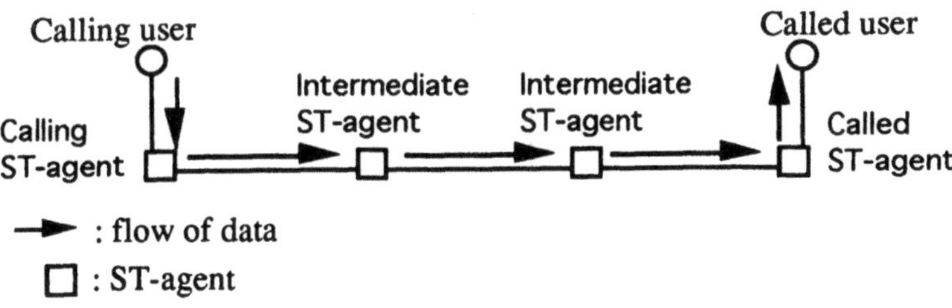

Fig 5.2 :Point-to-Point Simplex ST-II stream

In ST-II, the first step is to set a route along which resources are allocated and to activate ST-agents (figure 5.2).

5.1.1 Triangular Negotiation for a Bounded Target

In this type of negotiation, slightly different from the negotiation for a contractual value of § 2.2.2, the calling user introduces, in the request primitive, two values of a QoS parameter, the target and the lowest quality acceptable. The service provider is not allowed to change the value of the lowest quality acceptable. Here, the service provider is free, as long as it does not weaken it below the lowest quality acceptable, to weaken the target value before presenting the new value of the target and the unchanged value of the lowest quality acceptable to the called user, through an indication primitive. It will be the privilege of the called user to take the final decision concerning the selected value. This selected value of the QoS will be returned by the called user in the response primitive. This selected value will be included without change by the service provider in the confirm primitive. At the end of the negotiation, the three actors have agreed on the value of this QoS parameter (figure 5.3).

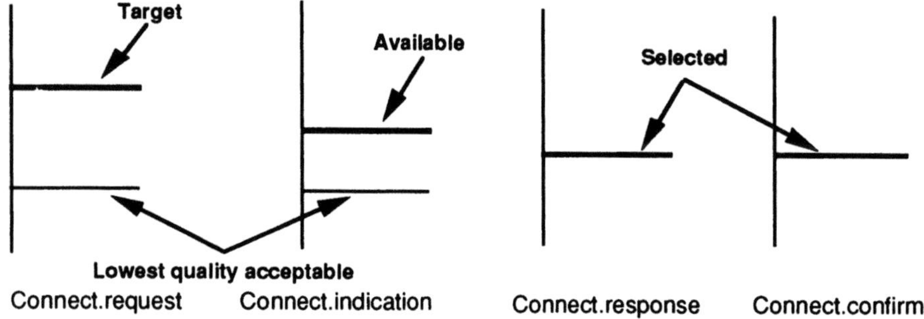

Fig. 5.3 Triangular Negotiation for a Bounded Target

The service provider may have to reject the request if it does not agree to provide a QoS in the requested range. The called user may also reject the connection attempt if it is not satisfied with the range of values proposed in the indication primitive.

With respect to the target value introduced by the calling user, the only possible modification introduced by the negotiation is the weakening of the target but limited by the lowest quality acceptable value.

The class 2 of the ISO Network Service [ISO 8348] is based on this scheme which is also used in ST-II [Top 90] and [FRV 93]. In this last paper, is also introduced the idea of a time-specified negotiation to ease the reservation problem.

5.1.2 The Flow Specification

ST-II uses a data structure named flow specification (abbreviated as flowspec). This flowspec is used to carry the target and the minimal values of the QoS parameters requested by the calling user. It is also used to carry values introduced by the service provider. The flowspec is passed from ST-agent to ST-agent as the path is defined (fig 5.2). Every ST-agent along the path has to reserve enough resources based on the requested values, to adjust the target values of the negotiation scheme of §5.1.1 and to estimate its contribution to cumulative QoS parameters such as minimum, mean and maximum transit delay. The global information about the path contained in the flowspec is presented to the called user in the CONNECT.indication. The called user will then have to decide whether the performance values are sufficient before accepting the connection.

From a service point of view, a ST-II stream is nothing else than a connection (which can be point-to-point or point-to-multipoint). The fact that ST-II uses a fixed route to send its data is hidden from the service user and is only a protocol matter. The state information is maintained by all the ST agents on the path from the calling to the called user. This information is necessary in all these agents to co-ordinate their activities, and allows users to join or leave the ST-II stream dynamically or when the stream must be reconfigured (i.e. routed over other ST-agents). This reconfiguration might cause the selection of a new route and the allocation of new resources. If some users of the stream are no longer reachable after the reconfiguration, the calling user will be indicated.

The meaning of "resource reservation" from a service user point of view is more subtle. In an ideal world, when all necessary resources are firmly reserved for a connection, that means that the connection is assured that the requested QoS will be fulfilled. In the real world, a pathological situation may always arise, in which case the QoS will not be achieved. This means that a guaranteed QoS will always be provided *but the case of pathological behaviour*.

5.1.3 Firm or Pre-emptive Reservation

In the case of firm reservation, all the resources allocated to a path will be used only for the data transfer on that path. From the calling user view point, it is the best solution but such a solution may prevent a later CONNECT.request to be accepted. The usual solution to this problem is to introduce a priority or a cost concept. The calling user has to specify, in its CONNECT.request, the precedence of its connection. The precedence parameter is used as a measure of the relative importance of the connection. During the connection set-up, a stream being established can steal resources to a stream of a lower precedence. If a new stream steals only a fraction of the resources of a lower precedence stream, the service provider will have to indicate the new values of the QoS parameters to the calling user of the "victim" stream.

When pre-emptive reservation is possible, it is difficult to attach a semantics of *guarantee* to the QoS of a path which is not of the highest precedence. In a more recent flow specification [Par 92], the pre-emptive reservation is not allowed.

5.1.4 Statistical Reservation

Up to now, we always discussed the QoS from the service user perspective and attached to one connection or to one stream. There is also a service provider perspective which takes a more global view. In the present case, the service provider may be tempted to try to satisfy the highest possible number of CONNECT.request. To do so, the service provider may allocate the resources on a statistical basis or using an overbooking scheme.

In this case, it is not only in pathological situations, associated with a very low probability of occurrence, that the QoS may not be achieved. In this case, the probability of not achieving the QoS may be high, reflecting the fact that the requirement is very likely not to be met. To distinguish both cases, we may qualify differently the QoS value when it has to be associated with a probability of achievement that is far from 1. It is what has been done in [CSZ 92] and [Par 92] where a distinction is made between a guaranteed service and a predicted service.

5.1.5 Incomplete Reservation

Up to now, we have assumed the possibility for an ST-agent to allocate resources on a firm basis in the best case, or on a statistical basis or following an over allocating scheme otherwise.

An internet service provider is based on the concatenation of relay functions and one of these relays may be completely unable to allocate resources and to guarantee any throughput on a segment between two ST-agents. Therefore, with incomplete reservation, it is impossible to speak of guaranteed QoS and even a predicted service will be difficult to quantify.

Incomplete reservation along a fixed path may however provide a better service than IP but we are back to a best effort semantics.

5.2 What if?

If the semantics of best effort is associated with the QoS value resulting from a negotiation for information exchange, the service provider has no obligation and nothing will be done if the agreed value is not achieved. The service provider may not even be aware of the situation as it is not obliged to monitor the achieved QoS values.

If the semantics of guaranteed QoS is requested by the service user, the service provider has to reject the request if it does not agree to provide a QoS in the requested range.

Once the request has been accepted, as the result of a negotiation for a bounded target or of a negotiation for a contractual value, the service provider will guarantee the negotiated QoS value and, unless (unavoidable and unpredictable) pathological cases, will provide the requested service. Here, **the service provider has an obligation of results.** This should be the only meaning of a guaranteed QoS and with this strong semantics, there is no need for monitoring, the service provider being certain, at the end of the negotiation, of its ability to provide the requested value[3].

6 The QoS enhancements in OSI95

If a guaranteed QoS is not possible to achieve, it does not mean that the only choice left is the best effort. There is a room for an enhanced QoS semantics. If the service

---

[3]Without the obligation of results and without the monitoring, we are back to the best effort.

provider is not able, for any reason, to guarantee the requested QoS, it may try_ to provide the requested QoS value but will have **an obligation of behaviour** if it does not succeed. In this case, the service provider will have to monitor the achieved QoS value.

When the service provider notices that the achieved QoS value is weaker than the negotiated one, it will have to take some action, such as aborting the connection if it has been instructed to do so by the service user.

The service provider may have been instructed not to abort, but to indicate to the service user(s) that it cannot maintain the selected value, and leave to the service user(s) the responsibility of aborting.

It is on this basis that, in OSI95, a new semantics for the performance QoS parameters has been introduced. In this semantics, a parameter is seen as a structure of three values, respectively called "compulsory", "threshold" and "maximal quality", all these values being optional. Each one has its own well-defined meaning, and expresses a contract between the service users and the service provider [DBL 93].

This means, and this is the most fundamental difference with the best effort, that the service provider is now subject to some well-defined duties, known by each side. In other words, the rules of the game are clear.

The existence of a contract between the service users and the service provider implies that, in some cases, the service users are also subject to well-defined duties also derived from the application environment.

Depending upon the type of facilities, these QoS parameters will be subject to a negotiation or not. In case of a negotiation, some rules related to the rights of strengthening or lowering (weakening) the QoS values are defined for each type of value and for each participant in the negotiation. This has to be done to keep the final result consistent with the meaning of the parameter value and is part of the needed admission control of a new connection, admission control which will be based on the QoS on one side and on the source characterisation on the other side. This connection admission control will have to be completed by an admission control on a connection to enforce the contract.

7 The Compulsory QoS Value

The idea behind the introduction of a "compulsory" QoS value is the following one: *when a compulsory value has been selected for a QoS parameter of a service facility, the service provider will monitor this parameter and abort the service facility when it notices that it cannot achieve the requested service.*

It must be clear that no obligation of results is linked to the idea of compulsory value. The service provider **tries** to respond to the requested service facility and, by monitoring its execution, it will

- either execute it completely without violating the selected compulsory value of the performance parameter;
- or abort it if the selected compulsory value of the performance parameter is not fulfilled.

This is, by the way, the semantics associated with the error control in TP4 [ISO 8073]. The connection is released after a number of unsuccessful transmission attempts.

7.1 Compulsory QoS versus Guaranteed QoS

The guaranteed QoS has a stronger semantics. When a guaranteed QoS value has been selected for a parameter of a service facility, the service provider will execute completely the service facility without violating the selected guaranteed value of the performance parameter.

The compulsory concept reflects the fact that, in some environments (e.g. a lightly loaded LAN), the compulsory QoS value may be achieved without resource reservation. Of course, the same LAN, which does not provide any reservation mechanism or any priority mechanism, may, when heavily loaded, prevent the service provider from reaching the compulsory QoS value and oblige it to abort the execution of the requested service facility.

7.2 QoS Parameters and Information Parameters

The introduction of compulsory QoS values implies that the service provider will have a more difficult task to fulfil. It is therefore not surprising that the service user may have to provide the service provider with more information about the characteristics of the elements associated with the request in order to facilitate the decision of rejection or acceptance of the request. Requesting a throughput of 2 Mb/s with SDUs of 10 Kbytes is different from requesting a throughput of 2 Mb/s with SDUs of 40 bytes.

Hence, the introduction of the concept of compulsory QoS requires the introduction, in the primitives associated with a request, of additional parameters. These additional parameters may be designated as information parameters to distinguish them from the QoS parameters proper. Information parameters will be used for instance for source characterisation.

Values of QoS information parameters may also have to be introduced to control the negotiation process to preserve the semantics associated with the negotiated value.

7.3 Negotiation of a Compulsory QoS value

The negotiation for a compulsory QoS value will follow the negotiation scheme for a contractual value introduced in § 2.2.2. When a service user introduces a compulsory QoS value for a performance parameter to be negotiated, the only possible modification is the strengthening of this compulsory value. In particular, it is absolutely excluded for the service provider to modify this value in order to relax the requirement

As the calling service user may not be interested in an unlimited strengthening of the proposed compulsory QoS value. It introduces therefore in the request primitive, a second value which fixes a bound indicating to what extent the proposed compulsory QoS value may be strengthened (fig 2.7).

When the service provider analyses the request of the calling service user, it has to decide whether it rejects it or not (it can already do so as it knows that the request could only be strengthened). In the latter case, it has to examine the bound of strengthening. This bound may be made poorer (brought closer to the compulsory value) by the service provider, before issuing the indication primitive to the called service user, in such a way to give, to the called service user, the range of compulsory values acceptable by both the calling service user **and** the service provider.

The service provider does not have to strengthen the compulsory QoS value which must be seen as the expression of the requirements of the service users.

After receiving the indication primitive, the called service user may accept or reject the request. If it accepts it, it may modify (strengthen) the compulsory QoS value up to the value of the bound and return it in its response. In this case the negotiation is completed and the service provider may confirm the acceptance of the request and provide the final selected compulsory QoS value to the calling service user.

If the negotiation is successful, the bound is of no interest anymore (the bound is an example of information values mentioned earlier) and the selected compulsory QoS value reflects now the final and global request to the service provider from both service users.

8 The Threshold QoS Value

Some service users may find that the solution of aborting the requested service facility, when one of the compulsory QoS values is not reached, is a little too radical. They may prefer to get information about the degradation of the QoS value.

To achieve that we introduced a "threshold" QoS value with the following semantics: *when a threshold value has been selected for a QoS parameter of a service facility, the service provider will monitor this parameter and indicate to the service user(s) when it notices that it cannot achieve the selected value.*

This threshold QoS value may be used without an associated compulsory value. In this case, the behaviour of the service provider is very similar to the one it has to adopt with a compulsory value. The main difference is that, instead of aborting the service facility when it notices it is unable to provide the specified value, it warns either or both users depending of the service definition. If the service provider is able to provide a QoS value better than the threshold value, everything is fine.

8.1 Threshold QoS versus Best Effort QoS

If the threshold QoS is used without any compulsory QoS, the main difference between the threshold and the best effort is that in the former case, the service provider has the obligation to monitor the parameter and to indicate if the threshold value is not reached.

8.2 Threshold and Compulsory QoS values

It is possible to associate, with the same QoS parameter, a threshold and a compulsory QoS values with, of course, the threshold value "stronger" than the compulsory one.

If the performance parameter degrades slowly and continuously, an indication will be delivered to the service users before the abortion of the service facility. Until such a threshold indication occurs, the service user knows that the service facility is not endangered by the current parameter value.

8.3 QoS Parameters and Information Parameters

Here also, as in the case of compulsory QoS values, information QoS parameters may also have to be introduced to facilitate the decision of rejection or acceptance of the request and to preserve the semantics associated with the negotiated value.

8.4 Negotiation of a Threshold QoS value

The negotiation procedure of a threshold value is similar to the negotiation procedure of a compulsory value. Here also the only possible modification is the strengthening of the threshold value. Here also the calling service user introduces, in the request primitive, an information parameter which fixes a bound indicating to what extent the proposed threshold QoS value may be strengthened.

If a compulsory and a threshold value are associated with the same QoS parameter, there exists a set of order relationship between the compulsory, the threshold and their bounds values which must be verified in the request primitive and maintained during the negotiation.

9 The Maximal Quality QoS Value

In most cases, if the service provider is able to offer a "stronger" value of the QoS parameter than the threshold, the service user will not complain about it. But it could happen that the service user wants to put a limit to a "richer" service facility.

A called entity, for instance, may want to put a limit to the data arrival rate or a calling entity may want, for cost reasons, to prevent the use of too many resources by the service provider.

Such a parameter may be useful to smooth the behaviour of the service provider. Introducing a maximal quality QoS value on a transit delay, i.e. fixing a lower bound to the transit delay values will reduce the transit delay jitter and facilitate the resynchronization at the receiving side.

To achieve that we introduced a "maximal quality" QoS value with the following semantics: *when a maximal quality value has been selected for a QoS parameter of a service facility, the service provider will monitor this parameter and avoid occurrence of interactions with the service users that would give rise to a violation of the selected value.*

It is possible to associate, with the same QoS parameter, a maximal quality, a threshold and a compulsory QoS values with, of course, the maximal quality "stronger" than the threshold value, itself "stronger" than the compulsory value.

9.1 Negotiation of a Maximal Quality QoS value

If a service user introduces a maximal quality QoS value for a performance parameter, the only possible modification is the weakening of this maximal quality QoS value. This value can be weakened during the negotiation by the service provider that indicates by this way the limit of the service it may provide and by the called service user.

If the maximal quality is the only value associated with a given QoS parameter, no bound will be introduced in the request primitive and the negotiation will result in the selection of the weakest of the maximal quality values of the service users and the service provider following the negotiation scheme of § 2.2.1.

If the maximal quality value and a compulsory or/and a threshold values are associated with the same QoS parameter, there exists an order relationship between the maximal quality value and the bound value on the threshold (or the bound value on the compulsory value if no threshold value is specified) which must be verified in the request primitive and preserved during the negotiation.

10 About an Example of Negotiated QoS Parameter

As an example of a negotiated QoS parameter, let us consider the throughput on a direction of transfer of a connection with the associated compulsory, threshold and maximal quality values and the throughput definition of the figure 2.4.

The compulsory is the smallest value and represents the throughput that the service provider must be able to offer to the sending and to the receiving users. If at anytime during the whole lifetime of the connection, the service provider happens to be unable to maintain this throughput a DISCONNECT.indication will occur. If the traffic on the connection is dedicated to a video channel, the compulsory value may represent the throughput necessary to maintain the synchronisation between the sending and the receiving users.

However, in the most general case, two particular situations have to be considered. Both are dealing with the responsibilities of the users and of the service provider when the selected values are not matched.

At the sending SAP, the compulsory value may be violated because no DATA request occurred in due time owing to the behaviour of the sending user only (i.e. the service provider was ready to accept a DATA request in due time). This does not imply a failure of the service provider but only results from the behaviour of the sending user. In this case, it may be reasonable to believe that the service provider does not have to issue a DISCONNECT indication.

If at the receiving SAP, no DATA indication occurs due to the receiving user, the compulsory QoS value will also be violated. Here however it seems reasonable to recognise that the compulsory value does not only constrain the service provider but also constrains the receiving user. Its acceptance of the compulsory value implies that it believes to be able to support the compulsory value. Therefore a DISCONNECT indication seems appropriate[4]. Notice that the constraints on the service provider on the input side and on the service user on the receiving side allow the use of a rate control at both ends of the connection.

It is possible, for the threshold value to apply the same rules as for the compulsory value, issuing, of course, a REPORT indication instead of a DISCONNECT indication. However, taking account of the warning characteristic associated with the threshold value, the service provider may issue a REPORT indication to the service users as soon as the threshold value is violated and without a different attitude depending of the responsible entity.

The maximal quality QoS value is the highest value of the three. The service provider, by controlling the occurrence of the interactions with the service users, prevents the sending user's throughput and the receiving user's throughput from crossing the limiting value.

An interesting point to mention is that the combination of the compulsory and maximal quality values helps to make more precise the model of the service provider. The internal queues may build up at a maximum pace linked to the difference between the maximal quality and the compulsory values of the throughput.

11 Related Work

Very few related work exist on user-oriented QoS and have been mentioned already. In this section, we mainly focus on related work on network-oriented QoS.

If a layered architecture may be based on a recursive definition as expressed in figure 5.1 and if the QoS service model of figure 2.1 may be applied to the network layer and even to the subnetwork layer, the model reaches its limits when the service provider is a bearer service. Such a bearer service may not be able to negotiate all the QoS performance parameters but is associated with "network-oriented" QoS allowing the bearer service user to evaluate the user-oriented QoS. An interesting discussion about BER, Cell Loss Ratio, Cell Insertion Ratio, Cell Delay Variation and End-to-end Transfer Delay in an ATM based B-ISDN may be found [ATV 91] where the relation between these parameters and the application requirements are discussed.

The idea of compensating the delay jitter, in packet networks, by an additional storage in the end node in order to restore a synchronous behaviour has been proposed and used for the voice packet in PARC in the early eighties. This idea has been extended to video in [DRB 87] and implemented in the BWN[5] project [DHH 88]. This idea of reducing or eliminating the transit delay jitter by an additional storage at the

---

[4]This asymmetrical behaviour may not be acceptable in every application environment

[5]BWN is an ESPRIT I project on a Broad Site Wideband Communication Network

end of the service provider may be achieved by a maximal quality QoS on the transit delay (section 9).

This idea of trying to provide a constant delay is also integrated in a proposed solution for a particular class of multimedia applications called *play-back* applications [CSZ 92]. By providing additional storage at the end node, it is possible to fix a play-back time for each packet and, if none arrives after the play-back time, to provide a synchronous service. In [CSZ 92], the play-back time is integrated in the scheduling of the node of an internet, packets which are not in danger to miss their play-back point being delayed in favour of packets which are.

The idea of play-back applications may be linked to a guaranteed service for intolerant and rigid clients or to a predictive service for tolerant and adaptive ones. In the former case, we will have an *a priori* bound for the play-back time and this will lead to a limited utilisation of the nodes. In the later case, it is possible to use a *post facto* bound and achieve a better utilisation of the nodes. However, this requires adaptive clients which may endure reduction and even temporary disruption of service as the result of admission of new clients or of modification of traffic charges of existing ones. The approach of [CSZ 92] favoured the predictive service in contrast to the Tenet approach [Fer 93] which favoured the guaranteed service. It is once again the conflict between the service provider view and the service user view.

The workshop on "Quality of Service Issues in High Speed Networks" held at Murray Hill on April 23-24, 1992 [Kes 92] demonstrated the increase of interest in the issue as well as the diversity of the notion of QoS even if the workshop put an emphasis on the QoS related to the network service. The basic question of the workshop was: how to provide different QoS to the diversity of the applications ? It appears however that the ability of providing bounds for the throughput is necessary, bounds on delay are desirable and bounds on delay jitter debatable. The scheduling discipline is perceived as the key policy [Kes 92]. This workshop confirmed that the performance characteristics of the ATM layer are not easily mapped in the QoS perceived by the user of the service. This workshop was also an opportunity to compare the approach of several projects such as Xunet II, TeraNet [LaP 91] and plaNET/Orbit.

In a recent paper [Kur 93], the author, after reviewing the present situation, discusses the challenges and the open issues involved in providing QoS guarantees to sessions in a high-speed wide area network. If the goal of providing end-to-end QoS guarantees on a session is clear, it is less obvious how a service provider can take advantage of the resource gains offered by the statistical multiplexing of sessions. For every new request, the call admission control must be able not only to provide the guaranteed end-to-end QoS for the new session but also to determine the performance impact of admitting this session on the already-accepted sessions of the network. The performance of the service provider, which previously was linked to the design and the dimensioning of the network is now becoming a real-time problem to be treated by the call admission control. The characterisation of the source is an essential part of the decision process and it is not surprising that the policing of the source is an absolute necessity. What is more surprising is the fact that each element of the network may contribute to deeply modify the characteristics of the source not always in the direction of reducing the difficulty of providing the QoS. The paper discusses the problem of bounds on delay comparing the solutions of queue scheduling which achieve absolute bounds and therefore are able to offer guaranteed QoS and approximate methods which are providing what has been defined earlier as predictive service.

12 Conclusion

After having introduced the QoS paradigm, the principle of the QoS parameter definition and a new taxonomy to distinguish different types of negotiation of the QoS, we discussed the limitation of the best effort QoS and the resources reservation constraints associated with the guaranteed QoS.

We presented in this paper an enhancement of the QoS semantics able to match the communication requirements of the new application environment. This enhanced semantics is based on compulsory, threshold and maximum values and involves a new negotiation scheme aiming at the definition of the contract between the service user and the service provider. This work is at the origin of the OSI95 Transport Service [Dan 92a, 92b], [DBL 92a, 92b], [BLL 92a, 92b, 93]. The table below summarises the behaviour of the service provider within the various semantics discussed.

Table 1 : Behaviour of the service provider within the different QoS semantics

	Best Effort	Maximum Quality	Threshold	Compulsory	Guaranteed
Obligation of Results	NO	YES	NO	NO	YES
Monitoring	NO	YES	YES	YES	NO
Disconnect if achieved value is weaker	NO	N/A	NO	YES	N/A
Indicate if achieved value is weaker	NO	N/A	YES	NO	N/A

N/A : Not applicable

This work has been also presented to the standardisation committees [DBL 93], [ISO 8010] and we hope that it will contribute to the specification of new communication services.

References

[Alm 92]   P. Almquist, **Type of Service in the Internet Protocol Suite**, Network Working Group RFC 1349, July 1992

[AHS 90]   D.P. Anderson, R.G. Herrtwich, C. Schaefer, **SRP : A Resource Reservation Protocol for Guaranteed-Performance Communication in the Internet**, TR-90-006 International Computer Science I,nstitute, February 1990, 26 p.

[ATV 91]   M.E. Anagnostou, M.E. Theologou, K.M. Vlakos, **Quality of service requirements in ATM-based B-ISDNs**, Computer Comm., Vol. 14, N° 4, May 1991, pp. 197-204

[BLL 92a]   Y. Baguette, L. Leonard, G. Leduc, A. Danthine, **Belgian National Body Contribution - Four Types of Enhanced Transport Services and their LOTOS Specifications,** *SC6 plenary meeting, San Diego, July 8-22, 1992,* ISO/IEC JTC1/SC6 N7323, May 1992, 58 p. (OSI95/ULg/A/22/TR/P, May 1992, SART 92/12/09).

[BLL 92b]   Y. Baguette, L. Leonard, G. Leduc, A. Danthine, O. Bonaventure, **OSI95 Enhanced Transport Facilities and Functions,** OSI95/Deliverable ULg-A/P, Dec.1992, 277 p. (SART 92/25/05)

[BLL 93]   Y. Baguette, L. Leonard, G. Leduc, A. Danthine, **The OSI95 Connection-Mode Transport Service,** *The OSI95 Transport Service with Multimedia Support on HSLAN's and B-ISDN,* A. Danthine, ed., Springer Verlag, pp 181-198.

[Bra 89]   R. Braden, Ed., **Requirements for Internet Hosts - Communication Layers,** Network Working Group RFC 1122, Oct. 1989

[CCITT I.350] CCITT, **General Aspects of Quality of Service and Network Performance in Digital Networks, including ISDN,** CCITT Recommandation I.350 Blue Book Fasc.III.8 (1988)

[CSZ 92]   D.D. Clark, S. Shenker, L. Zhang, **Supporting Real-Time Applications in an Integrated Services Packet Network : Architecture and Mechanism,** SIGCOMM'92, Communications Architectures and Protocols, Baltimore, August 17-20, 1992, pp. 14-24

[Dan 92a]   A. Danthine, **A New Transport Protocol for the Broadband Environment,** *IFIP Workshop on Broadband Communications,* Estoril, January 20-22, 1992, A. Casaca, ed., Elsevier (North-Holland), 1992, pp. 337-360.

[Dan 92b]   A. Danthine, **Esprit Project OSI95 - New Transport Services for High-Speed Networking,** *3rd Joint European Networking Conference,* Innsbruck, May 11-14 1992, also in: *Computer Networks and ISDN Systems 25 (4-5),* Elsevier Science Publishers (North-Holland), Amsterdam, November 1992, pp. 384-399.

[Dan 93]   A. Danthine, **The Networking Environment of the Nineties and the Need of New Standards,** *The OSI95 Transport Service with Multimedia Support on HSLAN's and B-ISDN,* A. Danthine, ed., Springer Verlag, pp 1-12.

[DBL 92a]   A. Danthine, Y. Baguette, G. Leduc, **Belgian National Body Contribution - Issues Surrounding the Specification of High-Speed Transport Service and Protocol,** *SC6 interim meeting, Paris, February 10-13, 1992,* ISO/IEC JTC1/SC6 N7312, January 1992 58 p.(OSI95/ULg/A/15/TR/P/V2, 46 p., January 1992, SART SART 92/04/05).

[DBL 92b]   A. Danthine, Y. Baguette, G. Leduc, L. Leonard, **Belgian National Body Contribution - The Enhanced Connection-Mode Transport Service of OSI95,** *SC6 plenary meeting, San Diego, July 8-22, 1992,* ISO/IEC JTC1/SC6 N7759, June 1992 16 p. (OSI95, OSI95/ULg/A/24/TR/P, 16 p., June 1992, SART 92/14/05).

[DBL 92c]   A. Danthine, Y. Baguette, G. Leduc, L. Leonard, **The OSI95 Connection-mode Transport Service - The Enhanced QoS,** *IFIP Conf. on High Performance Networking,* Liège, December 16-18, 1992, in: A. Danthine, O. Spaniol, eds., C14 High Performance Networking, Elsevier Science Publ. (North-Holland), Amsterdam, 1993, pp. 235-252.

[DBL 93]   A. Danthine, Y. Baguette, L. Leonard, G. Leduc, **An Enhancement of the QoS Concept,** ISO/IEC JTC1/SC6/N8010, 01-1993, 16p. (OSI95/ULg/A/28/TR/P, Jan. 1993)

[Dee 92]   S. Deering, **Simple Internet Protocol (SIP) Specification,** Internet Working Draft, Nov. 1992

[Dee 93]   S. Deering, **SIP: Simple Internet Protocol,** IEEE Network Magazine, Vol. 7 No. 3 (May 1993)

[DHH 88]   A. Danthine, B. Hauzeur, P. Henquet, C. Constantinidis, D. Fagnoule, V. Cornette, **Corporate Communication System by Lan Interconnection,** in Information Technology for Organisational Systems, H.-J. Bullinger et al (Ed.) Elsevier Science Publishers (North Holland) pp. 315-326

[DRB 87]   M. De Prycker, M. Ryckebusch, P. Barry, **Terminal Synchronization in Asynchronous Networks,** Proc. ICC'87, Seattle, June 1987, pp. 800-807

[Fer 92]   D. Ferrari, **Real-Time Communication in an Internetwork,** Journal of High Speed Networks,Vol. 1, N° 1, 1992 pp. 79-103

[Fra 93]   P. Francis, **A Near-Term Architecture for Deploying Pip,** IEEE Network Magazine, Vol. 7 No. 3 (May 1993)

[FRV 93]   D. Ferrari,   J. Ramaekers,   G. Ventre, **Client-Network Interactions in Quality of Service Communication Environments**, *IFIP Conference on High Performance Networking*, Liège, December 16-18, 1992, in: A. Danthine, O. Spaniol, eds., C14 High Performance Networking, Elsevier Science Publishers (North-Holland), Amsterdam, 1993, pp. 235-252.

[ISO 8010]   **Liaison Statement to ISO/IEC JTC 1/SC 21 on Qualities of Service (QoS) Work,** ISO/IEC JTC1/SC6 N8010, 17 March 1993, 16 p.

[ISO 8072]   ISO/IEC JTC1, **Transport service definition for Open Systems Interconnection,** ISO/IEC JTC1/SC6 N7734 (ISO DIS 8072 - DIS ballot terminates on April 1993).

[ISO 8073]   International Standard ISO/IEC 8073, **Connection oriented transport protocol specification** [Third edition , 1992-12-15]

[ISO 8348]   ISO/IEC JTC1, **Network Service Definition for Open Systems Interconnection,** ISO/IEC JTC1/SC6 N7558, 21 Sep. 1992

[Kes 92]   S. Keshav, **Report on the Workshop on Quality of Service Issues in High Speed Networks,** Computer Communication Review, Vol. 22, N° 5, October 1992, pp. 74-85

[Kur 93]   J. Kurose, **Open Issues and Challenges in Providing Quality of Service Guarantees in High-Speed Networks,** ACM Computer Communication Review, Vol. 23, N° 1, January 1993, pp. 6-15

[LaP 91]   A.A. Lazar, G. Pacifici, **Control of Resources in Broadband Networks with Quality of Service Guarantees,** IEEE Communication Magazine, Vol. 29, N° 10, October 1991, pp. 66-73

[LiG 90a]   T.D.C. Little,   A. Ghafoor, **Synchronization and Storage Models for Multimedia Objects**, IEEE Journal on Selected Areas in Communications, Vol. 8, N° 3, April 1990, pp. 413-427

[LiG 90b]   T.D.C. Little,   A. Ghafoor, **Network Considerations for Distributed Multimedia Object Composition and Communication**, IEEE Network Magazine, Vol. 4, N° 6, November 1990, pp. 32-49

[Par 92]   C. Partridge, **A proposed Flow Specification**, Network Working Group RFC 1363, Sep. 1992

[Pos 81a]   J. Postel, Ed., **Internet Protocol - Darpa Internet Protocol Specification**, Network Working Group RFC 791, Sep. 1981

[Pos 81b]   J. Postel, Ed., **Transmission Control Protocol - Darpa Internet Protocol Specification**, Network Working Group RFC 793, Sep. 1981

[RaD 91]   G. Ramamurthy, R.S. Dighe, **A Multidimensional Framework for Congestion Control in B-ISDN**, IEEE Journal on Selected Areas in Communications, Vol. 9, N° 9, December 1991, pp. 1440-1451

[Top 90]   C. Topolcic, Ed., **Experimental Internet Stream Protocol, Version 2 (ST-II)**, Network Working Group RFC 1190, Oct. 1990

[Tsu 93]   P. Tsuchiya, **PIP Near-term Architecture**, Internet Working Draft, Feb. 1993

[WoK 90]   G.M. Woodruff, R. Kositpaiboon, **Multimedia Traffic Management Principles for Guaranteed ATM Network Performance**, IEEE Journal on Selected Areas in Communication, Vol. 8, N° 3, April 1990, pp. 437-446

A Method for Applying LOTOS at an Early Design Stage and Its Application to the ISO Transport Protocol

Guy Leduc
Research Associate of the National Fund for Scientific Research (Belgium)
Université de Liège, Institut d'Electricité Montefiore, B 28, B-4000 Liège 1, Belgium
Email: leduc@montefiore.ulg.ac.be

We propose a method applicable to the design of large and abstract LOTOS specifications. More precisely, this method explains how to generate a constraint-oriented specification which is an adequate abstract and modular starting point of a complex design process leading to implementation. It is illustrated on a substantial part of the ISO Transport Protocol class 4 which is considered as a complex and stable case study. Having proved the feasibility of the method on this protocol, it was then used to specify (parts of) the new transport protocol TPX of OSI95.

Keywords: LOTOS, Transport Protocol, Formal Specification, Constraint-oriented style.

1 Introduction

The purpose of this method is to give the necessary guidelines for applying LOTOS [ISO 8807] at an early design stage of distributed systems. At this stage, the specification is required to be abstract, structured and easily modified. *Abstract* means that the specification is free from implementation details: it just describes what the behaviour of the system is. *Structured* means that the specification is decomposed into well-defined sub-parts dealing with specific aspects of the system. This is helpful for understanding the global behaviour. *Easily modified* means that a local change will not require major changes in the whole specification. This characteristic is important at an early design stage because many functions are changed in or added to or removed from the system.

The method presented here does not cover the transformation process of such an initial specification into a more implementation-oriented specification. These transformations are often called refinement transformations, and have to preserve the semantics of the specification. The transformations that are discussed in the previous paragraph (i.e. addition, removal and change of a function) are of a different nature, and usually affect the design: the semantics of the specification will generally change completely.

To clarify these ideas, we will first present in section 2 our view of the full design process composed of specification, validation and implementation activities, with a special emphasis on LOTOS.

Section 3 presents several LOTOS specification styles [VSv 91] and their purpose. Among them we find the constraint-oriented style, which is the main style used throughout the rest of the report.

Section 4 presents the method and section 5 explains the application of this method on a large case study, viz. the ISO transport protocol entity (class 4) [ISO 8073]. The LOTOS specification is presented in [Led 92c]. This specification has been based on the LOTOS specification of IS 8073 [ISO 10024]. The syntax and static semantics of this specification have been checked by the ESPRIT II / LOTOSPHERE Toolset [vEi 92] and some parts have been simulated by SMILE [EeW 93]. The specification has not been validated further up to now.

2 Design Process

This section is not a survey of all the known results on design with formal description techniques. Instead it gives a brief overview of the design activities and shows how they are related.

The design of a (concurrent) system is a complex activity that starts from very abstract or general requirements on the structure and/or the behaviour of the system, and ends up with an actual implementation of them. General requirements on the structure are for instance the need to fit in an architectural model such as the OSI Reference Model [ISO 7498] for open systems. Requirements on the behaviour may be purely functional or include some performance aspects.

The design process is composed of several activities that may be divided into two categories, viz. synthesis and analysis. Synthesis is any type of creative activity dealing mainly with the conception, the formalization and the implementation of a system. The purpose of analysis is to validate (to some extent) the design at different stages. Figure 2.1 presents a simplified view of the synthesis activities in the design process. Stages are represented by shaded circles, and synthesis activities by white rectangles.

Conception remains an informal activity that usually ends up with an *informal* or semi-formal *specification* of some desired features of the system. The first interesting synthesis activity where FDTs play a central role is the *formalization* of these ideas, and the achievement of a - preferably abstract - *formal specification*. This constitutes the first main phase of the design process. Here the structuring features of the FDT are important, as well as its abstraction facilities. These two criteria, viz. abstraction and structure, are the bases of the classification of several specification styles in LOTOS [VSv 91] which will be presented in section 3. Ideally, these activities of conception and formalization should be carried out hand in hand to conceive better protocols and services. Appropriate methods for using FDTs are essential at this level. The main objective of this report is to present such a method and apply it on a well-defined case study, viz. ISO 8073 (class 4).

The second main phase of the design process is the generation of a valid implementation of the specification. The greater the degree of abstraction of both the semantic model of the FDT and the specification itself, the greater the gap between specification and implementation is. Very often, the implementation cannot be derived directly from the formal specification. It may be so, for instance, because the FDT has a non-constructive character (e.g. logic) or because the specification is highly non-deterministic, or does not have an appropriate structure or has no structure at all. The solution to this problem is a more progressive approach, going stepwise towards an implementation. Figure 2.1 only shows the last step, referred to as the *formal implementation specification*. The successive transformations leading from an intermediate stage to the next one are synthesis activities. They may be automatic or not. A transformation however is supposed to preserve somehow the properties of the system. Examples of transformations may be a change in the structure of the specification to get closer to an implementation structure, the resolution of some non determinism or the translation into another FDT. After several transformations, the specification may be such that an automatic generation of some programming code is possible, at least for parts of the specification. This stepwise methodology is found in Software Engineering Standards where it is referred to as the Waterfall model. More details can be found in [QAP 92].

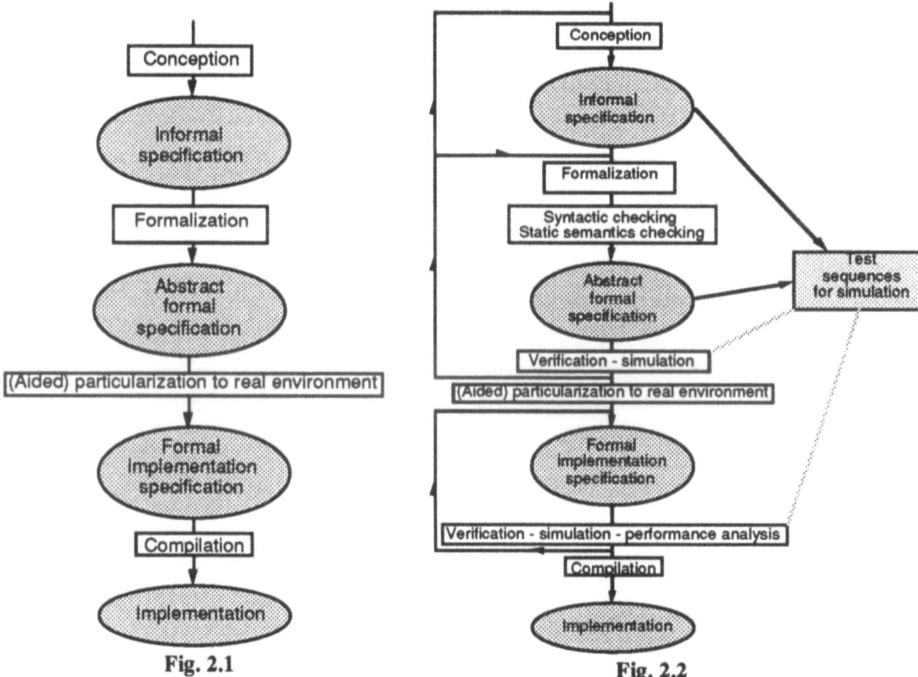

<table>
<tr><td align="center">**Fig. 2.1**</td><td align="center">**Fig. 2.2**</td></tr>
<tr><td align="center">Synthesis activities in the design process</td><td align="center">Analysis and synthesis activities in the design process</td></tr>
</table>

Figure 2.2 introduces analysis activities. They are also represented by white rectangles. Moreover some useful related objects such as the *simulation sequences* (defined later) are represented by shaded rectangles.

Some analysis activities are generally required at the level of the abstract formal specification. Firstly, the specification must have a well-defined meaning, which requires correct syntax and static semantics. It turns out in practice that many conceptual errors are detected at this early stage simply because they often lead to syntactic discrepancies for instance. However, some additional analyses are required to give more confidence and/or insight into the formal specification. Not all these analyses may be formal since the ultimate goal is to be convinced that the specification reflects the initial informal ideas. Some FDTs that have a constructive character may help in this informal validation, since the specification may often be animated or simulated. The designer may thus experiment with the specification and look at how it reacts in specific situations. FDTs with no constructive character, such as logic, usually express the specification with distinct requirements or properties. As far as each informal requirement has been translated into a formal one, this may need less (or no) machinery to be convinced that the specification is an appropriate model of the intended system.

Of course, many FDTs have been designed to support some formal reasoning about the specification, referred to as *verification*. This activity largely depends upon the FDT that is used to describe the protocol. However, the basic idea is the same: in order to verify whether a formal description is "correct", it is mandatory to define formally the meaning of this term. Besides the classical absence of pathological situations such as deadlocks, things are not correct per se, but with respect to something else usually more simple. Therefore, a formal proof of correctness starts with TWO formal descriptions: the first one is the specification that has to be verified (e.g. a protocol), whereas the second one is a simpler reference specification (e.g. a service) that the first one must satisfy. This satisfaction relation between two specifications may have different forms depending on the FDT used. When both specifications are expressed in the same FDT based on a labelled transition system (LTS) model (e.g. CCS or LOTOS), the verification consists mainly of checking some equivalence relation between the two specifications. When the second specification is expressed in an appropriate logic, model-checking techniques are used. For property-oriented specifications, some proofs of consistency may also be carried out at this stage. Other verification techniques are based on the exhaustive generation of all the reachable global states of a system description to check for the absence of pathological situations. This technique has been used with descriptions based on finite-state transition systems.

Many proof techniques cannot be applied directly to the formal specification itself, but require a preliminary translation into a tractable internal form (e.g. a labelled transition system, or a finite state automaton) that allows verification. However, for realistic specifications, this translation faces the well-known state explosion problem. Note, however, that it is now possible to analyse systems of up to 5.10^8 states [Hol 92].

During the second phase of the design process, i.e. the transformations towards the implementation-oriented specification, there may exist rules that guarantee that some properties are preserved if the transformations have a specific form. In this case, no validation[1] of the transformation is necessary. Otherwise, some verification or at least some simulation should be carried out. The simulation may be based on sequences of events (simulation sequences) generated from the initial specification [MAQ 92]. The nature of the proof is closely related to the computation model used. This proof usually consists of verifying that the behaviours described by the two design stages

---

[1] The term validation is used is a broader sense than verification. Validation covers any activity that contributes to give more confidence in the design, whereas verification necessitates the proof or guarantee of a certain property.

are equivalent, or that the behaviours of the less abstract specification constitute a subset of (or satisfy more properties than) the behaviours of the more abstract one. When a specification can be (semantically) mapped onto a Labelled Transition System, one may apply all the proof techniques that exist for these LTSs. These techniques are mainly based on the existence of adequate equivalences and other relations based on a concept of observation and/or implementation. There exist several reasonable ways to observe systems [dNi 87, vGl 90, Led 91]: observation equivalence [Par 85, Mil 89], testing equivalence or some related preorders [dNi 84, BHR 84, BSS 87, Hen 88] have been defined for that purpose. A general framework for dealing with this transformation or implementation process has been defined in [Led 91, Led 92b]. Examples of transformation rules are given in [Lan 90, Mas 92].

Ideally, a performance analysis ought to be carried out at several intermediate stages of the design to quantitatively support some design decisions and their associated transformation steps. However, the difficulty is to bridge the gap between the formal specification and an adequate performance model.

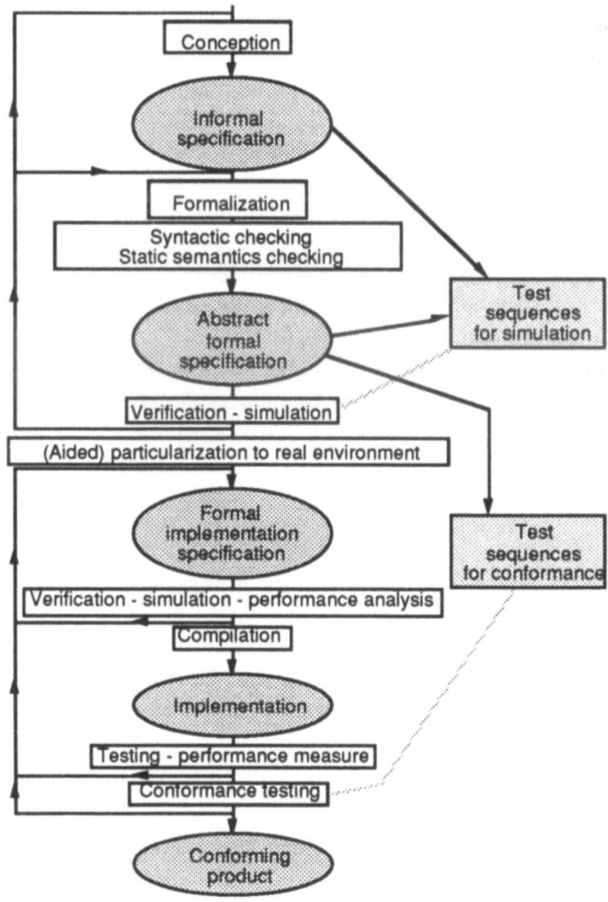

Fig. 2.3 The design process including testing

Figure 2.3 finally illustrates the complete design process with a third phase denoted testing phase. It consists of experimenting with the system implementation integrated in its real environment. This is particularly useful when the initial specification is a standard. In this case some conformance to the standard is usually required to ensure interoperability between implementations. We will only discuss dynamic conformance that is expressed by test suites. These suites describe how the implementation under test will be stimulated and how it should normally react. Ideally, these test suites should be derived from the formal specification of the standard. If the tests succeed, the implementation is claimed to conform to the standard. Note that the simulation sequences introduced in figure 2.2 may already be conformance test sequences. In LOTOS, a testing theory has been proposed in [Bri 88a], extended and brought into play in [WBL 91, Led 92a]. Conformance testing is now widely accepted but the theory is incomplete here too, e.g. the generation of test sequences and the estimation of their coverage are difficult topics.

Most, if not all, FDTs do not totally support this design process. Some are more appropriate to expressing very abstract specifications rather than implementation-oriented ones, or the converse. Some have underlying models that provide a rigorous mathematical basis for the analysis or transformation of a specification. Others are limited to informal validations.

FDTs such as Estelle and LOTOS have proved their abilities to specify very complex protocols and services of the OSI RM [ISO 10023, ISO 10024]. However, they fail to provide concise and readable specifications, especially in the Estelle case. The second main phase of the design process is more critical for all the techniques. The applicability of LOTOS, for instance, has been investigated in the Pangloss/ESPRIT project [Bog 89, Mar 91] and in the Lotosphere/ESPRIT project [vKv 90, QAP 92].

For realistic specifications of respectable size, tools play an essential role in the design process to automate some activities or assist the designer in others. Prototype tools for the standardised languages Estelle (e.g. [CoS 92, Bud 92, ADG 93]), LOTOS (e.g. [GHL 88, MaM 89, QPF 89, Fer 89, MaV 90, Gar 90, GaS 90, vEi 92, EeW 93]) and SDL (e.g. [SSR 89, Atl 90, Hol 92]) are now available.

3 Specification Styles

In practice, it is necessary to produce specifications that are comprehensible and to specify their functionality efficiently. However, depending on the purpose of the specification and its level in the design process described in section 2, different specification styles may be used. In [VSv 91] four specification styles have been identified and described. The two basic criteria used for classifying these styles are structure and abstraction.

Along the *structure* axis, two description types are defined, viz. structured and unstructured:

- The *structured* description type is defined as a composition of basic subsystems influencing each other.
- The *unstructured* description type is defined as a whole without any decomposition in terms of subsystems influencing each other.

Note, in the above definitions, the importance of the expression "influencing each other". This means that the term "structure" has to be considered as a *spatial* structure instead of a *temporal* structure. A temporally structured behaviour description is for instance a *sequential* composition of several behaviours (or subsystems) B1, B2 and B3. In LOTOS, we write B1 >> B2 >> B3. In a temporal structure, the subsystems *cannot* influence each other (except by fixing an initial state). This temporal structure is important but not considered in this report. Instead we focus on the spatial structure. A spatially structured behaviour description is for instance the parallel composition of several behaviours (or subsystems) B1, B2 and B3. In LOTOS, we write for instance[2] B1 ‖ B2 ‖ B3. In a spatial structure, the subsystems can influence each other (except in the special case of interleaving: ⦀).

A (spatial) structure should not be confused with an implementation architecture even if a parallel operator is used. Structure is used as a way to present complex behaviours appropriately to allow a global understanding. In this respect, any complex system will require a structured description to be apprehended by a person. For instance, when in logic we write a specification of an object as the conjunction of several properties (e.g. property1 and property2 and property3), we use a structuring operator that is the logical "and" operator. However this does not introduce any architectural constraints in the design. In LOTOS, the parallel composition operator may be used as a kind of "and" operator without inducing an implementation architecture.

Let us summarise the presentation of the "structure" axis as follows. Structure is understood as spatial structure, and does not mean "implementation architecture". This last aspect is precisely one of the criteria leading to the other classification axis: the abstraction axis.

Along the *abstraction* axis, two description types may also be defined, viz. extensional and intentional types:

- The *extensional* description type is defined solely in terms of its external observable behaviour, viz. in terms of the interactions between the system and its environment.
- The *intensional* description type is defined with respect to internal (unobservable) state variables, or in terms of parts that interact internally (in an unobservable way).

Whereas the structure had a priori nothing to do with implementation details, the abstraction axis explicitly refers to them. A description will be considered extensional (i.e. abstract) if it does not contain internal details. By contrast, a description will be intensional (i.e. concrete) if it goes beyond the description of only the external behaviour as seen by an external user.

Despite the careful definition of the term "structure", the reader might still be convinced that structure and abstraction are not independent criteria because a structure inevitably influences an implementer and therefore goes against abstraction. We hope to further clarify the distinction between structure and abstraction after a detailed classification of the styles. Indeed, it will turn out that a specification style may be extensional and structured, or intensional and unstructured.

In LOTOS, these description types are easily related to the use of specific operators:

- In the extensional type, no "hide" operator is allowed. This means no internal interaction point are allowed.

---

[2] There are of course other parallel composition operators in LOTOS.

- In the unstructured type, no "parallel composition" operator is allowed.

Using these two orthogonal criteria, one naturally derives four specification styles:

- Intensional and unstructured (no parallel composition): the *state-oriented* style in which the system is seen as a single resource whose internal state space is explicitly defined. This style, therefore, presents only observable interactions - not spatially structured - and a representation of the global state space that these interactions manipulate.
- Intensional and structured: the *resource-oriented* style in which both observable and internal interactions are presented. The temporal ordering relationship of the observable interactions is defined by the parallel composition of separate resources that hides the internal interactions. Each resource is defined by a temporal ordering of external and internal interactions or of merely internal interactions.
- Extensional and unstructured (no hide, no parallel composition): the *monolithic* style in which only observable interactions are presented. Their temporal ordering relationship is defined as a collection of alternative sequences of interactions.
- Extensional and structured (no hide): the *constraint-oriented* style in which only observable interactions are presented, but their temporal ordering relationship is composed as the conjunction of separate constraints (like the conjunction operator in logic). Each constraint is defined on the subset of the interaction set that is relevant to it. An in-depth presentation of this style may be found in [Bri 89].

As defined, these styles are mutually exclusive. If the description is unstructured, there is a single process, which is state-oriented or monolithic. Note that in practice these two styles are often combined together, which leads to a model of an extended automaton. If the system description is structured (constraint-oriented or resource-oriented), it is possible to use a different style to specify each of the basic components of the system specification. For instance, each component of a resource-oriented specification may be specified in any of the four styles. This recursive use of the styles stops when all the basic processes are unstructured.

The usefulness of the styles will be further explained in the next section.

4 Method Guidelines

4.1 Use of Specification Styles in the Design Process

In the design process presented in section 2, the initial (abstract) LOTOS specification is successively transformed to reach a final (implementation-oriented) LOTOS specification that can be efficiently compiled. The four specification styles are very useful to support these transformations. Their characteristics make them suitable at specific stages of the design.

The starting point of the design is an abstract and structured specification. Abstraction is needed because this specification must be as tolerant as possible regarding possible implementations. This is particularly crucial for the specification of a standard. Structure is needed because the object to be described is very often so complex that it cannot be globally apprehended by a person. These two elements lead us to an extensional and structured style, i.e. the constraint-oriented specification style, as the basic style for the initial LOTOS specification. The whole method explained in this report deals only with this initial abstract and structured LOTOS specification. There are three reasons for this.

First, this initial specification is the *reference* specification for all possible implementations. It is the only one that can be input to standardization bodies like ISO.

Second, the design of this abstract specification is far from being a trivial task. It needs the support of a method, which is today mostly learned by experience only.

Third, a methodology for covering the whole design trajectory (i.e. from the initial specification down to a implementation specification) needs much more resources, and was outside the scope of OSI95. Such an implementation methodology can be found in the Lotosphere / ESPRIT project [QAP 92].

The objective of our work may now be made more precise. It consists of finding for LOTOS a method for tackling the inherent difficulty of specifying large systems in an abstract way suitable for ISO. This goes beyond the selection of only a specification style, which would be the constraint-oriented style in this case. The method will explain the procedure to be followed by a LOTOS designer who wants to use the constraint-oriented style.

The main problem, which renders the specification process quite complex, originates from the fact that LOTOS does not fully support the constraint-oriented style. A solution to this problem would consist of enhancing LOTOS to allow it to support perfectly the constraint-oriented style. Some specific problems and known means to extend LOTOS are briefly described at the end of section 4.2. Even if this basic research would probably be the right way to go in the long term, it is beyond the scope of OSI95. Therefore, we will keep LOTOS and the constraint-oriented style "as is", and focus our objective on a method that is intended to help LOTOS designers to circumvent the difficulties faced when using the constraint-oriented style on large case studies.

Going down towards implementation, abstraction becomes less important whereas implementation details become highly desirable: adding architectural details is the main purpose of the implementation process, which is the second phase of the implementation process presented in section 2.

In this activity, the first constraint-oriented specification will become progressively more concrete in various ways. For instance, in a software system, choices have to be made about modules, interfaces, data structures, algorithms, ... Since the gap between a "high-level" specification and a "low-level" implementation is usually large, some "intermediate", less and less abstract, specifications are produced. The first constraint-oriented specification is thus likely to be replaced by a resource-oriented specification whose basic components will be in turn specified in the constraint-oriented style or any unstructured style. The process may then continue downward in the structure, and transform the constraint-oriented subsystems into adequate resource-oriented ones. The process stops when all the basic modules are unstructured.

4.2 Guidelines for Using the Constraint-Oriented Style

The constraint-oriented style is intrinsically related to the parallel composition operator that allows the synchronisation of a certain (possibly infinite) number of processes
at interaction points. These interaction points are necessarily external because no hide
operator in allowed. A system is thus described as a collection of sub-processes interconnected together as illustrated in figure 4.1.

The first design decision is thus this decomposition in terms of subsystems. This
problem is far from being simple and we will try to give in this section a general
method to solve this difficult problem. Our method contains several phases denoted:

a) Identification of the main functions
b) Identification of the data structures updated or simply referred to by the functions
c) Grouping of functions according to data structures
d) Mapping of (groupings of) functions onto LOTOS processes (i.e. constraints)
e) Refinement step: For each LOTOS process defined at the end of step d), the whole
    process may be applied again, which leads to a recursive method.

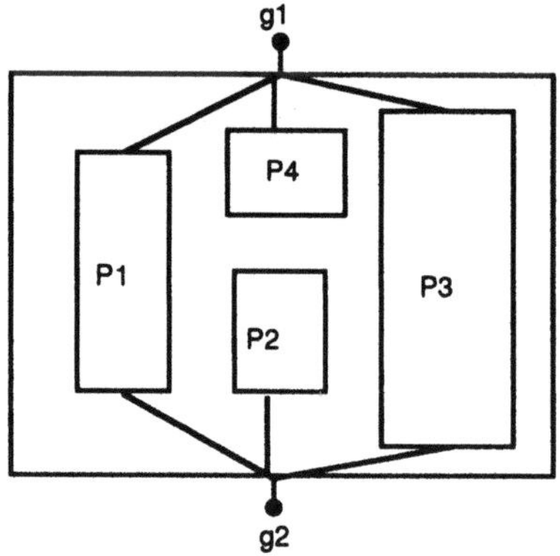

Fig. 4.1 An example of a constraint-oriented specification structure

a) Identification of The Main Functions

The first step consists of isolating the main functions of the specification. There is no
need to define precisely all the elementary functions. Instead, it will be better to stay
first at a high level of design of the specification. The precise substructure of these
functions will be taken into account later, during step e (refinement step).

b) Identification of the Data Structures Updated or Simply Referred to by the Functions

The purpose of this step is to prepare the mapping of the functions onto LOTOS processes. This mapping has to respect some well-defined principles to circumvent some
known deficiencies of the constraint-oriented style, as further explained at the end of
section 4.2.

The main task consists of enumerating *all* the data structures used in the specification, and associating them with the functions. During this association process, we consider only the data structures that the function *updates* (i.e. read-write access). The data structures which the function simply *refers to* (i.e. read-only access) will be considered in the next step.

c) Grouping of Functions According to Data Structures

When a single function updates a data structure, it is easy to model the function as the manager of this data structure. By contrast, when several functions may update the same data structure, we have to be careful. Two possibilities exist:

a) We merge the functions into one process. This has the drawback of losing the separation of concern between processes.
b) We extract from each function the part dealing with the update and we isolate the management of the data structure in a new and distinct process. This process is such that it includes all the basic functionalities extracted from each function. This approach is not always feasible when the update of the data structure is deeply intertwined within the respective functions.

d) Mapping of (Groupings of) Functions onto LOTOS Processes (i.e. Constraints)

Now we consider the data structures that are *referred to* by the function (i.e. read-only access).

From the previous classification, we have groups of functions that will be associated with LOTOS processes. We still need to know whether some of these processes must have read access to the data structures localised in other processes. Again, to answer this question we have to take every process one by one and analyse its behaviour. If such read access is needed, we will have to provide a way to allow this access in the LOTOS specification. There are several possibilities:

a) We merge each process that needs read access with the process handling the data structure. Again, this has the drawback of losing the separation of concern between processes.
b) We add an attribute to all interactions occurring at a suitable interaction point common to both processes. This attribute will be the value of the data structure at the time of the interaction, and will be imposed by the process that knows it.

e) Refinement Step

After the first three steps, we have a constraint-oriented LOTOS specification whose structure reflects the separation of concerns between the main functions. For the functions that, at this stage, remain too complex to be specified in a single unstructured LOTOS process, the whole method may be applied again in a similar way. This leads naturally to a recursive method.

4.2.1 Known Deficiencies of LOTOS to Support the Constraint-Oriented Style

Ideally, every function should be mapped onto a distinct process (or constraint). This allows a clear separation of concerns between functions that can be explained independently from each other. However this ideal separation cannot always be carried out as far as we would like. For instance in LOTOS, even if the functions may be specified independently, they have sometimes to refer to some knowledge handled by another process. In other words, processes sometimes need to exchange knowledge about their context (or state). The problem then originates from the fact that there is no concept of shared variable in LOTOS. Therefore a process has to communicate with the others to get some information. The only basic hypothesis is that processes that refer to such a variable should have at least one common interaction point with the owner of the variable. If it is not the case, this interaction point has to be (artificially) added to the process to allow it to get information.

The following example will illustrate this. Suppose process B is modelling the constraints on the allowed numbering of successive TPDUs. B is responsible for the updates of a context variable such as the number of the last TPDU sent on a connection. The other processes have no way to access this variable. Thus if another process C needs this value to perform its function correctly (e.g. a retransmission of TPDUs up to the last number), there are two solutions: either C has observed all the TPDUs sent at a given interaction point and has registered permanently the highest number, or C has to exchange information with the first process B. The first solution implies that the updating function of the last TPDU number is performed in two processes; this is cumbersome and counter-intuitive. The exchange of information, which constitutes the second solution, is only possible at external interaction points since there are no internal ones in a constraint-oriented specification. Therefore, the environment of the system has to participate in these interactions; this is clearly undesirable.

Of course, since our method is based on the constraint-oriented style, it will suffer from these limitations. We will try however to minimize their impact. The ideal solution would be extend LOTOS according to the section 4.2.2 for instance.

4.2.2 Known LOTOS Extension to Solve this Problem

This problem could be overcome for the larger part if it was possible to hide partially observable actions to the external environment, i.e. to hide some attributes of the observable actions that only concern the communication among the subsystems but not the environment. In this case, by adding suitable attributes to the observable actions (e.g. the value of some variables that are of interest for several functions), we allow an additional exchange of knowledge between the subsystems. If a suitable construct may hide these attributes to the environment, then the goal is achieved.

A formal approach that would solve this problem has been proposed in [Bri 88b]. The underlying semantic model of LOTOS has however to be changed to incorporate the notion of a composite event. A composite event is an event composed of more elementary actions that have to occur all at the same instant (i.e. a composite event remains atomic) to say that the composite event occurs (at that instant). The idea then comes as a hiding operator that may hide some actions of a composite event. The way to map a LOTOS event into a composite event in the semantic model is quite straightforward: it suffices to split an event into as many actions as there are attributes. For instance, the LOTOS event $g?x{:}int?n{:}nat$ would be mapped onto the composite event $<g_1?x{:}int, g_2?n{:}nat>$ where g_1 and g_2 are the elementary interaction points in the semantic model. The internal event i would be simply mapped onto the empty event $<>$. Then, it is possible to hide only one of the g_i's; in this case this consists in removing the g_i component from all the composite actions. Removing all attributes leads of course to classical hiding, and to the empty event, which is the model of i.

Another problem that this approach solves is the possibility to specify the successful termination of a process when a specific event occurs. In LOTOS, the *exit* construct, which is used to model termination, has some known shortcomings. For instance, consider the process that is specified as *(P [> Q) >> R*, if P terminates successfully, then Q is disabled. However, suppose P is defined as *P := a; b; exit*, then we could think that what we have specified is that P has terminated successfully when b has occurred. This is wrong, after b has occurred, the global process is described by *(exit [> Q) >> R* (which is equivalent to *i; R [] Q*), and Q may still occur. This is only the internal event modelling the relay between *exit* and R that disables Q. In other words, it is impossible to model that Q is disabled as soon as event b has occurred; which is unfortunate. The composite event solves this problem easily. It suffices to replace the b action (which is the composite event $<b>$) by a composite event $<b, \delta>$ where δ is the well-known termination action in the semantic model of LOTOS. Any composite event containing δ as a component becomes a termination event.

5 Application of the Method to ISO 8073 (Class 4)

In this section we illustrate the method presented in section 4 on a large case study, viz. a transport entity as specified in ISO 8073 class 4, referred to as TP4 in the sequel. According to ISO 8073, if class 4 is supported by a transport entity, then it is required that class 2 be supported too. In CCITT X224 describing the same transport protocol, this static conformance requirement is different: this is class 0 which is required. Nevertheless, in our specification we consider that our transport entity only supports class 4.

5.1 The Model of a TP4 Entity

According to the Reference Model [ISO 7498], a transport protocol entity interacts with transport service users by means of transport service primitives exchanged at some local TSAPs, and with the network service provider by means of network service primitives at some local NSAPs.

In LOTOS, this architecture will be modelled as a system with two external interaction points, viz. t and n, modelling respectively the (set of all local) TSAPs and the (set of all local) NSAPs. TP4 [ISO 8073] is connection-oriented at both the transport and network service interfaces. Therefore each SAP is in turn composed of CEPs. Primitives exchanged at service interfaces will then have the following structure:

t ?ta:TAddress ?tcei:TCEI ?tsp:TSP

n ?na:NAddress ?ncei:NCEI ?nsp:NSP

where ta and tcei (resp. na and ncei) identify the TSAP and TCEP (resp. NSAP and
NCEP) where the transport (resp. network) service primitive occurs, and tsp (resp.
nsp) is a primitive proper.

5.2 Application of the Method

Step a: Functions Performed by a TP4 Entity

In class 4, all basic functions are supported except the error release - which is used
only in Classes 0 and 2 to release a transport connection on the receipt of an N-
DISCONNECT or N-RESET indication - and all TPDUs are supported except the RJ
(Reject) TPDU.

We list hereafter all the functions supported by a TP4 entity. The function names are
those provided in the standard, and the numbers between parentheses refer to the
clauses of the standard where the functions are described.

- Assignment to network connection (§ 6.1)
  - Initial assignment of a transport connection to a (set of) network connection(s)
  - Creation of a network connection
  - New assignment in relation to splitting
- TPDU transfer (§ 6.2)
  - TPDU -> NSP
  - NSP -> TPDU
- Segmenting and reassembling (§ 6.3)
  - Segmenting (TSDU -> set of TPDUs)
  - Reassembling (set of TPDUs -> TSDU)
- Concatenation and separation (§ 6.4)
  - Concatenation (set of TPDUs -> NSDU)
  - Separation (NSDU -> set of TPDUs)
- Connection establishment (§ 6.5, § 12.2.2.2.b)
  - Initiator function
  - Responder function
  - Negotiation (classes, TPDU size, format, checksum, QoS parameters, expedited
    data)
  (other functions strongly related to the connection establishment are dealt with sep-
  arately, e.g. assignment to network connections, transport connections reference al-
  location, mapping of transport addresses onto network addresses)
- Connection refusal (§ 6.6)
- Normal release (§ 6.7)
  - Explicit variant normal release only (§ 6.7.5)
  - Release of a network connection
- Association of TPDUs with transport connections (§ 6.9)
  - Identification of TPDUs (§ 6.9.4.1)
  - Association of individual TPDUs (§ 6.9.4.2)
- Data TPDU numbering (§ 6.10)

- Expedited data transfer (Network normal data variant only) (§ 6.11, § 12.2.3.4)
- Reassignment after failure (this function is provided using procedures other than those used in § 6.12, refer to § 12.2.2.2.a)
- Retention until acknowledgement of TPDUs (§ 6.13)
- Resynchronization (this function is provided using procedures other than those used in § 6.14, refer to table 23, line NRSTind)
- Multiplexing and demultiplexing (§ 6.15)
  - Multiplexing
  - Demultiplexing
- Explicit flow control (§ 6.16, § 12.2.3.6, § 12.2.3.9)
  - Explicit flow control proper (sending part) (§ 12.2.3.9)
  - Explicit flow control proper (receiving part) (§ 12.2.3.9)
  - Use of flow control confirmation parameter (§ 12.2.3.9)
- Checksum (§ 6.17)
  - Compute checksum
  - Verify checksum
- Frozen references (§ 6.18, § 12.2.1.1.6)
- Retransmission on time-out (§ 6.19, § 12.2.1.2.i, § 12.2.1.1.4, § 12.2.1.1.1, § 12.2.3.1.2)
- Resequencing (§ 6.20, § 12.2.3.5)
- Inactivity control (§ 6.21, § 12.2.3.3, § 12.2.3.1.1)
- Treatment of protocol errors (§ 6.22)
- Splitting and recombining (§ 6.23)
  - Splitting
  - Recombining
- Sequencing of received AK TPDUs (§ 12.2.3.7)
- Procedures for transmission of AK TPDUs (§ 12.2.3.8, § 12.2.1.1.3, § 12.2.3.1.2)
  - Transmission of AK TPDUs (§ 12.2.3.8.1, § 12.2.1.1.3, § 12.2.3.1.2)
  - Sequence control for transmission of AK TPDUs (§ 12.2.3.8.2)
  - (Optional) retransmission of AK TPDUs after CDT set to zero (§ 12.2.3.8.3, § 12.2.3.1.2)
  - (Optional) retransmission procedures following reduction of the upper window edge (§ 12.2.3.8.4, § 12.2.3.1.2)
- Encoding of TPDUs (§ 13)
  - Encoding of CR TPDU (§ 13.3)
  - Encoding of CC TPDU (§ 13.4)
  - Encoding of DR TPDU (§ 13.5)
  - Encoding of DC TPDU (§ 13.6)
  - Encoding of DT TPDU (§ 13.7)
  - Encoding of ED TPDU (§ 13.8)
  - Encoding of AK TPDU (§ 13.9)
  - Encoding of EA TPDU (§ 13.10)
  - Encoding of ER TPDU (§ 13.12)
- Mapping of transport addresses onto network addresses (§ 5.3.1.2 e)
- Transport connection identification (§ 5.3.1.3.e + local mapping on TCEP)

- Network connection identification (local mapping on NCEP: implicit in the standard)
- Options supported by the transport entity (implicit in the standard)
- Provision of transport connections (implicit in the standard)
- Provision of network connections (implicit in the standard)
- Transport connection reference allocation and release (implicit in the standard)
- Backpressure at the transport service interface (missing in the standard)
  Backpressure means that the transport entity may decide not to accept a T-DATA.request for some local reasons. This function is necessary for reasons of conformance with the transport service definition where this phenomenon may appear. In the transport entity, application of this backpressure is of course a local matter (e.g. an implementation may use backpressure when it would overflow its local memory resources, due to flow control or retention at the protocol level).

These functions may be classified as follows:
1. Functions whose scope is a transport connection, i.e. functions which are associated with a transport connection and are completely independent from any other existing transport connection or network connection. There will be as many instances of these functions as there are transport connections handled by the entity.

- Segmenting and reassembling
- Connection establishment
- Connection refusal
- (Explicit variant) normal release
- Data TPDU numbering
- Expedited data transfer
- Retention until acknowledgement of TPDUs
- Explicit flow control
- Checksum
- Retransmission on time-out
- Resequencing
- Treatment of protocol errors
- Sequencing of received AK TPDUs
- Procedures for transmission of AK TPDUs
- Encoding of TPDUs
- Transport connection identification
- Backpressure at the transport service interface

2. Functions whose scope is a network connection, i.e. functions which are associated with a network connection and are completely independent from any other existing network connection or transport connection. There will be as many instances of these functions as there are network connections handled by the entity.

- TPDU transfer
- Concatenation and separation
- Normal release (Part: Release of a network connection)

- Association of TPDUs with transport connections (Part: Identification of TPDUs)
- Resynchronization
- Network connection identification

3. Functions whose scope is the overall transport entity, i.e. functions which cannot be classified in the first two categories either because they are associated with a relation between transport connections and network connections, or because they refer to sets of transport or network connections. There will be one instance of these functions in the transport entity.

- Assignment to network connection
- Association of TPDUs with transport connections (Part: Association of individual TPDUs)
- Reassignment after failure
- Multiplexing and demultiplexing
- Frozen references
- Inactivity control
- Splitting and recombining
- Mapping of transport addresses onto network addresses
- Options supported by the transport entity
- Provision of transport connections
- Provision of network connections
- Transport connection reference allocation and release

In addition, the relation between the transport connections and the network connections is a *many-to-many* relation, due to multiplexing and splitting.

The more general structure we can therefore think of is the one presented in figure 5.1. The process declaration is as follows:

```
process TPEntity [t,n] (tas:Taddresses, nas:Naddresses,
                        tpeo: TPEOptions) :noexit :=
    (TCs [t,n] (...)  |[n]|  NCs [n] (...))
    |[t,n]|
    Entity_functions [t,n] (...)
endproc
```

where tas (resp. nas) is the set of transport (resp. network) addresses supported by this transport entity, and tpeo is the set of options supported by the entity (e.g. supported classes, roles, optional procedures, maximal TPDU size supported for each supported class).

In this structure the data flow between t and n is handled by process TCs, and split between the instances of TC. Process Entity_functions is not used to transfer information between t and n; its purpose is to add constraints which can only be expressed at the entity level.

This structure is however unable to take account of the concatenation function. Indeed, the many-to-many relation between transport and network connections becomes even more complex due to segmentation and concatenation functions, and especially their intertwining with multiplexing and splitting functions. The essence of the problem is the presence of *concatenation and separation* functions in the transport entity specification, which clearly enforces the identification of a third important data structure (i.e. the TPDU) in addition to the unavoidable TSDU and NSDU. In order to understand this point, it is important to know that a TSDU may be segmented into several TPDUs (possibly sent on *different* network connections), and several TPDUs (possibly from *different* transport connections) may be concatenated into a single NSDU. Therefore, a process TC alone cannot directly deal with a NSDU which is potentially associated with several TCs. TC cannot go further than dealing with a TPDU. Instead, NC is the only place where NSDUs may be dealt with.

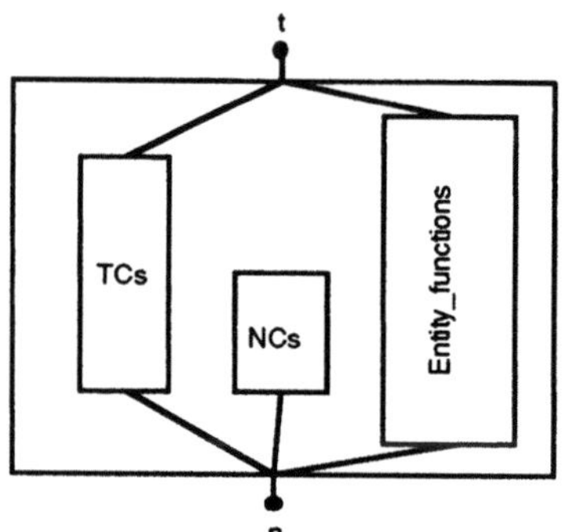

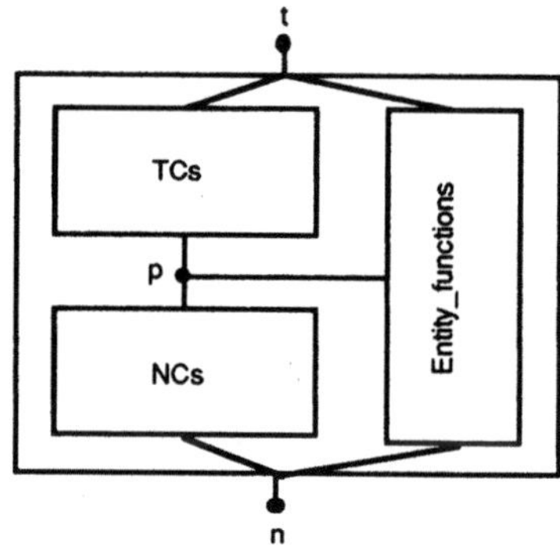

Fig. 5.1 Naïve structure of a TPEntity   **Fig. 5.2** TPEntity

Therefore, if we want to preserve the separation of concerns between functions related to transport and network connections, the only solution is to introduce an internal interaction point between TCs and NCs. This interaction point will be used to transfer the TPDUs between TCs and NCs, so that NCs are able to concatenate TPDUs or separate NSDUs when necessary. This internal interaction point will be denoted p.

We have also to add the interaction point p to process Entity_functions in order to remain as general as possible, and to allow all three processes to synchronize on p.

The more general structure we can therefore think of is the one presented in figure 5.2. The process declaration is as follows:

```
process TPEntity [t,n] (tas:Taddresses, nas:Naddresses,
                        tpeo: TPEOptions) :noexit :=
  hide p in
      (TCs [t,p] (…)    |[p]|   NCs [p,n] (…))
      ||
      Entity_functions [t,p,n] (…)
endproc
```

It is likely that some parameters of TPEntity will not be referred to by some subprocesses, e.g. tas by NCs. To decide on that, it is necessary to check which function refers to which parameter.

Step b: Identification of Data Structures

Instead of doing so, we propose to directly generalize this process as explained in point b of the method (section 4.2). We first enumerate *all* the data structures (not only the TPEntity parameters) used by a transport entity, and associate them with the functions. During this association, a distinction will be made between the data structures simply *referred to* by the function (i.e. read-only access) and those *updated* by the function (i.e. read-write access).

The reason why we propose this generalization is that we already foresee that these data structures will be needed when going downwards in the refinement of the transport entity structure (i.e. when TCs, NCs and Entity_functions will have to be described).

From our understanding of the informal text, we have been able to extract the following data structures of a TP4 entity. For each of them, we have also listed the TP4 functions that are responsible for updating them (Table 5.1).

Many of these data structures are explicit in the standard and are therefore easy to find. We realized however that some other data structures (e.g. Local Transport Connection References) are needed to write a complete specification in LOTOS. We did not follow some specific guidelines to find these data structures. Past experience of protocol specification is probably the central criterion here.

Among these data structures we find some "timers" (e.g. inactivity timer). We use this term in the following sense: the time *variable* whose value keeps track of the passing of time. This is clearly a data structure. By contrast, the function of updating this value is done by a timer *process*(e.g. inactivity control).

Step c: Grouping of Functions or Isolation of Data Structure Handling

When several functions may update the same data structure, some problems arise. We presented the two ways of tackling them in section 4.2, together with their advantages and shortcomings:

i) We merge the functions together into one process.
ii) We extract from each function the part dealing with the update and we isolate the management of the data structure in a new and distinct process. This solution is likely to require that the remaining function cores will still need to know the value of the isolated data structure. This read-access problem will be dealt with in part d of the method where it best fits. It will imply the presence of extra attributes to the interaction points common to the processes concerned. Each new attribute will convey the value of a data structure.

Table 5.1 List of data structures and corresponding functions that update them

Data structure	Updated by (function)
At the entity level	
Local TSAPs (free and in use)	Provision of transport connections
Local NSAPs (free and in use)	Provision of network connections
Local Transport Connection references (Free, in use, frozen)	Transp. connect. ref. alloc. and release, frozen ref.

Supported options	(Static)
Mapping table: Remote TSAPs -> Remote NSAPs	(Static)
Many-to-many relation (TCs <-> NCs)	Assign. netw. connect., reassign. after failure, multiplexing, splitting
Inactivity timer	Inactivity control
Reference timer	Frozen ref.
At the transport connection level	
My Transport Connection Id	Transport connection identification
Negotiated TPDU size	Segmentation [3]
Negotiated format	Data TPDU numbering, Expedited data transfer, Transmission of AK TPDUs [4]
Negotiated use of checksum	Checksum [4]
Queue of Incoming Transport Service Primitives to be (segmented and) sent as TPDUs	Segmentation
Queue of retained TPDUs after sending	Retention until acknowledgement of TPDUs
Queue of Incoming TPDUs to be (reassembled and) delivered as Transp. Service Primitives	Reassembling, resequencing
Role (initiator, responder)	Connection establishment
Credit of last AK received in sequence	Explicit flow control (sending part)
Credit of last AK sent	Explicit flow control (receiving part)
Highest number of DT sent	Data TPDU numbering
Highest number of DT received in sequence	Resequencing
Highest number of ED sent	Expedited data transfer
Highest number of ED received	Expedited data transfer
Highest number of AK received in sequence	Sequencing of received AK TPDUs
Highest number of AK sent	Transmission of AK TPDUs
Highest number of EA received	Expedited data transfer
Highest number of EA sent	Expedited data transfer
Retransmission timer	Retransmission on time-out
Window timer	Transmission of AK TPDUs
Maximum number of transmissions	Retransmission on time-out
At the network connection level	
My Network Connection Id	Network connection identification
Queue of TPDUs to be (concatenated and) transmitted as Network Service Primitives	Concatenation, TPDU transfer (TPDU -> NSP)
Queue of Network Service Primitives to be (separated and) delivered as TPDUs	Separation, TPDU transfer (NSP -> TPDU)

A third solution is a variant of (i). Instead of merging the functions, it would consist of adding a new internal interaction points in each group to allow some synchronisation between the functions that update the same data structure. This is what has been done at the first design stage (Figure 5.2). However, we want to use this solution as a last resort only to stay as close as possible to a pure constraint-oriented style.

Some basis for choosing between solutions (i) and (ii) above may be the nature of the data structure which is common to the respective functions. Let us suppose that the data structure is specified as a data type and not as a LOTOS process. When sev-

---

[3] It may seem counter-intuitive to consider that the data structure is updated by these processes. This is indeed a particular case: being updated only once, at connection establishment, it is more natural to associate the data structure directly with the processes which will refer to it later on.

eral functions refer to a complex data structure (i.e. data type) like a queue, we recommend solution (i) because solution (ii) would lead to the addition of a new attribute for conveying the state of a complex data type (viz. the queue) repeatedly. In our opinion, this way of doing would make the specification less understandable. For example, the functions "reassembling" and "resequencing", which share the same queue of incoming TPDUs, are better to be merged.

Another reason for merging functions according to solution (i) is that, after having looked carefully at these functions, they do not appear any more as separate concerns. For example, the two parts of the function "TPDU transfer" can simply be merged respectively with the functions "concatenation" and "separation".

Let us review the listed data structures and their localization within functions according the general idea explained above.

According to the previous discussion we recommend that the functions grouped together hereafter be merged into a single process:

- Reassembling + resequencing
- Assignment to a NC + Reassignment after failure + Multiplexing + Splitting
- Frozen references + Transport Connection references allocation and release
- Concatenation + TPDU transfer (emission)
- Separation + TPDU transfer (reception)

The "negotiated format" data structure does not necessitate a merge between "Data TPDU numbering", "Expedited data transfer" and "Transmission of AK TPDUs" because these values are only "updated" once (viz. during the connection establishment).

Step d: Mapping of (Groupings of) Functions onto LOTOS Processes

Now that this classification has been achieved, we have groups of functions which will be associated with LOTOS processes. We still need to know whether some of these processes must have a read-access to the data structures localized in other processes. Again, to answer this question, we have to take every process one by one, and analyse its behaviour. If such read-access is needed, we will have to provide a means to allow this access in the LOTOS specification. There are several possibilities:

a) We merge each process which needs read-access with the process handling the data structure. Again, this has the drawback of losing the separation of concern between processes.

b) We add an attribute to all interactions occurring at a suitable interaction point common to both processes. This attribute will be the value of the data structure at the time of the interaction, and will be imposed of course by the process which knows it. There are two sub-cases.

b1) This common interaction point is external, i.e. t or n. In this case, the attribute is visible at the system interface and, since the external environment has to participate in these interactions, the specification of the environment itself has to include these attributes in each of its offers. This is unacceptable because the environment has to be designed independently: changing the structure of interactions at n (resp. at t) would imply a redesign of the network (resp. transport) service specification.

b2) This common interaction point is internal, i.e. p. In this case, this seems the perfect solution since the external environment ignores everything.

Therefore, if p is a common interaction point of both processes, solution b2) will be selected. Otherwise, we have to apply solution a) because b1) is not applicable without a partial hiding operator.

Let us review the listed data structures and the processes which need a read-access to them (Table 5.2).

Table 5.2 List of data structures and functions that refer to them

Data structures	Referred to by (function)
Local TSAPs (free and in use)	-
Local NSAPs (free and in use)	-
Local Transport Connection references (Free, in use, frozen)	-
Supported options	-
Mapping table: Remote TSAPs -> Remote NSAPs	-
Many-to-many relation (TCs <-> NCs)	Association of individual TPDUs
Inactivity timer	-
Reference timer	-
My Transport Connection Id	-
Negotiated TPDU size	-
Negotiated format	-
Negotiated use of checksum	-
Queue of Incoming Transport Service Primitives to be (segmented and) sent as TPDUs	-
Queue of retained TPDUs after sending	Retransmission on time-out
Queue of Incoming TPDUs to be (reassembled and) delivered as Transport Service Primitives	-
Role (initiator, responder)	-
Credit of last AK received in sequence	Retention until acknowledgement
Credit of last AK sent	Resequencing, Transmission of AK TPDUs
Highest number of DT sent	-
Highest number of DT received in sequence	Transmission of AK TPDUs
Highest number of ED sent	-
Highest number of ED received	-
Highest number of AK received in sequence	Explicit flow control (sending part, use of conf. param.)
Highest number of AK sent	Explicit flow control (receiving part), Resequencing
Highest number of EA received	-
Highest number of EA sent	-
Retransmission timer	-
Window timer	-
Maximum number of transmissions	-
My Network Connection Id	-
Queue of TPDUs to be (concatenated and) emitted as Network Service Primitives	-
Queue of Network Service Primitives to be (separated and) delivered as TPDUs	-
Role (owner, non-owner)	-

By applying some criteria explained in phase (c) of the method, we recommend that the functions grouped together hereafter be merged into a single process:

- Association of individual TPDUs merged with
  (Assignment to a NC + Reassignment after failure + Multiplexing + Splitting)
- Retransmission on time-out merged with Retention until acknowledgement

For other functions, we recommend not to merge and therefore to add the following attributes to all interactions at p:

- Highest number of AK sent (i.e. the lower bound of the receiving window)
- Credit of last AK sent
- Highest number of AK received in sequence (i.e. the lower bound of the transmit window)
- Highest number of AK received in sequence
- Highest number of DT received in sequence

The "Highest number of AK sent" together with the "Credit of last AK sent" define the Receiving window. The "Highest number of AK received in sequence" together with the "Highest number of AK received in sequence" define the Transmit window.

Figures 5.3, 5.4, and 5.5 illustrate a TC process, a NC process, and the Entity_functions process according to the above study.

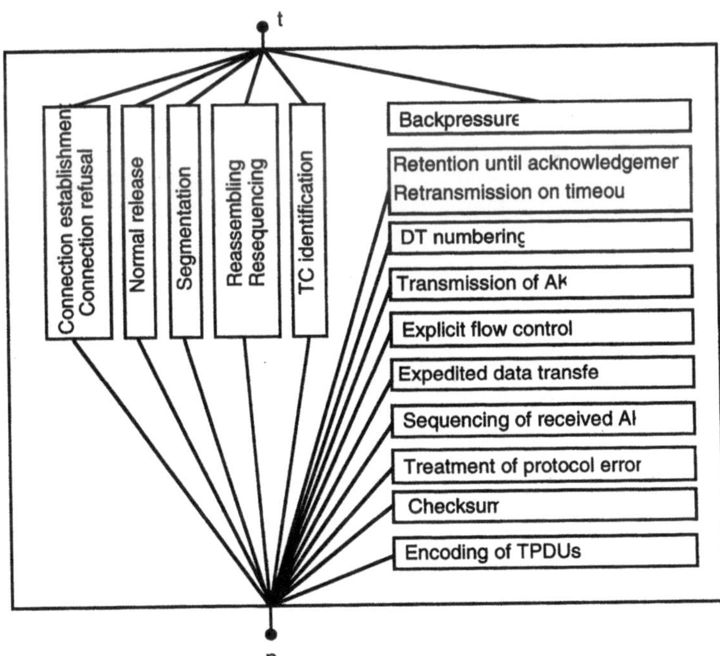

Fig. 5.3 A TC process

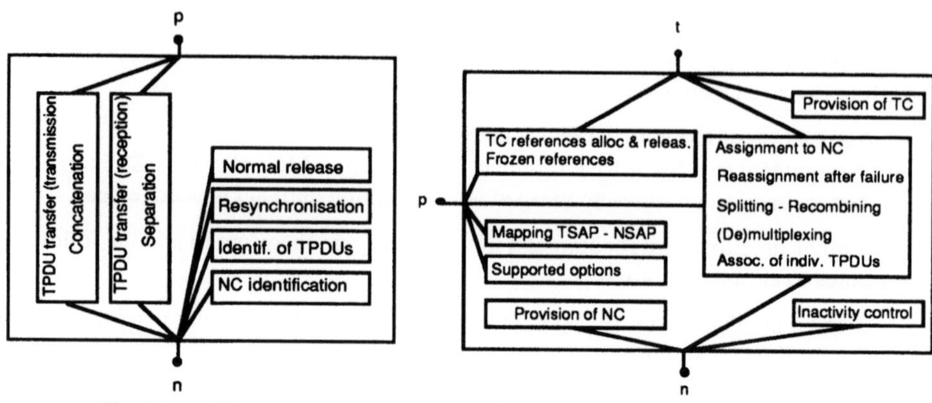

Fig. 5.4 An NC process   **Fig. 5.5** The Entity_functions process

There remains an important step in the method. We have introduced an internal interaction point p where all the TPDUs are passing. The TPDU is the main attribute of actions occurring at p, but we have seen that additional attributes are necessary to allow an exchange of knowledge between processes attached at p.

We will see now that it is useful to go further and add another set of attributes to the p actions in order to avoid the duplication of functions in several processes.

Many functions (processes) refer explicitly to the internal structure of the TPDU (i.e. the values of the various fields of the TPDU) in order to behave correctly. Usually, a process does not need to know the whole TPDU but only a useful abstraction of it, e.g. if it is correct, in sequence, received in the window or a duplicate. However, two problems may occur regarding the computing of such abstractions:

- The processes are unable to compute these abstractions because they have only a partial knowledge of the context of the transport entity. They know only the data structures managed by them, and the values of the attributes coming with the TPDU. This information may well be insufficient to perform the appropriate function. This may be solved if suitable abstractions are computed by the processes which have the information to do so, and then add new attributes to pass their value to all other processes.
- When the same abstractions of a TPDU are needed by several processes which have enough information to compute them, we have to avoid that they be computed in all of them. This may be avoided if only one performs this computation and passes the value in a new attribute.

It is rather easy to list the useful abstractions of a TPDU as done hereafter. Note that some of them are only applicable to a subset of the TPDUs (e.g. DT (Data), AK (Acknowledgement of DT), ED (Expedited Data), EA (Acknowledgement of ED)):

- its direction of flow (either up or down)
- if it is a duplicate or not
- if it is correct or in error (this boolean value may be generalized and replaced by a set containing all the errors associated with a passing TPDU, the set being empty if the TPDU is correct)

- if it is (received) in sequence
- if its number is in the receiving window
- if it is kept or not on reception

These attributes have to be computed by only one process. Let us review (Table 5.3) which process may compute which attribute, and then select the appropriate one. If some attributes cannot be computed by any process, we have to identify which information is missing in which process, and then add it as a new attribute.

Table 5.3 List of attributes at "p" and functions that compute them

Attribute	May be computed by (function)
Direction	Segmentation (for sending non-duplicated CR (Connect Request), DT, ED, DR (Disconnect Request))
	Retransmission on time-out (for sending duplicated CR, CC (Acknowledgement of CR), DT, ED, DR)
	Connection establishment + refusal (for sending CR, CC, DR)
	Normal release (for sending DR, DC (Acknowledgement of DR))
	Transmission of AK (for sending AK)
	Expedited data transfer (for sending EA)
	TPDU transfer (reception) + separation (for receiving any TPDU)
Duplication at emission	Segmentation (for non-duplicated CR, DT, ED, DR)
	Connection establ. + refusal (for non-duplicated CR, CC, DR)
	Normal release (for non duplicated DR, DC)
	Retransm. on time-out (for duplicated CR, CC, DT, ED, DR)
	Transmission of AK (for AK) (but no duplication is possible)
	Expedited data transfer (for EA)
Duplication at reception	Connection establ. + refusal (for CR, CC, DR)
	Connection release (for DR)
	Resequencing (for DT)
	Sequencing of received AK TPDUs (for AK)
	Expedited data transfer (for ED, EA)
Errors at reception	Encoding of TPDUs (for encoding errors)
	Checksum (for bad checksums)
	Treatment of protocol errors
Reception in sequence	Resequencing (for DT)
	Expedited data transfer (for ED)
	Sequencing of AK TPDUs (for AK)
	Expedited data transfer (for EA)
In the receive window	Explicit flow control (for DT)
Kept or not	Resequencing (for DT)

Taking account of these new attributes at gate "p", there may be some redundancy with previously selected attributes. This redundancy has to be carefully analysed and removed. In this example, the attribute "Credit sent in the last AK" may be removed for the following reason: it was referred to by "Resequencing" and "Transmission of AK TPDUs" to check whether a received DT TPDU was in the receive window. Now that the special attribute "In the Receive window" exists, which directly informs any process about this fact, the credit becomes useless.

Step e: Refinement of Processes

In this section the term "refinement" has the following meaning which may slightly differ from its usual sense. Consider the function "Expedited Data Transfer" which has been specified as a single LOTOS process up to now. When looking closer at this function in the informal text, one comes to the conclusion that this function can be split into several sub-functions such as "Numbering of ED", "Transmission of EA", "Reception of EA" and "DT blocking". This decomposition can be done for other functions in a similar manner, e.g. the process "Explicit Flow Control" will be split for instance into "Explicit Flow Control on transmission", "Explicit Flow Control on reception", and "use of flow control confirmation parameter".

Thus decomposition into sub-functions is what we call refinement in this section. During this refinement process, the constraint-oriented style remains the sole style used; that is, a function is specified as a collection of processes (sub-functions) combined with a parallel operator, without introducing additional (internal) interaction points. Therefore, the sub-processes are synchronized via the existing interaction points, and each of them introduces only the constraints associated with its sub-function. The constraints resulting from this composition are simply the union of the constraints of each sub-process.

This refinement may create some difficulties which may lead to the addition of some attributes to the existing interaction points. The reason is that the whole method explained up to now has to be reapplied to each process. That means that if sub-processes have to access some data structures whose scope is restricted to another sub-process, they have to exchange knowledge via suitable attributes in their interactions. Since we have the requirement that no additional interaction point is allowed, the attributes should be added to existing interaction points.

Let us consider the refinement of "Expedited data transfer" and apply our method. In order to do that, reconsider all the data structures whose updates are the responsibility of this process (Table 5.4).

Table 5.4 List of attributes that are computed by the function "expedited data transfer"

Attribute	**May be updated by (function)**
Negotiated format	Expedited data transfer
Highest number of ED sent	Expedited data transfer
Highest number of ED received	Expedited data transfer
Highest number of EA received	Expedited data transfer
Highest number of EA sent	Expedited data transfer

Suppose that this process is refined into the four processes "Numbering of ED", "Transmission of EA", "Reception of EA" and "Suspension of DT flow". We have to assign each data structure to one of these sub-processes: the one which updates it (Table 5.5).

Then we analyse whether the other sub-processes need read access to values of data structures not updated by themselves (Table 5.6).

Table 5.5 Refinement of the function "expedited data transfer" and new updates of attributes

Attribute	May be updated by (function)
Negotiated format	ED numbering, EA transmission
Highest number of ED sent	ED numbering
Highest number of ED received	EA transmission
Highest number of EA received	EA reception
Highest number of EA sent	EA transmission

Table 5.6 Refinement of the function "expedited data transfer" and new needs for references

Attribute	May be referred to by (function)
Highest number of ED sent	Suspension of DT flow
Highest number of ED received	-
Highest number of EA received	Suspension of DT flow
Highest number of EA sent	-

We see that "Suspension of DT flow" needs to know the values of "Highest number of ED sent" and "Highest number of EA received". Therefore, these values have to be added to all interactions at gate p (which is the sole gate of this process) in order to allow the needed exchange of knowledge between sub-processes.

The shortcoming of this approach is that all the event offers of other functions have to be changed to include these new attributes as well. This problem was already encountered during the first decomposition and is due to the lack of a partial hiding operator in LOTOS.

As a characterization of our design we propose the final set of attributes at gate "p" and their meaning.

`p ? tr: TPid`	The transport connection reference given by the protocol
`? ni: NId`	The network connection identification
`? d: Dir`	The direction in which the ETPDU parameter passes (either *Send* or *Receive*)
`? c: Copy`	The notification of duplication (either *New* or *Dupl*)
`? err: TPerr`	The set of protocol errors detected on the passing ETPDU parameter
`? IsInRWindow: Bool`	The notification of presence in the receive window
`? rw: TPDUWindow`	The receive window, i.e. the pair (highest number of AK sent, highest number of AK sent + credit sent in the last AK)
`? sw: TPDUWindow`	The sending window, i.e. the pair (highest number of AK received in sequence, highest number of AK received in sequence + credit received in the last AK in sequence)
`? kon: KeptOrNot`	The notification of storage or deletion of the passing ETPDU parameter
`? IsInSeq: Bool`	The notification of being in sequence with previous ETPDU parameters
`? MaxDTRecInSeq: TPDUNumber`	The highest number of DT TPDU received in sequence
`? etpdu: ETPDU`	The passing TPDU in its encoded form

The outline structure of the resulting specification is depicted on figures 5.2, 5.3, 5.4 and 5.5.

6 Conclusion

In this report, a method for applying the constraint-oriented style to large LOTOS specifications has been presented and illustrated on a case study of relatively high complexity, viz. the ISO transport protocol class 4. Our approach allows us to circumvent some limitations of the constraint-oriented style in LOTOS but not all of them. For instance, the LOTOS specifications of all the main modules of TP4 are constraint-oriented, except the topmost module which is resource-oriented. Nevertheless, the specification has a cleaner structure than the official LOTOS specification of IS 8073 [ISO 10024, KLR 93] which uses much more the resource-oriented style, and thereby introduces several additional internal interaction points and specific associated synchronisations in processes. To go further, LOTOS, despite its powerful parallel composition operator, would need some improvements to support the constraint-oriented style better. The operators which ought to be more flexible are mainly the parallel composition and hiding operators. Nevertheless, we think that when LOTOS is applied according to the proposed method, it offers enough flexibility to specify objects as complex as a transport protocol entity or service. This has been further examined in the OSI95 project on the TPX protocol [BLL 92].

A substantial part of ISO 8073 class 4 has been specified according the method presented in the previous section. The commented LOTOS specification size is 5100 lines (2700 lines of abstract data types and 2400 lines of processes). However, our specification is far from being complete. In particular, process NC has been left unspecified, as well as a large part of the process Entity_functions. The main work has focused on process TC, which includes the main transport functions that are almost independent from the nature of the underlying network service. This choice is motivated by the fact that this method is intended to be applied in a second step to specify TPX which is a transport protocol relying on a connectionless-mode network service provider. In such a case, processes NCs and Entity_functions are much simpler.

References

[ADG 93]   B. Algayres, L. Doldi, H. Garavel, Y. Lejeune, C. Rodriguez, **VESAR: a Pragmatic Approach to Formal Specification and Verification,** in: *Computer Networks & ISDN Systems 25 (7)* (1993).

[Atl 90]   M. Atlevi, **SDT - A Real Time CASE Tool for the CCITT Specification Language SDL,** in: S. T. Vuong, ed., *Formal Description Techniques II* (North-Holland, Amst. 1990) 37-41.

[BHR 84]   S.D. Brookes, C.A.R.Hoare, A.W.Roscoe, **A Theory of Communicating Sequential Processes,** *Journ. ACM*, Vol. 31, No. 3, July 1984, 560-599.

[BLL 92]   Y. Baguette, L. Léonard, G. Leduc, A. Danthine, O. Bonaventure, **OSI95 Enhanced Transport Facilities and Functions,** OSI95/Deliverable ULg-A/P, 11-12-1992, pp. 277.

[Bog 89]   K. Bogaards, **LOTOS Supported System Development,** in: K.J. Turner, ed., *Formal Description Techniques* (North-Holland, Amsterdam, 1989) 279-294.

[Bri 88a]   E. Brinksma, **A Theory for the Derivation of Tests,** in: S. Aggarwal, K. Sabnani, eds., *Protocol Specification, Testing and Verification, VIII* (North-Holland, 1988) 63-74.

[Bri 88b]   E. Brinksma, **On the Design of Extended LOTOS, a Specification Language for Open Distributed Systems,** Doctoral Dissertation, Twente University of Technology, Department of Informatics, Enschede, The Netherlands, 1988.

[Bri 89]   E. Brinksma, **Constraint-oriented Specification in a Constructive Formal Description Technique,** in: J.W. de Bakker, W.-P. de Roever, G. Rozenberg, eds., *Stepwise Refinement of Distributed Systems - Models, Formalisms, Correctness,* LNCS 430 (Springer-Verlag Berlin Heidelberg New York, 1989), 130-152.

[BSS 87]   E. Brinksma, G. Scollo, C. Steenbergen, **Process Specifications, their Implementations and their Tests,** in: G.v. Bochmann, B. Sarikaya, eds., *Protocol Specification, Testing and Verification, VI* (North-Holland, Amsterdam, 1987) 349-360.

[Bud 92]   S. Budkowski, **Estelle Development Toolset (EDT),** in: *Computer Networks & ISDN Systems 25* (1) (1992) 63-82.

[CoS 92]   J.-P. Courtiat, P. de Saqui-Sannes, **ESTIM : an Integrated Environment for the Simulation and Verification of OSI Protocols Specified in Estelle,** in: *Computer Networks & ISDN Systems 25* (1) (1992) 83-98.

[dNi 87]   R. De Nicola, **Extensional Equivalences for Transition Systems,** *Acta Informatica 24,* (Springer - Verlag, Berlin Heidelberg, 1987) 211-237.

[EeW 93]   H. Eertink, D. Woltz, **Symbolic Execution of LOTOS Specifications,** in: M. Diaz, R. Groz, eds., *Formal Description Techniques V* (North-Holland, Amst., 1993) 295-310.

[Fer 89]   J.-C. Fernandez, **Aldébaran : A Tool for Verification of Communicating Processes,** Technical Report SPECTRE, Laboratoire de Génie Informatique - Institut IMAG, 1989.

[Gar 90]   H. Garavel, **Compilation of LOTOS Abstract Data Types,** in: S. T. Vuong, ed., *Formal Description Techniques II* (North-Holland, Amsterdam, 1990) 147-162.

[GaS 90]   H. Garavel, J. Sifakis, **Compilation and Verification of LOTOS Specifications,** in: L. Logrippo, R.L. Probert and H. Ural, eds., *Protocol Specification, Testing and Verification X,* (North-Holland, Amsterdam, 1990) 379-394.

[GHL 88]   R. Guillemot, M. Haj-Hussein, L. Logrippo, **Executing Large LOTOS Specifications,** in: S. Aggarwal, K. Sabnani, eds., *Protocol Specification, Testing and Verification, VIII* (North-Holland, Amsterdam, 1988) 399-410.

[GuL 89]   R. Guillemot, L. Logrippo, **Derivation of Useful Execution Trees from LOTOS Specifications by Using an Interpreter,** in: K.J. Turner, ed., *Formal Description Techniques,* Elsevier Science Publishers B.V. (North-Holland, Amsterd., 1989) 311-324.

[Hol 92]   Gerard J. Holzmann, **Practical Methods for the Formal Validation of SDL Specifications,** *Computer Communications,* vol. 15, n°2, March 92.

[ISO 7498] ISO-TC97/SC16/WG1, **Information Processing Systems - Open Systems Interconnection - Basic Reference Model,** IS 7498, 1984.

[ISO 8073] ISO-TC97/SC6/WG4, **Information Processing Systems - Open Systems Interconnection - Connection-mode Transport Protocol Specification,** IS 8073, 1986.

[ISO 8807]   ISO/IEC-JTC1/SC21/WG1/FDT/C, **Information Processing Systems - Open Systems Interconnection - LOTOS, a Formal Description Technique Based on the Temporal Ordering of Observational Behaviour,** IS 8807, February 1989.

[ISO 10023] ISO/IEC-JTC1/SC6/WG4, **Information Technology - Telecommunications and Information Exchange between Systems - Formal Description of ISO 8072 in LOTOS,** TR 10023, July 1992.

[ISO 10024] ISO/IEC-JTC1/SC6/WG4, **Information Technology - Telecommunications and Information Exchange between Systems - Formal Description of ISO 8073 (Classes 0,1,2,3) in LOTOS,** TR 10024, 1992.

[KLR 93]   H. Kremer, J. v.d. Lagemaat, A. Rennoch, G. Scollo, **A Critical Synthesis of a Standardization Experience,** in: M. Diaz, R. Groz, eds., *Formal Description Techniques V* (North-Holland, Amsterdam, 1993) 231-246.

[Lan 90]   R. Langerak, **Decomposition of Functionality: a Correctness Preserving LOTOS Transformation,** in: L. Logrippo, R. Probert, H. Ural, eds., *Protocol Specification, Testing and Verification X*, (North-Holland, Amsterdam, 1990) 229-242.

[Led 91]   G. Leduc, **On the role of Implementation Relations in the Design of Distributed Systems,** in: *Coll. des Publications de la Faculté des Sciences Appliquées de l'Université de Liège, n° 130* (Liège, 1991), Thèse d'agrégation de l'enseignement supérieur, 283 p.

[Led 92a]   G. Leduc **Conformance Relation, Associated Equivalence, and New Canonical Tester in LOTOS,** in: B. Jonsson, J. Parrow, B. Pehrson, eds, *Protocol Specification, Testing and Verification XI* (North Holland, 1992); updated in: Rept. No. S.A.R.T. 91/05/13, Université de Liège, Dept. Systèmes et Automatique, B28, B-4000 Liège 1, Belgium, August 1991.

[Led 92b]   G. Leduc, **A Framework Based on Implementation Relations for Implementing LOTOS Specifications,** in: Computer Networks & ISDN Systems 25 (1) (1992) 23-41.

[Led 92c]   G. Leduc, **A Methodology for the Design of Large LOTOS Specifications and its Application to ISO 8073,** OSI95/Deliverable ULg-3/P/V3, 09-1992, 89p. (SART 92/19/05)

[MaM 89]   J. Mañas, T. de Miguel, **From LOTOS to C,** in: K.J. Turner, ed., *Formal Description Techniques* (North-Holland, Amsterdam, 1989) 79-84.

[MAQ 92]   T. Miguel, A. Azcorra, J. Quemada, J. Mañas, **A Pragmatic Approach to Verification, Validation and Compilation,** in: *Proceedings of the Third Lotosphere Workshop & Seminar*, Pisa, Sept. 1992,

[Mar 91]   F. Marso, **LOTOS as an Aid in the Design of Computer Communication Systems by Stepwise Refinement,** Doctoral dissertation, University of Liège, Dept. Systèmes et Automatique, B28, Liège, Belgium, Sept. 1991.

[Mas 92]   T. Massart, **A calculus to Define Correct Transformations of LOTOS Specifications,** in: G. Rose, K. Parker, eds, *Formal Description Techniques IV* (North Holland, Amsterdam, 1992) 281-296.

[MaV 90]   E. Madelaine and D. Vergamini, **Auto : A Verification Tool for Distributed System Using Reduction of Finite State Automata Networks,** in: S. T. Vuong, ed., *Formal Description Techniques II* (North-Holland, Amsterdam, 1990) 61-66.

[Mil 89]   R. Milner, **Communication and Concurrency,** (Prentice-Hall Int., London,1989).

[Par 85]   D. Park, **Concurrency and Automata on Infinite Sequences,** *Theorical Computer Science, 5th G1-Conference*, Springer Verlag, 1985.

[QAP 92]   J. Quemada, A. Azcorra, S. Pavón, **The LOTOSPHERE Design Methodology,** in: *Proceedings of the Third Lotosphere Workshop & Seminar*, Pisa, Sept. 1992,

[QPF 89]   J. Quemada, S. Pavón, A. Fernández, **Transforming LOTOS Specification with LOLA - The Parametrised Expansion,** in: K.J. Turner, ed., *Formal Description Techniques* (North-Holland, Amsterdam, 1989) 45-54.

[SSR 89]   R. Saracco, J.R.W. Smith, R.Reed, **Telecommunications Systems Engineering using SDL,** (North-Holland, Amsterdam, 1989).

[vEi 92]   P. van Eijk, **The LOTOSPHERE Integrated Tool Environment Lite,** in: G. Rose, K. Parker, eds, *Formal Description Techniques IV* (North Holland,Amsterd., 1992) 471-474.

[vGl 90]   R.J. van Glabeek, **The Linear Time - Branching Time Spectrum,** in: J.C.M. Baeten, J.W. Klop, eds., CONCUR '90, *Theories of Concurrency: Unification and Extension*, LNCS 458 (Springer - Verlag, Berlin Heidelberg New York, 1990) 278-297.

[vKv 90]   P. van Eijk, H. Kremer, M. van Sinderen, **On the Use of Specification Styles for Automated Protocol Implementation from LOTOS to C,** in: L. Logrippo, R.L. Probert and H. Ural, eds., *Protocol Specification, Testing and Verification X*, (North-Holland, Amsterdam, 1990) 157-168.

[VSv 91]   C.A. Vissers, G. Scollo, M. van Sinderen, E. Brinksma, **Specification Styles in Distributed Systems Design and Verification,** in: *Theoretical Computer Science*, 89:179-206, 1991.

[WBL 91]   C.D. Wezeman, S. Batley, J.A. Lynch, **Formal Methods to Assist Conformance Testing - A case study,** in: J. Quemada, J. Mañas, E. Vazquez, eds., *Formal Description Techniques III* (North-Holland, Amsterdam, 1991) 147-174.

The OSI95 Connection-Mode Transport Service

Yves Baguette[1], Luc Léonard[1], Guy Leduc[2] and André Danthine
[1] Research Assistant of the Belgian National Fund for Scientific Research (F.N.R.S.)
[2] Research Associate of the F.N.R.S.
Université de Liège, Institut d'Electricité Montefiore B28, B-4000 Liège 1, Belgium
Email: baguette@montefiore.ulg.ac.be

We present the main features of the connection-mode service developed as a part of the entire OSI95 Enhanced Transport Service. This connection-mode service results from modifications and enhancements to the standard ISO/IEC Connection-Mode Transport Service. Some of the enhancements are just mentioned in this paper for they are addressed in detail in companion papers.

Keywords: OSI95, Transport Service, Connection-Mode, QoS Semantics, Error Management, Transport Connection Release, QoS Re-negotiation, Out-of-Band Data Transfer.

1 Introduction

This paper deals with the main innovations of the OSI95 Enhanced Connection-Mode Transport Service in comparison with the ISO/IEC Connection-Mode Transport Service [ISO 8072].

These main innovations are:

- the ability to open a uni-directional Transport Connection (TC);
- a modification of the set of the Quality of Service (QoS) parameters used, with the introduction of an additional transit delay jitter performance QoS parameter but also the suppression of other performance QoS parameters (such as the residual error rate, the transfer failure probability, ...) that are deemed unmanageable;
- the definition of a new semantics for the performance QoS parameters, consisting in the introduction of the concepts of compulsory QoS and threshold QoS;
- more flexibility of the error control (i.e. the error detection and recovery) on a TC;
- the widening of the TC release facilities to allow the graceful or abrupt release of either direction of data transfer separately from the reverse one on a TC;
- the introduction of a QoS re-negotiation facility;
- the introduction of an out-of-band data transfer facility associated with a TC.

The new semantics for the performance QoS parameters will not be explained in detail here since a specific companion paper [DBL 93] is devoted to this very important issue. Similarly, another specific companion paper [BaL 93] deals with the substantial widening of the TC release facilities consisting in the introduction of a graceful TC release facility and in the enhancement of the standard abrupt TC release facility. All the other aforementionned innovations are examined thoroughly in the remainder of this paper.

2 TC Establishment Facility and Uni-directional TCs

The OSI95 Enhanced Connection-Mode Transport Service is still offered through the classical sequence of four T-CONNECT primitives (request, indication, response and confirm) to establish a TC, allowing a full negotiation of the characteristics of the TC before the beginning of the data transfer.

A significant innovation in comparison with the ISO/IEC Connection-Mode Transport Service, however, is the ability to open uni-directional TCs. This is achieved by means of a new parameter in the T-CONNECT primitives: the "directions to open" parameter. The parameter indicates the direction(s) of data transfer that will be opened on the TC if the TC set-up request is accepted, by selecting one of three possible options:

  a) "bidirectional TC": both directions of data transfer have to be opened;
  b) "direct unidirectional TC": only the direction of data transfer from the calling OSI95 TS user to the called OSI95 TS user (referred to as the calling-to-called direction) has to be opened;
  c) "reverse unidirectional TC": only the direction of data transfer from the called OSI95 TS user to the calling OSI95 TS user (referred to as the called-to-calling direction) has to be opened.

The "directions to open" parameter is not negotiated at all. The decision on which direction(s) is (are) to be opened on a TC is taken by the calling OSI95 TS user only.

Additionally, a new kind of parameter in the T-CONNECT primitives is the TSDU size range parameter. There is a separate TSDU size range parameter for each direction of data transfer that is to be opened. The TSDU size range parameter associated with one direction of data transfer permits the negotiation, between the OSI95 TS users and the OSI95 TS provider, of a TSDU size range (i.e. a minimum and a maximum TSDU size) for this direction. The negotiated TSDU size range will apply to all normal TSDUs submitted by the corresponding sending OSI95 TS user via T-DATA requests. The TSDU size range for either direction, which is negotiated in parallel with the QoS for this direction, is a very useful information for the OSI95 TS provider and the called OSI95 TS user to determine more precisely their ability to provide the requested QoS.

3 Modification of the Set of QoS Parameters

The QoS parameters used in the ISO/IEC Connection-Mode Transport Service include parameters which express TS performance (cf. table 3.1) and two parameters which express other TS characteristics, namely the TC protection and the TC priority.

Phase	Performance criterion:	
	Speed	Accuracy / Reliability
TC establishment	TC establishment delay	TC establishment failure probability (misconnection or TC refusal on the part of the TS provider)
Data transfer	Throughput Transit delay	Residual error rate (corruption, duplication or loss) Resilience of the TC Transfer failure probability
TC release	TC release delay	TC release failure probability

Table 3.1 The performance QoS parameters of the standard ISO/IEC Transport Service

For the OSI95 TS, we have kept the TC protection and TC priority parameters, without any change in their semantics up to now.

Among the performance QoS parameters, we have kept the throughput and transit delay parameters only, but with the new semantics of compulsory and threshold QoS. All other performance QoS parameters of the current ISO/IEC standard have been removed because they have been deemed unmanageable in practice.

We have introduced a new interesting performance QoS parameter, viz. the transit delay jitter, with here also the new semantics of compulsory and threshold QoS. Additionally, several QoS parameters have been introduced to deal with error control. The error control QoS parameters are discussed separately in the next chapter.

For each open direction of data transfer of a TC, the throughput is defined as *the ratio of the size (in octets) of a TSDU submitted via a T-DATA request primitive to the time elapsed until the occurrence of the next T-DATA request primitive relating to the same direction on the TC*. The definition of the throughput has been modified in comparison with the one in the current ISO/IEC standard : it relates now to a form of "instantaneous" throughput which is measured at each invocation of the TSDU transfer facility.

After a QoS negotiation (see [DBL 93] for more details on the negotiation scheme), the meaningful values regarding the throughput for an open direction of data transfer are the minimum compulsory value, the threshold value and the maximum value. The following relation must always be verified: $0 \leq$ negotiated minimum compulsory throughput value $\leq$ negotiated threshold throughput value $\leq$ negotiated maximum throughput value. According to the semantics of compulsory and threshold QoS, the behaviour of the OSI95 TS provider will be time-dependent, as shown by table 3.2 where T_0 is the time at which the last T-DATA request primitive occurred, $\Delta t_{max} = L\,/$ minimum compulsory throughput, $\Delta t_{thres} = L\,/$ threshold throughput, $\Delta t_{min} = L\,/$ maximum throughput and L is the size of the last submitted TSDU in the non-isochronous case (Figure 3.1):

Time interval considered:	Behaviour of the OSI95 TS provider:
In $] T_0, T_0 + \Delta t_{min} [$,	the OSI95 TS provider is not allowed to accept a T-DATA request interaction.
In $[T_0 + \Delta t_{min}, T_0 + \Delta t_{thres} [$,	the OSI95 TS provider is allowed to accept or not a T-DATA request interaction.
In $[T_0 + \Delta t_{thres}, T_0 + \Delta t_{max} [$,	the OSI95 TS provider should be able to accept a T-DATA request interaction at any time, otherwise it has to report its inability to both OSI95 TS users as soon as possible.
In $[T_0 + \Delta t_{max}, \infty [$,	the OSI95 TS provider should be able to accept a T-DATA request interaction at any time, otherwise it has to shut down the TC as soon as possible.

Table 3.2 Constraints on the OSI95 TS provider as regards the throughput

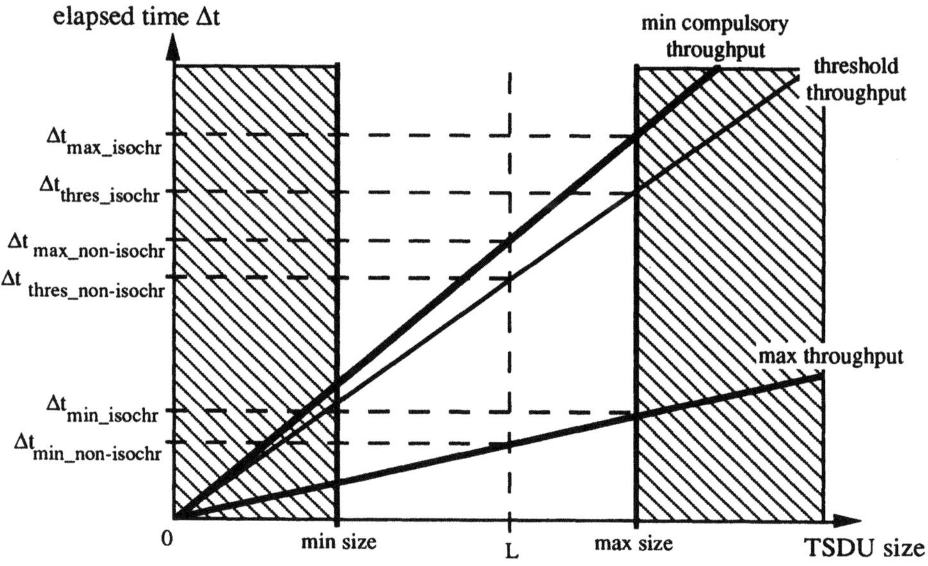

Fig. 3.1 Δt_{min}, Δt_{thres} and Δt_{max} for a non-isochronous or an isochronous traffic

If a sending OSI95 TS user generates an isochronous traffic, that is if it produces TSDUs (possibly of variable length) at a fixed rate, this sending OSI95 TS user may want the OSI95 TS provider to be able to accept the submission of TSDUs at the same rate characterized by the constant time interval Δt_{isochr} between the submission of successive TSDUs whatever their length. For this purpose, the OSI95 TS user may negotiate with the OSI95 TS provider a minimum compulsory (resp. threshold) throughput value given by the ratio of the maximum size authorized for the TSDUs to the constant time interval Δt_{isochr}. With such a minimum compulsory (resp. threshold) throughput value, if T_0 is the time at which the last T-DATA request interaction occurred, the sending OSI95 TS user is assured that the OSI95 TS provider will be ready to accept another T-DATA request interaction at the latest at time $T_0 + \Delta t_{isochr}$ in the case where the TSDU submitted in the interaction at time T_0 is of maximum size and even earlier in the case where the TSDU is of smaller size, or will disconnect (resp. report to the OSI95 TS users) otherwise. However, by forcing the OSI95 TS provider to be able to accept a new T-DATA request primitive before $T_0 + \Delta t_{isochr}$ when the last TSDU is not of maximum size, an unnecessary constraint is imposed on it. In fact, for an actually isochronous traffic, forcing the OSI95 TS provider to be able to offer a T-DATA request interaction every Δt_{isochr} units of time, whatever the size of the TSDU submitted in the previous interaction, is a sufficient constraint. This is why for each open direction of data transfer, the sending OSI95 TS user informs the OSI95 TS provider of the type of traffic it intends to submit, by selecting either of two traffic type options: "non-isochronous" or "isochronous".

The traffic type indicator is not negotiated at all: the traffic type option is imposed by an OSI95 TS user for its output direction of data transfer on a TC. If the traffic type option is fixed to "non-isochronous", then the Δt_{max}, Δt_{thres} and Δt_{min} are calculated as explained previously, i.e. by taking the size L of the previous submitted TSDU into account (Figure 3.1). By contrast, if the traffic type option is fixed to "isochronous", then the Δt_{max}, Δt_{thres} and Δt_{min} are simply taken equal to Δt_{max_isochr}, Δt_{thres_isochr} and Δt_{min_isochr} (Figure 3.1) as if all the TSDUs were of maximum size.

For each open direction of data transfer of a TC, the transit delay is defined as *the time elapsed between the occurrence of a T-DATA request primitive at a TSAP and the occurrence of the corresponding T-DATA indication primitive at the peer TSAP*, just like in the current ISO/IEC standard. This definition means that a measure of the transit delay is associated with each invocation of the TSDU transfer facility.

After a QoS negotiation, the meaningful values regarding the transit delay for an open direction of data transfer are the maximum compulsory value and the threshold value. The following relation must always be verified: $0 \leq$ negotiated threshold transit delay value $\leq$ negotiated maximum compulsory transit delay value. The maximum compulsory value for the transit delay is such that it will never be exceeded at any invocation of the TSDU transfer facility while it applies: if the OSI95 TS provider cannot deliver a T-DATA indication anymore without violating this value, it has to shut down the TC as soon as possible (via T-DISCONNECT primitives). The threshold value for the transit delay is such that every time the OSI95 TS provider exceeds this QoS value for a particular invocation of the TSDU transfer facility while this value applies, it has to report this fact to the OSI95 TS users as soon as possible (via T-REPORT indication primitives) but the TC is not shut down.

For each open direction of data transfer of a TC, the transit delay jitter is defined as *the difference between the longest and the shortest transit delays observed on this direction since the successful completion of the last QoS negotiation on the sending OSI95 TS user's side*. According to this definition, a measure of the transit delay jitter is thus associated with each invocation of the TSDU transfer facility, but this measure of the transit delay jitter depends upon the transit delays observed for all invocations of the TSDU transfer facility (including the current one) since the successful completion of the last QoS negotiation on the sending OSI95 TS user's side.

After a QoS negotiation, the meaningful values regarding the transit delay jitter for an open direction of data transfer are the maximum compulsory value and the threshold value. The following relation must always be verified: $0 \leq$ negotiated threshold transit delay jitter value $\leq$ negotiated maximum compulsory transit delay jitter value. The role of these two values for the transit delay jitter is similar to the one of the corresponding values for the transit delay.

4 Flexibility of the Error Control on a TC

The ISO/IEC Connection-mode Transport Service is always error-free during the data transfer phase, in the sense that there are normally no losses, no corruptions of contents and no duplications of transferred TSDUs except for what is tolerated by the residual error rate QoS parameter.

Even if the connection mode has to provide a full error recovery capability to many applications, and to bulk data transfer applications in particular, such a capability may be quite unsuitable in case of multimedia applications for example. In several multimedia applications indeed, the receiving end is able to manage some losses but has stringent transit delay and transit delay jitter requirements, so that a full error recovery capability at the transport service level based on retransmissions at the protocol level is not viable. Our approach in OSI95 has been to permit a tune-up (which may lead ultimately to a suppression) of the error control [BLL 92]. This has been done by means of a new QoS parameter: the error control policy.

4.1 Description of the Possible Error Control Policies

An error control policy is negotiated for either open direction of data transfer on a TC regardless of and separately from the negotiation of the error control policy for the reverse direction (if the TC is bi-directional). The error control policy for a direction of data transfer consists in the combination of a TSDU corruptions policy and a TSDU losses policy [Bag 93a].

The TSDU losses policy is specified qualitatively by selecting one of two options:
    a) losses of (parts of) TSDUs not accepted;
    b) losses of (parts of) TSDUs accepted but indicated.

The TSDU corruptions policy is also specified qualitatively by selecting one of three options:
    a) corruptions of contents in (parts of) TSDUs not accepted;
    b) corruptions of contents in (parts of) TSDUs accepted but indicated;
    c) corruptions of contents in (parts of) TSDUs accepted and not indicated.

The six possible combinations result in six different error control policies. Table 4.1 shows the error rates that are meaningful in each of the six error control policies.

Error control policy →		**LOSSES**	
		Losses not accepted	**Losses accepted but indicated**
C O R R U P T I O N S	Corruptions not accepted	① — —	④ Lost TSDU error rate
	Corruptions accepted but indicated	② Corrupted TSDU error rate	⑤ Corrupted TSDU error rate Lost TSDU error rate
	Corruptions accepted and not indicated	③ — —	⑥ Lost TSDU error rate

Table 4.1 The six possible error control policies and the corresponding meaningful error rates

When losses (resp. corruptions of contents) of (parts of) TSDUs are not accepted, all losses (resp. corruptions of contents) have to be recovered. When losses (resp. corruptions of contents) of (parts of) TSDUs are accepted provided they are indicated, these losses (resp. corruptions of contents) do not need to be recovered, within the limits expressed by the lost (resp. corrupted) TSDU error rate requirements. They do not need to be recovered but, of course, they may be. However, they necessarily need to be detected so that they can be indicated. By contrast, when corruptions of contents in (parts of) TSDUs are accepted without having to be indicated, these corruptions of contents do not need to be recovered or even detected but, of course, they may be.

Error control policy (1) is the classical full error recovery policy (i.e. no errors at all are tolerated). When this policy is adopted for an open direction of data transfer, the OSI95 TS provider has to preserve unchanged the boundaries and the contents of all submitted TSDUs. That is, any TSDU delivered to the receiving OSI95 TS user via a T-DATA indication shall have the same number of octets and the same value for each octet as the TSDU received from the sending OSI95 TS user in the corresponding T-DATA request.

With error control policy (2), all losses of (parts of) TSDUs have to be recovered, whereas corruptions of contents in (parts of) TSDUs do not need to be recovered, within the limits expressed by the corrupted TSDU error rate requirements, but need to be detected. With error control policy (3), all losses of (parts of) TSDUs have to be recovered, whereas corruptions of contents in (parts of) TSDUs do not need to be recovered or even detected. With error control policy (4), losses of (parts of) TSDUs do not need to be recovered, within the limits expressed by the lost TSDU error rate requirements, but need to be detected, whereas all corruptions of contents in (parts of) TSDUs have to be recovered. With error control policy (5), losses and corruptions of contents of (parts of) TSDUs do not need to be recovered, within the limits expressed respectively by the lost and corrupted TSDU error rate requirements, but both need to be detected. Finally, with error control policy (6), losses of (parts of) TSDUs do not need to be recovered, within the limits expressed by the lost TSDU error rate requirements, but need to be detected, whereas corruptions of contents in (parts of) TSDUs do not need to be recovered or even detected.

In the case where a whole submitted TSDU is lost through the OSI95 TS provider and the loss is not recovered by the OSI95 TS provider, a zero-length TSDU shall be delivered to the receiving OSI95 TS user instead. In all other cases (i.e. when a non-zero-length TSDU is delivered), the TSDU delivered to the receiving OSI95 TS user via a T-DATA indication shall have the same number of octets as the TSDU received from the sending OSI95 TS user in the corresponding T-DATA request, but the value of some octets may have been altered by the OSI95 TS provider:

- If a part of a submitted TSDU (not the whole TSDU) is lost through the OSI95 TS provider, and if the loss is not recovered by the OSI95 TS provider, then each octet of this part in the delivered TSDU will have a value that is either any value or a specific default value (called dummy character value), the value being provided by the OSI95 TS provider regardless of the correct value of the octet in the submitted TSDU. In the worst case, all the octets of this part will have an incorrect value in the delivered TSDU.

- If a part of a submitted TSDU (possibly the whole submitted TSDU) is corrupted through the OSI95 TS provider, and if the corruption of contents is not recovered by the OSI95 TS provider, then at least one octet of this part in the delivered TSDU will have a value different from the correct value in the submitted TSDU. In the worst case, all the octets of this part will have an incorrect value in the delivered TSDU. In other words, a number of octets between 1 and the length of this part inclusive have an incorrect value in the delivered TSDU.

As already said above, when error control policy (4) applies, losses of (parts of) TSDUs do not need to be recovered, within the limits expressed by the lost TSDU error rate requirements, but need to be detected, whereas all corruptions of contents in (parts of) TSDUs have to be recovered. One possible way to recover from corruptions of contents of (parts of) TSDUs in such a case is to consider and process the corrupted parts as if they had been lost. Other ways are conceivable however.

The error control policy for either direction of data transfer on a TC is negotiated between the OSI95 TS users only. This is because the choice of the error control policy must depend upon the needs of the application only, and must not be influenced by the OSI95 TS provider which has to be able to provide full error recovery if required. The error control policy proposed by the calling OSI95 TS user in the T-CONNECT request (and replicated unchanged in the T-CONNECT indication) may only be strengthened by the called OSI95 TS user in the T-CONNECT response (which means that, in table 4.1, the square corresponding to the policy finally selected by the called OSI95 TS user may not be on the right or under the square corresponding to the policy proposed by the calling OSI95 TS user). This introduces a partial order among the error control policies.

4.2 Role of the Other Error Control QoS Parameters

In OSI95, besides the error control policy, there are four other error control QoS parameters: the lost TSDU error rate, the corrupted TSDU error rate, the dummy replacement indicator and the dummy character. Each of these parameters is specified for either open direction of data transfer on a TC regardless of and separately from its specification for the reverse direction (if the TC is bi-directional) [Bag 93a].

The successful completion of a QoS (re-)negotiation corresponds to the occurrence of a T-CONNECT confirm or a T-RENEGOTIATE confirm (see section 5) on the calling side, and to the occurrence of a T-CONNECT response or a T-RENEGOTIATE response (see section 5) on the called side. All the TSDUs submitted via T-DATA requests by a same sending OSI95 TS user between two successive successful completions of a QoS (re-)negotiation make up a sequence of TSDUs. The TSDUs of a sequence can be numbered conceptually in the order in which they are submitted to the OSI95 TS provider with consecutive integers starting from zero. Then, a sequence of TSDUs can be partitioned into successive N-subsequences of TSDUs. Given an integer $N \geq 1$, an N-subsequence of TSDUs is a series of N TSDUs that belong to a same sequence, that have consecutive increasing numbers and such that the number of the first TSDU of the series is a multiple of the integer N. A TSDU error rate is expressed as a ratio of N1 to N (N1 and N being integers and verifying $0 \leq N1 \leq N$ with $N \geq 1$) where N1 is the number of TSDUs in error in a series of N TSDUs.

More precisely, for an open direction of data transfer on a TC, the lost (resp. corrupted) TSDU error rate is defined as the number of TSDUs, out of an N-subsequence (resp. N'-subsequence) of TSDUs submitted by the sending OSI95 TS user to the OSI95 TS provider, which are delivered to the receiving OSI95 TS user with an indication that (parts of) their contents has been lost (resp. corrupted). It has to be stressed that the size N of the N-subsequences of TSDUs intervening in the measure of the lost TSDU error rate may be different from the size N' of the N'-subsequences of TSDUs intervening in the measure of the corrupted TSDU error rate. As previously shown in table 4.1, for either open direction of data transfer, the lost (resp. corrupted) TSDU error rate parameter is meaningful only if the error control policy negotiated for this direction specifies that losses (resp. corruptions of contents) of (parts of) TSDUs are accepted but have to be indicated.

The semantics adopted for the lost and corrupted TSDU error rates is the one adopted for the performance QoS parameters (i.e. the throughput, the transit delay and the transit delay jitter), giving the possibility to negotiate compulsory and threshold QoS values. The sole difference in comparison with the performance QoS parameters is in the negotiation scenario: the OSI95 TS provider may not intervene at all in the negotiation, so that the proposal of the calling OSI95 TS user in the T-CONNECT request as regards the lost and corrupted TSDU error rates is replicated unchanged in the T-CONNECT indication.

Moreover, for either open direction of data transfer on a TC, the dummy replacement indicator parameter is meaningful only if the error control policy negotiated for this direction specifies that losses of (parts of) TSDUs are accepted provided they are indicated. Either of two dummy replacement options may be selected by the receiving OSI95 TS user: "replacement with dummy character" or "no replacement with dummy character". The dummy replacement indicator is not negotiated at all: the dummy replacement option is imposed by an OSI95 TS user (in the T-CONNECT request for the calling OSI95 TS user and in the T-CONNECT response for the called OSI95 TS user) for its input direction of data transfer on a TC.

If the dummy replacement option is fixed to "replacement with dummy character" by the receiving OSI95 TS user, then, in every TSDU received with an indication of loss, all lost octets in the zone of the TSDU affected by errors have a default value (called the dummy character value) provided by the OSI95 TS provider regardless of the correct value of the octets in the submitted TSDU. By contrast, if the dummy replacement option is fixed to "no replacement with dummy character", then, in every TSDU received with an indication of loss, each lost octet in the zone of the TSDU affected by errors may have any value provided by the OSI95 TS provider regardless of the correct value of the octet in the submitted TSDU.

Lastly, for either open direction of data transfer, the dummy character parameter is meaningful only if the dummy replacement indicator is meaningful and if the dummy replacement option is set to "replacement with dummy character". Any dummy character value between 0 and 255 (8-bit unsigned value) may be selected by the receiving OSI95 TS user. The dummy character is not negotiated at all: its value is imposed by an OSI95 TS user for its input direction of data transfer on a TC.

4.3 Use of Transfer Status Parameters in the T-DATA indication

Two transfer status parameters, viz. the error type indicator and the affected zone, have been inserted in the T-DATA indication to cope with error control [Bag 93a].

The possible error type indicator options in a T-DATA indication depend upon the error control policy negotiated for the corresponding direction of data transfer:

a) with error control policies (1) & (3), the sole possible error type indicator option is "no errors signalled in the TSDU";

b) with error control policies (4) & (6), two error type indicator options are possible: "no errors signalled in the TSDU" or "losses signalled in the TSDU";

c) with error control policy (2), two error type indicator options are possible: "no errors signalled in the TSDU" or "corruptions of contents signalled in the TSDU";

d) with error control policy (5), four error type indicator options are possible: "no errors signalled in the TSDU", "losses signalled in the TSDU", "corruptions of contents signalled in the TSDU" or "both losses and corruptions of contents signalled in the TSDU".

Any TSDU delivered to the receiving OSI95 TS user with the error type indicator indicating "losses signalled in the TSDU" or "both losses and corruptions signalled in the TSDU" increments by 1 the number of (partly) lost TSDUs pertaining to the current N-subsequence of TSDUs. In the same way, any TSDU delivered to the receiving OSI95 TS user with the error type indicator indicating "corruptions signalled in the TSDU" or "both losses and corruptions signalled in the TSDU" increments by 1 the number of (partly) corrupted TSDUs pertaining to the current N'-subsequence of TSDUs.

The affected zone parameter in a T-DATA indication is not meaningful when the associated error type indicator indicates "no error signalled in the TSDU", or when the whole submitted TSDU is considered to be lost and a zero-length TSDU is delivered instead (which is necessarily accompanied by an error type indicator that indicates "losses signalled in the TSDU"). In all other cases, this parameter shall delimit the zone affected by errors, by giving a pair of integers which are the number of the first octet of the zone and the number of the last octet of the zone (the octets of the submitted TSDU are supposed to be numbered consecutively starting from 1).

Even when the corrupted and lost parts of a TSDU are not adjacent, the affected zone in a delivered TSDU is a continuous block of octets. The affected zone is the smallest continuous block of octets containing all corrupted and lost parts of the TSDU. In the example of figure 4.1, the affected zone is determined by the pair of integers (b, e), despite the fact that there are parts of the TSDU whose integrity is preserved between the octets numbered b and e. If it is deemed useful, the transfer status information could be refined in a future definition, in order to give details about the corrupted, lost and safe parts in the affected zone.

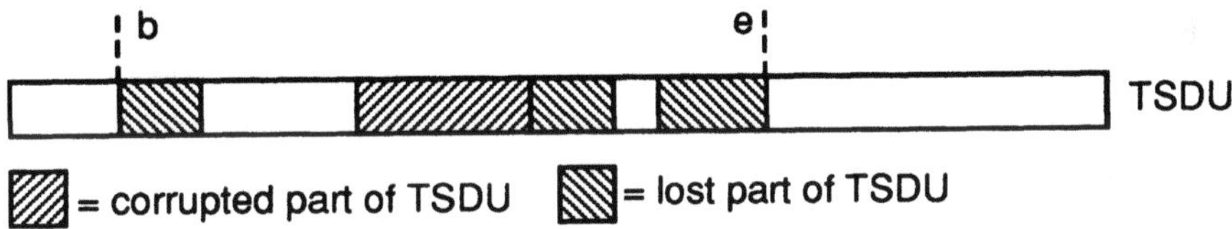

Fig. 4.1 Affected zone parameter in the transfer status

5 QoS Re-negotiation Facility

In the ISO/IEC Connection-Mode Transport Service, the QoS is negotiated once and for all at the TC establishment. This means that the negotiated QoS always applies for the whole TC lifetime.

To respond to certain requirements formulated in [LCH 93], we have decided to introduce a QoS re-negotiation facility in our OSI95 Enhanced Connection-Mode Transport Service definition.

5.1 Who Will Be Allowed to Start a Re-negotiation ?

The introduction of such a facility poses a first problem: who will be allowed to initiate a re-negotiation ? If both users of a given TC are permitted to start a re-negotiation, we will face a major difficulty when they invoke the facility at the same time. Let us remind that, as regards the TC establishment, colliding TC set-up requests lead to the establishment of distinct TCs. Bypassing the difficulty in a similar way is of course impossible in case of re-negotiation.

An obvious solution to the problem is to suppress it by precluding one of the two OSI95 TS users to initiate a re-negotiation on the TC. Therefrom comes another question: which user should have the right to start a re-negotiation whereas the other one would not have this right ? This could be decided during the TC set-up phase, through a form of negotiation by way of the T-CONNECT primitives. The sole OSI95 TS user authorized to initiate a re-negotiation could also be always the same one, either the calling or the called one. At first glance, it would be preferable to elect the calling OSI95 TS user, i.e. the initiator of the TC establishment, as the only possible initiator of a QoS re-negotiation on the TC. Whichever the OSI95 TS user authorized to start the re-negotiation, the other one could anyway use the out-of-band data transfer facility (see section 6) to indicate to its peer its wish to have a re-negotiation.

Another solution to the aforementioned problem is to allow both OSI95 TS users to initiate the re-negotiation, and thus to accept the possibility of colliding re-negotiation requests, while giving the precedence to one of them. In case of collision, the re-negotiation request from the OSI95 TS user that has not the precedence is simply discarded and the other request is processed normally. The different alternatives, proposed in the context of the previous solution to choose the only OSI95 TS user that should be authorized to start the re-negotiation, remain conceivable to choose the OSI95 TS user that should have the precedence.

In the current version of our connection-mode service, we have decided that only the initiator of the TC set-up, i.e. the calling OSI95 TS user, will have the right to start a QoS re-negotiation.

5.2 Sequence of Primitives for a Successful Re-negotiation

The sequence of OSI95 TS primitives for QoS re-negotiation has been inspired by the classical four-primitive sequence used for TC set-up. Nevertheless, because of the wish not to suspend the TSDU transfers over the TC while a QoS re-negotiation takes place, a fifth primitive named T-NEW-QOS indication has been added (Figure 5.1).

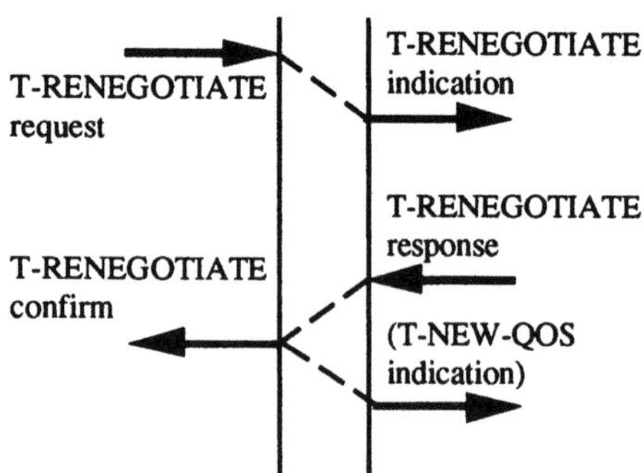

Fig. 5.1 Time sequence diagram for a successful re-negotiation

On the calling side, both the output and input traffics switch from the old QoS to the new one upon receipt of the T-RENEGOTIATE confirm. On the called side, the output traffic switch from the old QoS to the new one upon invocation of the T-RENEGOTIATE response but the additional T-NEW-QOS indication is necessary to indicate when the QoS of the input traffic changes.

The QoS re-negotiation primitives are used to re-negotiate the QoS as well as the TSDU size range in either or both open direction(s) of data transfer on an established TC. If the established TC is bidirectional and both directions of data transfer are still considered as open by the calling OSI95 TS user, this calling OSI95 TS user may choose to re-negotiate the QoS and the TSDU size range in either or both open direction(s). In all other cases, the re-negotiation will apply to the only direction still considered as open by the calling OSI95 TS user. The calling OSI95 TS user may never initiate a QoS re-negotiation when it considers both directions of data transfer as closed or closing[1].

The parameters of the T-RENEGOTIATE primitives are thus identical to those of the T-CONNECT primitives, except that the unnecessary address parameters are removed and that the "directions to open" parameter becomes a "directions to re-negotiate" parameter. The relations which have to be verified between the parameters of the different T-RENEGOTIATE primitives are summarized in Table 5.1.

	PP-ECOTS primitives:			
	T-RENEG…	T-RENEG…	T-RENEG…	T-RENEG…
Parameters:	request	indication	response	confirm
Directions to re-negotiate	M	M	M	M
Quality of service {…}	M	M(N)	M(N)	M(=)
TSDU size range	M	M(N)	M(N)	M(=)
OSI95 TS user-data	U	C(=)	U	C(=)

Keys: {…} : generic denomination gathering several parameters;

     M : presence of this parameter or of this group of parameters in the primitive is mandatory;

     U : inclusion of this parameter or of this group of parameters in the request or response primitive is a choice made by the OSI95 TS user;

     C : presence of this parameter or of this group of parameters in the indication or confirm primitive is conditional, depending upon its inclusion in the preceding request or response primitive;

    (=) : the value or option of the parameter in the primitive is identical to that of the corresponding parameter in the preceding primitive;

    (N) : each value or option of the parameter in the primitive is obtained from the values or options of the corresponding parameter in the preceding primitive in accordance with the negotiation rules;

Table 5.1 Relations between the parameters of the different T-RENEGOTIATE primitives

---

[1] When an OSI95 TS user has invoked the graceful TC release facility (see [2] for more details) to release its output direction of data transfer gracefully, then, while it is waiting for the confirmation of the graceful release, it does not regard this direction as open any more, neither does it regard this direction as closed yet. Its output direction is considered to be in an intermediate state between the open and closed states, referred to as the closing state.

The T-NEW-QOS indication has no parameter. As its sole function is to indicate to the called OSI95 TS user when the QoS and TSDU size range characteristics relating to the calling-to-called direction of data transfer switch from the old ones to the new ones, this T-NEW-QOS indication shall be delivered only if the calling-to-called direction appears in the "directions to re-negotiate" parameter of the T-RENEGOTIATE confirm and if the called OSI95 TS user does not consider this direction as closing or closed before the delivery of the primitive.

5.3 Possible Sequences of Primitives for an Unsuccessful Re-negotiation

The QoS re-negotiation of any direction of data transfer has three possible outcomes: it succeeds, it is rejected by either the OSI95 TS provider or the called OSI95 TS user, or it is stopped if it becomes meaningless.

The QoS re-negotiation procedure may fail either due to the inability of the OSI95 TS provider to accept the TC with the new requested characteristics or due to the unwillingness or the inability of the called OSI95 TS user to accept the TC with the new requested characteristics. These two cases are described in figures 5.2 & 5.3. When the QoS re-negotiation involves both directions, the rejection of the proposal made by the calling OSI95 TS user for one direction will inescapably cause the rejection of the whole re-negotiation. When a QoS re-negotiation is rejected, the TC is of course maintained with its old QoS and TSDU size range characteristics.

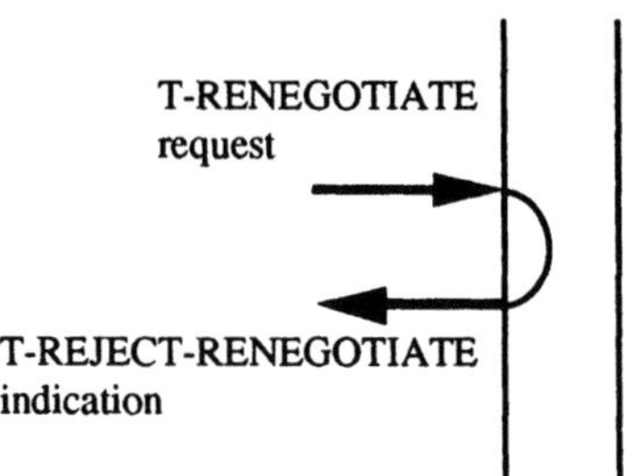

Fig. 5.2 Time sequence diagram for the rejection of a re-negotiation attempt by the OSI95 TS provider

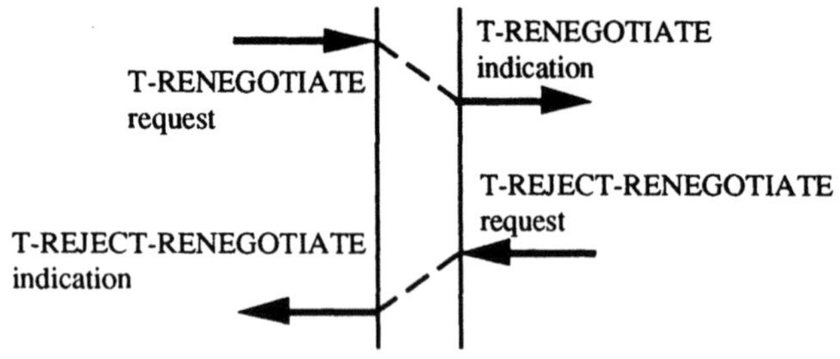

Fig. 5.3 Time sequence diagram for the rejection of a re-negotiation attempt
by the called OSI95 TS user

The parameters of the T-REJECT-RENEGOTIATE primitives, and the relations which have to be verified between them, are summarized in table 5.2.

The reject reason is one of the following:

a) remote OSI95 TS user initiated;

b) OSI95 TS provider initiated due to any internal reason which may be of transient or permanent nature.

	PP-ECOTS primitives:	
Parameters:	T-REJECT-RENEG... request	T-REJECT-RENEG... indication
Reject reason		**M**
OSI95 TS user-data	U	C(=)

Keys: M : presence of this parameter or of this group of parameters in the primitive is mandatory;

U : inclusion of this parameter or of this group of parameters in the request or response primitive is a choice made by the OSI95 TS user;

C : presence of this parameter or of this group of parameters in the indication or confirm primitive is conditional, depending upon its inclusion in the preceding request or response primitive;

(=) : the value or option of the parameter in the primitive is identical to that of the corresponding parameter in the preceding primitive;

Blank : the parameter is absent.

Table 5.2 Relations between the parameters of the T-REJECT-RENEGOTIATE primitives

Any of the time sequence diagrams in figures 5.1, 5.2 & 5.3 may terminate before completion if the QoS re-negotiation is stopped. A QoS re-negotiation is stopped in the case where all direction(s) of data transfer indicated initially in the directions to re-negotiate parameter of the T-RENEGOTIATE request is (are) no more considered as open by the OSI95 TS user which should have issued or received the next expected primitive of the sequence.

The calling OSI95 TS user may not initiate a first QoS re-negotiation before the T-CONNECT confirm primitive has been issued. Moreover, the calling OSI95 TS user may not initiate a subsequent QoS re-negotiation until a T-RENEGOTIATE confirm or a T-REJECT-RENEGOTIATE indication primitive has been received in response to the previous T-RENEGOTIATE request primitive, indicating that the previous re-negotiation is successful or rejected, or until it considers the direction(s) of data transfer involved in the previous re-negotiation as closing or closed.

6 Out-of-Band Data Transfer Facility

In addition to the transfer of normal TSDUs, the OSI95 Enhanced Connection-Mode Transport Service provides for an exchange of out-of-band TSDUs between the OSI95 TS users. Like the exchange of normal TSDUs, the exchange of out-of-band

TSDUs may not take place before the successful completion of the TC set-up phase. Unlike the exchange of normal TSDUs, the exchange of out-of-band TSDUs is allowed even when the corresponding direction of data transfer is not considered as open. Transfers of out-of-band TSDUs can always occur in both directions simultaneously as long as the TC remains established, i.e. as long as both directions of data transfer of the TC are not considered as closed.

Since the interest of transmitting out-of-band TSDUs over TCs with no acknowledgement procedure at all seems quite limited to us, we have elected an out-of-band TSDU transfer facility which is very similar to the Enhanced Acknowledged Connectionless-mode Transport Service defined in detail in [Bag 93b]. In fact, the out-of-band TSDU transfer facility differs from this acknowledged connectionless-mode service only by the fact that the transmitted out-of-band TSDUs are automatically associated with an established TC by means of the TCEP (Transport Connection Endpoint) identification mechanism. This is why the T-OOB-DATA primitives, unlike the T-ACKDATA primitives (cf. [Bag 93b]) do not need any address parameter: all addressing information is provided in the T-CONNECT primitives, and the other connection-mode primitives makes use of the TCEP identification mechanism to identify the TC to which they pertain.

The QoS associated with an out-of-band TSDU is specified by the sending OSI95 TS user regardless of the QoS negotiated on the TC. The QoS associated with an out-of-band TSDU may thus be lower or higher than that for the TSDUs transferred via the T-DATA primitives. In particular, the monitoring of the compulsory and threshold QoS negotiated on the TC does not apply to the out-of-band TSDU transfers. An out-of-band TSDU is conveyed by the OSI95 TS provider exactly in the same way as would be an acknowledged connectionless-mode TSDU submitted with the same QoS in the Enhanced Acknowledged Connectionless-mode Transport Service.

With respect to error control, the OSI95 TS provider always preserves the integrity and the boundaries of the submitted out-of-band TSDUs. On the other hand, it may lose some submitted out-of-band TSDUs, may deliver out-of-band TSDUs in an order different from that in which they have been submitted and may duplicate out-of-band TSDUs. Out-of-band TSDUs may also pass or be passed by normal TSDUs.

Clearly, an identification mechanism has to be provided at the OSI95 TS interface if an OSI95 TS user and the OSI95 TS provider need to distinguish between several invocations of the out-of-band data transfer facility relating to a same TC. For this purpose, the TAEP (Transport Association Endpoint) identification mechanism which is also used by the OSI95 Acknowledged and Request/Response Connectionless-Mode Transport Services (cf. [Bag 93b]) is provided locally. Here however, unlike what is done in the connectionless-mode services where the TAEP identification mechanism is used at the TSAP level, this identification mechanism is used at the TCEP level!

Even though it has not exactly the same properties, the out-of-band data transfer facility may be an interesting alternative to the expedited data transfer facility. A recommended use of the out-of-band data transfer facility is in support of the QoS re-negotiation. With the current QoS re-negotiation facility, only the OSI95 TS user that initiated the TC may start a re-negotiation. When the other OSI95 TS user wants to re-negotiate the QoS on the TC, it could inform the OSI95 TS user that has the right to re-negotiate and provide this OSI95 TS user with the relevant information by way of the out-of-band channel corresponding to the TC.

7 Conclusions

A lot of modern applications (multimedia applications for instance) may not be completely satisfied with the capabilities currently available at the transport service interface in the networks, especially for what regards the multiple QoS aspects. In this light, the modifications and enhancements in the field of QoS has been a very important facet of our work on a new connection-mode transport service within the OSI95 project. The removal of QoS parameters deemed unmanageable in practice, the change in the definition of the throughput, the introduction of a transit delay jitter, the definition of a new semantics for the performance QoS parameters, the introduction of specific QoS parameters to deal with error control, the introduction of a TSDU size range information parameter and finally the definition of a QoS re-negotiation facility all aim at better meeting the needs of modern applications.

The ability to open uni-directional TCs, as well as bi-directional TCs, in the OSI95 Connection-Mode Transport Service reflects our choice to adopt the concept of "simplex virtual crcuit" (corresponding to one direction of data transfer) as the basic building block for a TC. By the way, this choice has had consequences on the design of the TC release facilities (see [BaL 93] for the details).

Lastly, despite the fact that it has not exactly the same properties, the ouf-of-band data transfer facility has taken the place of the classical expedited data transfer facility in the current version of the OSI95 Connection-Mode Transport Service. The requirements of the applications in the field of priority data transfer need further investigations however, and therefore it remains an open question to know whether the out-of-band data transfer facility (if kept in a next version) should remain a substitute for or should come in addition to the expedited data transfer facility.

References

[Bag 93a]   Y. Baguette, "Enhanced Transport Service Definition (Informal specification in English - Version 1)", Internal ref.: S.A.R.T. 93/03/15 (also circulated as ISO/IEC JTC1/SC6/WG4 N 822), Université de Liège, Institut Montefiore B28, B-4000 Liège, Belgium, Feb. 5, 1993.

[Bag 93b]   Y. Baguette, "The OSI95 Connectionless-Mode Transport Services", *The OSI95 Transport Service with Multimedia Support on HSLAN's and B-ISDN*, A. Danthine, ed., Springer Verlag, 212-224

[BaL93]   Y. Baguette, G. Leduc, "The Connection Release Facilities in the OSI95 Transport Service", *The OSI95 Transport Service with Multimedia Support on HSLAN's and B-ISDN*, A. Danthine, ed., Springer Verlag, 199-211

[BLL 92]   Y. Baguette, L. Léonard, G. Leduc, A. Danthine, O. Bonaventure, "OSI95 Enhanced Transport Facilities and Functions", Technical report for ESPRIT II project 5341 (OSI95), Ref.: OSI95/Deliverable ULg-A/P/V1, Université de Liège, Institut Montefiore B28, B-4000 Liège, Belgium, Dec. 1992.

[DBL 92]   A. Danthine, Y. Baguette, G. Leduc, "Issues Surrounding the Specification of High-Speed Transport Service and Protocol", Technical report for ESPRIT II project 5341 (OSI95), Ref.: OSI95/ULg/A/15/TR/P/V2 (also submitted as ISO/IEC JTC1/SC6 N 7312 through the Belgian National Body at the SC6 interim meeting, Paris, France, Feb. 10-13, 1992), Université de Liège, Institut Montefiore B28, B-4000 Liège, Belgium, Jan. 1992.

[DBL 93]   A. Danthine, O. Bonaventure, G. Leduc, "The QoS Enhancements in OSI95", *The OSI95 Transport Service with Multimedia Support on HSLAN's and B-ISDN*, A. Danthine, ed., Springer Verlag, 125-150

[ISO 8072]   ISO/IEC JTC1: "Information Processing Systems - Open Systems Interconnection - Transport Service Definition", ISO 8072, 1986.

[ISO 8807]   ISO/IEC JTC1: "Information Processing Systems - Open Systems Interconnection - LOTOS - A Formal Description Technique Based on the Temporal Ordering of Observational Behaviour", ISO 8807, 1989.

[LCH 93]   H. Leopold, A. Campbell, D. Hutchinson, N. Singer, "Distributed Multimedia Communication System Requirements", *The OSI95 Transport Service with Multimedia Support on HSLAN's and B-ISDN*, A. Danthine, ed., Springer Verlag, 64-81

[LeD 93]   G. Leduc, A. Danthine, "On the Provision of a Fast Connect Facility in a Connection-Mode Transport Service", *The OSI95 Transport Service with Multimedia Support on HSLAN's and B-ISDN*, A. Danthine, ed., Springer Verlag, 225-238

The Connection Release Facilities in the OSI95 Transport Service

Yves Baguette[1] and Guy Leduc[2]
[1] Research Assistant of the Belgian National Fund for Scientific Research (F.N.R.S.)
[2] Research Associate of the F.N.R.S.
Université de Liège, Institut d'Electricité Montefiore B28, B-4000 Liège 1, Belgium
Email: baguette@montefiore.ulg.ac.be

We explain in detail how the work carried out on the transport connection release aspects in the framework of the OSI95 Connection-Mode Transport Service has led to the introduction of a graceful transport connection release facility and, later on, to an enhancement of the existing ISO/IEC abrupt transport connection release facility.
   Keywords: OSI95, Transport Service, Connection-Mode, Graceful Release, Abrupt Release.

1 Introduction

The sole Transport Connection (TC) release facility provided by the current ISO/IEC Connection-Mode Transport Service [ISO 8072] is an abrupt (and thereby possibly destructive) one. At present, in the OSI Reference Model [ISO 7498], the concept of orderly (or graceful) release is introduced at the session service level. So it is up to the peer session entities to make sure that there are no more in-transit data on the TC to which is assigned the session connection that is to be gracefully released before requesting the abrupt termination of this TC.

Some people even think that there is no objective reason for providing an orderly connection release in any layer except the application layer (cf. Question 6 in [ISO SD9]). They argue that peer application entities could ensure, by way of an exchange of application protocol control information, that there are no more data in transit on the presentation connection and could then safely perform an abrupt termination of this presentation connection. This, of course, suggests that an orderly release facility be provided in the Common Application Service Element (CASE) and that all other layers below implement an abrupt connection termination only.

We do not share this opinion, neither do we agree with the ideas that the existing orderly connection release facility is the most appropriate one and that the session layer is the best place to introduce it. We have therefore defined a graceful connection release facility *significantly different from what exists for the moment in the ISO/IEC standards*, and we have provided this facility in the connection-mode part of the OSI95 Transport Service (OSI95 TS) [BLL 93]. This decision has been based on several arguments which we deem of sufficient weight in favour of such a new facility at the transport level.

First, from a functional point of view, we think that the orderly release of a connection may be considered as a concern of the transport layer. In fact, the transport layer deals with most of the OSI aspects relating to the provision of a reliable data transfer service and, in our opinion, the graceful connection release may be regarded as a way to achieve reliability (unlike the abrupt connection release which is not reliable by itself in the sense that it may cause the loss of all data possibly still in transit on the connection).

Second, if a graceful TC release facility is introduced, the invocation of the orderly connection release facility at any upper service interface will merely result in the invocation of the same facility at the underlying service interface down to the transport service interface. The end-to-end PDU exchanges required to complete an orderly presentation connection release will thus be accomplished in the transport layer. No PDU processing will be needed above the transport layer, relieving the upper layers of that burden.

Third but not least, it is clear that today we have mainly two different network infrastructures, the one based on OSI and the one based on DoD Internet. From an essentially political point of view, an interesting issue is whether an enhanced transport service has to be strictly aimed at the OSI stack or instead has to be designed also to allow a rather smooth migration from the TCP and UDP transport services to this enhanced transport service. TCP offers a graceful connection release facility. One major reason for this is probably that there exists no session service on top of TCP, so that a graceful connection release, if required and not provided directly at the application level, can only be provided at the transport level. A graceful TC release facility does not appear as essential within the strict OSI framework, but its introduction would obviously ease the migration from the TCP world.

Of course, in the OSI95 TS, the graceful TC release facility is intended to be a complement to the abrupt TC release facility which is already defined in the ISO/IEC standard, and not a substitute for this abrupt TC release facility.

The remainder of the paper is organized as follows. Section 2 discusses the successive steps of the rationale that has led us to the definition of the graceful connection release facility we have ultimately adopted in the OSI95 TS. Section 3 examines the possible interferences that may occur between the graceful TC release facility and the abrupt TC release facility in the OSI95 TS, and proposes an enhancement of the standard ISO/IEC abrupt TC release facility to cope properly with these interferences. Section 4 describes this enhancement in a more detailed manner. The conclusions of the paper are drawn in the last section.

2 What Kind of Graceful TC Release Facility for OSI95 ?

2.1 Discussion of the ISO/IEC Orderly Session Connection Release Facility

The starting point of our study has been the orderly connection release facility currently in use at the ISO/IEC Basic Connection-Oriented Session Service level [ISO 8326]. This facility is always based on the classical 4-primitive (i.e. request, indication, response and confirm) scheme. However, there are two variants to release a session connection in an orderly manner, depending upon the availability of the release token as negotiated during the session connection establishment.

If it has been negotiated not to use the release token, either of the two session service users may invoke a S-RELEASE request, and the other session service user cannot refuse the release. A S-RELEASE request means that the requestor asks for the orderly release of the session connection (which implies that it has no more data to send in the context of the session and thus will not invoke any S-DATA request any more). The S-RELEASE request gives rise to a S-RELEASE indication at the peer SSAP. The acceptor of the S-RELEASE indication knows that it will not receive any S-DATA indication any more, but may still invoke S-DATA requests. When it has no more data to send, the acceptor of the S-RELEASE indication will respond positively with a positive S-RELEASE response, which gives rise to a positive S-RELEASE confirm at the peer SSAP. When a positive S-RELEASE confirm comes back, the requestor of the release knows that the other session service user has been correctly informed of its request and also that the session connection orderly release is successfull. Moreover, a specific solution has been developed to deal with the possible problem of a collision between S-RELEASE requests.

If it has been negotiated to use the release token, only the session service user that owns the token may invoke a S-RELEASE request, but the other session service user may refuse the release and continue the session connection without loss of data. In this case, the acceptor of the S-RELEASE indication issues a negative S-RELEASE response and remains in the data transfer phase. The negative S-RELEASE response gives rise to a negative S-RELEASE confirm at the peer SSAP, which causes the requestor of the orderly session connection release to go back to the data transfer phase. The acceptor of the S-RELEASE indication may also behave as in the case where no release token is available, i.e. may wait until it has no more data to send before responding with a positive S-RELEASE response.

We think that either of the two variants of the orderly session connection release presents a certain drawback. In the first variant, the requestor may not get a confirmation of the orderly release of its output direction of data transfer until the acceptor of the T-RELEASE indication decides to also release the reverse direction by issuing a positive T-RELEASE response. This means that if the acceptor is not ready to perform the orderly release immediately and if something wrong happens before it becomes ready, the requestor will not have any chance to know whether the orderly release of its output direction of data transfer has been successful. In the second variant, we do not see the interest of putting the requestor back in the data transfer phase when the acceptor of the T-RELEASE indication responds negatively. In fact, the requestor has clearly no more data to send since it has invoked a T-RELEASE request. Putting the requestor back in the data transfer phase will therefore have as sole effect to force it to try and try again (until success or until the release token is passed to the other session service user) the orderly session connection release.

The aforementioned drawbacks come from the fact that the positive S-RELEASE confirm has a double role: it confirms the orderly release of the requestor's output direction of data transfer and indicates the orderly release of the reverse direction as well. There is no way for the acceptor of the S-RELEASE indication to confirm the orderly release of the requestor's output direction of data transfer without closing the reverse direction. So, if the acceptor is not willing to perform the orderly release of its output direction immediately, it has to either delay the confirmation of the orderly release of the requestor's output direction (first variant) or refuse the orderly release of the requestor's output direction (second variant). We judge both alternatives unacceptable.

2.2 Graceful Release of Either Direction Separately from the Other One

This explains why we have chosen to define a new graceful connection release facility which relies on an idea previously exploited at the transport level by some well-known non-OSI protocols, e.g. TCP [RFC 793] and more recently XTP [PEI 3.6].

This idea is that either of the directions of data transfer of a connection should be treated quite separately from the other one as for its graceful release. Such an idea conforms perfectly to our view of a TC in OSI95. Indeed, in OSI95, the building block for the concept of TC is the simplex virtual circuit (corresponding to one direction of data transfer). Thus, a TC may be uni-directional, in which case it is made up of a single simplex virtual circuit, or bi-directional, in which case it has to be regarded as two simplex virtual circuits rather than as a single duplex virtual circuit. The new graceful connection release facility we have designed applies to a single direction of data transfer, without any consideration for the current state of the reverse direction if the connection is bi-directional.

2.3 Design of the OSI95 Graceful TC Release Facility

The sequence of OSI95 TS primitives we have first envisaged to gracefully release a single direction of data transfer on a TC is the following one: a T-RELEASE request generated by one of the two users of a TC gives rise to a T-RELEASE indication issued to the peer OSI95 TS user [DBL 92]. The invocation of a T-RELEASE request by one of the two users of a TC means that it *wants to gracefully release* its output direction on this TC (which implies that the requestor has no more data to send and thus will not invoke any T-DATA request any more). The reception of a T-RELEASE indication by one of the two users of a TC means that its input direction on this TC *has been released gracefully* owing to the invocation of a T-RELEASE request by the remote OSI95 TS user.

This first scenario raises a few questions. For instance, may an OSI95 TS user assume a TC to be closed (so that no OSI95 TS primitive pertaining to this TC is allowed at the local TSAP any more) immediately after both the issuance of a T-RELEASE request and the reception of a T-RELEASE indication, in any order ? The questions that arise are due to the fact that a T-RELEASE request just means that the graceful release of the output direction *is requested* and not that it *is completed*. When completed, there is no confirmation of the graceful release to the requestor. Let us remember here that the impossibility to confirm the orderly release of one direction of data transfer as long as the reverse direction is not also closed in an orderly manner has been identified as a main drawback of the ISO/IEC orderly session connection release facility. By choosing to apply our graceful release facility to a direction of data transfer rather than to a connection, we have paved the way for a solution to the problem of confirmation but we have not actually solved the problem yet.

The obvious solution to this problem is to use a confirmed service facility, i.e. to introduce a form of confirmation at the OSI95 TS level by means of an additional T-RELEASE confirm primitive. A T-RELEASE confirm simply confirms to the requestor the graceful release of its output direction, i.e. confirms the delivery of a T-RELEASE indication to the remote OSI95 TS user. Figure 2.1 illustrates the sequences of OSI95 TS primitives finally elected for the graceful release of a single direction or both directions of data transfer of a TC [DBL 92]. With this new graceful TC release scenario, an OSI95 TS user may consider a TC to be completely closed (i.e. closed in both directions) immediately after the receipt of both a T-RELEASE indication and a T-RELEASE confirm, in any order.

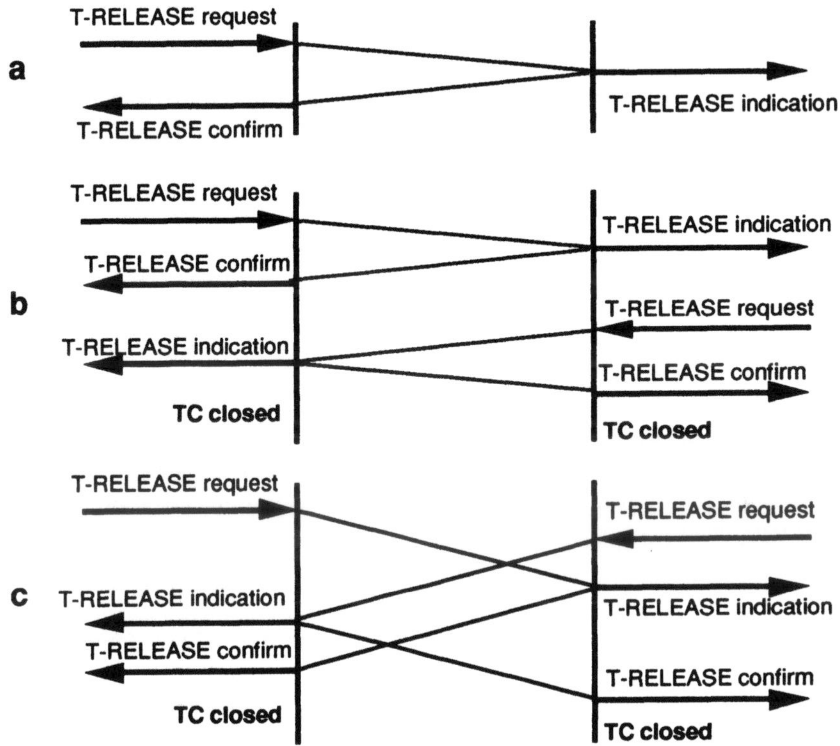

Fig. 2.1 Graceful TC release facility:
a. release of a single direction of data transfer of a TC;
b. successive releases of both directions of a TC;
c. interleaved releases of both directions of a TC

2.4 Parameters of the T-RELEASE Primitives

The T-RELEASE request may have a TSDU parameter. The T-RELEASE indication may have a TSDU parameter and transfer status parameters used for error control purposes. The T-RELEASE confirm has never any parameter. The relations which have to be verified between the parameters of the T-RELEASE request and indication primitives are summarized in table 2.1.

Parameters:	Primitives:	
	T-RELEASE request	T-RELEASE indication
TSDU	U	C(E)
Transfer status {...}		C'

Keys: {...} : generic denomination gathering several parameters;
　　　　U : inclusion of this parameter or of this group of parameters in the request or response primitive is a choice made by the OSI95 TS user;
　　　　C : presence of this parameter or of this group of parameters in the indication or confirm primitive is conditional, depending upon its inclusion in the preceding request or response primitive;
　　　　C' : presence of this parameter or of this group of parameters in the indication or confirm primitive is conditional, depending upon the inclusion of another parameter (here, the TSDU parameter) in the preceding request or response primitive;
　　　　(E) : the contents of this parameter may differ from that of the corresponding parameter in the preceding primitive only within the limits authorized by the error control policy;
　　　　Blank : the parameter is absent.

Table 2.1 Relations between the parameters of the T-RELEASE request and indication primitives

When they convey a TSDU parameter, the T-RELEASE request and indication primitives may be considered as classical T-DATA request and indication primitives in regard to the transfer of the TSDU, but they carry additional information that concerns the graceful release of the corresponding direction of data transfer of the TC. In practice, all happens as if a T-RELEASE request or indication primitive conveying a TSDU were split up conceptually into a classical T-DATA request or indication primitive instantaneously followed by a T-RELEASE request or indication primitive without any parameter.

3 Interferences Between the T-RELEASE and T-DISCONNECT Primitives

3.1 Examples of Encountered Problems

Since the OSI95 TS is intended to provide the graceful TC release facility in complement to the abrupt TC release facility which is already defined in the ISO/IEC standard, it is important to see how the two facilities may interfere with one another.

Let us remind that the abrupt TC release facility currently offered by the ISO/IEC Connection-Mode Transport Service [ISO 8072] always closes both directions of data transfer of a TC together. Therefrom comes an interesting question: if an OSI95 TS user invokes a T-RELEASE request to gracefully release its output direction on a TC and then decides to abruptly release its input direction on this TC via a classical T-DISCONNECT request, when may the T-DISCONNECT request be issued without taking the risk of releasing both directions abruptly ? Since the OSI95 TS user is assured that its output direction on the TC has been closed gracefully as soon as it receives the T-RELEASE confirm, waiting for this primitive is a firm guarantee. By contrast, there cannot be any guarantee on the graceful release of the output direction if the T-DISCONNECT request is invoked prior to the receipt of the T-RELEASE confirm. In this latter case, as shown in figure 3.1, depending upon which one of the remote T-RELEASE indication and the remote T-DISCONNECT indication (prompted respectively by the T-RELEASE request and the T-DISCONNECT request) is issued first, the requestor's output direction is closed gracefully or abruptly. However, even when the output direction is closed gracefully, the graceful release cannot be confirmed.

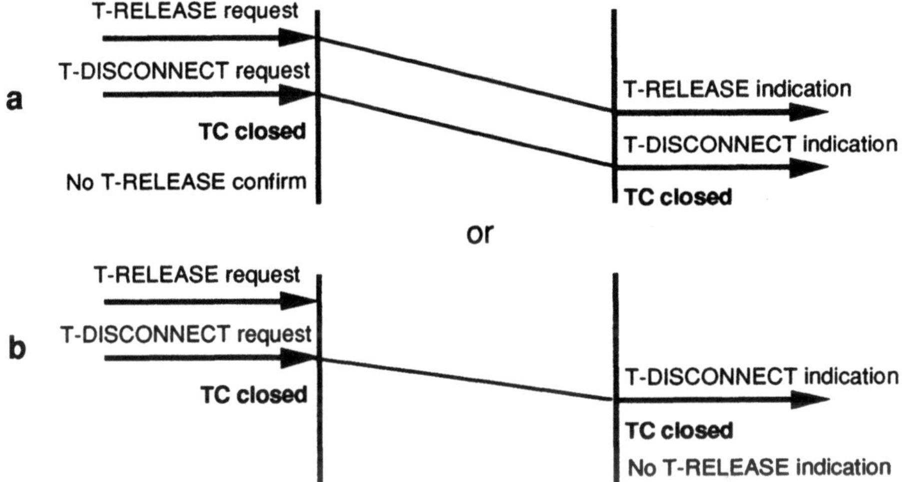

Fig. 3.1 Condition of competition for issuance between a T-RELEASE indication and a T-DISCONNECT indication:
    a. left user's output direction released gracefully (without confirmation to the requestor),
       left user's input direction released abruptly;
    b. both directions released abruptly

Now, what happens when an OSI95 TS user receives a T-RELEASE indication, indicating that its input direction on the TC has been released gracefully, and then decides to abruptly close its output direction ? The only way for the OSI95 TS user to close its output direction abruptly is to issue a T-DISCONNECT request. However, in this case, nothing prevents the corresponding T-DISCONNECT indication to occur at the remote TSAP before the expected T-RELEASE confirm (see figure 3.2). Clearly, it would be preferable for the remote OSI95 TS user to definitely get the confirmation of the graceful release of its output direction before receiving the T-DISCONNECT indication (Figure 3.2(a)). On the other hand, if the OSI95 TS provider was forced to always deliver the T-RELEASE confirm before delivering the T-DISCONNECT indication, we would no more strictly conform to the semantics of the T-

DISCONNECT primitives as specified in the current ISO/IEC Connection-Mode Transport Service (which we call the "pre-emptive right" semantics in the sequel).

This "pre-emptive right" semantics states indeed that once a T-DISCONNECT request has been invoked by a TS user, the TS provider should be in a position to issue the corresponding T-DISCONNECT indication to the peer TS user without any consideration for the pending actions, whatever they are. In the queue model describing in an abstract way the operation of a TC, this statement is formalized by the property that a Disconnect object takes precedence over any other object (see section 9 of [ISO 8072] for more details). Obviously, if we want the competition for issuance illustrated by figure 3.2, i.e. the competition for issuance between a T-RELEASE confirm prompted by a remote T-RELEASE indication on the one hand and a T-DISCONNECT indication prompted by a remote T-DISCONNECT request on the other hand, to always lead to the sequence of primitives (a), the Disconnect object must not keep precedence over the new ReleaseConfirm object in the queue model.

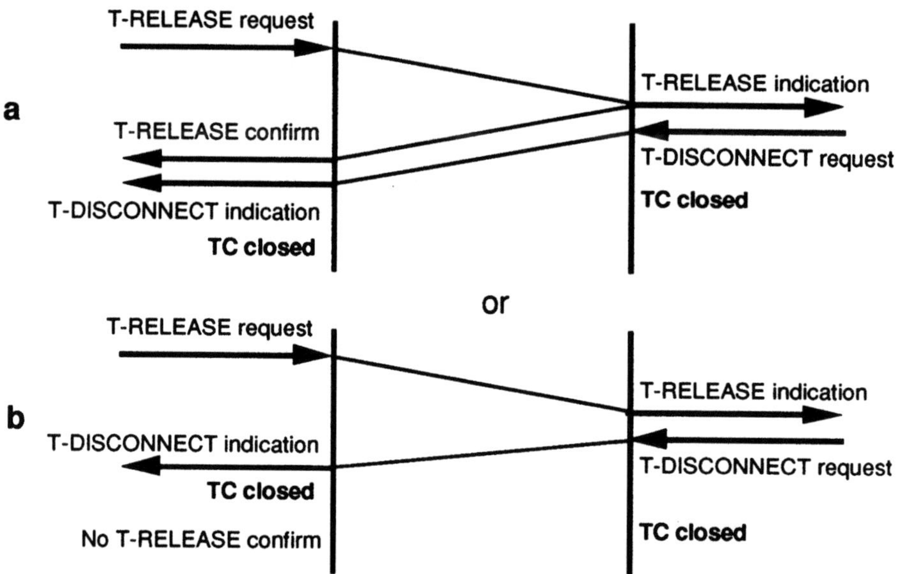

Fig. 3.2 Condition of competition for issuance between a T-RELEASE confirm and a T-DISCONNECT indication:
 a. left user's output direction released gracefully (WITH confirmation to the requestor), left user's input direction released abruptly;
 b. left user's output direction released gracefully (WITHOUT confirmation to the requestor), left user's input direction released abruptly

3.2 Investigation of an Attractive Solution

It is acceptable to modify the property of the Disconnect object in the queue model, by stating that a Disconnect object takes precedence over any other object *except the ReleaseConfirm object*, since this modification concerns a new object that does not exist in [ISO 8072] where no graceful TC release facility has been defined. However, for the OSI95 TS, we have preferred to investigate another solution to this problem of condition of competition for issuance between TC release primitives [BLL 92]. We find our solution more promising because we think it is elegant and because it does not require any change in the "pre-emptive right" semantics of the T-DISCONNECT primitives (in other words, it allows the Disconnect objet to keep precedence over any other object, without any exception, in the queue model).

This solution consists in an enhancement of the abrupt TC release facility to allow the abrupt closing of a single direction of data transfer on a TC. With such an enhanced facility, an OSI95 TS user that receives a T-RELEASE indication may invoke a T-DISCONNECT request (my output) to request the abrupt release of its output direction on the TC without any risk. Indeed, even if the corresponding T-DISCONNECT indication (your input) is issued at the peer TSAP prior to the T-RELEASE confirm, the TC will not be considered to be closed until the receipt of the T-RELEASE confirm, which suppresses the problem of condition of competition for issuance between these primitives (Figure 3.3).

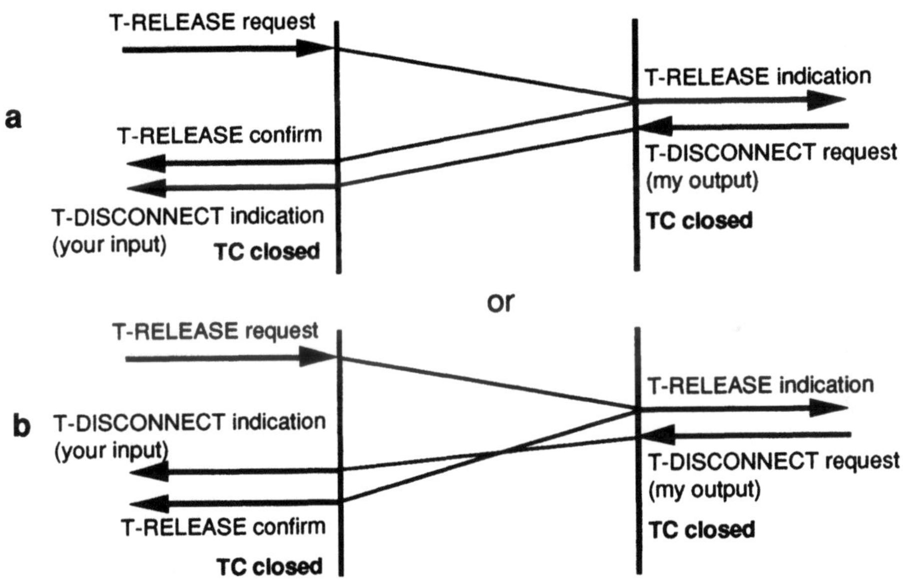

Fig. 3.3 Suppression of the problem of condition of competition for issuance shown in figure 3.2 thanks to the enhancement of the abrupt TC release facility: in both cases, the left user's output direction is released gracefully (WITH confirmation to the requestor) while its input direction is released abruptly

The enhancement of the abrupt TC release facility also presents an advantage in the situation shown in figure 3.1. The OSI95 TS user that decides to abruptly release its input direction on the TC, after the invocation of a T-RELEASE request to gracefully release its output direction on this TC, has to wait for the T-RELEASE confirm before issuing a usual bi-directional T-DISCONNECT request if it wants to be sure that its output direction is actually closed gracefully. But if it uses a uni-directional T-DISCONNECT request (my input) instead, it may issue this primitive before the receipt of the T-RELEASE confirm without jeopardizing the graceful release of its output direction (Figure 3.4).

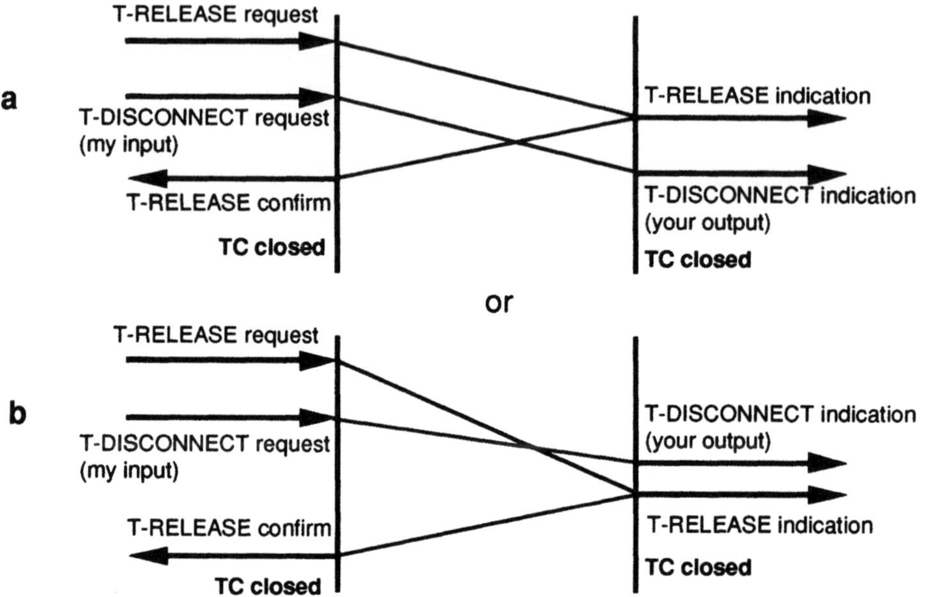

Fig. 3.4 Suppression of the problem of condition of competition for issuance shown in figure 3.1 thanks to the enhancement of the abrupt TC release facility: in both cases, the left user's output direction is released gracefully (WITH confirmation to the requestor) while its input direction is released abruptly

Additionally, the enhanced abrupt TC release facility gives the possibility to abruptly close one direction on a TC while still keeping the reverse direction open.

4 Enhancement of the Abrupt TC Release Facility for OSI95

As explained in the previous section, the proposed enhancement of the ISO/IEC abrupt TC release facility is the introduction of a new parameter for the T-DISCONNECT primitives: a "directions to close" parameter[1]. The enhanced abrupt

---

[1] In the OSI95 TS, another enhancement has been made to the ISO/IEC abrupt TC release facility in order to deal with the aspects relating to the semantics of the compulsory Quality of Service. This other enhancement is the introduction of special compulsory QOS report parameters in any T-DISCONNECT indication that is generated by the OSI95 TS provider to shut down the TC due to a compulsory QOS violation. The use of the compulsory QOS report parameters will not be detailed.

TC release facility may thus be used to release either or both direction(s) of data transfer of a TC abruptly. If an established TC is uni-directional, the release of its only open direction causes a complete TC release. If an established TC is bi-directional, the TC release completes when both open directions are released, together or separately.

Let us give some details about the new parameter. The parameters of the T-DISCONNECT primitives in the OSI95 TS, and the relations that have to be verified between them, are summarized in table 4.1.

	Primitives:	
	---	---
Parameters:	**T-DISCONNECT** request	**T-DISCONNECT** indication
Disconnect reason		M
Directions to close	M	M
Compulsory QOS report {...}[2]		P
OSI95 TS user-data	U	C(=)

Keys: {...} : generic denomination gathering several parameters;

      M : presence of this parameter or of this group of parameters in the primitive is mandatory;

      U : inclusion of this parameter or of this group of parameters in the request or response primitive is a choice made by the OSI95 TS user;

      C : presence of this parameter or of this group of parameters in the indication or confirm primitive is conditional, depending upon its inclusion in the preceding request or response primitive;

      P : presence of this parameter or of this group of parameters in the OSI95 TS provider-initiated indication primitive is conditional, depending upon the reason why the OSI95 TS provider generates the primitive;

      (=) : the value or option of the parameter in the primitive is identical to that of the corresponding parameter in the preceding primitive;

      Blank : the parameter is absent.

Table 4.1 Relations between the parameters of the T-DISCONNECT request and indication primitives

The value of the "directions to close" parameter in a T-DISCONNECT primitive obeys the following rules:

1) When an OSI95 TS user invokes a T-DISCONNECT request primitive prior to the successful completion of the TC set-up on its side (in order to abandon the TC set-up attempt on the calling side or to reject the TC set-up attempt on the called side), every direction of data transfer whose opening has been requested shall appear in the "directions to close" parameter.

2) In a T-DISCONNECT indication primitive that is initiated by the remote OSI95 TS user and that occurs prior to the successful completion of the TC set-up, every direction of data transfer whose opening has been requested shall appear in the "directions to close" parameter.

3) When an OSI95 TS user invokes a T-DISCONNECT request primitive after the successful completion of the TC set-up on its side, a direction of data transfer may appear in the "directions to close" parameter only if this direction is not yet considered as closed by the OSI95 TS user.

---

[2] Refer to the footnote on the previous page.

4) In a T-DISCONNECT indication primitive that is initiated by the remote OSI95 TS user and that occurs after the successful completion of the TC set-up, every direction of data transfer that appeared in the corresponding T-DISCONNECT request primitive and that is also not yet considered as closed by the OSI95 TS user to which the indication primitive is issued shall appear in the "directions to close" parameter.

5) An OSI95 TS provider-initiated T-DISCONNECT indication primitive always releases the TC completely. If the OSI95 TS provider initiates the T-DISCONNECT indication primitive prior to the completion of the TC set-up, every direction of data transfer whose opening has been requested shall appear in the "directions to close" parameter. If the OSI95 TS provider initiates the T-DISCONNECT indication primitive after the completion of the TC set-up, every direction of data transfer that is not yet considered as closed by the OSI95 TS user to which the indication primitive is issued shall be indicated in the "directions to close" parameter.

5 Conclusions

In the framework of the Connection-Mode Transport Service developed within the OSI95 project, we have advocated important additions in the field of TC release.

The most important addition is the definition of a new graceful TC release facility. A bi-directional TC in OSI95 is viewed as made up of two simplex virtual circuits. In this line, our investigations have led us to a scenario of graceful TC release where each direction of data transfer on a TC is released quite independently of the reverse one. The graceful release of the requestor's ouput direction of data transfer is confirmed to this requestor, and is thus based on a 3-primitive sequence.

By examining how the new T-RELEASE primitives can interfere with the T-DISCONNECT primitive of the existing ISO/IEC abrupt TC release facility, we have highlighted some problems due to conditions of competition for issuance between these primitives.

We have proposed an elegant solution which respects the "pre-emptive right" semantics of the T-DISCONNECT primitives in the current ISO/IEC standard. This solution consists mainly in adding a parameter to the T-DISCONNECT primitives in order to allow the abrupt release of a single direction of data transfer on a TC.

We deem that the combination of the new graceful TC release facility with the enhanced abrupt TC release facility will be able to cover most of the needs of today's transport service users.

References

[BLL 92]   Y. Baguette, L. Léonard, G. Leduc, A. Danthine, O. Bonaventure, "OSI95 Enhanced Transport Facilities and Functions", Technical report for ESPRIT II project 5341 (OSI95), Ref.: OSI95/Deliverable ULg-A/P/V1, Université de Liège, Institut Montefiore B28, B-4000 Liège, Belgium, Dec. 1992.

[BLL 93]   Y. Baguette, L. Léonard, G. Leduc, A. Danthine, "The OSI95 Connection-Mode Transport Service", *The OSI95 Transport Service with Multimedia Support on HSLAN's and B-ISDN*, A. Danthine, ed., Springer Verlag, 181-198

[DBL 92]   A. Danthine, Y. Baguette, G. Leduc, "Issues Surrounding the Specification of High-Speed Transport Service and Protocol", Technical report for ESPRIT II project 5341 (OSI95), Ref.: OSI95/ULg/A/15/TR/P/V2 (also submitted as ISO/IEC JTC1/SC6 N 7312 through the Belgian National Body at the SC6 interim meeting, Paris, France, Feb. 10-13, 1992), Université de Liège, Institut Montefiore B28, B-4000 Liège, Belgium, Jan. 1992.

[ISO 7498]   ISO/IEC JTC1: "Information Processing Systems - Open Systems Interconnection - Basic Reference Model", ISO 7498-1, 1984.

[ISO 8072]   ISO/IEC JTC1: "Information Processing Systems - Open Systems Interconnection - Transport Service Definition", ISO 8072, 1986.

[ISO 8073]   ISO/IEC JTC1: "Information Processing Systems - Open Systems Interconnection - Connection-Oriented Transport Protocol Specification", ISO 8073, 1992.

[ISO 8326]   ISO/IEC JTC1: "Information Processing Systems - Open Systems Interconnection - Basic Connection-Oriented Session Service Definition", ISO 8326, 1987.

[ISO SD9]   ISO/IEC JTC1/SC21 (Information Retrieval, Transfer & Management for OSI): "Standing Document 9 - Approved Commentaries on the Basic Reference Model for Open Systems Interconnection", ISO/IEC JTC1/SC21 N 5217, June 19, 1990.

[RFC 793]   Network Information Center: "Transmission Control Protocol", J.B. Postel, ed., NIC Standard RFC-793, Sept. 1981.

[PEI 3.6]   Protocol Engines, Inc.: "XTP Protocol Definition - Revision 3.6", Ref.: PEI 92-10, Jan. 11, 1992.

The OSI95 Connectionless-Mode Transport Services

Yves Baguette
Research Assistant of the Belgian National Fund for Scientific Research (F.N.R.S.).
Université de Liège, Institut d'Electricité Montefiore B28, B-4000 Liège 1, Belgium
Email: baguette@montefiore.ulg.ac.be

We present the connectionless-mode services of the OSI95 Transport Service. Three services are thus examined: an unacknowledged connectionless-mode service, an acknowledged connectionless-mode service and a request/response connectionless-mode service.

Keywords: OSI95, Transport Service, Connectionless-Mode, Unacknowledged Service, Acknowledged Service, Request/Response Service

1 Introduction

This paper covers the three types of connectionless-mode services that have been defined as parts of the entire OSI95 Transport Service (OSI95 TS) [BLL 92]:

- an unacknowledged (or basic) connectionless-mode service, which is an updated version of the standard ISO/IEC Connectionless-Mode Transport Service;
- a new acknowledged connectionless-mode service;
- a new request/response connectionless-mode service.

2 The Unacknowledged Connectionless-Mode Transport Service

2.1 Service Primitives and Parameters

This service is simply an updated version of the ISO/IEC Connectionless-Mode Transport Service [ISO 8072/1] (Figure 2.1).

Fig. 2.1 Normal sequence of T-UNITDATA primitives

The only improvement regards the semantics of the transit delay performance Quality of Service (QoS) parameter in the T-UNITDATA primitives. This semantics has been indeed modified, with the replacement of the old concept of "best-effort" QoS by the concept of compulsory QoS [DBL 93], as explained below.

The parameters of the T-UNITDATA primitives in the OSI95 TS, and the relations that have to be verified between them, are summarized in table 2.1.

	Primitives:	
Parameters:	T-UNITDATA request	T-UNITDATA indication
Calling address	M	M(=)
Called address	M	M(=)
QoS {...}	M	M
TSDU	M	M(=)

Keys: {...} : generic denomination gathering several parameters;
  M : presence of this parameter or of this group of parameters in the primitive is mandatory;
  (=) : the value or option of the parameter in the primitive is identical to that of the corresponding parameter in the preceding primitive;

Table 2.1 T-UNITDATA primitives and parameters

2.2 Details on some QoS Parameters

The only QoS parameters used by the OSI95 Unacknowledged Connectionless-Mode Transport Service are the transit delay, the TC protection and the TC priority (the last two ones without any change in their semantics for the moment). The residual error probability included in the ISO/IEC standard has been removed because we deem that it is unmanageable and thus useless.

Let us remind that the transit delay is defined as *the time elapsed between the occurrence of a T-UNITDATA request at a TSAP and the occurrence of the corresponding T-UNITDATA indication at the peer TSAP*. Of course the transit delay has a sense only when the T-UNITDATA indication occurs.

The single value specified for the transit delay parameter of the T-UNITDATA request in the OSI95 TS is a maximum compulsory value which is replicated unchanged in the T-UNITDATA indication. This maximum compulsory value is such that it shall never be exceeded. If the OSI95 TS provider has not been able to deliver the TSDU at the latest at time $T_0 + \Delta t_{unack\_transit\_max}$, where T_0 is the time of occurrence of the T-UNITDATA request and $\Delta t_{unack\_transit\_max}$ is the maximum compulsory transit delay, the T-UNITDATA indication may no more be issued (so that no transit delay is actually measured) and the TSDU must be considered as lost.

In our opinion, adding the possiblity to specify a threshold value for the transit delay parameter would be a nonsense. In fact, it may be expected that, in most cases, the calling OSI95 TS user be more interested than the called OSI95 TS user in being informed of a threshold value violation. Unfortunately, it is not possible to envisage a threshold violation report to the calling OSI95 TS user without changing ipso facto the unacknowledged connectionless mode into an acknowledged connectionless mode.

3 The Acknowledged Connectionless-Mode Transport Service

3.1 Purpose of a Connectionless-Mode Service with Acknowledgement

With an unacknowledged connectionless-mode service, the calling service user has of course no means to be informed by the service provider of the outcome of its service invocation. So it has no chance to know definitely whether its submitted SDU has been delivered correctly, except by relying on the protocol between the service users to get the information. Such a chance can be given directly at the service level if a new service is defined where the correct delivery of the SDU to the called service user gives rise to the issuance of an acknowledgement to the calling service user.

Unfortunately, since this new service has to be provided by operating above unreliable communication channels, it may happen that the acknowledgement cannot be delivered to the calling service user whereas its SDU has been transferred correctly to the called service user. In other words, the delivery of the acknowledgement to the calling service user has to be regarded as a firm guarantee on the success of the delivery of the SDU to the called service user, but the absence of acknowledgement (which may lead to the delivery of a negative acknowledgement after a while) does not necessarily mean that the delivery of the SDU to the called service user has failed.

At the OSI95 TS level, this type of new service, that merely ensures that the successful delivery of the TSDU is acknowledged correctly when no particular problem takes place (which is expected to be the most likely case), may appear attractive to some OSI95 TS users because it meets the needs of the application. More information on the kinds of applications which this new service is meant for is given in section 3.4.

3.2 Service Primitives and Parameters

This acknowledged service is an extension of the unacknowledged one [Dan 92][DBL 92]. The T-UNITDATA primitives are renamed T-ACKDATA request and T-ACKDATA indication. A third primitive denoted T-ACKDATA confirm is needed that confirms to the calling OSI95 TS user the delivery of the T-ACKDATA indication by the OSI95 TS provider to the called OSI95 TS user. Another primitive denoted T-REJECT-ACKDATA indication is also added that is issued to the calling OSI95 TS user instead of the T-ACKDATA confirm in case of failure.

Furthermore, the set of QoS parameters is slightly modified with the introduction of a maximum service completion delay QoS parameter, and the semantics of the transit delay performance QoS parameter is enriched with the concept of threshold QoS [DBL 93]. Here indeed, it is possible to envisage a threshold violation report to the calling OSI95 TS user in the T-ACKDATA confirm in addition to the same threshold violation report to the called OSI95 TS user in the T-ACKDATA indication.

The T-ACKDATA confirm contains the following parameters:
- the responding transport address parameter;
- threshold report parameters that indicates whether the measured transit delay value was greater than the specified threshold value and that gives that measured value when it is actually greater[1].

The T-REJECT-ACKDATA indication only contains a reason parameter.

The parameters of the T-(REJECT-)ACKDATA primitives in the OSI95 TS, and the relations that have to be verified between them, are summarized in table 3.1.

	Primitives:			
Parameters:	T-ACKDATA request	T-ACKDATA indication	T-ACKDATA confirm	T-REJECT -ACKDATA indication
Calling address	M	M(=)		
Called address	M	M(=)		
Responding address			M	
QoS {...}	M	M		
TSDU	M	M(=)		
Threshold report {...}		M	M(=)	
Reason				M

Keys: {...} : generic denomination gathering several parameters;
M : presence of this parameter or of this group of parameters in the primitive is mandatory;
(=) : the value or option of the parameter in the primitive is identical to that of the corresponding parameter in the preceding primitive;

Table 3.1 T-(REJECT-)ACKDATA primitives and parameters

3.3 Details on some QoS Parameters

The QoS parameters used for the OSI95 Acknowledged Connectionless-Mode Transport Service are a service completion delay, the transit delay, the TC protection and the TC priority (the last two ones without any change in their classical semantics for the moment).

The transit delay is obviously defined as *the time elapsed between the occurrence of a T-ACKDATA request at a TSAP and the occurrence of the corresponding T-ACKDATA indication at the peer TSAP*. Of course the transit delay has a sense only when the T-ACKDATA indication occurs.

---

[1] The choice to communicate the measured transit delay value only when it is greater than the threshold value permits a strict alignment with the concept of threshold QoS as it is perceived in connection mode. In connection mode indeed, the actual measured value is reported only if it violates the threshold value since no threshold report occurs otherwise. Here however, the measured transit delay value could be systematically included in the threshold report in the T-ACKDATA confirm without putting any additional burden on the ETS provider. If this interesting alternative was adopted ultimately, we should question the interest of keeping a threshold value violation flag in the threshold report as the comparison between the threshold value and the actual measured value could be made readily by the calling ETS user itself.

As said above, two values are specified in the transit delay parameter of the T-ACKDATA request: a threshold value and a maximum compulsory value. These values are replicated unchanged in the T-ACKDATA indication. The following relationship must always be verified between them: $0 \leq$ imposed threshold transit delay value $\leq$ imposed maximum compulsory transit delay value. The maximum compulsory value is such that it shall never be exceeded. If the OSI95 TS provider has not been able to deliver the TSDU at the latest at time $T_0 + \Delta t_{ack\_transit\_max}$, where T_0 is the time of occurrence of the T-ACKDATA request and $\Delta t_{ack\_transit\_max}$ is the maximum compulsory transit delay, the T-ACKDATA indication may no more be issued (so that no transit delay is actually measured) and the TSDU must be considered as lost. Clearly, in such a case, a T-REJECT-ACKDATA indication shall be issued to the calling OSI95 TS user. The threshold value is such that if it is exceeded this fact shall be reported to the called OSI95 TS user by the threshold report parameters of the T-ACKDATA indication and to the calling OSI95 TS user by the same parameters of the T-ACKDATA confirm provided this primitive is issued.

The service completion delay, as for it, is defined as *the time elapsed between the occurrence of a T-ACKDATA request at a TSAP and the occurrence of the corresponding T-ACKDATA confirm at the same TSAP*. Of course the service completion delay has a sense only when the T-ACKDATA confirm occurs.

The introduction of a service completion delay aims at limiting on the time axis the possibilities of interactions associated with a given service invocation on the calling side. For this purpose, a single maximum compulsory value is specified for the service completion delay parameter of the T-ACKDATA request and is replicated unchanged in the T-ACKDATA indication. The maximum compulsory service completion delay value is such that it shall never be exceeded. If the OSI95 TS provider has not been able to issue the T-ACKDATA confirm at the latest at time $T_0 + \Delta t_{ack\_service\_max}$, where T_0 is the time of occurrence of the T-ACKDATA request and $\Delta t_{ack\_service\_max}$ is the maximum compulsory service completion delay, the T-ACKDATA confirm primitive may no more be issued (so that no service completion delay is actually measured) and a T-REJECT-ACKDATA indication shall be issued instead as soon as possible after time $T_0 + \Delta t_{ack\_service\_max}$. Obviously the maximum compulsory service completion delay value must be greater than or equal to the maximum compulsory transit delay value described above.

3.4 Possible Sequences of Service Primitives

Either a T-ACKDATA confirm or a T-REJECT-ACKDATA indication shall be issued to the calling OSI95 TS user in reply to its T-ACKDATA request. This means that whenever no T-ACKDATA confirm is returned, a T-REJECT-ACKDATA indication has to be returned instead. The reason parameter of the T-REJECT-ACKDATA indication may only take one of the three following values:

1) OSI95 TS provider initiated due to an exceeded maximum compulsory service completion delay;
2) OSI95 TS provider initiated due to an exceeded maximum compulsory transit delay;

3) OSI95 TS provider initiated due to any other internal reason which may be of
   transient or permanent nature.

Figure 3.1 illustrates the three possible scenarios of the OSI95 Acknowledged
Connectionless-Mode Transport Service.

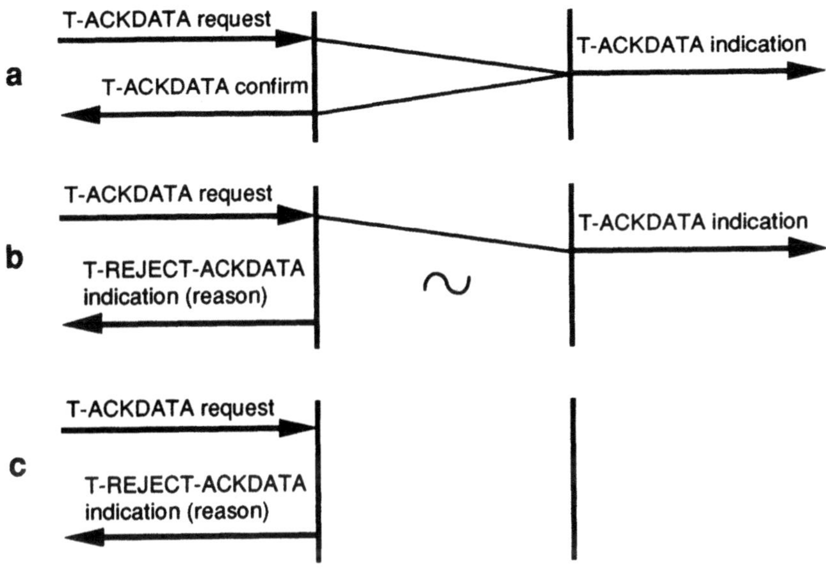

Fig. 3.1 Possible sequences of T-(REJECT-)ACKDATA primitives

Figure 3.1(a) (successful scenario) shows that the receipt of a T-ACKDATA
confirm at the calling TSAP gives the firm confirmation of the delivery of the TSDU
to the called OSI95 TS user. By contrast, as illustrated by figure 3.1(b), it is possible
that the OSI95 TS provider delivers a T-REJECT-ACKDATA indication to the
calling OSI95 TS user whereas the transfer of the TSDU has been successful. This
will happen for instance when the maximum compulsory service completion delay is
exceeded. So no definite conclusion can be drawn by the calling OSI95 TS user as for
the success of the TSDU transfer when the OSI95 TS provider issues a T-REJECT-
ACKDATA indication in response to a T-ACKDATA request, except if the reason
parameter in the T-REJECT-ACKDATA indication is the reason 2) indicating without
any doubt that the T-ACKDATA indication cannot have been issued. This explains
why the T-REJECT-ACKDATA indication should be generally considered as an
absence of confirmation rather than as a negative confirmation.

The OSI95 Acknowledged Connectionless-Mode Transport Service is therefore
intended for applications that
- either can tolerate duplicated TSDUs on the called side because the processing of
  the duplicate of a TSDU does not have any prejudicial consequence on the future
  behaviour of the called OSI95 TS user (such a TSDU is then said to be nil-potent),
- or provide in the protocol between the OSI95 TS users a mechanism to detect (and
  thus eliminate) the duplicates.

In fact, in both cases, the receipt of a duplicate of a TSDU by the called OSI95 TS user, resulting from the fact that the calling OSI95 TS user has re-submitted a TSDU whose previous submission has led to the scenario of figure 3.1(b), does not entail problems.

With the current ISO/IEC Connection-Mode Transport Service [ISO 8072], the TS provider may never give by itself the guarantee that TSDUs submitted via T-DATA requests to be sent on a given TC will actually be delivered to the peer user of the TC, since the TC release is always abrupt. It is therefore up to the TS users to make sure that there are no more data in transit on the TC before asking for its unconditional and therefrom abrupt release. But here, with the OSI95 Acknowledged Connectionless-Mode Transport Service, the OSI95 TS provider can give, via a T-ACKDATA confirm, the guarantee that the submitted TSDU has actually been delivered. Otherwise, like what happens in connection mode when a T-DISCONNECT request is invoked or a T-DISCONNECT indication is received after a T-DATA request is invoked, there may be an uncertainty as for the correct delivery of the submitted TSDU when a T-REJECT-ACKDATA indication is received in response to a T-ACKDATA request.

The OSI95 Unacknowledged Connectionless-Mode Transport Service provider may in general perform any of the following actions: discard submitted TSDUs, duplicate submitted TSDUs and change any order of independent submitted TSDUs into a different order of delivered TSDUs. The possibility of duplication means that a single T-UNITDATA request may give rise to several consecutive T-UNITDATA indication at the called TSAP. Obviously, the OSI95 Acknowledged Connectionless-Mode Transport Service provider may lose submitted TSDUs (nevertheless, this will be reported by T-REJECT-ACKDATA indications) or may change any order of independent submitted TSDUs into a different order of delivered TSDUs. What about the duplication of submitted TSDUs? Considering that, as explained above, the OSI95 Acknowledged Connectionless-Mode Transport Service is really useful only for applications that can cope with duplicated TSDUs (since, due to the scenario of figure 3.1(b), the behaviour of the calling OSI95 TS user may lead to the generation of duplicates), it is needless to strive to avoid the generation of duplicates by the OSI95 TS provider. This is why we authorize that a single T-ACKDATA request gives rise to several consecutive T-ACKDATA indications at the called TSAP. However, a single T-ACKDATA confirm is allowed at the calling TSAP in any case.

3.5 Service Invocation Identification Mechanism

The T-ACKDATA confirm or the T-REJECT-ACKDATA indication is obviously issued at the same TSAP as the T-ACKDATA request (i.e. at the calling TSAP).

Clearly, an identification mechanism has to be provided at the OSI95 TS interface if an OSI95 TS user and the OSI95 TS provider need to distinguish between several invocations of the Acknowledged Connectionless-Mode Transport Service at a same TSAP. We have chosen to provide a TAEP (Transport Association Endpoint) identification mechanism locally. This TAEP identification mechanism has a role similar to that of the TCEP (Transport Connection Endpoint) identification mechanism used locally to make the distinction between several TCs at a same TSAP in connection mode. All T-ACKDATA request, T-ACKDATA confirm and T-REJECT-ACKDATA indication have then to make use of the TAEP identification mechanism at the calling TSAP to identify the invocation of the service to which they apply. In other words, the function of the TAEP identification mechanism utilized at a given TSAP is to relate a T-ACKDATA confirm or T-REJECT-ACKDATA indication primitive with the corresponding T-ACKDATA request primitive. This implicit identification, which is fully consistent with the OSI RM [ISO 7498], is not shown as a parameter of the aforementioned primitives and should not be confused with the address parameters.

We strongly favour this approach against the approach that would consist in relying on an explicit user-managed parameter to identify the service invocation.

4 The Request/Response Connectionless-Mode Transport Service

4.1 Purpose of a Connectionless-Mode Service with Request/Response

An acknowledged connectionless-mode service does not allow the called service user to send back a SDU in reply to the SDU sent by the calling service user. Such a generic two-SDU exchange could be performed by invoking twice the acknowledged (or even the unacknowledged) connectionless-mode service (i.e. one invocation for each service user), but a service that automatically associates the two SDUs pertaining to a same request/response operation would be preferable from the service user's standpoint.

If the two SDUs of a request/response exchange are associated automatically, a specific acknowledgement at the service level is no more necessary to confirm the correct delivery of the SDU sent in request since the correct delivery of the associated SDU sent in response can play this role of acknowledgement. From there on, if we agree with the idea that the service user which is first and foremost concerned with the success of the complete request/response exchange is the calling service user rather than the called (and also responding) service user, we may accept not to acknowledge at the service level the correct delivery of the TSDU sent in response. Indeed, as the correct delivery of the TSDU sent in response constitutes the only acknowledgement of the correct delivery of the TSDU sent in request, the calling service user is expected to issue its request again when no response is delivered correctly.

At the OSI95 TS level, this type of new service may appear attractive to some OSI95 TS users because it meets the needs of the application, in particular for certain client/server applications. More information on the kinds of applications which this new service is meant for is given in section 4.4.

4.2 Service Primitives and Parameters

This request/response service is a further extension of the acknowledged one [Dan 92][DBL 92]. The three T-ACKDATA primitives are renamed T-REQRES request, T-REQRES indication and T-REQRES confirm. A fourth primitive denoted T-

REQRES response is needed that allows the called OSI95 TS user to submit a response TSDU[2]. Another primitive denoted T-REJECT-REQRES indication plays the same role as the T-REJECT-ACKDATA indication in the acknowledged service.

Moreover, the set of QoS parameters is modified with the removal of the transit delay performance QoS parameter. In fact, whereas the actual goal of an invocation of the acknowledged service is the delivery of the TSDU to the called OSI95 TS user, it is the receipt of the response TSDU by the calling OSI95 TS user in reply to the its request TSDU that must be considered as the actual goal of an invocation of the request/response service. This implies that the ability to separately control the transit delay of the request TSDU should not really be a useful service feature from the OSI95 TS users' point of view.

The T-REQRES response and confirm contain the following parameters:
- the responding transport address parameter;
- the response TSDU parameter.

The T-REJECT-REQRES indication only contains a reason parameter.

The parameters of the T-(REJECT-)REQRES primitives in the OSI95 TS, and the relations that have to be verified between them, are summarized in table 4.1.

	Primitives:				
Parameters:	T-REQRES request	T-REQRES indication	T-REQRES response	T-REQRES confirm	T-REJECT -REQRES indication
Calling address	M	M(=)			
Called address	M	M(=)			
Responding address			M	M(=)	
QoS {...}	M	M	M	M(=)	
TSDU	M	M(=)	M	M(=)	
Reason					M

Keys: {...} : generic denomination gathering several parameters;
　　　M : presence of this parameter or of this group of parameters in the primitive is mandatory;
　　　(=) : the value or option of the parameter in the primitive is identical to that of the corresponding parameter in the preceding primitive;

Table 4.1 T-(REJECT-)REQRES primitives and parameters

4.3 Details on some QoS Parameters

The QoS parameters used for the OSI95 Request/Response Connectionless-Mode Transport Service are the service completion delay, the TC protection and the TC priority (the last two ones without any change in their classical semantics for the moment).

---

[2] From now on, the TSDU in the T-REQRES request and indication primitives will be called the request TSDU, and the TSDU in the T-REQRES response and confirm primitives will be called the response TSDU.

The service completion delay is defined as *the time elapsed between the occurrence of a T-REQRES request at a TSAP and the occurrence of the corresponding T-REQRES confirm at the same TSAP*. Of course the service completion delay has a sense only when the T-REQRES confirm occurs.

The introduction of a service completion delay aims at limiting on the time axis the possibilities of interactions associated with a given service invocation on the calling side but also on the called side. For this purpose, a single maximum compulsory value is specified for the service completion delay parameter of the T-REQRES request and is replicated unchanged in the T-REQRES indication. The maximum compulsory service completion delay value is such that it shall never be exceeded. If the OSI95 TS provider has not been able to issue the T-REQRES confirm at the latest at time $T_0 + \Delta t_{reqres\_service\_max}$, where T_0 is the time of occurrence of the T-REQRES request and $\Delta t_{reqres\_service\_max}$ is the maximum compulsory service completion delay, the T-REQRES confirm primitive may no more be issued (so that no service completion delay is actually measured) and a T-REJECT-REQRES indication shall be issued instead as soon as possible after time $T_0 + \Delta t_{reqres\_service\_max}$. A T-REJECT-REQRES indication shall also be issued as soon as possible to the called OSI95 TS user if this user has already received the T-REQRES indication and has not responded yet.

4.4 Possible Sequences of Service Primitives

Either a T-REQRES confirm or a T-REJECT-REQRES indication shall be issued to the calling OSI95 TS user in reply to its T-REQRES request. This means that whenever no T-REQRES confirm is returned, a T-REJECT-REQRES indication has to be returned instead. The reason parameter of the T-REJECT-REQRES indication may only take one of the two following values:

1) OSI95 TS provider initiated due to an exceeded maximum compulsory service completion delay;
2) OSI95 TS provider initiated due to any other internal reason which may be of transient or permanent nature.

On the called side, following the issuance of a T-REQRES indication, either a T-REQRES response shall be invoked by the OSI95 TS user or a T-REJECT-REQRES indication shall be issued by the OSI95 TS provider.

Figure 4.1 illustrates the four possible scenarios of the OSI95 Request/Response Connectionless-Mode Transport Service.

Figure 4.1(a) (successful scenario) shows that the receipt of a T-REQRES confirm at the calling TSAP gives the firm confirmation of the delivery of the request TSDU to the called OSI95 TS user. By contrast, as illustrated by figure 4.1(b)&(c), it is possible that the OSI95 TS provider delivers a T-REJECT-REQRES indication to the calling OSI95 TS user whereas the transfer of the request TSDU has been successful. This may happen for instance when the maximum compulsory service completion delay is exceeded. So no definite conclusion can be drawn by the calling OSI95 TS user as for the success of the request TSDU transfer when the OSI95 TS provider issues a T-REJECT-REQRES indication in response to a T-RESREQ request. Anyway, the calling OSI95 TS user is normally not interested in the success of the transfer of its request TSDU when this success does not result in the successful receipt of the response TSDU. The calling OSI95 TS user must however be aware of the fact that its request TSDU may have been delivered correctly to the called OSI95 TS user even when it does not receive the response TSDU.

The OSI95 Request/Response Connectionless-Mode Transport Service is therefore intended for request/response-oriented applications that

- either can tolerate duplicated request TSDUs on the called side because the processing of the duplicate of a request TSDU does not have any prejudicial consequence on the future behaviour of the called OSI95 TS user (such a request TSDU is then said to be nil-potent),
- or provide in the protocol between the OSI95 TS users a mechanism to detect (and thus eliminate) the duplicates.

In fact, in both cases, the receipt of a duplicate of a request TSDU by the called OSI95 TS user, resulting from the fact that the calling OSI95 TS user has re-submitted a request TSDU whose previous submission has led to the scenario of figure 4.1(b) or (c), does not entail troubles.

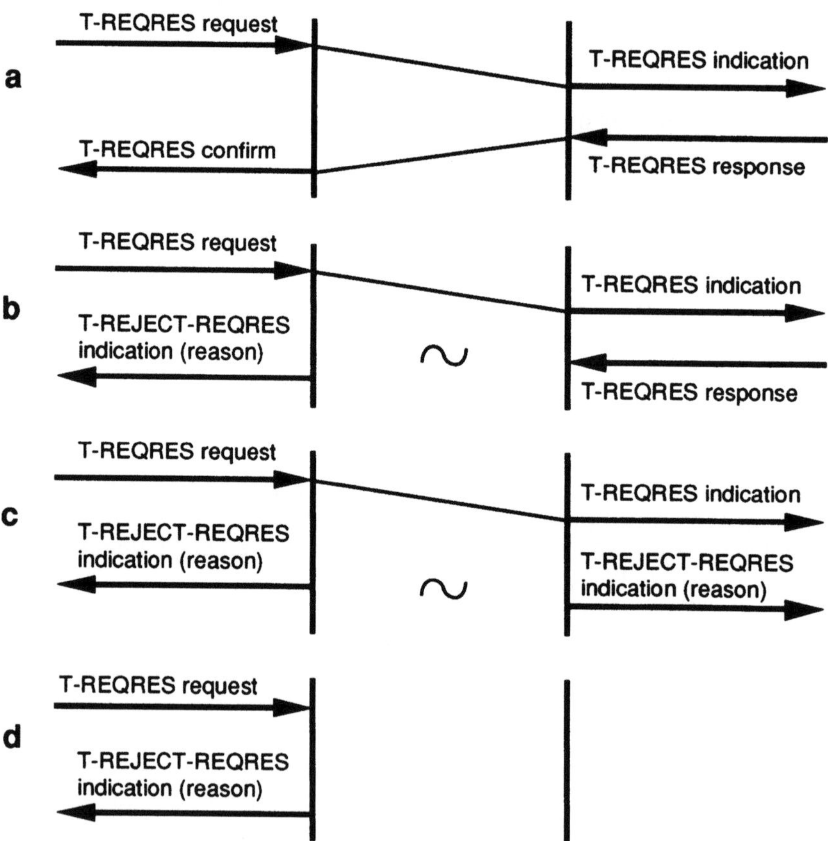

Fig. 4.1 Possible sequences of T-(REJECT-)REQRES primitives

The OSI95 Unacknowledged Connectionless-Mode Transport Service provider may in general perform any of the following actions: discard submitted TSDUs, duplicate submitted TSDUs and change any order of independent submitted TSDUs into a different order of delivered TSDUs. Obviously, the OSI95 Request/Response Connectionless-Mode Transport Service provider may lose submitted TSDUs (nevertheless, this will be reported by T-REJECT-REQRES indications) or may change any order of independent submitted TSDUs into a different order of delivered TSDUs. What about the duplication of submitted TSDUs? Considering that, as explained above, the OSI95 Request/Response Connectionless-Mode Transport Service is really useful only for applications that can cope with duplicated request TSDUs (since, due to the scenario of figure 4.1(b) or (c), the behaviour of the calling OSI95 TS user may lead to the generation of duplicates of requests), it is needless to strive to avoid the generation of duplicates by the OSI95 TS provider. This is why we authorize that a single T-REQRES request gives rise to several consecutive T-REQRES indications at the called TSAP. Each T-REQRES indication issued at the called TSAP has to be followed by an associated T-REQRES response or T-REJECT-REQRES indication at this called TSAP. However, a single T-REQRES confirm is allowed at the calling TSAP in any case.

4.5 Service Invocation Identification Mechanism

Again, an identification mechanism has to be provided at the OSI95 TS interface to associate a T-REQRES confirm (or a T-REJECT-REQRES indication) with the corresponding T-REQRES request, as well as to associate a T-REQRES response (or a T-REJECT-REQRES indication) with the corresponding T-REQRES indication. We have chosen to provide a TAEP (Transport Association Endpoint) identification mechanism locally, as already explained in section 3.5.

We strongly favour this approach against the approach that would consist in relying on an explicit user-managed parameter (i.e. a "transaction identifier") to identify the service invocation.

5 Conclusions

In the framework of the OSI95 Transport Service, we have defined three types of connectionless-mode services.

The unacknowledged connectionless-mode service is merely an updated version of the ISO/IEC standard [ISO 8072/1].

By contrast, the acknowledged and the request/response connectionless-mode services are quite new ones and may be regarded as two successive extensions of the unacknowledged connectionless-mode service.

The acknowledged connectionless-mode service is a confirmed service. Its successful scenario is therefore based on a 3-primitive sequence. The service provider issues a confirmation primitive to the calling service user to confirm the issuance of the indication primitive to the called service user.

In the request/response connectionless-mode service, the service-provider-generated confirmation primitive disappears in favour of a called-service-user-generated response primitive which gives rise to a confirmation primitive to the calling service user. The successful scenario of this request/response connectionless-mode service is therefore based on a classical 4-primitive sequence.

In order to distinguish between several invocations of the acknowledged or the request/response connectionless-mode service at a TSAP, we have decided to provide a TAEP (Transport Association Endpoint) identification mechanism locally at the OSI95 TS interface. This concept of TAEP is similar to the concept of TCEP (Transport Connection Endpoint) used in connection mode. Let us stress however that, unlike the well-established concept of connection endpoint, the concept of association endpoint has never been defined in the OSI RM [ISO 7498] which only allows for basic (unacknowledged) services in connectionless mode.

References

[BLL 92]   Y. Baguette, L. Léonard, G. Leduc, A. Danthine, O. Bonaventure, "OSI95 Enhanced Transport Facilities and Functions", Technical report for ESPRIT II project 5341 (OSI95), Ref.: OSI95/Deliverable ULg-A/P/V1, Université de Liège, Institut Montefiore B28, B-4000 Liège, Belgium, Dec. 1992.

[Dan 92]   A. Danthine, "A New Transport Protocol for the Broadband Environment", in: A. Casaca, ed., IFIP International Workshop on Broadband Communication (Estoril, Portugal, Jan. 20-22, 1992), Elsevier Science Publishers B.V. (North-Holland), Amsterdam, 337-360.

[DBL 92]   A. Danthine, Y. Baguette, G. Leduc, "Issues Surrounding the Specification of High-Speed Transport Service and Protocol", Technical report for ESPRIT II project 5341 (OSI95), Ref.: OSI95/ULg/A/15/TR/P/V2 (also submitted as ISO/IEC JTC1/SC6 N 7312 through the Belgian National Body at the SC6 interim meeting, Paris, France, Feb. 10-13, 1992), Université de Liège, Institut Montefiore B28, B-4000 Liège, Belgium, Jan. 1992.

[DBL 93]   A. Danthine, O. Bonaventure, G. Leduc, "The QoS Enhancements in OSI95", *The OSI95 Transport Service with Multimedia Support on HSLAN's and B-ISDN*, A. Danthine, ed., Springer Verlag, 125-150

[ISO 7498]   ISO/IEC JTC1: "Information Processing Systems - Open Systems Interconnection - Basic Reference Model", ISO 7498-1, 1984.

[ISO 8072]   ISO/IEC JTC1: "Information Processing Systems - Open Systems Interconnection - Transport Service Definition", ISO 8072, 1986.

[ISO 8072/1] ISO/IEC JTC1: "Information Processing Systems - Open Systems Interconnection - Transport Service Definition - Addendum 1: Connectionless-Mode Transmission", ISO 8072 ADD1, 1986.

On the Provision of a Fast Connect Facility in a Connection-Mode Transport Service

Guy Leduc[1] and André Danthine
[1] Research Associate of the Belgian National Fund for Scientific Research (F.N.R.S.)
Université de Liège, Institut d'Electricité Montefiore, B 28, B-4000 Liège 1, Belgium
Email: leduc@montefiore.ulg.ac.be

A "fast connect" facility in a connection-mode service allows the calling service user to send service data units without waiting for the round trip delay associated with the successful establishment of the connection. In this paper we investigate how such a fast connect facility can be added to a connection-mode transport service. Starting from the eXpress Transfer Protocol (XTP) which is at the origin of the concept, we analyse the High Speed Transport Service (HSTS) and study how it can be improved by the introduction of a listening interaction at the called side. After having criticized these two transport facilities, we finally propose three possible models of the fast connect facility. The first one is the 4-primitive connection establishment from the Broadband Transport Service (BTS) and the last ones are two 3-primitive connection establishments.

Keywords: Transport Service, Connection-mode, Fast Connect

1 Introduction

In today's high-speed networks, the 'throughput × round-trip delay' (denoted T × RTD) product may easily be of the order of several hundred Kbits or even several Mbits. Such values of the T × RTD product imply that an opportunity for sending a sizeable amount of data will be lost if data can only be sent when the negotiation of the Transport Connection (TC) is complete. The figures 1a and 1b depict the classical 4-primitive exchange of a connection establishment in case of success and refusal respectively. The ISO Transport Service ([ISO 8072]) is the standard example of such a TC establishment.

To allow the Transport Service (TS) users to send data without waiting for a RTD, the eXpress Transfer Protocol (XTP) [PEI XTP] introduced the "fast connect" idea. We would have liked to explain this idea on XTP directly. However, as there is unfortunately no official description of the service provided by XTP, this would have led us to a description of the protocol mechanisms, whereas we want to refrain from giving protocol details in this paper that deals essentially with service issues. The fast connect idea may nevertheless be explained on the associated High Speed Transport Service (HSTS) [ISO HSTS]. The HSTS definition originates from the effort to standardize XTP in ISO: the XTP group, which wanted to promote XTP, realized that a clear service definition was needed prior to any attempt to standardize the protocol.

226   Guy Leduc and André Danthine

HSTS is a service rather close to the ISO Connection-mode Transport Service ([ISO 8072]), and is based on the classical sequence of four connection establishment primitives. The fast connect idea appears as the ability for the calling TS-user to submit T-DATA.request primitives before receiving the T-CONNECT.confirm (fig. 2a). If the connection is refused (fig. 2b) then the data in transit will simply be discarded.

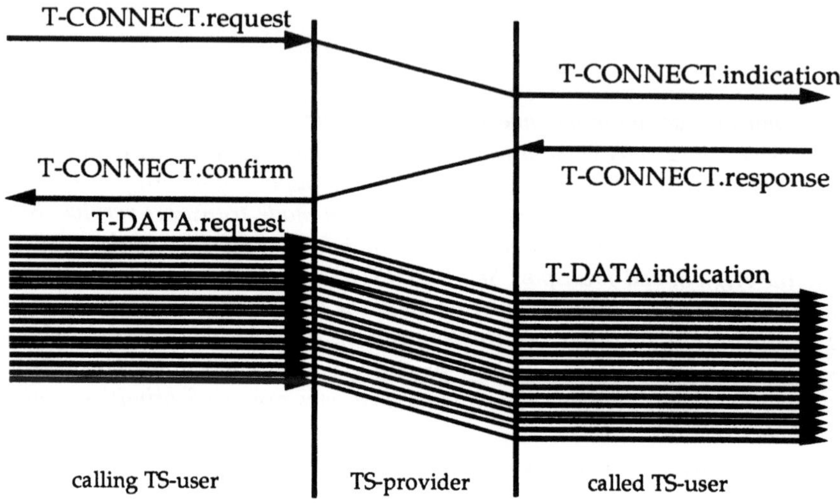

Fig.1a: Connection establishment in ISO 8072 - Successful case

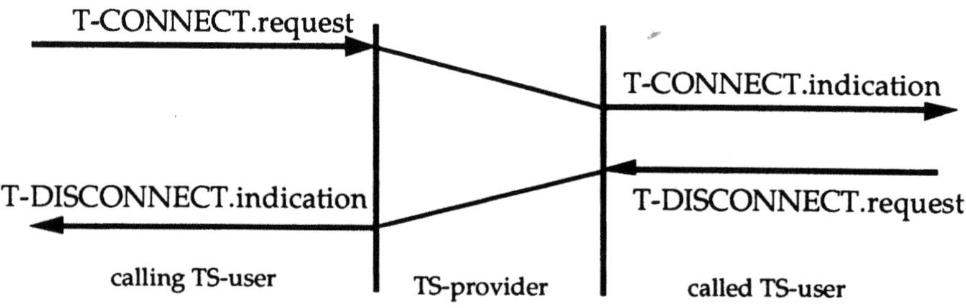

Fig.1b: Connection establishment in ISO 8072 - Unsuccessful case

This 4-primitive establishment was also considered in our group as an interesting first service description of the fast connect facility [Dan 92, DBL 92]. And from these figures 2a and 2b, we started to draw two main conclusions. First, the negotiation, which normally takes place during the 4-primitive exchange of the TC establishment, cannot be achieved any more in the fast connect scenario. This is simply because the data transfer may begin before the end of the TC establishment. Stated otherwise, the negotiation is restricted to a "take-it-or-leave-it" negotiation. This means that, during the establishment, neither the TS-provider nor the called TS-user can change the QoS (Quality of Service) parameters specified in the T-CONNECT.request: they have to take it "as is" or leave it. This is a price to pay for the fast connect service. Moreover, this important change in the semantics requires the provision of new specific TC set-up primitives for the fast connect facility, or equivalently the addition of an explicit fast connect option in the T-CONNECT.request primitive. We also drew a second conclusion about the fast connect. In the refusal case, a very large amount of data in transit will be thrown away, which leads to a wasting of the provider resources. This is certainly not expected to occur too often. Therefore we made the second hypothesis that such "fast connections" are supposed to be accepted in most cases. This is why we came to consider that a "fast connect" service facility is only meaningful when the two TS-users are in a same and thus well-known environment.

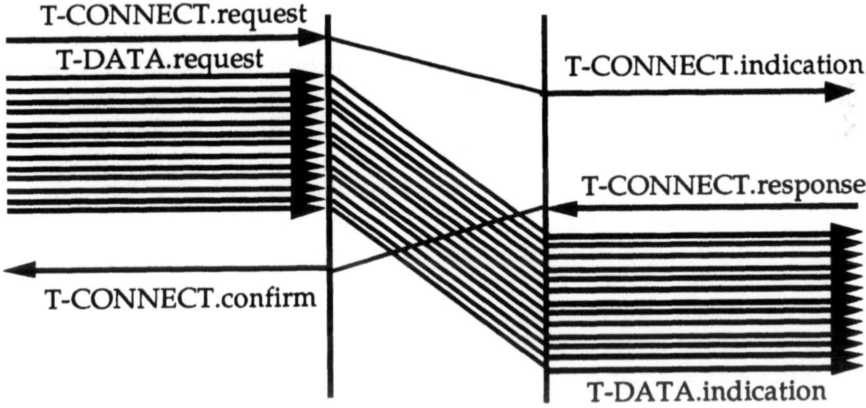

Fig.2a: Fast connect in HSTS - Successful case

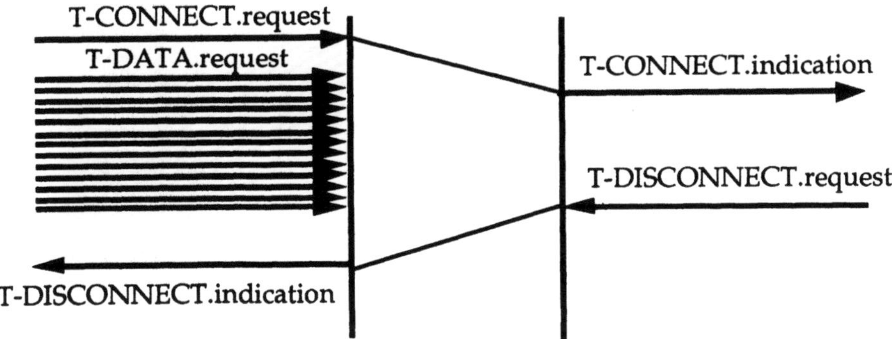

Fig.2b: Fast connect in HSTS - Unsuccessful case

In summary we may argue that the fast connect facility and the classical connection establishment facility with full negotiation are complementary. To avoid confusion between these two service facilities in the sequel, we will use distinct TS-primitives to refer to them. For a fast connect, the four establishment primitives will be denoted T-FastCONNECT.request, T-FastCONNECT.indication, T-FastCONNECT.response and T-FastCONNECT.confirm.

2 An Improvement of the Fast Connect by Way of a Listening Concept

After having clearly defined under which conditions a fast connect facility was meaningful, we tried to improve the scenario of figure 2a to overcome the following problem.

On the called side, it is only after a complete interaction with the called TS-user (by way of the T-FastCONNECT.indication and then the T-FastCONNECT.response primitives) that the TS-provider is able to deliver the beginning of the stream of data to the called TS-user. The delay associated with this interaction between the TS-provider and the called TS-user means that the amount of data that the TS-provider will have to store before the delivery may be very important, which will interfere with the resources available in the TS-provider. Figures 3a and 3b are refinements of figures 2a and 2b. They better illustrate (i) that the data are stored in the called transport entity (fig. 3a), and (ii) that Network Service (NS) provider resources are wasted in the unsuccessful case (fig 3b).

Even if the second effect is unavoidable, we considered that the solution proposed by HSTS in figure 2a needed improvements to overcome the first undesired effect.

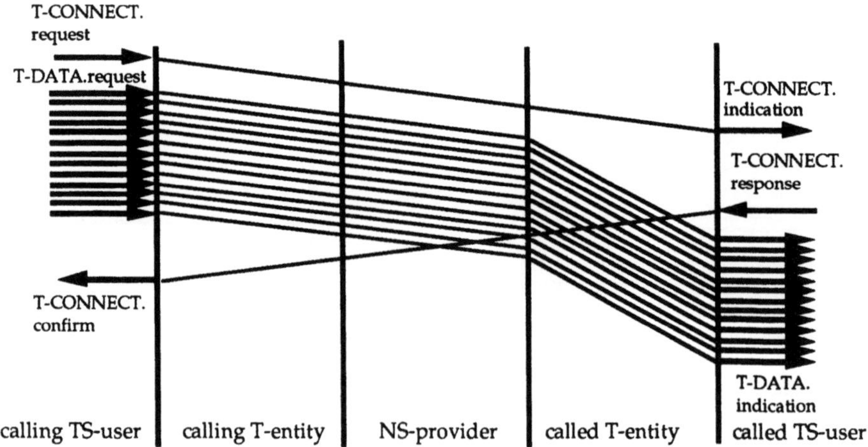

Fig. 3a: Refinement of the Fast connect in HSTS - Successful case

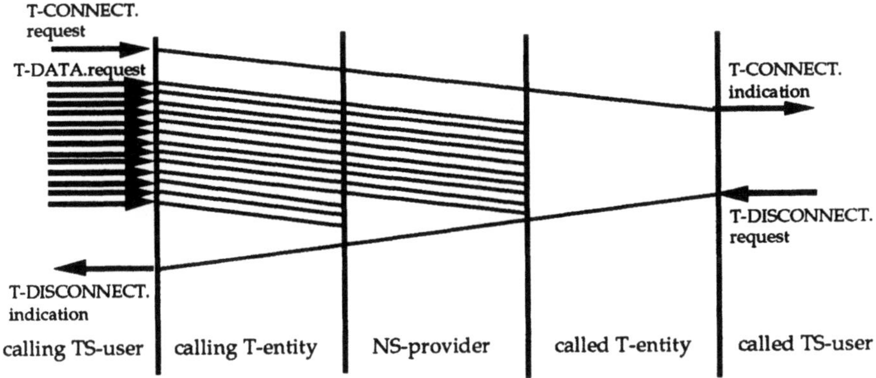

Fig. 3b: Refinement of the Fast connect in HSTS - Unsuccessful case

2.1 The Listening Interaction

Coming back to XTP, from which the fast connect idea originates, and looking at how this problem is solved in the implementations, we realize that in XTP implementations a user has the ability to pass some information to its local XTP entity to inform it on its acceptation conditions of incoming fast connections. Thanks to this interaction, denoted "listening interaction", the XTP entity may decide on the acceptation or refusal of an incoming fast connection. In the refusal case, the XTP entity discards the incoming data packets and sends back a disconnect packet. In the acceptation case, the XTP entity informs its local user, sends back an acceptation packet to the source entity, and forward immediately the incoming data packets to its user.

Our first idea in OSI95 [Dan 92, DBL 92] has been to derive from these implementation considerations an abstract service facility in terms of an ISO-like sequence of TS-primitives. A first proposal is depicted in figure 4a. A T-LISTEN.request is used to model the listening interaction of XTP. Its parameters are composed of QoS and addressing requirements. We may consider it as a pending request to any incoming demand matching its QoS and addressing requirements. When the TS-provider receives a T-FastCONNECT.request, it can thus check whether the requested QoS and addressing parameters are matching those of the T-LISTEN.request. If the matching is successful, the TS-provider delivers to the service user a T-FastCONNECT.indication which may be immediately followed by the stream of TSDUs through T-DATA-indications. If the matching is not possible (fig. 4b), the connection will not be established, the TSDUs sent following the T-FastCONNECT.request will be lost and the provider will issue a T-DISCONNECT.indication to the calling TS-user.

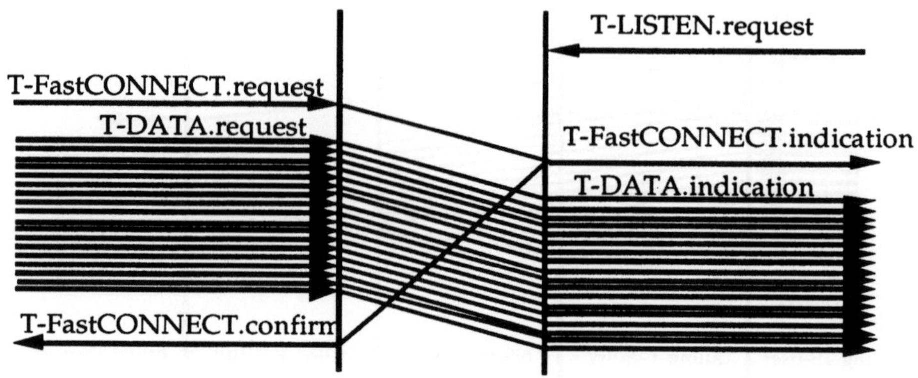

Fig. 4a: Fast connect with listening interaction - Successful case

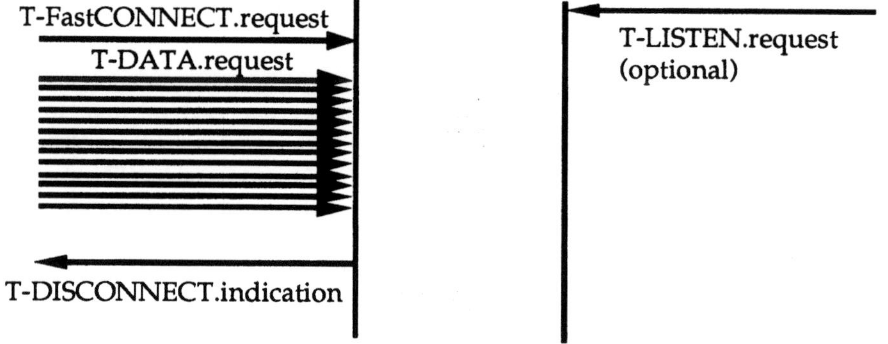

Fig. 4b: Fast connect with listening interaction - Unsuccessful case

2.2 Shortcomings of the Solution Based on the Listening Interaction

At first glance the listening interaction seems to avoid at the called side the delay associated with the decision on the acceptation or refusal of the incoming connection. This is wrong. This delay is still present but it is spent in the TS-provider rather than in the TS-user. Indeed, the delay between the T-FastCONNECT.request and the T-FastCONNECT.indication in fig. 4a is comparable to the delay between the T-FastCONNECT.request and the T-FastCONNECT.response in fig. 2a. In the first case, the TS-provider has to evaluate a certain predicate which is a comparison between the requested QoS from the T-FastCONNECT.request and the required QoS from the T-LISTEN.request. In the second case, the TS-user has to evaluate a similar predicate between the QoS from the T-FastCONNECT.indication and its own requirements. There is no reason to consider that the TS-provider will decide faster than the TS-user.

Therefore, as the delay associated with a decision at the called side is unavoidable, we simply consider this issue as an implementation matter, which has only to do with the design of a 'performant' local service interface. At an abstract level, this is a non-problem that we will not consider any more in the sequel.

Furthermore, the T-LISTEN.request primitive is a very special primitive. First, in contrast to all other TS-primitives, it is not associated with a given connection. It occurs on a TSAP but no TCEP is created until an acceptable T-FastCONNECT.indication is issued on this TSAP. This induces some unclear situations. The following questions give an idea of the additional problems that cannot be ignored.

- Does the TS-user have to re-issue a new T-LISTEN.request after a successful T-FastCONNECT.indication ?
- Is it possible that several T-LISTEN.requests be pending at the same TSAP ? If so, what is the semantics of the "union" of T-LISTEN.requests ?
- How does the TS-user retract a pending T-LISTEN.request ? In particular, how does it identify them if several such requests are pending at the same TSAP ?
- How shall we define the called TS-user requirements to be included in the T-LISTEN.request ? The fact that the T-LISTEN.request is not associated with a given connection adds some non trivial intertwinings between the addressing requirements and the QoS requirements.

All the previous questions indicate that the T-LISTEN.request primitive should not be considered as a TS-primitive, but instead as a local management primitive. Indeed, its sole purpose is *not* end-to-end but purely local. It just tries to improve the design of the local TS-interface. Being local in scope, and of a management nature, one may even wonder if it is correct to consider the T-LISTEN.request as an OSI primitive. To further substantiate this point, it suffices to remark that, whatever the answers to the above questions are, the service does not change conceptually. However, the presence of the T-LISTEN.request requires for completeness that these questions be answered in the service specification. This is unfortunate because abstraction is lost when particular implementations are described explicitly.

In conclusion, the fast connect facility based on the T-LISTEN.request does not solve anything as regards the potential bottleneck at the called side, and leads to an important loss of abstraction.

Furthermore, we realized that neither HSTS nor the variant based on the listening interaction have made consistent hypotheses to address the fast connect facility correctly. Indeed, opposite hypotheses on the behaviour of the TS-provider and on that of the called TS-user have been made: the TS-provider was supposed to be able to accept "immediately" the T-DATA.requests that follow a T-FastCONNECT.request, whereas the called TS-user was *not* supposed to be able to accept "immediately" the T-DATA.indications after a T-FastCONNECT.indication. These two cases are however fully comparable. The TS-provider, like the called TS-user, has to create a context before it can accept incoming TSDUs. It has also to check the QoS parameters of the T-FastCONNECT.request to avoid accepting TSDUs when it knows that the requested QoS cannot be achieved. And so on.

So if we harmonize our hypotheses and abstract away from local and/or implementation-oriented considerations, we come to three possible well-defined scenarios to express the fast connect service facility: a relaxed 4-primitive establishment and two variants of a 3-primitive establishment. We will present them and compare them in the remaining part of this paper.

3 Fast Connect as a Relaxed 4-primitive Establishment

In this scenario, depicted on figures 5a and 5b, the classical 4-primitive establishment applies, but T-DATA.requests may occur between the T-FastCONNECT.request and the T-FastCONNECT.confirm on the calling side, and *similarly* T-DATA.indications may occur between the T-FastCONNECT.indication and the T-FastCONNECT.response on the called side. This fast connect scenario has been first proposed in the Broadband Transport Service (BTS) of the RACE / CIO project [HeD 92].

If the called TS-user does not accept the connection, it *may* not accept the T-DATA.indications and *shall* use the T-DISCONNECT.request to release the connection. This case is illustrated on figure 5b.

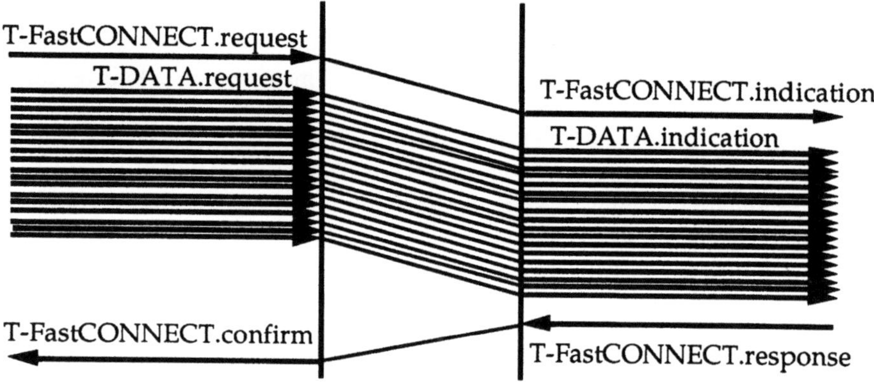

Fig.5a: Fast Connect in BTS - Successful case

As regards the QoS associated with the TS-primitives, BTS does not distinguish between the classical establishment and the fast connect. In both cases, the T-FastCONNECT.response and the T-FastCONNECT.confirm have QoS parameters and nothing precludes the called TS-user to respond with different QoS. By contrast we restrict the negotiation to a take-it-or-leave-it in which the called TS-user cannot change the QoS vector present in the T-FastCONNECT.indication. Therefore, this vector becomes useless in the T-FastCONNECT.response and T-FastCONNECT.confirm primitives and will simply be removed.

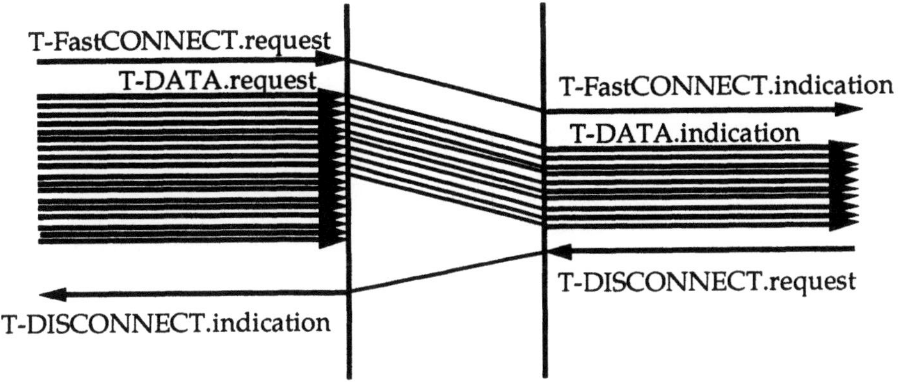

Fig.5b: Fast Connect in BTS - Unsuccessful case

4 Fast Connect as a 3-primitive Establishment

Noticing that the T-FastCONNECT.response has almost no purpose, we may even get rid of this primitive, and thereby propose a 3-primitive establishment as depicted in figures 6a and 6b, if we take the following two-step argument into account:

- The T-FastCONNECT.response in the relaxed 4-primitive establishment is reduced to a very simple interaction because no QoS parameter vector is needed. The same reasoning applies to the T-FastCONNECT.confirm primitive.
- If we want to optimize the successful establishments with respect to the unsuccessful ones, we may consider that a T-FastCONNECT.response is implicit in the T-FastCONNECT.indication, and thereby completely remove the need for this response primitive.

The scenario depicted in figure 6a shows the successful establishment. The refusal scenario is depicted in figure 6b. When the called TS-user decides to refuse the connection, it may not accept the T-DATA.indications and shall use the T-DISCONNECT.request to release the connection.

Note that in this case, the calling TS-user may first receive the T-FastCONNECT.confirm before the T-DISCONNECT.indication. Actually, the refusal scenario per se disappears, but the more general release scenario remains and can be used for this purpose. This is aligned with the idea that such "fast connections" are not supposed to be refused. Acceptation is implicit or taken by default.

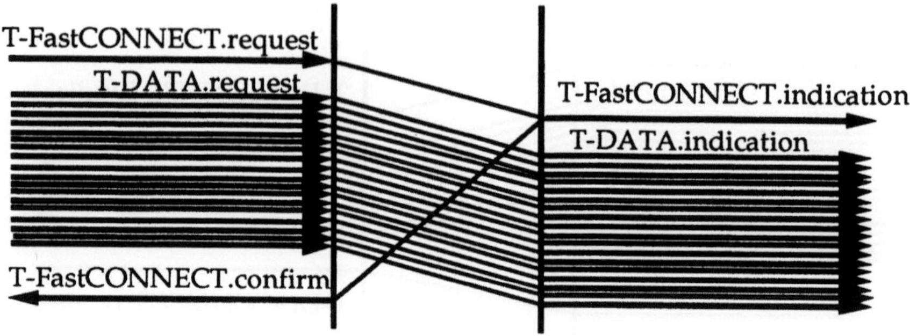

Fig.6a: Fast connect as a 3-primitive establishment - Successful case

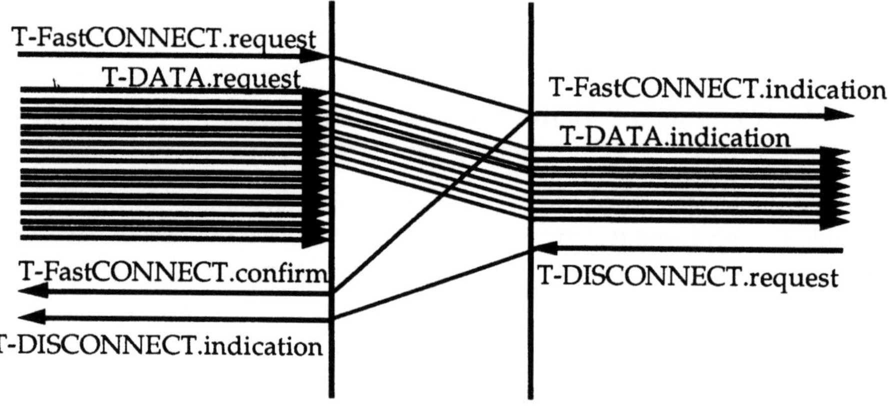

Fig.6b: Fast connect as a 3-primitive establishment - Unsuccessful case

5 A Variant of the 3-primitive Establishment

In the previous 3-primitive establishment, the implicit a priori acceptation of the connection by the called TS-user leads to a less elegant scenario when the called TS-user wants to refuse the connection (which is supposed to be an exception). Indeed the TS-user first accepts the connection as always and then has to disconnect it. This scenario may be improved if the TS-user has a means to inform the TS-provider of its acceptation or refusal *when the T-FastCONNECT.indication occurs*. This will be explained in detail after a reminder of the concept of a *service primitive*.

5.1 The Service Primitive

It is important to keep in mind that a service primitive is an *interaction* between two entities (the service user and the service provider). This interaction occurs at the SAP to which both entities are "attached". This abstraction has several consequences:

- When a service primitive occurs, both entities are aware of it simultaneously.
- A service primitive cannot occur if one entity does not agree on its occurrence. This idea is used in ISO standards to describe interface flow control. For example, when a TS-provider is momentarily blocking the T-DATA.requests, or when a TS-user is blocking the T-DATA.indications.
- When a service primitive occurs, a negotiation of values may take place instead of the more classical value passing. For example, a connection endpoint identifier (cep_id) is negotiated by both entities on the occurrences of a T-CONNECT.request as well as of a T-CONNECT.indication. Of course, the way this negotiation is made is implementation-dependent and is therefore beyond the scope of the standard.

5.2 The New 3-primitive Establishment

Along the same principles, pieces of information may be passed in both directions when a primitive occurs. For example, suppose that a certain parameter of the T-FastCONNECT.indication primitive (denoted 'verdict' for example) is not determined by the TS-provider but instead by the TS-user. By way of this parameter the TS-user informs the TS-provider of its acceptance or refusal of the QoS values proposed unilaterally by the TS-provider. The figures 7a and 7b respectively depict the successful establishment and the refusal of the fast connection.

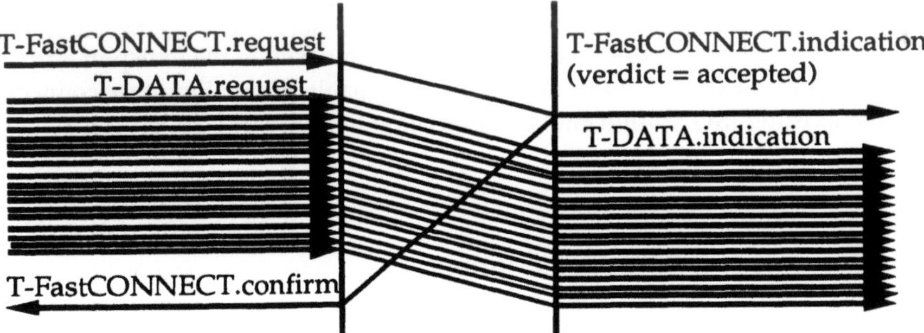

Fig.7a: Fast connect as a new 3-primitive establishment - Successful case

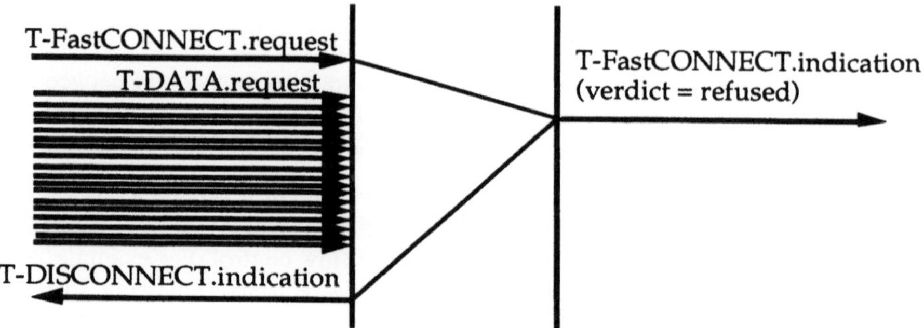

Fig.7b: Fast connect as a new 3-primitive establishment - Unsuccessful case

The verdict parameter determined by the TS-user plays the role of the T-FastCONNECT.response primitive. In other words, two primitives (T-FastCONNECT.indication and T-FastCONNECT.response) are merged into a single *atomic* primitive. This fast response included in the T-FastCONNECT.indication is only interesting if the TS-user has at its disposal (that is in its local context) all the information required to return the verdict immediately. Let us consider three cases:

1) The called TS-user is an application entity which sits directly on top of the TS-provider. In this case the TS-user can obviously determine the verdict by evaluating some predicate based on the other T-FastCONNECT.indication parameters and its local context.
2) The called TS-user is an intermediate entity (e.g. a session entity if a complete OSI stack is present). There are two subcases:
   - This entity has been informed of the acceptance conditions by the above layers. We are back in case 1.
   - This entity has not been informed of the acceptance conditions. It shall always accept the connection and thus fix the verdict to 'accepted'. The decision will be taken in higher layers, but the data flow is not blocked in the transport entity.
3) There may also exist TS-users that are never capable of handling fast connections (e.g. the existing ISO session entity). In this case the TS-user shall always return the verdict "refused".

5.3 An Example of LOTOS Description

For information we provide hereafter a simplified LOTOS description that formalizes the above ideas. It consists of two processes (a TS-provider and a called TS-user) that are synchronized on the gate `dest_tsap` to interact via a T-FastCONNECT.indication primitive. The specification illustrates that the TS-provider determines the calling address and the QoS parameters, whereas the called TS-user determines the acceptation verdict parameter. Depending on the verdict the resulting behaviour of the TS-provider is also described (either it issues a T-FastCONNECT.confirm or a T-DISCONNECT.indication at the `source_tsap`).

```
specification fast_connect_example [source_tsap,dest_tsap] :noexit
behaviour     TS_provider [source_tsap,dest_tsap]
              |[dest_tsap]|
              TS_user [dest_tsap] (...)
where
process TS_provider [source_tsap,dest_tsap] :noexit :=
      source_tsap?tsp1: T_FastCONNECT_request;
      dest_tsap?tsp2: T_FastCONNECT_indication
            [(QoS(tsp2) eq QoS(tsp1))
              and (calling_of(tsp2) eq calling_of(tsp1))
              and (dest_sap is_valid_value_for called_of(tsp1))];
      (([verdict(tsp2) = accepted] ->
            source_tsap!T_FastCONNECT_confirm;
            Data_transfer [source_tsap,dest_tsap])
        []
        ([verdict(tsp2) = refused] ->
            source_tsap!T_DISCONNECT_indication;
            stop))
endproc (* TS_provider *)
process TS_user [dest_tsap] (u: user_context) :noexit :=
      dest_tsap?tsp: T_FastCONNECT_indication
          [(Acceptance_predicate(calling_of(tsp),QoS(tsp),u)
                implies (verdict(tsp) eq accepted))];
      Receive_data [dest_tsap]
      []
      dest_tsap?tsp: T_FastCONNECT_indication
          [(not (Acceptance_predicate(calling_of(tsp),QoS(tsp),u))
                implies (verdict(tsp) eq refused))];
      stop
endproc (* TS_user *)
endspec
```

5.4 The Responding Address Issue

There remains a pending issue regarding the 3-primitive scenario. How can the TS-provider fill in the responding address field of the T-FastCONNECT.confirm primitive if no T-FastCONNECT.response primitive exists to provide this responding address to the TS-provider ?

According to us, the TSAP at which the T-FastCONNECT.indication primitive actually occurred is known by the TS-provider and the responding address *is precisely* the address of this TSAP. So, the TS-provider knows the responding address, and a T-FastCONNECT.response is not necessary to provide this address.

By contrast, the called address that is present in the T-FastCONNECT.indication primitive does *not* in general identify the TSAP at which the T-FastCONNECT.indication is occurring. Two obvious counter-examples are when the called address is a generic or a group address.

6 Conclusion

A fast connect facility is incompatible with a full negotiation. Instead it can only provide the limited "take-it-or-leave-it" negotiation. Moreover the refusal of a fast connection has worse consequences than the refusal of a classical connection due to the resources wasted in the TS-provider. For these reasons, we argued on the complementarity of the fast connect facility and the existing connection-mode TC establishment facility.

Several models of a fast connect facility have been presented and the last three ones have been retained. They all are defensible but, according to us, the last one is the best one. It combines soundness with simplicity and abstraction: 3 primitives for a take-it-or-leave-it negotiation is necessary and sufficient. It also preserves all the nice properties of a classical connection: the called TS-user can refuse the connection (not true in the first version of the 3-primitive establishment), and no T-DATA.indications occur at the called side when the connection is refused (not true in BTS and not applicable in the first 3-primitive establishment).

Acknowledgements

We would like to thank Yves Baguette and Olivier Bonaventure for many valuable discussions and for their in-depth reviewing of this paper.

References

[Dan 92]   A. Danthine, **A New Transport Protocol for the Broadband Environment**, in: A. Casaca, ed., Broadband Communications, Elsevier Science Publishers (North-Holland), Amsterdam, 1992, 337-360.

[ISO HSTS] Source: USA, Proposed working draft for high-speed transport service definition, ISO/IEC JTC1/SC6/N7070, 1991-11-28.

[ISO 8072] ISO-TC97/SC6/WG4, Information Processing Systems - Open Systems Interconnection - Transport Service Definition, IS 8072, 1986-06-15.

[DBL 92]   A. Danthine, Y. Baguette, G. Leduc, **Issues Surrounding the Specification of High-Speed Transport Service and Protocol**, Rept. no. OSI95/ULg/A/15/TR/P/V2, University of Liège, Institut Montefiore B28, B-4000 Liège, Belgique, Jan. 1992, also in: ISO/IEC JTC1/SC6/N7312, 11-05-1992.

[HeD 92]   L. Henckel, S. Damaskos, **Multimedia Communication Platform: Specification of the Broadband Transport Service**, Rept. no. R2060/GMD/CIO/DS/P/001/b1, GMD-FOKUS, Hardenbergplatz 2, D-1000 Berlin 12, Germany, Dec. 1992.

[PEI XTP]   Protocol Engines Inc., **XTP Protocol Definition - Revision 3.6**, PEI 92-10, 1900 State Street, Santa Barbara, CA 93101, Jan. 1992.

The LOTOS Specification of the Enhanced Transport Service

Luc Léonard[1]
[1] Research Assistant of the Belgian National Fund for Scientific Research (F.N.R.S.)
Université de Liège, Institut d'Electricité Montefiore, B 28, B-4000 Liège 1, Belgium
Email: leonard@montefiore.ulg.ac.be

This paper introduces the LOTOS specification of the OSI95 Enhanced Transport Service. It discusses the problems encountered during its writing, such as the ones caused by its validation, or by the lack of a 'Timed' LOTOS. But the interest is not only in the specification itself. It also lies in the way this specification was developed: not after, but in parallel with the design of the Service it had to describe. The reasons for the choice of this way of working, its advantages (and drawbacks) are presented.

Keywords: LOTOS, specification, OSI95, Transport Service.

1 Introduction

This paper introduces the LOTOS specification of the OSI95 Enhanced Transport Service, which can be found in the Appendix. The interest is not only in the specification itself, which gives a precise, unambiguous description of the Service, and can be considered as a (very early) prototype, but lies also in the way this specification was developed, and in the way it helped in the very complex design phase of the Service.

Unlike what usually happens, for instance with the LOTOS specification of the ISO Transport Service, this specification is not simply the translation, in a formal language, of an already existing detailed description in English. It did not come after the design phase of the Service, but in fact, was developed in parallel. This means that its evolution followed the design of the Service. The advantages (and disadvantages) of this procedure are discussed in the next point. One of them is the quite important amount of time our method required. This explains why the English description of the transport services have not reached their final ISO-like form yet. As we have never been in favour of quick and dirty descriptions which always lead subsequently to difficult problems during the implementation phase, we have also decided to limit our design to the parts which have been thoroughly studied and formalized. This is why multicasting has not been included.

A few comments will also be made about the problems encountered during the writing of the specification, such as the ones caused by the lack of a 'Timed' LOTOS, or by the validation of the specification.

2 Discussion of the Method

One could wonder about the interest of trying to develop a LOTOS specification in parallel with the design of the service. This takes much more time indeed: the specifier has no precise idea of the final form of its specification. He cannot plan the easiest way to realise it, but on the contrary, he is forced to accompany the evolution of the system and thus, he must proceed to regular updates, not only to insert new developments, but also to carry out the (frequent) modifications of already defined, and specified parts, causing changes that are usually much easier to realise in the fertile imagination of the Service designers than in its specification.

But on the other side, although the work of the specifier was made harder, proceeding this way also appeared to be of great interest.

The need for formal techniques came from the lack of precision of the usual languages. A formal description in LOTOS is clear and without ambiguity. Thus, when translating into LOTOS an informal, and sometimes quite vague, description, as the one the designers have of their system while developing it, the specifier will probably face ambiguities or even incoherences, he will ask answers for details the designers very often did not even think about. Our way of working, where the ideas were first formalised before being adopted definitively, was thus of great help for them. It helped them getting faster a clearer view of their own ideas, and pointed out troubles that could thus be solved earlier. It brought more rigour and finally led to a more precise and detailed definition of the service.

As an illustration of the problems that could be avoided by using formal methods, let us briefly report on our experience with existing ISO standards. The work on the LOTOS methodology and its related specification of ISO's Transport Protocol (IS 8073) in LOTOS [Led 93] has led recently to the detection of some ambiguities or errors in this standard. These problems have been collected in a list of 10 defect reports (numbered 143 to 152) submitted to ISO/SC6/WG4. For this reason, Belgium also cast a negative vote on DIS 8073 prior to the San Diego meeting of ISO/IEC JTC1/SC6 in July 1992. A solution to the defect nr. 150 was proposed in San Diego and was circulated for ballot [ISO 6N7619]. Some other defect reports (nr. 143, 145, 147, 149 and 152), being too complex, remained however unresolved by the editors' group. They have been circulated for comments in [ISO 6N7609]. Defect nr. 148 was also unresolved but no conclusion was drawn in San Diego. Finally, some others (nr. 144, 146, 151) were rejected as defects, because the editors' group considered that the standard is clear and correct on the matter. Considering the discussion we had however, it is likely that some potential implementers will still have difficulties to understand the standard correctly. These 10 defects came after 142 other ones from which 4 were also still unresolved in San Diego, including the old defect nr. 76 (May 1987). If we add to these defects, some other open issues, like the non alignment between the Transport Service and the Transport Protocol on the equality between the called and responding transport addresses, we think we have enough illustrated the kind of problems that will be with us in ISO standards for many years to come.

3 The Constraint-Oriented Style

The specification presented here corresponds to the first design phase of the Service. This means that it has to remain as abstract as possible, just describing the observable behaviour of the Service, that must be seen as a black box. Details about an internal structure that could provide the Service must not be given, this would lead to over-specification. Also, as we said above, due to the method we used, the specification had to be frequently updated, either to integrate new elements or to modify already existing ones. This required it to be as flexible as possible, in order that any small novelty did not cause a complete rewriting. In [Led 93] the constraint oriented style is explained, and some arguments are given of its adequacy in such an early design phase, for it is abstract, structured and modular.

One must not think, however, that specifying in this style is kind of holidays. The reality is not so simple: the attempt to decompose a process into independent constraints and to structure it as a conjunction of parallel processes interacting only on external gates, what is the philosophy of the constraint-oriented style, faces the limitations of the LOTOS formalism. It is not always possible to separate in different processes logical constraints one first thought to be independent. In fact, this desire is often in conflict with the ban on the hiding operator, that could help continuing the decomposition, but with a loss of abstractness and a risk to make subsequent modifications less easy. This sometimes leads to quite complex monolithic processes. It may also happen that conceptually inoffensive small modifications in fact deeply change the nature of the constraints and finally require very important rewriting.

But anyway, although it is not perfect this style still appears to be the most convenient in such a situation and to simplify the work a lot by comparison with other ways of specifying.

4 A Brief Description of the Specification

As announced the specification is written in the constraint-oriented style. It describes the behaviour of a never terminating service provider. Omitting some less important processes, its general structure can be represented as on the first figure, next page.

Any PP-ECOTS, PP-EACLTS, PP-ERCLTS is described by a given process, a potentially infinite number of such processes being available. A single process suffices to provide a potentially infinite number of PP-EUCLTS. All these processes are composed in parallel and interleave independently of each other.

Most of these processes are quite simple, except the ones providing a PP-ECOTS, whose definition requires the main part of the specification. At a first level of decomposition, the structure of such a process is described by figure 4.2. The plain lines indicate interacting (sub-)processes, the dotted lines correspond to (sub-)processes interleaving independently.

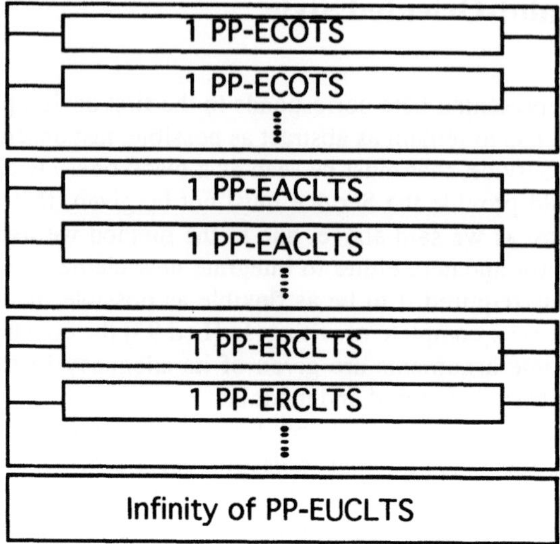

Fig. 4.1 general structure of the specification

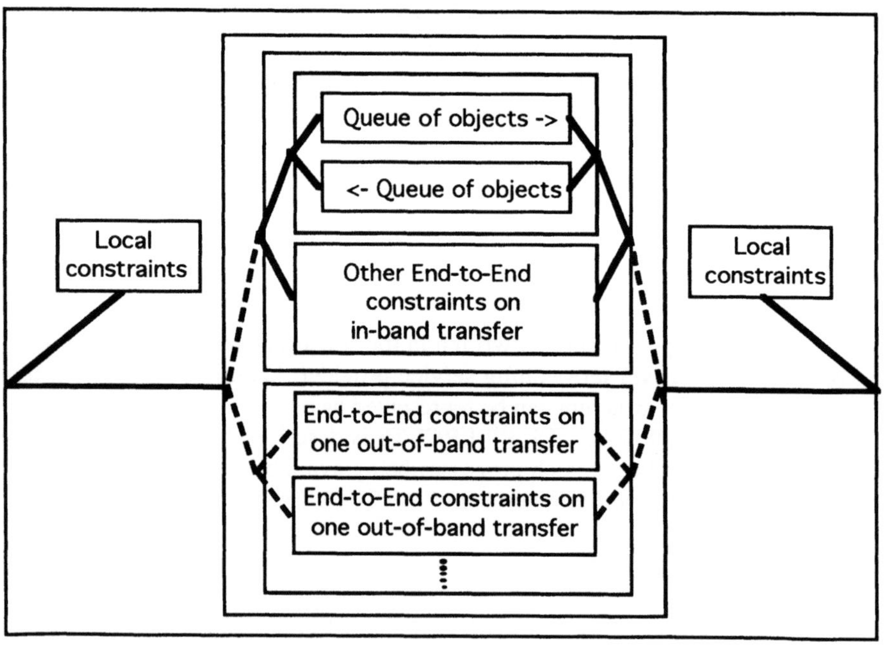

Fig. 4.2 general structure of a process describing a PP-ECOTS

Local constraints relate the behaviour at a TCEP to the history of the TSPs executed at that TCEP. End-to-End constraints relate the behaviour at a TCEP to the history of TSPs at the other TCEP. End-to-End constraints are separated into in-band and out-of-band constraints. Out-of-band constraints provide a potentially infinite number of out-of band transmissions, each one described by a given process. In-band constraints are made of two main parts. The first one expresses the link between the histories at both TCEPs, taking care of the non-determinacy in this relation. The second part is made of various (sub-)processes, most of them describing the influence of the qualities of service parameters on the transmission of data.

Some general figures about the specification to finish:

- Resource consumed by this LOTOS work: ± 200 man.days
- Size (non commented): ± 2200 lines of Abstract Data Types
                         ± 2200 lines of Processes
- Number of (sub-)processes: 102

5 The Lack of a Timed LOTOS

A particular problem arouse from the fact that LOTOS does not allow (at least in a realistic way), the description of time constraints, when some of the mechanisms of the service were fundamentally based on such features. We adopted the solution of simplifying the description by expressing all the possible behaviours while omitting to describe the way the time influenced the choice among them, this choice being usually solved as an internal, non deterministic event. This solution however presents two main deficiencies. The first and most obvious one is of course the lack of information that the specification provides about the actual behaviour of some mechanisms of the Service. The second one is more subtle: the loss of information about the influence of time allows some sequences of events to happen in a normally impossible order, due to their relative timings. This could be hardly avoided in the case these events are accomplished by concurrent processes, for adding special synchronisation to prevent such sequences would be extremely heavy and complicated and would lead to a loss of abstraction

Within the OSI95 framework, TIC has developed a 'timed' LOTOS formalism, that is presented in this book with its application to the 'Tick-Tock' case-study, specially designed to assess it. ULg is also developing its own 'timed' LOTOS which is currently applied to the case-study

6 Validation

This specification has been developed with the help of the LOTOSPHERE toolset: LITE. It has been syntactically and semantically checked. It has not been completely simulated, using SMILE, because this happened to be extremely time and resource consuming. As an example, just obtaining the possible first two steps for only one

Connection process (see Appendix) required around five hours and 280 MB of swap-space on a standalone SPARC1+. However, even if just some smaller sub-processes were partially simulated, this helped discovering some mistakes.

References

[ISO 6N7609] Source: ISO/IEC JTC1/SC6, For comments - Various defect reports - 8072/008, 8073/143, 8073/145, 8073/147, 8073/149 and 8073/152, ISO/IEC JTC1/SC6/N7609, 1992-08-07.

[ISO 6N7619] Source: ISO/IEC JTC1/SC6, Defect reports 8073/141, 8073/142 and 8073/150 to ISO/IEC DIS 8073 ISO/IEC JTC1/SC6/N7619, 1992-08-07.

[Led 93]   G. Leduc , **A method for applying LOTOS at an early design stage and its application to ISO's transport protocol**, *The OSI95 Transport Service with Multimedia Support on HSLAN's and B-ISDN*, A. Danthine, ed., Springer Verlag, 151-180

ATM and the New Communication Environment

We have already stressed that the last ten years have seen a tremendous change of the communication environment in their performance as well as in the service capability.

The stability of the transport service has prevented to offer, at the transport level, services available at network and sub-network levels, such as the multicast and broadcast services available in LANs and the synchronous service with guaranteed average throughput available in FDDI I.

The decision of the CCITT in 1988, to select the asynchronous transfer mode (ATM) for the B-ISDN, has introduced a new lower level communication service able to offer a guaranteed throughput over an end-to-end connection, associated with a great flexibility in the selection of the value of this throughput. With the existing transport service, it will not be possible to offer such a service which will be of great interest for the networking of the multimedia-based application.

The foreseen importance of the ATM-based network in the LAN as well as in the WAN has triggered the research in OSI95 in two directions.

The first one is related to the integration of the B-ISDN services in the OSI Reference Model. A large number of possible solutions are presently discussed in various forums and standardisation bodies to achieve the integration of ATM in the existing communication architecture. Studies are also going on to offer, in the wide area, existing or enhanced services such as connectionless service with sequence preservation.

Another approach is to bring the service offered by the ATM layer at the transport level. The OSI95 transport service is able, through its enhanced QoS, to offer, to the transport service users, access to these enhanced services.

This approach requires to specify a connection-oriented (sub-)network service on top of the ATM bearer service. One way to do it has been specified and implemented by ORL with its MSNA (Multi Service Network Architecture). The main characteristics of this architecture are to support the multiplexing by the VCI which are linked to the applications and to have the SAR (Segmentation and Reassembly) as the top sub-layer of the network layer.

It is possible to envisage for the multimedia application, direct access to such a network service, but it is also possible to have, for the data communication as well as for the multimedia-based application and its video streams, a unique transport layer based on an enhanced transport service such as OSI95 .

B-ISDN Services and their Relation with the OSI Protocol Reference Model

Joost Boerjan and Robert Peschi
Alcatel Bell Telephone Research Centre, F. Wellesplein 1, B-2018 Antwerpen, Belgium

This contribution results from work undertaken as part of the Esprit Project OSI95. The goal of the task, making an assessment of the services delivered by B-ISDN, was to provide a clear idea of the kind of new service that the enhanced transport facility TPX may expect from the lower layers. In this paper we describe the various groups of services which are available from a B-ISDN. Furthermore, the relation between these services and those defined within the OSI Reference Model has been indicated.
Keywords: B-ISDN Services, AAL, ATM, B-ISDN Protocol Reference Model

1 Introduction

The goal of this paper is to provide a description of the various services available from the broadband integrated services digital network (B-ISDN). Secondly, once we have characterised the various services, we will establish a relation between these B-ISDN services and the services defined within the scope of the OSI Reference Model.

The OSI Reference Model is a layered model with a clear distinction between the service of a layer and the underlying protocol used to provide this service. If, in the context of this paper, we are talking about services, the term "service" designates the service offered by a protocol or in other words, the so called "layer service". In other contexts the word often refers to the concept of a telecommunication service (e.g. telephony) which is a more application oriented notion.

1.1 The OSI and B-ISDN Protocol Reference Models

At the time of the definition of the B-ISDN Protocol Reference Model (B-ISDN PRM) [I.321] no explicit relation with the OSI PRM [ISO 7498] was established. As a result of this both models differ from each other in more than one respect. While this is true one should realise that their respective scopes are differing as well. The OSI PRM, as originally envisaged, establishes a model for end to end communication between data terminals. Later contributions have tried to enlarge the scope of the model. The Esprit Project OSI95 went ahead in this direction by specifying enhanced transport services, aiming towards the integration of new network services and multimedia applications in the OSI model. The B-ISDN PRM model, on the other hand, provides a model for describing the transfer of information using the Asynchronous Transfer Mode technique. Although the B-ISDN model is independent from the OSI PRM, a few principles of the OSI PRM, such as the layering principle, were adopted. In other aspects, the models are different. E.g. the B-ISDN PRM makes use of several planes to describe the communication. Besides, a split is made between the control plane and the user plane for describing respectively the connection control (establishment, release, etc.) phase and the user data transfer. The figure below shows the relation between the various layers and planes of the B-ISDN PRM.

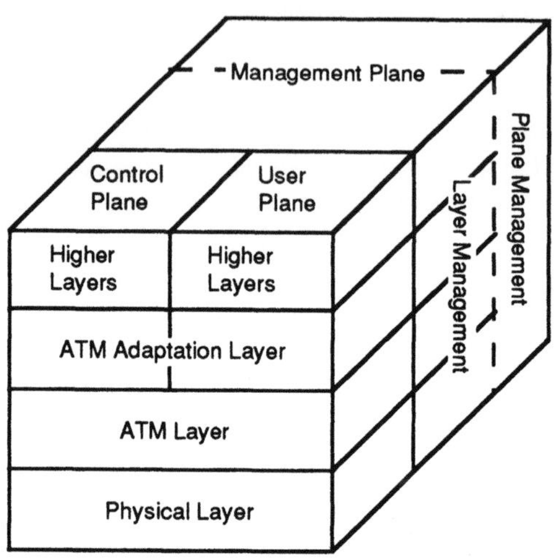

Fig. 1.1 The B-ISDN Protocol Reference Model

In the OSI95 project we have aimed at establishing a relation between the B-ISDN PRM and the OSI PRM. This is a useful goal because it allows the smooth integration of this new communication technology into the existing communication architecture requiring minimal changes. To be clear with respect to our goals; we did not aim at "integrating" B-ISDN in the OSI PRM. Rather, we have tried to establish a relation between some of the layer services defined within the B-ISDN PRM ([I.363], [I.364]) and layer services defined within the OSI PRM. It is clear that not all of the B-ISDN services can be integrated into the OSI PRM in its present form. In order to exploit all the capabilities of the new network technologies, the definition of the OSI PRM layer services would have to be modified. For the transport layer, the OSI95 project has provided such an enhanced service definition [Dan'92a]. A similar enhanced specification of the network layer was outside the scope of the project. However, several aspects of the network service definition requiring enhancement were indicated. These are mainly concerned with multicast facilities, isochronicity and resource reservation aspects as will be discussed further on in this paper.

Also, in order to accommodate all the new communication paradigms, like e.g. distributed processing, into the model, it may be required to go beyond simple modification of the OSI layer services and consequently, the research community also discusses new protocol architectures [Leo'92].

At present the discussion whether to go on with OSI or to start from scratch with new models continues. Many diverging ideas are being forwarded with on the one side people wanting to accommodate the "new" into the existing, but suitably adapted models and on the other side people who want to discard the "old" and start building again from scratch while benefiting from past experience. It is amusing to note that this contraposition is not limited to the field of communication; it almost seems to be an age-old theme.

1.2 The ATM Technique

Before we go ahead with a detailed description of the B-ISDN services in the next section, it may be useful, for a better understanding, to briefly remind some important characteristics of ATM. An in depth discussion of ATM can be found in [MDP'91a]. As is well known, ATM is a connection oriented technique based on the concept of virtual channel connections. ATM cells are routed through the switching network on the basis of the Virtual Channel Identifier (VCI). The ATM service user requests for a connection with a certain bandwidth through an out-of-band signalling procedure with the service provider. The service provider will then check all along the communication path and at the receiving entity whether enough resources are available. If this is the case, the resources will be allocated to the connection until connection release. During a call, the user may negotiate changes to the allocated bandwidth via signalling although the details of this procedure remain for further study. In other words, ATM performs network resource scheduling. This system works well as far as every user complies with the contract negotiated with the network provider prior to the communication instance. The network enforces this compliance by a policing procedure which drops cells in excess of the negotiated contract [MDP'91a].

The ATM service is enhanced by the ATM Adaptation Layer (AAL). Several types of AAL have been defined catering to users with differing service requirements. In the next section, this will be discussed in depth. Two essential functions of the AAL are the segmentation of user information into the ATM cell format and the detection and handling of lost or errored information. Depending on the type of AAL, other functions, such as e.g. jitter absorption for the AAL Type 1, are executed as well. Though the purpose of this paper is to describe services rather than protocols, it is helpful to note that the AAL is sublayered in a Segmentation and Reassembly sublayer (SAR) and a Convergence Sublayer (CS) as in the figure below.

CS sublayer (AAL)
SAR sublayer (AAL)
ATM layer
ATM Physical layer

Fig.1.2 B-ISDN Protocol Stack

The ATM Adaptation Layer works on end-to-end basis within the ATM network. An example is the end-to-end error recovery of lost and errored cells which may be carried out by the AAL. The ATM network may, however, be interconnected with other networks in an interworking configuration. Of course, in that case, the AAL does not perform end-to-end functions between one user connected on one network and another user connected on the ATM network. For these end-to-end functions, a higher layer protocol may be required.

1.3 Multiplexing in B-ISDN

The idea of B-ISDN is that various types of services can be transported by a single digital network. This requires multiplexing of information of very different types and of diverging service requirements. This objective can be attained by using the ATM technique which offers, as one of its main advantages, a very flexible multiplexing solution to its users. The different streams of user information, carried by separate virtual channels, are multiplexed at the cell level, independently of the type of information carried by the cells.

A multimedia user requires this multiplexing of streams of differing information types. Some of the information streams may be delay and delay jitter sensitive, others won't. A synchronisation between some of the streams may also be required. For this user, there are now several alternatives for the location, within the protocol stack, of these multiplexing and synchronisation functions. One of the alternatives is to do the multiplexing of the various media at the ATM level. The approach is attractive because of the flexibility of ATM multiplexing. Some mechanisms are required in order to keep the differential delays between continuous media streams travelling to the same destination over different paths bounded.

Another alternative is to multiplex the media components somewhere at the higher layers, into a single stream before entering the network. A single VC is then used for transport of the multiplexed stream.

It is not clear at this time whether one of these approaches will be preferred. It may well be possible that both multiplexing schemes are used simultaneously. The choice will also be influenced by the charging principles practised by the network operators. If network operators charge by the number of VPI/VCIs used, ATM layer multiplexing will be penalised. As a consequence, different exploiting practices may develop for the local ATM and the wide area ATM.

2 B-ISDN Services

In the section below we will present an overview of the groups of B-ISDN services and, where possible, indicate the relationship with the OSI PRM [Boe'92]. For the sake of clarity we have grouped the B-ISDN services in three main groups; the connection oriented time constrained services, the non time constrained connection oriented services and the connectionless service.

2.1 B-ISDN Connection Oriented Time Constrained Services

B-ISDN offers services oriented towards the transport of user information with time constraints. Typical examples here are the transport of voice and video. The essential characteristic of these services is that the relation in time between the instances at which primitives are submitted to the service interface at the sending side is exactly copied at the interface at the receiving side. These services could also be described as real-time services. In other words a user of such a service experiences strictly bounded values for the delay and the delay jitter of the transmitted service data units (SDUs). This kind of time constrained service is delivered by the AAL Type 1 and the AAL Type 2. It is intended that the AAL Type 2 would cater more specifically to variable bit rate video services but few things can be said about the AAL Type 2 as standardisation hasn't progressed a great deal yet. Consequently, the AAL Type 2 will not be discussed in this paper.

The AAL Type 1 offers four constant bit rate (CBR) services to the AAL user; circuit transport (asynchronous circuits, e.g. G.703 plesiochronous digital hierarchies and synchronous circuits, e.g. I.231 circuit mode bearer service), video signal transport, voice-band signal transport and high-quality audio transport (still under consideration). These four services are offered through 4 different Convergence Sublayer (CS) protocols which make use of a common Segmentation and Reassembly (SAR) sublayer (see figure below).

CS 1.1	CS 1.2	CS 1.3	CS 1.4
SAR			
ATM layer			

Fig. 2.1 Protocol stack for B-ISDN Time Constrained Services

The four CS protocols differ mainly with respect to the QOS offered to the AAL user. The main QOS aspect in this respect is the way errors are treated by the CS. Detected errors may be notified to the user, concealed or corrected by a Forward Error Correcting method depending on the requirements of the service user. All error handling methods must be compatible with the time constraints of the service specification which excludes e.g. the use of retransmission schemes to deal with lost or errored cells.

The AAL Type 1 service is connection oriented which eases the burden of complying with the synchronous service constraints as connection resources can be allocated prior to the transfer of user data. The most important functions of the AAL Type 1 protocol are the handling of cell delay variation by means of a buffer, the recovery of timing information and, as mentioned before, error handling.

Once the service has been described, the question rises whether and at what level this service would fit into the OSI RM. It is clear, however, that the OSI RM, as such, does not generally include aspects of time relations between primitives. This is not amazing as the OSI RM was originally designed for modelling data communications. It is only during the last years that more and more typical data environments also start to make use of applications in which timing is important. This is clearly illustrated by the recent moves towards multimedia applications. As the OSI RM does not include timing aspects, the timing feature of the AAL Type 1 service cannot be explicitly used by the current higher OSI layers. But as more and more applications do require some kind of synchronicity between sender and receiver, it would be interesting to introduce timing aspects in the OSI RM.

In fact, it is not entirely correct to say that the OSI Reference Model does not include timing aspects, since at the lowest level of the OSI Model we find signals travelling on a physical medium. Depending on the transmission technique this transport can respect a timing relation between source and destination which has also been reflected in the definition of the OSI Physical Layer Service. The OSI Physical Service [CCITT X.211] [ISO/IEC 10022] knows two types of transmission; either a synchronous or an asynchronous type of transfer of Physical Layer PDUs. In synchronous transmission, SDU synchronisation is supported by the Physical Service.

In a sense the AAL Type 1 service describes the behaviour of a physical wire for synchronous transmission. This makes the AAL Type 1 service comparable to the synchronous service at the OSI Physical layer. Depending on further evolution in standardisation, it remains to be seen whether the AAL Type 1 could not also be considered as offering a datalink service as a result, e.g. of PDU delimiting, structure information and error recovery using forward error Correction (FEC).

Since no OSI Datalink/LLC layer presently preserves the synchronous character of the synchronous Physical layer service, one will have to define in OSI, a new connection oriented Datalink/LLC service transparent to time, possibly including error recovery mechanisms provided they are compatible with the synchronous constraint. Depending on CCITT, the AAL Type 1 service could be considered as such a new LLC service.

From a modelling point of view, explicit timing primitives may not be necessary. If the higher layer applications do not need direct access to the network clock, it may suffice to introduce another kind of QOS semantics into the service definition in order to model the synchronous behaviour. With QOS semantics as they are defined today, the user specifies a QOS value desired for the transmission of PDUs. The network provider tries to offer the required QOS with a best effort way of proceeding. By introducing, in the service definition, guaranteed and compulsory QOS values, the behaviour of a synchronous service with its strict bounds on delay and delay jitter could also be modelled. The idea behind the concept of guaranteed and compulsory QOS [DAN'92a] is that the corresponding protocol should, at call set-up, take the necessary actions for meeting the service requirements such as allocation of network resources in a deterministic way. If the service provider can't meet the specified QOS requirements it should refuse the connection request.

2.2 The Non Time-Constrained Connection Oriented Service

We have chosen this unfamiliar title to name a group of B-ISDN connection oriented services which have no explicit timing features. Quite often this group is referred to as the connection oriented *data* services group. However, this terminology may be slightly confusing as these services are also used for the transmission of information with time constraints but where the handling of the timing properties is done at a higher layer, e.g. when multimedia multiplexing is done in the higher layers instead of at the ATM level.

Traditionally, the connectionless transmission mode has been popular in LAN environments for datacom applications. However, for high speed transfers and multimedia applications, more and more people consider a connection oriented service in order to avoid problems of congestion that may arise with bursty traffic at high bit rates.

Describing the B-ISDN non time constrained services is probably most easily done starting from the architectural framework.

AAL-3/4	AAL 5	Null
ATM		
ATM PHY		

Fig. 2.2 Protocol stack for the B-ISDN Non Time-Constrained Connection Oriented Service

There are three AAL options for providing non time constrained connection oriented services; the AAL Type 3/4, the AAL Type 5 or a null AAL. Note that the AAL Type 3/4 can also be used for the support of the connectionless service in B-ISDN as will be discussed in section C.

The null AAL, which does nothing else but mapping the ATM service primitives to the AAL service primitives, is useful in those cases that the higher layer user would like to access the bare ATM layer service directly [MDP'91a]. This option will not be further discussed in this paper. The bare ATM layer service corresponds to an OSI Physical Layer service with, however, extensive multiplexing facilities [MDP'91b]. The ATM layer service offers the capability of realising virtual wires (i.e. virtual connections), which merely exist for the duration of the connection, in a very dynamic way.

The AAL Type 3/4 and the AAL Type 5 layers have been subdivided in three sublayers as indicated in the figure below.

SSCS 1	SSCS 2	Null	SSCS 1	SSCS 2	Null
AAL 3/4 CPCS			AAL-5 CPCS		
AAL-3/4 SAR			AAL-5 SAR		
ATM layer					

Fig. 2.3 Sublayering of Protocol Stack for the Non Time-Constrained Connection Oriented Service

The service of the Common Part Convergence Sublayer (CPCS) protocol is an unassured service which means that integral SDUs may be lost or corrupted without being corrected by retransmission. On top of this CPCS sublayer, various Service Specific Convergence Sublayer (SSCS) protocols are currently being defined. These SSCS protocols can be used to build an assured service on top of the unassured CPCS service by providing error correction with a suitable retransmission protocol and congestion control. In the assured service every SDU is delivered with exactly the data content that the user sent.

The SSCS protocols may also provide emulations of existing services. An example of such an SSCS protocol of which the specification has already progressed substantially is the Frame Relaying SSCS [ETSI FR]. This SSCS offers an emulation of the frame relaying service (non-assured) [I.233]. The different SSCS protocols can be either supported by the AAL Type 3/4 CPCS or the AAL Type 5 CPCS as their respective services are identical.

Although the service of the AAL Type 5 and the AAL Type 3/4 CPCS sublayer is the same, two different protocols are used to support it. Originally only one CPCS protocol existed. The AAL Type 5 CPCS and SAR featuring less functionality were proposed later. The main difference lies in the CS frame delineation mechanism of the SAR and the fact that the AAL Type 5 CPCS does not support multiplexing of several CPCS connections over a single ATM connection. Also, the AAL Type 5 CPCS has no provisions for allowing simultaneous interleaved transmission of several CPCS PDUs or in other words, on a single ATM connection, cells of different messages can not be found intermingled. This makes the AAL Type 5 protocol more simple. The original goal of the AAL Type 5 was also to make it easier to implement than the AAL Type 3/4 but this promise still has to materialise. The fact that the AAL Type 5 does not allow PDU interleaving at the CPCS layer makes it less suitable for the support of a connectionless service, like SMDS, as will be discussed in the next section.

Whereas, the AAL Type 3/4 does support multiplexing, from the viewpoint of system architecture, multiplexing of connections at this level of the protocol stack seems redundant. The impression is reinforced when one considers the extensive multiplexing facilities available at the ATM layer. It is for this reason that the AAL Type 5 does not provide the multiplexing facility at the CPCS layer. In the case of a network, constituted solely by ATM islands, the absence of connection multiplexing in the AAL offers the interesting possibility of direct mapping of ATM connections to higher layer connections. It is then the application which has direct access to the ATM multiplexing facility. This approach is attractive and can easily be implemented in Local ATM environments. On the wide area side, the multiplexing level at the CPCS layer may be useful (not desirable), if the network operators would charge the user by the number of ATM connections he uses.

At the level of the AAL, the B-ISDN non time constrained connection oriented service is of the same nature as the OSI Datalink Service. Some of the SSCS protocols even offer an emulation of existing datalink services. The Frame Relaying SSCS, e.g., supports the core OSI Datalink Service. The primitives of the Frame Relaying SSCS and the OSI Datalink are exactly similar as far as the data transfer part is concerned.

2.3 The Connectionless Service in B-ISDN

As mentioned before, one of the services available from B-ISDN is the connectionless service. The connectionless service from B-ISDN may be considered as a superset of the connectionless service which one can normally find in the OSI model. The typical LLC Type 1 service e.g. offers at its service interface two primitives to the higher layer user; UNITDATA-Request and UNITDATA-Indication. These primitives take as parameters a source address, a destination address and a priority parameter. The connectionless service for broadband as it is being defined in CCITT [Rec. I.364] and ETSI [ETR 53203] and which is inspired by the Bellcore SMDS specification [BCR '91], is slightly more complex and offers new features which did not previously exist. Examples of these new features are access class enforcement (see further on), source and destination address screening, source address validation. Address screening ensures that a user cannot receive data from certain users and that a user can send data to certain users only. These latter features allow for the constructing of closed user groups which is useful for oa. supporting the concept of virtual LANs [DVAN'92].

The parameters of the connectionless service in B-ISDN are, over a public network, the usual source and destination address (group addresses allowed) and parameters specifying the desired QOS for the SDU transfer. The number of QOS parameters has been increased when compared with e.g. the LLC Type 1 service, however, details still remain for further study.

Noteworthy here is that there is also a parameter specifying service characteristics with respect to the missequencing of frames. Whereas normally, a connectionless service does not guarantee correct in sequence delivery of the datagrams, a typical MAC service offered by a LAN shows a very low probability of missequencing. Consequently, for SMDS, a requirement on the missequenced PDU ratio has been specified (individually addressed packets only). In CBDS, two classes have been specified; Class 1 exhibits a negligible rate of missequenced PDUs and Class 2 does not guarantee sequence preservation. Class 1 does conform nicely with typical LAN characteristics whereas Class 2 allows for more flexible routing mechanisms responding e.g. to real time information on traffic load, which might potentially break the correct datagram sequence.

In fact the whole concept of the traditional connectionless service is evolving. E.g., if we look at the connectionless service offered by SMDS, we may observe that SMDS offers strict guarantees, not only on PDU sequence order but also on delay and delay jitter. E.g. SMDS guarantees that 95% of all packets from one network access to the other (at 45 Mbit/s) will experience a delay smaller than 20 ms. The idea, of course, is to make connectionless networks more suitable for the transport of e.g. multimedia and voice traffic. This is also reflected in recent research work on scheduling algorithms that allow some kind of organisation and guarantees of network resources.

Even although the service is connectionless, access to the Cl network, at subscription time, is associated with a negotiation of the bandwidth needs between the user and the network operator prior to the first datagram being sent. A number of access classes have been defined and Access Class Enforcement function monitors user's traffic with regard to the peak bit rate and the sustained information rate (long term average of the information rate for bursty traffic).

The connectionless service in B-ISDN needs some extra attention in this B-ISDN services overview. This is because ATM by itself, is connection oriented and provisioning a connectionless service must thus be handled in a special way. Several alternatives have been considered ranging from semi-permanent user end-to-end paths, sleeping connections to connectionless servers [TVLA'91]. We finally opted for providing the connectionless service with special connectionless service functions located in so called connectionless servers. Connectionless traffic is routed between these connectionless servers interconnected by semi-permanent ATM connections. These interconnected connectionless servers can be modelled as a virtual overlay network on top of ATM [VANL'91] [DDEL'92]. The approach is attractive in the sense that the ATM connections between the servers can be adapted to the traffic needs. As well the topology of the meshed network as the bandwidth of individual ATM connections constituting the overlay network can be adjusted to form a very flexible overlay network.

The connectionless servers are designed so that connectionless frames do not have to be reassembled in the server before routing to the next server; from the moment the first cell constituting a frame has arrived it is immediately routed towards the next server.

From the viewpoint of the protocol stack, the connectionless service in B-ISDN is offered by the so called Connectionless Network Access Protocol (CLNAP). The protocol's main functions are routing of the connectionless frames based on the final destination address (Rec. E.164 addressing) and QOS selection. The CLNAP protocol is supported by the AAL Type 3/4 providing an unassured service. The CLNAP could also be considered as another SSCS protocol but CCITT explicitly stated that routing and addressing for the connectionless service takes place in a layer above the AAL. And, thus, as a consequence, the Service Specific Convergence sublayer (SSCS) of the AAL Type 3/4 will in this case be a null sublayer.

		CLNAP
SSCS 1	SSCS 2	Null
AAL 3/4 CPCS		
AAL-3/4 SAR		
ATM layer		

Fig. 2.4 Protocol Stack for the B-ISDN Connectionless Service

As to the relationship with services defined within the OSI RM, the CLNAP service can be situated at the level of the MAC service. Due to the increased number of QOS parameters and the extra features on address screening, the CLNAP service does not exactly map to the MAC service but both services can clearly be modelled at the same level.

The CLNAP service allows for easy bridging between the connectionless service in B-ISDN and IEEE 802 LANs and MANs as described below. In this way high speed catenets can be formed. We will consider LANs or MANs interconnected through an SMDS or CBDS network on top of ATM by using the remote bridging technique [WROZ'92], [DVAN'92].

In this approach, frames from a LAN or a MAN are encapsulated into an SMDS packet at the sending side. Then, the SMDS packets are transferred over the SMDS network and at the remote side, the SMDS packets are decapsulated to recover the LAN or the MAN frame. The SMDS network simply acts as a transparent means of interconnection.

Note that in principal, this technique is only suitable for interconnecting homogeneous networks. Heterogeneous networks can also be bridged (e.g. a LAN to a MAN). However, this may require extensive format conversion as well as complete address translation.

There are basically two approaches to remotely bridge LANs over networks such as SMDS. The first is the point-to-point link approach, the second is the virtual LAN approach. Hybrid approaches are possible, combining both point-to-point links and virtual LANs. The ideas, summarised below, have been described in greater detail in [WROZ'92] and [DVAN'92].

If two stations know each others SMDS address, we can say that there is a logical point-to-point link between them. These logical links can be used to set up a mesh of paths between SMDS Remote bridges, just as in traditional point-to-point remote bridging approaches. The spanning tree protocol will be used to ensure that in catenets of interconnected networks, frames cannot make loops.

The other approach to remote bridging over SMDS uses the concept of virtual LANs. A group address can be established on the SMDS network, of which all SMDS remote bridges can be a member. The SMDS network will take care that one copy of frames sent to this group address will be delivered to all SMDS remote bridges in the group. This simulates the operation of a real shared-medium LAN (e.g. an Ethernet), where frames are also always seen by all stations on the LAN. The group address can thus be seen as a virtual LAN. Obviously, if all frames are sent to the group address, this will lead to a lot of superfluous traffic on the SMDS network, and thus to a higher cost to the user. Many frames could probably be directed to only one other remote bridge, without the need to send to the group address by performing self-learning of remote addresses. Both techniques, the Virtual LAN approach and the point-to-point approach can be combined in order to minimise traffic on the SMDS network.

3 Conclusion

In this paper three main groups of B-ISDN services have been described; the time constrained connection oriented services, the non-time constrained connection oriented services and the broadband connectionless service. For each group, where possible, a relation with services defined within the OSI Reference Model was indicated.

References

[BOE'92]   Boerjan, J. , **ATM LLC Assessment**, *Esprit Project 5341/OSI95 Deliverable, BELL-1*, November 1992

[Lanc-3]   J. Boerjan, A. Campbell, G.Coulson, F.Garcia, D.Hutchison, H.Leopold And N. Singer, **TPX Service Evaluation**, *Esprit Project 5341/OSI95 Deliverable Lanc-3*, November 1992

[ORL'92]   D.J. Greaves, D. Mc Auley, **ATM Network Services for Workstations**, *The OSI95 Transport Service with Multimedia Support on HSLAN's and B-ISDN*, A. Danthine, ed., Springer Verlag, 260-278

[DAN'92a] A. Danthine, Y. Baguette, G. Leduc, L. Léonard, **The OSI95 Connection-Mode Transport Service - The Enhanced QOS**, *4th IFIP Conference on High Performance Networking*, Liege, December 1992

[DAN'92b] A. Danthine, OSI95 - **New Transport Services for High Speed Networking**, *3rd Joint European Networking Conference*, Innsbruck, May 11-14, 1992

[DAN'92c] A. Danthine, **A new Transport Protocol for the Broadband Environment**, *IFIP Workshop on Broadband Communications* - Estoril - Jan 1992.

[LEO'92]   H. Leopold, A. Campbell, D. Hutchison, N. Singer, **Towards an Integrated Quality of Service Architecture (QOS-A) for Distributed Multimedia Communications**, *Proceedings 4th IFIP conference on High Performance Networking*, Liege, December 1992.

[MDP'91a] Martin De Prycker, **Asynchronous Transfer Mode, Solution for Broadband ISDN**, *Ellis Horwood Series in Computer Communications and Networking*, 1991.

[MDP'91b] M. De Prycker, R. Peschi, T. Van Landegem, **B-ISDN and the OSI Protocol Reference Model**, *Proceedings Third IFIP Conference on High Speed Networking*, Berlin, March 1991.

[MDP'93]   M. De Prycker, R. Peschi, T. Van Landegem, **B-ISDN and the OSI Protocol Reference Model**, *IEEE Network*, March 1993

[VANL'91]   T. Van Landegem, R. Peschi, **Managing a connectionless Virtual Overlay Network on Top of an   ATM Network**, *Proceedings ICC'91*

[DDEL'92]   D. Deloddere, P. Reynders, P. Verbeeck, **Architecture and Implementation of a Connectionless Server for B-ISDN**, *Proceedings ISS'92*, Yokohama

[WROZ'92]   W. Rozenblad, B.Li, R. Peschi, **Interconnection of LANs/802.6 Customer Premises Equipments, (CPEs) via SMDS on Top of ATM: a case description**, *4th IFIP Conference on High Performance Networking, Liege*, December 1992

[DVAN'92]   D. Van Achter, **Routing approaches for remote LAN bridging over SMDS using the spanning tree**, *Interworking '92, Berlin*

[ISO 7498]   ISO Standard 7498, **Reference Model for Open Systems Interconnection**

[I.321]   CCITT Recommendation I.321, **B-ISDN Protocol Reference Model and its application** (1991)

[I.363]   CCITT Recommendation I.363, **B-ISDN ATM Adaptation Layer (AAL) specification** (July 1992)

[I.364]   CCITT Recommendation I.364, **Support of Broadband Connectionless Data Service on B-ISDN** (1993)

[I.233]   CCITT Recommendation I.233, **ISDN Frame Mode Bearer Services (FMBS) - ISDN Frame Relaying Bearer Service**

[ETS 300]   European Telecommunication Standard ETS 300 217 ETSI, **Connectionless Broadband Data Service** (1992)

[ETR 53203] European Telecommunication Report, ETSI TC-NA, Network Aspects; **CBDS over ATM** (Dec 1992)

[ETSI FR]   Draft European Telecommunication Report (DETR), **Support of Frame Relaying Bearer Service in B-ISDN**, ETSI/STC- NA5(92) 4, November 1992

[BCR'91]   Bellcore Technical Reference, TR-TSV-000772, May 1991, **Generic System Requirements in support of Switched Multi-Megabit Data Service.**

[IVS'92]   CCITT Study Group XVIII - Report R 103, **Integrated Video Services (IVS) Baseline Document**

List of abbreviations

AAL	ATM Adaptation Layer
ATM	Asynchronous Transfer Mode
B-ISDN	Broadband ISDN
CBR	Constant Bit Rate
CLNAP	Connectionless Network Access Protocol
CPCS	Common Part Convergence Sublayer
FDDI	Fibre Distributed Data Interface
FEC	Forward Error Correction
IDU	Interface Data Unit
LAN	Local Area Network
LLC	Logical Link Control
MAC	Medium Access Control
MAN	Metropolitan Area Network
PDU	Protocol Data Unit
PRM	Protocol Reference Model
QOS	Quality of Service
SAP	Service Access Point
SAR	Segmentation and Reassembly

SDU	Service Data Unit
SMDS	Switched Multimegabit Data Service
SRTS	Synchronous Residual Time Stamp
SSCS	Service Specific Convergence Sublayer
VBR	Variable Bit Rate

ATM Network Services for Workstations

D. J. Greaves and D. McAuley
Olivetti Research Ltd, Cambridge, UK.
Email: dm, djg@cam-orl.co.uk

Workpackage B of OSI95 was titled 'New Communications Techniques' and was an evaluation of how to make use of the new communications techniques which offer a guaranteed quality of service, including ATM and B-ISDN. This document discusses the provision of the ATM networking services to application programs running on general purpose computing equipment which is connected to an ATM network.
    Keywords: ATM, Adaptation, Protocol, Interface.

1 Overview

Section 2 of this document asserts that ATM may either be integrated into end systems or else be used as an encapsulation technology. Section 3 describes how and why ATM technology should be used in the local area. Section 4 describes the essential component of service common to all ATM systems. Section 5 describes the Multi-Service Network Architecture (MSNA) approach to ATM integration. Section 6 describes a set of logical link control primitives for access to ATM. Section 7 describes connection set up and control within MSNA. Section 8 describes the design space for ATM host interface hardware. Greater detail for each topic may be found in the related OSI95 'deliverable' documents.

2 How May we Use ATM ?

A new transport protocol may either use existing network layer services or may be defined in terms of new network layer services. The work within Workpackage A of OSI95 which developed the TPX transport protocol specification has adopted the former approach, using the standardised OSI connectionless network service [BLL92].

The MSNA (multi-service network architecture) approach, described herein and in [MAC-89] takes the alternative route of offering a new network layer service. Within MSNA, ATM virtual circuits are brought into the end systems and terminated at or above the network layer service access point. This opens the way for a transport protocol which treats the ATM layer as a connection-oriented network layer service.

Recently, within the Internet Engineering Task Force and also within the highly regarded discussion group *atm@sun.com*, there have been proposals for using network layer addresses as the termination points for ATM virtual circuits. This is in keeping with MSNA.

2.1 Encapsulation versus Integration

ATM may be used as the bearer service for encapsulation of existing protocol stacks. For example, the Bellcore SMDS service offers encapsulation and, in the future, will operate over ATM. In addition, a draft RFC 'Multiprotocol over ATM Adaptation Layer 5' by Juha Heinanen specifies encapsulation formats for virtually all existing protocol stacks.

Task B3 of project OSI95 has produced interfaces for ATM end systems and therefore has had the chance to introduce ATM specific lower-layer protocols into the end systems. The MSNA approach to this has been introduced in OSI95/ORL/Deliverable-1.

Encapsulation of older protocols, such as IP, cannot offer new services. For example, IP does not support a large set of quality of service options and those which are specified are not associated with guarantees or always implemented. This situation has been described by Bob Metcalfe as "the worst of both worlds" because the overheads of an ATM subsystem are present but the advantages of ATM are not being realised. In order to support new applications, such as real-time traffic, it is sensible to concentrate on full integration of ATM with the existing protocol stacks. We leave the field of encapsulation to those who are unable to modify their end systems. This is addressed within OSI95 in [BoP93].

3 ATM in the Local Area

There is no point supporting multiple classes of quality of service within the international ATM network or local ATM backbones unless these separate qualities of service can be presented to the individual application programs loaded into an end workstation. This consideration is addressed in an approach where the ATM virtual circuit is terminated inside individual applications and there is no multiplexing of multiple applications onto a common virtual circuit.

The arguments which convinced the CCITT to recommend ATM as the solution for Broadband-ISDN also operate in the local area for private multiservice networks. Research work to establish this point has been carried out within OSI95 at Olivetti Research Ltd. and elsewhere. Increasingly, computer manufacturers are promoting the use of ATM techniques in privately owned and operated digital networks.

The attractive properties of ATM itself include:

- Support of a mixture of traffic types, including fixed, variable rate, and bursty traffic.

- Low jitter owing to short cell size and reduced switching delay due to cut through of multi-cell blocks.[1]

The fact that B-ISDN has adopted ATM as its transfer mode brings two further, consequential advantages:

- Increased availability of special purpose VLSI devices.
- Opportunity for ease of interoperability between private and public networks.

These advantages can only be fully realised if the private networks use the same size cell *payload* as the public networks, namely 48 bytes. However, the header size and *header format* are not important in terms of the service offered to the end systems, since in ATM, headers can be manipulated by each switching entity, while the payloads are passed unaltered from one point to another (indeed two header formats are defined within B-ISDN). This implies that network manufacturers can use an optimum cell header for the number of virtual circuits and particular media characteristics that their equipment supports. An example is a current Olivetti Research ATM radio project, where sequence numbers and a MAC layer response field are put in the cell header.

It is not universally agreed that 48 bytes is the optimum payload size for a general purpose network. On the other hand, if one agrees that a fixed-size packet (or cell-based) network has superior real-time performance to one which supports variable length packets, then it is clear that, at least, all components of the network infrastructure should support the same cell size. This implies that private ATM networks should employ the same 48 byte payload size as the CCITT's B-ISDN.

4 The ATM Service

The definition of 'ATM' from the CCITT recommendations is: 'the use of a fixed-length cell as the primary means of information transfer where the periodicity of cells is not known by the receiver in advance, but it is indicated by a circuit identifier in the cell header.'

More specifically, any equipment which can offer the CCITT ATM layer service is a suitable component for a private ATM network. The ATM service is the ability to transparently ship ATM cell payloads along a virtual circuit from one ATM-layer SAP to another while preserving order. A result of our work is the realisation that the implementor of an ATM interface for a host is free to use any techniques which meet the ATM-layer specification. Naturally he will also benefit if the lower and higher layers of his implementation are a CCITT or ATM Forum standard, but he may have good reason for using alternative techniques.

Private ATM networks may be interconnected over the public B-ISDN network without an in-band processing overhead provided the B-ISDN service is accessible at

---

[1]Cut-through switches are a class where the start of a message may already have left the switch on the appropriate output port, before the end of the message has been received at the input.

the ATM layer. The cost of a virtual circuit is likely to be greater within the global public network than in a private network. Therefore translation between signalling and addressing formats, along with other management functions, will probably have to be implemented at the gateway.

5 The MSNA Approach to Multiplexing.

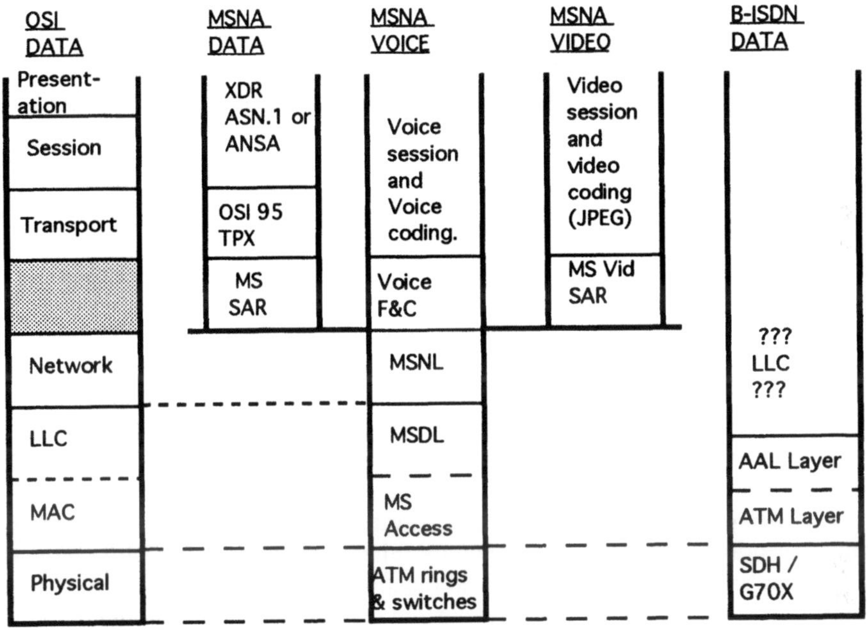

OSI DATA	MSNA DATA	MSNA VOICE	MSNA VIDEO	B-ISDN DATA
Present-ation	XDR ASN.1 or ANSA	Voice session and Voice coding.	Video session and video coding (JPEG)	
Session				
Transport	OSI 95 TPX			
	MS SAR	Voice F&C	MS Vid SAR	
Network		MSNL		??? LLC ???
LLC		MSDL		AAL Layer
MAC		MS Access		ATM Layer
Physical		ATM rings & switches		SDH / G70X

Fig. 5.1 Functional mapping of MSNA to the OSI reference model and CCITT B-ISDN reference model.

The Multi-Service Network Architecture (MSNA) protocols have been used at Olivetti Research Ltd. and at the University of Cambridge Computer Laboratory. Recently the MSNA system has been extended using 34 Mbit links from the new UK academic network 'SuperJANET' to University College London. Although these sites internally employ and are interconnected by a variety of networks, the MSNA platform offers a homogeneous 'ATM Internet'. The aims of MSNA include the provision of transparent inter-operation between the ATM and non-ATM style networks to support access to devices and machines situated on Ethernet and other non-ATM subnetworks. MSNA is designed for efficient implementation in software, with or without hardware support. It makes efficient use of transmission bandwidth, allowing many types of transaction to fit into a single ATM cell. Management and connection procedures are conducted 'out-of-band', allowing implementation of data paths in hardware. Most importantly, despite providing interoperation between different classes of computer network, it attempts never to compromise the capabilities of the underlying network technologies.

Current CCITT work, regarding the integration of ATM with the OSI 7 layer model, can effectively hide ATM specific features below an LLC service interface. The current 802.6 MAN protocol stack takes a similar approach. As will be shown, the MSNA approach is radically different, especially with regard to the semantics of its link layer connection and its elevated position of the segmentation and reassembly function. This is shown in Figure 5.1.

The most basic MSNA function is the interconnection of MSNA service access points (MSAPs). These are identified by a 64 bit MSNL (Multi-service Network Layer) address. They are basically a concatenation of a conventional network layer address and an application port number. Owing to the separation of control and data within ATM, these addresses need never appear in the header field of a protocol data unit. They are communicated in the data fields of management PDUs. A design goal for MSNA was , as far as possible, to eliminate multiplexing outside the physical layer.

5.1 'Layered Multiplexing Considered Harmful'

Many conventional protocol implementations, including OSI and TCP/IP, offer provision for multiplexing at multiple points in their layered architecture. By *multiplexing*, the process of tagging component SDUs from several streams with a different stream identifier and then merging the resulting streams is implied. One aim of MSNA was to avoid unnecessary multiplexing. Tennenhouse summarised the motivation for this approach in his short paper *'Layered Multiplexing Considered Harmful'* [TEN-89]. The unnecessary and harmful effects of layered multiplexing in a connection-oriented ATM environment are reiterated here in this section. However, multiplexing and demultiplexing are vital functions, if only to enable multiple streams to share one network, and so multiplexing and demultiplexing are inevitably required at certain points in the architecture. MSNA attempts to confine this function to a single point: specifically the ATM switching layer.

To give a feel for the argument, we quote from Tennenhouse's paper:

> 'The principal advantage of layered architectures is that they provide for the *step-by-step enhancement of communications services* [ISO-1]. In theory, each service boundary between adjacent layers identifies a stage in the enhancement process. In order to minimise the duplication of functionality across layers, the network architect should *collect similar functions into the same layer*.[2]

> 'In the case of the multiplexing function, this principle has largely been ignored. For example the OSI architecture presently provides for

---

[2]Principle P4 guiding OSI layer determination [TEN-2] (Tennenhouse's footnote).

multiplexing within six of the seven layers of the protocol stack.[3] It is claimed that the extensive duplication of multiplexing functionality across the middle and upper layers is harmful and should be avoided.'

Let us consider, as an example, the layers of multiplexing present when conventional TCP/IP is run over an Ethernet. Inevitably there is multiplexing of packets from different sources and destinations over the physical medium of the Ethernet. The corresponding de-multiplexing function is performed on a MAC layer address basis by the interface hardware. Please note that there is only one MAC layer address for a station and that this is fixed for the lifetime of the station. (As will be shown, a different approach is possible in an ATM system.) Conventional LANs of this nature, the Ethernet being no exception, generally include an additional source address field in the MAC layer header. Note that the source address field is not used by the media-access hardware and that the destination address field has no further role after the message had been received by the hardware. In an ATM architecture, these fields can be compressed into the virtual circuit identifier field.

Encapsulated in the MAC data field is the IP (Internet Protocol) header. This is usually 20 bytes in length and contains a 32 bit IP source address and a 32 bit IP destination address. Encapsulated in the IP data field there is a TCP header. This contains a 16 bit source port and a 16 bit destination port addresses (NSAPs). These identify the network layer service access points. Still further levels of layered addressing can often be found: for instance some RPC systems multiplex their own logical connections over a single NSAP. If we only consider multiplexing in and below the transport service layer, by examining the TCP and IP headers, we find

$$2 \times (48+32+16) = 192 \text{ bits } (24 \text{ bytes})$$

of addressing associated with every message. This is just to support a single end-to-end TCP byte stream.

As Tennenhouse stated, a justification for layered protocols is provided by the step-by-step enhancement principle. However layers of *multiplexing* below a particular service layer boundary are intended to be completely hidden and therefore to offer no perceived enhancement of service. Indeed, hidden multiplexing can sometimes manifest itself as *crosstalk*, causing interference between the multiplexed streams. By 'crosstalk' we mean that the service within one logical instance of a protocol stack within a machine experiences interference, mainly in the form of processor service jitter, resulting for messages arriving for other instances of the protocol stack.

On the other hand, worthwhile enhancements of service is provided by certain of the algorithms at end and intermediate network points and their associated fields in the headers. An example is flow control. Equivalent functionality to these services must be found or preserved in a new protocol architecture which is avoiding multiplexing.

---

[3] The presentation layer is the sole exception. However, presentation address selectors have been incorporated into the naming and addressing scheme on the grounds of *architectural consistency* (Tennenhouse's footnote).

5.2 Layered Protocols in a Connection-Oriented Environment

Given that avoiding multiplexing allows us to delete the sub-addressing fields in the protocol headers, the remaining information bearing fields can be divided into two categories: those which are of constant value (the same value for every message sent) and those whose value varies each time. Protocol identifiers are examples of the first category and checksums are examples of the second. In a connection-oriented environment, fixed circuit parameters, including quality of service requests and protocol identifiers can be properties of the connection and then need not be sent every time. This corresponds to 'VC-based multiplexing' in Juha Heinanen's draft RFC.

These connection attributes can be permanent properties of a permanent circuit, requested properties of a dynamically set up circuit, or negotiated properties, dependent on the networking resources available. MSNA uses connection-oriented *lightweight virtual circuits*, and the MSNA architecture supports any of these approaches to circuit establishment. A connection-oriented system has advantages for delay-sensitive traffic in a multi-media environment, since the probability of adjacent messages between the same source and destination being routed along different paths is much reduced. Fixed routing is almost certainly required for jitter sensitive, multi-media services, and therefore is be provided in MSNA to achieve its multi-service goal.

Returning to the Ethernet/IP/TCP protocol combination; a brief calculation shows that if it were possible to delete the addressing and constant value fields, the header length would be reduced from 52 bytes to about 15 bytes.[4] This is of great interest considering the ATM cell size of 48 bytes.[5] As stated, at least one level of addressing is always required, and when MSNA is run over an ATM substrate, MSNA uses the cell VCI[6] in the cell header for this purpose.

Through the deletion of the fixed fields from the low-level protocol headers, and by concentrating the multiplexing function at a single point just above the media access layer, it is clear that the size of the smallest network layer service PDU, including a

---

[4]This estimate is obtained by counting the following fields only. These are the IP time-to-live (1 byte) and length (2 bytes) fields and the TCP source and acknowledgment sequence number fields (4 bytes each), and the checksum and window fields (2 bytes each).   The IP fragmentation facility has been ignored in this account since MSNA supports fragmentation using MSSAR above the MSNL layer.

[5]Cells of length 32 bytes have been used by several of the older ATM networks constructed in Cambridge. When appropriate, the length of a cell to be used on a circuit is an MSNL liaison attribute. It may be argued that the 52 to 15 byte comparison is not strictly valid, since the figure of 52 bytes includes the two 48 bit Ethernet MAC addresses, where if all of the multiplexing is concentrated into the VCI, the cell header length should also be counted. The discrepancy arises since the Ethernet MAC layer address cannot be used as flexibly as a directly interpreted VCI on an ATM network.

[6] In this document, we make no distinction between VCI and VPI (virtual circuit and virtual path identifiers), and since a general principle of MSNA is that there should only be one multiplexing mechanism, a distinction is not desireable. However, an item of hardware, such as an ATM switch or host interface is free to partition the header field in any convenient way (e.g. to ease fast lookup). Naturally, the MSDL entity which allocates VCI values for a particular device must know what range of VCI fields the device can handle.

few bytes of actual data, can be reduced to a single cell. When RPC and transport layer services are overlaid, including optional segmentation and reassembly protocols, many types of useful application traffic still map into single cell messages.

To summarise, the principle advantages of the points outlined so far are: the reduction of protocol header overhead on the network, the reduction of processing time needed to generate and check headers, and the confinement of crosstalk between multiplexed traffic streams.

5.3 Further Aspects of Layered Multiplexing.

When intermediate layers of multiplexing are eliminated, essentially we are left with multiple instances of low-level protocol stacks, one for each virtual circuit. These, of course, do not need to be the same, but may be chosen to suite the traffic type in use. For high speed operation, the individual protocol stacks can be implemented in one monolithic section, rather than separate, layered software modules. All state variables of a connection can then be held in one activation record, rather than in several different places, reducing the number of time-consuming look up operations per message. Monolithic software can usually be faster than layered software and can be written in modular form given appropriate compiler technology.

Further benefits accrue when we consider the implementation of a protocol stack in a multi-threaded environment. When a single processor is servicing multiple client processes or threads, scheduling decisions must be performed efficiently, fairly and with regard to minimising context switching rate, and therefore overhead. Inefficiencies arise if incoming messages have to be partially demultiplexed before their recipient process can be identified. This also precludes exact resource accounting for scheduling purposes. MSNA avoids this by eliminating intermediate multiplexing points, thereby enabling the correct recipient thread to be identified as soon as a message arrives and requiring no subsequent context swaps as a 'port' or other sub-addresses are uncovered.

As stated before, the use of separate instances of a protocol stack for each logical connection also enables heterogeneous protocol stacks to be used if desired. An example is that not every connection will require segmentation and reassembly information. An obvious case is where only single cell messages are to be used on a connection. This may be the case for acknowledgement cells going in the reverse direction to a forward circuit, or for multicast messages which are placed in a single cell to ensure atomic reception. Using a null SAR layer obviously improves performance, both in terms of processing overhead and payload usage. Voice is another example: a time stamp is probably more appropriate than SAR information. There is no multiplexing of different SARs over an active liaison.

In a multi-processor or multi-media workstation, it is convenient for the hardware to demultiplex incoming messages according to which processor, or framestore, or whatever, they are destined for. For instance, some streams of traffic may require to be routed through decompression or de-encryption hardware, while others may not. Again, each of these types of traffic will generally treat the cell information field in a different way, and therefore require a separate protocol. When multiplexing is performed on a virtual circuit basis, streams may be routed in hardware to the appropriate resource, just using bits from the cell header.

Within a uni-processor system, it may be argued that a direct implementation of an entirely vertical protocol stack shifts multiplexing complexity from the networking software to the process scheduler. It is certainly true that the lower levels need to operate on multiple PDUs in parallel. For reception, this is necessary when multiple higher-level blocks are being reassembled in parallel, and it is also required on the transmit side when there is a rate limitation on particular VCIs, or to avoid head-of-line blocking in general. It is, after all, the scheduler's job to control multiplexing.

6 LLC Primitives for an ATM Interface

In this section we give a brief consideration to the formal primitives for an ATM interface which would approximate to the LLC level of the OSI reference model. The primitives offer service to both the in-band network layer and the out-of-band protocol layer for operation and maintenance.

We consider the connection control plane's protocols to be implemented below the offered primitives. This is appropriate since the complexity of connection establishment lies in end-point naming, which is above the LLC layer, and in bandwidth reservation databases, which (logically) lie outside the host: they lie in the ATM interconnection fabric.

For the in-band data, multiplexing may be avoided by using a null connection-oriented network layer on top of these LLC primitives. This is how many of our MSNA implementations work.

6.1 OSI LLC Situation

The ISO/IEC 10039 MAC service description defines an abstraction of the MAC definitions 802.3, 4, 5 and 7 [SC6-90]. Only connectionless service is specified in the 1990 draft. The MAC services provided in the abstraction are:[7]

- Independence from the underlying MAC and physical layer, except of course, in terms of quality of service.
- Transparency of transferred information, except there may be a byte limit on PDU size.
- Priority selection - the MAC service makes available to MAC service user a means to request the data at a specified priority.
- Addressing - the MAC service allows the MAC service user to identify itself and to specify the MSAP (MAC layer service access point) to which data is to be transferred.

The PDU transferred is termed the *unitdata object*. The MAC service is allowed to

---

[7] This list is slightly condensed from section 6, 'Overview of MAC service', from [SC6-90].

- Discard objects,
- Change the order of the objects, and
- Exhibits a negligible rate of object duplication, and
- Exhibits negligible reordering of objects of a given priority.

The relative rates of occurrence of the events are assumed to be known *a priori*. The service specification states that the receiver is not able to influence the speed of the transmitter, although it does not discuss rate control at the transmit side between the MAC and the MAC transmit side user. MSAP addresses are defined for broadcast and group addressing and it is stated that the received addresses must be the same as that transmitted.

6.2 An ATM LLC Primitive Set

In an example ATM LLC, in order to open a virtual circuit, we might use the primitive

        lid = vc_open(ADDR destsap, QOS qos);

where a negative value of lid indicates failure. We assume lid is a local identifier for a local virtual circuit control block (descriptor). The fields in this control block are established using signalling protocols. The most important value held is the VCI and VPI to be put in the header of all cells sent on this virtual channel.

To close a virtual circuit one may use

        vc_close(lid);

The quality of service field includes the maximum rate to be used on the virtual circuit and other parameters such as the expected average rate of cells and whether cells should be dropped or delayed when unavoidable.

Outgoing rate control will be parameterised within the QoS values supplied in the open primitive and implemented below the LLC layer by the interface or device driver.

In order to send data we may either use LLC primitives with the AAL underneath or we may have LLC primitives which have direct access to the ATM layer. In this example set of primitives, both types of access are available according to which primitive is invoked.

Transmission of a cell might be done as follows

        send_cell(lid, char *data);

where data is a pointer to a cell payload data unit (held in 48 bytes of consecutive memory), and receive might use

        rc = receive_cell(lid, char *data);

which is a blocking call which returns when one cell has been read. The return code is available for error reporting.

Where the primitives automatically perform segmentation and reassembly using a segmentation AAL, we must extend the two primitives just presented with the addition of a length field

        send_block(lid, int length, char *data);
        rc = receive_block(lid, int length, char *data);

where length is the number of data bytes to be transmitted or the buffer length available for reception. The return code is available to report errors and the actual size of the data unit received.

6.2.1 F4 and F5 Operation and Management Flow Messages

Cells of the F4 and F5 flows of a virtual channel are identified by a value in the cell header type field. Such cells may only be sent and received once their virtual circuit is established. Access to the F4 and F5 flows at the LLC boundary may be offered using primitives similar to those for normal data. The local management plane entities may either use these primitives or the ATM management may be considered to lie below the LLC boundary. In any case, the device driver's interface to the hardware is the same as for in-band data, except for the need to produce the alternative payload types.

6.2.2 Cell Loss Priority

The CCITT standards have not finalised how congestion control and cell loss priority should be handled by the ATM adaptation layer. It is reasonable that our LLC primitives ignore the CLP flag of received cells and set the outgoing CLP flag according to the underlying AAL standard. Therefore CLP does not appear in the primitives.

6.2.3 Congestion Indications

Received cells indicate congestion through values in the payload type field of the cell header. Congestion notification can be passed up to transport level entities (including VBR video compression entities, etc.) in order for them to modify their sending discipline. Alternatively, congestion may be considered an ATM management issue and the use of congestion indications can be restricted to modifying future connection acceptance and routing decisions.

Current CCITT working documents recommend that the congestion indications may be available at the AAL service layer but the mapping onto cells is not agreed. A sensible approach may be for the AAL congestion indication to simply reflect the congestion indication field of the end-of-message cell. A host interface may easily provide this and the LLC primitive return code may be extended in range to include it.

6.2.4 Generic Flow Control

When the interface is connected to media which employs generic flow control, the generic flow control protocol may be implemented entirely below the LLC layer, probably in the interface hardware. Lack of throughput for outgoing cells may be detected from an increase in the outgoing queue length and notification passed up through the congestion indication primitives.

7 MSNL and Connection Set Up

McAuley defined Multiservice Network Layer (MSNL) as follows [MAC-89]. The multi-service network layer is an interworking service which is based on the idea of MSNL *liaisons*. An MSNL liaison is a *lightweight* concatenation of MSDL associations. There are three important aspects of MSNL:

- it defines the MSNL addresses,
- it defines association and liaison set-up procedures,
- it does not multiplex its liaisons over the MSDL associations.

MSNL provides an out-of-band connection establishment mechanism. Since MSNL liaisons are not multiplexed over MSDL associations, MSNL does not require in-band protocol headers in the service units, so MSNL introduces no processing overhead on the data path and it provides the same data interface as an MSDL association.

Defining the MSNL connection as *'lightweight'* means that the resources allocated to the connection are neither to be thought of as valuable or permanent. As discussed in slightly more detail later, MSNL connections may be unilaterally de-allocated, or fail in other ways, in which case MSNL-layer software may provide re-establishment without explicit interaction with higher-layer software. On the other hand, MSNL users cannot always be expected to explicitly close connections (for instance they might be unexpectedly re-booted), and so the existence of garbage collection mechanisms is assumed.

An MSNL liaison is established between two MSNL SAPs (MSAPs). These are unique and are allocated from the 64 bit global address space.[8]

A host computer may have many MSAPs (loosely corresponding to the conventional idea of multiple ports), but on the other hand, there may be many microprocessors sharing a single MSAP, such as individual controllers on the ports of a fast-packet switch. In general, for ease of routing decisions when a connection is set up, it is beneficial if the structure of the 64 bit numbers is actually hierarchical. In [MAC-89], the division into separate, 32 bit, *identifier* and *port* fields is suggested. This optimises the typical case where multiple MSNL clients are situated at a single location (host). However, the individual client streams do not become multiplexed, owing to the separate liaisons for each client.

Setting up an MSNL liaison involves establishing a concatenation of MSDL association hops. MSDL does not have a mechanism for naming peers before an association is set up, so cannot directly perform this function. The MSNL address provides a naming mechanism, and so provides the basis for association and hence liaison set up.

---

[8] The initially adopted approach was to base 32 of the 64 bit address on IP addresses. This provided a convenient unique identifier space. Increasingly we have MSNL entities, such as ATM cameras, which do not have an IP address, so MSNL addresses have now become less tightly coupled to IP addresses.

7.1 Promiscuous MSNL

For start-of-day reasons, MSNL must be able to contact at least one management service before any connections can be established. (Permanently established virtual circuits do not require this, but then they do not have a 'start-of-day' case.) This management service is contacted using *meta-signalling* consisting of idempotent MSDL messages on one of a set of *well-known* VCIs. The receiver of these messages must be prepared to accept messages from any source. This is known as *promiscuous* MSNL. Messages sent on such a virtual channel are not part of an association.

This initial management entity may either be a fixed server offering the start-of-day services, or if the VCI is in fact a multicast address, this basic service may be provided in a distributed manner by a number of machines. Bridges and switches must, by default, route start-of-day virtual circuits from all ports where equipment might be connected, to an appropriate server. No state is retained for incoming messages on a promiscuous association. Such messages must be idempotent and fit into a single MSDL PDU (cell).[9]

Once a mechanism for the start-of-day problem is provided, arbitrary levels of indirection can be inserted before the service actually wanted is reached, although there is no advantage to excessive complexity. Of course, the most important services are bandwidth reservation and connection establishment.

At least one level of indirection is sensible for reliable connection establishment. A connection involves allocation of VCIs from VCI space, if not the reservation of other resources, such as buffer space and bandwidth. Allocation of resources cannot be done with idempotent semantics, hence the requirement for the establishment of a signalling connection. The last level of indirection in the chain generally becomes the only one frequently used, since sensible implementations are able to cache earlier results. In practice, this means that a station has signalling MSNL connections to management entities which are able to establish further connections to related MSNL addresses, or else provide further indirections to further management entities. In the current implementation, the hierarchical structure of the MSNL address is used along with mask fields to help with this. The concept of 'caching' the earlier enquiries, rather than retaining hard state relating to connection management entities, provides intrinsic adaptation to network reconfigurations. If a cached valued is found inadequate, the concept of an 'authoritative' inquiry can be used, which will propagate backwards through the network, ignoring cached responses, until the source is found.

7.2 Connection Closure and Time-Out

Liaison closure can either be performed explicitly by higher layers at the end points, or autonomously, when any intermediate entity wishes to reclaim resources from an

---

[9]CCITT draft recommendation I.311, 'B-ISDN General Network Aspects' defines the *meta-signalling channel*. This provides a start-of-day service which is available before signalling channels are allocated. As in MSNA, this channel is situated on a well known VCI (VCI=0x00001) and uses single cell messages.

old or under-used association. Entities performing a close have the opportunity to inform other participating entities of impending closure.

There is also the possibility that connections can be re-routed or re-established by the MSNL layer, while minimising interruptions to the provided service. An example is virtual circuit hand-over in a mobile ATM radio system. Further aspects of liaison set up, routing and closure are presented in [MAC-89], however, in this report it is only necessary to concentrate on the ATM and LLC specific features of MSNA.

8 Design Space for ATM Host Interfaces

This section considers the implementation space for an interface, concentrating on the trade off between protocol implementation in the host against within the interface.

An ATM host interface receives ATM cells from the physical layer and, in conjunction with the device driver, allows processing of their contents. It also transmits cells onto the physical layer.

This section limits itself to data-oriented applications and to systems with a conventional host bus architecture. Multi-media traffic carried over data adaptation layers, which is a likely situation for workstation videoconferencing, is covered implicitly.

8.1 Four Major Parameters

Four major parameters of a host interface are briefly discussed in this section; more subtle parameters are examined in subsequent sections.

1. How fast does it go? Three speeds are of primary interest:

- The cell rate on the physical medium (a function of the signalling rate and coding and framing efficiency).
- The data rate sustainable over the host bus. This relates to the percentage of cell slots on the physical medium that the host interface can fill or empty.
- The data rate delivered to host processes after processing by the operating system drivers, protocols and sockets.

Other speed issues include: the maximum rate at which full cell slots may be handled, performance degradation when transmitting and receiving simultaneously and performance degradation when incoming VCI's are highly interleaved.

2. Which physical layer? The physical layer may be optical or copper and may use one of several line codes and cell framing mechanisms. Three framing mechanisms are of primary interest:

- Pure ATM (i.e. non-SDH) using TAXI block coded data according to theT1.S1 specification as described in the Appendix to this document.

- Synchronous Digital Hierarchy (SDH) encapsulated ATM according to the CCITT packing functions and using the SDH scrambler.
- The ATM Forum block-coded, fibre channel-like encapsulation.

All of these can operate at various line speeds on different physical media with various connectors.

3. Which host bus? The interface normally connects into an expansion slot of the workstation. What type of bus and/or workstation is supported?

4. Which device drivers and operating systems are supported? A separate device driver is required for each operating system and each workstation.

Given these basic parameters, the next distinction between designs is whether the interface maintains any state between handling one cell and the next.

8.2 Interfaces Without Inter-Cell State

If the interface does not maintain state information between cells, then the interface requires processing support from the host CPU for each cell received or transmitted. The functions of such an interface are thus restricted to:

- generation, checking and correction of the per-cell header check (HEC),
- generation and checking of payload per-cell CRCs, as used in AAL-3 and AAL-4,
- outgoing peak rate control, determined by an interface timer control register
- possibly calculating multi-cell CRCs, as used in AAL-5, but with the host being required to load and save the running CRC residue register at the start and end of each cell.

Although host processing is required for each cell received and transmitted, this does not necessarily require a host context switch per cell, since the interface will buffer cells. An example of this approach is the ORL 'Yes V2' interface presented in OSI-95/ORL/Deliverable-2. Received cells are buffered until a cell with the 'push bit' (i.e. the ATM-layer user-user indication in the header) is set. The host then takes an interrupt and is able to process all received cells. Similarly for transmission, the host is able to write cells into the network interface at a higher rate than the actual line interface, allowing the processor to perform other operations while the interface is transmitting at full (or rate-controlled) speed.

Such an interface can operate using either programmed IO only or DMA in one of the following ways (in all cases we assume the host interface strips or generates the cell HEC, resulting in a host view of the ATM cell as a 32 bit header word and twelve payload words of 32 bits):

- Programmed IO is used for transferring both ATM payloads and ATM headers. No DMA is used.
- DMA of single complete cells, including a header word (32 bits) and 12 payload words.
- DMA to or from a host memory of a block of cells where the host area is formatted into a cell pattern, where every 13th word is an ATM header.

- A hybrid where programmed IO is used for cell headers and DMA is used for the payloads. Payload data is therefore transferred, by DMA, directly to or from the appropriate system buffer in the host. (This has proved most successful with the 'YES V2' interface.)

8.3 Minor Improvements to a Simple ATM Host Interface

A pair of minor improvements can be applied to an interface of the type described in the last section. These add some inter-cell state, but without implementing the *sorting* function described in the next section. However, they might result in a several times improvement in performance for many applications. These minor improvements include:

- **An outgoing VCI/VPI register.** The interface contains a 32-bit register which the CPU loads with a prototype cell header (4 bytes) to be prepended onto each transmitted payload. Payloads are taken by DMA from consecutive groups of twelve (32-bit) word memory blocks in host memory and transmitted without per-cell host processing. The interface sets flag fields in the cell header (such as 'push bit') as required.
- **A reception VCI/VPI register.** The host reads a cell header using programmed IO and uses the VCI/VPI to determine a buffer address in host memory. The VCI/VPI from the header is copied into an interface reception VCI/VPI register and the host writes the buffer address and length into interface registers. The interface then copies data from consecutive received cells held in its receive buffers, by DMA, into the host buffer, but only while the cell headers match the value in the register.

Together, these two improvements enable the host to send and receive blocks of data with low overhead, provided cells of separate blocks are not heavily interleaved on reception, or are required to be interleaved on transmission (for switch reasons, etc.). Support for adaptation can be provided through a logical extension of this method, where separate additional registers are added for each adaptation function which runs from one cell to another within a block. Example functions are sequence numbers, total length, and per-block CRC.

8.4 Interfaces Which Sort

A sorting interface contains state for each active association and is able to receive interleaved cells without host intervention. It sorts received cells into separate queues, one for each VCI/VPI. When an end-of-block cell arrives, the completed block is available in the queue, ready for host processor attention. Important design parameters of a sorting interface are:

- **Which fields within the cell are available as sorting tags?** A full implementation would support the VCI, the VPI, and for AAL-3 and AAL-4, the MID (multiplexing id).

- **How many different queues (distinct sorting tags) are supported?** This determines the number of active incoming virtual circuits for which sorting is supported. Other virtual circuits (e.g. for broadcast VCIs) may not need sorting.
- **Are there restrictions on the mapping between the fields which generate the sorting tag?** If associative look-up of the cell header (and MID field) is used, then there are unlikely to be any restrictions. If directly indexed lookup using a hash function of the VCI/VPI field is used then there will be certain restrictions, but the practical impact of these should have been minimised by the hardware designer.

8.4.1 DMA for a Sorting Interface

A sorting interface may keep the queues of cells either in its own local RAM or in the RAM of the host CPU. Examples of both alternatives are demonstrated in two of the commercial ATM host interface chipsets from Adaptive and Fujitsu.

If cells are reassembled in the interface RAM, then a complete block can be copied in a single DMA transfer to the appropriate host buffer. This results in block-sized host bus transfers unless they are artificially broken up into smaller bursts by additional hardware to achieve a finer-grain of bus sharing. Fine grain sharing is desirable to prevent cache starvation, etc., and is not uncommon in high performance bus interfaces. For example, existing Ethernet controllers will claim the bus for typically 16 words before allowing rearbitration.

Another consequence of reassembling cells in the interface RAM is that dynamic memory allocation is required within the interface, since it is too expensive to supply sufficient local RAM to support a maximal block for each sorting tag.

Alternatively, cells may be copied into host RAM as soon as they are received. This implies that the interface requires only sufficient buffering RAM to absorb variations in bus access latencies. It also now requires descriptor RAM to contain buffer pointers in host space. The memory management is provided by the host, where it can be integrated with operating system IPC and the allocation system for network IO buffers.

8.4.2 Cells on New Circuits

With either of the above approaches, cells either arrive on circuits for which memory has been allocated, or else must be handled otherwise. Cells on new circuits may simply be togged (thrown on ground) until the circuit has been properly set up by management software, or else held while an interrupt is generated. The host may then allocate buffers. Owing to pipelining in the interface, it may be necessary to suspend the receive side of the interface while the interrupt is serviced. The new cells can be copied after the handler has re-enabled the interface, but other circuits may have been held up while the interface is suspended. Without the suspend option, such cells are lost, but other circuits are uninterrupted. A further possibility is for the interface to reserve one descriptor for processing all unexpected cells.

Interfaces which copy data into host memory require that consistency is enforced for any host CPU data caches. This has an execution cost for processors which do not bus-snoop. If the data is held in uncached interface RAM and copied using processor block moves, then snooping is not required.

8.4.3 Support of Buffers not a Multiple of 48 Bytes

Operating system IO buffers may not be a multiple of 48 bytes long (or 44 as required when using AAL-3 or AAL-4), so if the interface is handling a list of buffer descriptors, it may need to arrange for the first section of a cell to go at the end of one buffer and the remainder to go at the start of the next.

Certain interfaces may restrict cells in host memory to be aligned only on 32-bit word boundaries. This is not likely for interfaces which support AAL-3 or AAL-4, since these have a more byte-oriented feel to them, but these might restrict alignment to 16-bit boundaries.

8.5 Choice of Adaptation Layers Supported

An interface may provide full or partial support for a number of standard and non-standard adaptation layers. The AAL-5 standard connection-oriented data adaptation layer is possibly the most important, but others may also be available.

Partial support implies that only certain CPU-intensive functions are performed by the interface hardware, whereas full support implies that the host access to the data is fully above the adaptation layer with little or no visibility of the underlying cell boundaries.

Implementations of an AAL may vary greatly in their ability to recover cleanly after a lost cell or other AAL fault. A suitable implementation is for errored PDUs to be togged along with the incrementing of an error counter.

8.6 Transmission Scheduling

An interface may be given a chain of buffers which describes more than one block of data to be transmitted on different virtual circuits. It must then decide in which order to send cells. ATM allows cells of separate virtual circuits to be freely mixed on transmission. This alters the quality-of-service received by each virtual circuit and the peak-rate of a given virtual circuit at the first switch.

A related control matter is whether outgoing rate control can be applied individually to each virtual circuit. This should be combined with interleaving of outgoing cells where possible. One commercial chipset provides 8 separate leaky buckets and requires each transmit circuit to be allocated to one of these. A further , identical rate control mechanism can be applied to clip the total output rate of the interface.

A full functionality interface should be able to pre-empt low-priority transmission work, already scheduled, with newly arrived higher-priority work.

8.7 Other Options Available in an ATM Host Interface

ATM host interfaces have to choose how to pack bytes received from the network into a 32-bit word. This gives either a big-endian or a little-endian interface. Big-endian implies that the first byte of a cell is put in bits 24 to 31 of a 32-bit word. The packing order may be an association-specific function or a per-interface control option. (The ORL OSI-95 ATM interface boards support both endians on a per-interface basis.)

Encryption is sometimes combined with a network interface.

Transport layer functions are sometimes combined with an interface. In particular, support for the TP-4, TCP or XTP checksums may be included.

Under certain adaptation layers, interfaces could use AAL information to re-order cells which are out of sequence.

8.7.1 Operation and Maintenance

The interface will offer certain loopback options to enable remote testing of the physical link or interface testing without a physical link.

The interface will also terminate certain O & M flows, such as the SDH BIP parity, and the interface may choose to autonomously reply to certain O & M cells (i.e. send a reply cell without consulting the host).

9 Remarks

ATM is a promising technology for both local and wide area networking. ATM must be an integrated part of the protocol suite if its full benefits are to be realised. The MSNA approach for this has been presented. Figure 5.1 showed a simple mapping between MSNA layers and corresponding OSI layers.

The basic in-band operations required for a host interface, including cell framing, buffering, header generation, CRC computation, etc. are well defined within ATM. Olivetti Research has constructed a set of ATM interfaces for a variety of machines and has run several applications over them. Many of these applications would benefit from a multi-media transport protocol layer implemented as an application layer library.

References

[MAC 89]   Dr Mc Auley, **Protocol Design For High Speed Networks**, University of Cambridge technical report 186. December 1989.

[BLL92]   Y. Baguette, L. Léonard, G. Leduc, A. Danthine, O. Bonaventure, **OSI95 Enhanced Transport Facilities and Functions**, OSI95/Deliverable ULg-A/P, 11-12-1992, 277 p. (SART 92/25/05)

[BoP93]   J. Boerjan, R. Peschi, **B-ISDN Services and their Relation with the OSI Protocol Reference Model**, *The OSI95 Transport Service with Multimedia Support on HSLAN's and B-ISDN*, A. Danthine, ed., Springer Verlag, 247-259

[SC6-90]   ISO / IEC JTC 1/ SC6 10039: 1990. LAN media access control service definition.

[TEN-90]   D. Tennenhouse, **Layered Multiplexing Considered Harmful**, In `Protocols for High Speed Networks'* H Rudin and R.Williamson (editors). Elsevier Science Publishers. IFIP WG 6.4 workshop 1989.

[ISO-1]   International Standards Organisation - Information Processing, International Standard 7498-1. Open Systems Interconnection - Basic Reference Model.

[ISO-2]   International Standards Organisation - Information Processing, International Standard 7498-3, OSI Reference Model - Part 3: Naming and Addressing.

Specific Studies

The congestion control is becoming more and more difficult with the increase of the transfer rate on a connection and one of the objective is to be sure that the mechanisms introduced in the transport protocol will not have the effect of worsening the situation and/or creating oscillating behaviour.

It was therefore important for OSI95 to study a congestion avoidance algorithm for its possible use in the OSI95 Transport Protocol to be implemented in the second phase of the project.

The possible hardware implementation of the transport protocol during the second phase of the project was also envisaged and it was necessary to study during this first phase, the problem associated with the VLSI support for the transport protocol.

Analysis of a Delay Based Congestion Avoidance Algorithm *

Walid Dabbous
INRIA, 2004, Route des Lucioles, BP 93, 06902, Sophia Antipolis, France

There has been considerable interest recently on flow control mechanisms where the data flow into a network is regulated based on feedback from the network. The window mechanism developed by Jacobson, in which sources use packet losses to adjust their window sizes, resulted in dramatic reduction of congestion in the Internet and has become an Internet standard. Recent studies have shown that Jacobson's algorithm gives rise to large-amplitude oscillations of window sizes and packet delays.

In this paper, we describe and evaluate the performance of a flow control mechanism in which sources use packet round trip delays to adjust their window sizes. The objective is to maintain packet delays at a target level close to the minimum possible delay, and yet not severely restrict throughput. Simulation and experimental results show that the mechanism reduces the oscillations of window sizes, and provides connections with low average delay and low delay jitter, thus making it suitable to support applications such as a video conference application currently under study, provided that some reservation mechanism is used in the intermediate gateways.

Keywords : Congestion control, datagram networks, high speed networks.

1 Introduction

In a computer network, packets generated by a source node are delivered to their destination by routing them via a sequence of intermediate nodes. If the source rates are increased without constraint, queues of packets waiting to be routed build up at bottleneck nodes, leading to high delay. Eventually, the buffering capacity of these nodes is exceeded and packets are dropped, resulting in low throughput as well. Flow control mechanisms attempt to maximize throughput and minimize delays and losses by imposing constraints on the flow of data into the network..

Consider a source-destination pair whose packets are routed via intermediate nodes. Let μ_i be the service rate offered by intermediate node i to packets of this source-destination pair. For two nodes i and j, μ_i and μ_j can differ for several reasons: they may be connected to links with different bandwidths; the total traffic through the nodes may differ because they support different source-destination pairs; their

---

* This paper has been published in "High Performance Networking IV", (C-14), A.Danthine and O.Spaniol (Editors), Elsevier Science Publishers B.V. (North Holland), and is reprinted here with the kind permission of IFIP.

hardware may differ; etc. If the network is in steady-state, the ideal flow control policy is to limit the source rate to the *bottleneck rate* $\mu = \min_i (\mu_i)$ [GKl80], or equivalently the interpacket gap to $1/\mu$; a higher rate would result in packet delay or loss, and a lower rate in underutilization.

In reality, however, the bottleneck rate changes with time because connections (i.e., source-destination pairs) are being set up and terminated, and because their sources do not maintain constant data rates. Recently, flow control mechanisms have been proposed where the source rate is dynamically regulated based on feedback from the intermediate nodes ([CLZ87], [Jac88], [KAS91], [MSe90], [RFS90]). These so-called end-to-end mechanisms attempt to adapt the source rate to the bottleneck rate in minimum time and with minimum packet delay or loss. Other mechanisms, referred to as gateway flow control mechanisms, determine which packets are discarded when the buffering capacity of a node is exceeded, and the order in which buffered packets are sent [MRa91]. Gateway control mechanisms determine the way in which packets from different sources interact with each other, which in turn affects the behaviour of end-to-end mechanisms. They are still the subject of active research ([Man90], [ZKe91]).

The rest of the paper is organized as follows. In the next section, we describe in more detail the issues concerning the congestion control problem, focusing on the analysis of the standard "slow start" algorithm. In section 3, we describe a delay based flow control mechanism in detail. In section 4, we study the performance of the delay based algorithm via simulation and compare it with slow start. Our results show that the mechanism provides low average delay and low delay variance in a homogeneous environment, i.e. a network in which all connections use the same flow control mechanism. In section 5, we describe modifications which improve the ability of the mechanism to respond to changes in network conditions, e.g. route changes. In section 6, further simulation results indicate that the mechanism would be adequate to support video applications in heterogeneous networks with so-called rate allocating gateways, such as Fair Queueing gateways [DKS89]. Section 7 concludes the paper.

2 Congestion Avoidance and Control

In this section, we focus on end-to-end mechanisms. One of the most important characteristics of such a mechanism is its feedback: What information does the source obtain about the state of the intermediate nodes. In the binary feedback scheme used in DECNET [RJa90], each data packet received by the destination has a bit indicating whether or not the packet encountered an intermediate node with average queue size greater than 1; this information is sent back to the source in acknowledgement packets. In the Internet TCP, the feedback consists of the arrival times (or lack of arrival) of acknowledgement packets [Jac88]. Other schemes rely on feedback information that includes average link utilization [RFS90], interarrival times between successive acknowledgement packets [KAS91], packet round trip delays [MSe90], [Jai89] etc.

The feedback information is used to regulate the flow of data from the sources into the network. In *rate-based* mechanisms, the source rate is regulated by adjusting interpacket gaps. Examples are found in the protocols NETBLT [CLZ87], VMTP [Che86], and XTP [Che89]. However, most protocols currently in operation ([Jac88], [Jai89]) use a *window-based* scheme, rather than the rate-based scheme. Here, a limit referred to as the *window size* is placed on the number of packets that can be outstanding at the source, but no constraint is placed on the rate at which these packets can be sent. Slow Start is a congestion avoidance and control scheme defined by Jacobson [Jac88]. In this scheme, the source increases the window when an acknowledgement is received, and decreases the window when a packet loss is detected. Specifically, the window adjustment mechanism has two phases, the slow-start or congestion recovery phase, and the linear increase or congestion avoidance phase. At connection set-up, the window is set to one (window sizes are expressed in terms of maximal packet sizes) and the slow-start phase begins. The window is increased by 1 whenever an acknowledgement is received, until it reaches a threshold value, at which point the algorithm switches into the congestion avoidance phase. There, the window is increased by 1/W whenever an acknowledgement is received, where W denotes the current window size. When a packet loss is detected, the window is set to 1 and the slow start phase resumes; the idea being that a packet loss means that the queues at the intermediate nodes are overflowing.

In Jacobson's scheme, queue sizes at the intermediate nodes are kept relatively large, in case of FCFS (First Come First Serve) service discipline. It is also intuitively clear, and it has been verified in numerous simulation (e.g. [SZC90]) and experimental studies (e.g. [Dab92]), that the evolutions of the window size as a function of time follow a cyclic pattern. In many cases, the amplitudes of the window oscillations turn out to be quite large.

The above observations imply that queueing delays, and hence packet delays, follow a cyclic behaviour with large maximum value and large variance. This makes Jacobson's algorithm unsuitable to support applications such as video image transmission (e.g. video conference) which require relatively low average delay and low delay jitter. Furthermore, we note that the periodic reduction of the window size to 1 following a packet loss will result in periodic blackouts of the video signal.

The goal of our study is to choose a congestion control algorithm suitable to be used by the TPX protocol currently under design within the OSI95 ESPRIT project [OSI95]. Therefore we will focus on transport level procedures and consider only end-systems mechanisms which we will compare in order to propose an adequate algorithm. Gateway congestion control algorithms should be studied in the framework of a general architectural design for high speed network layer extension. In most of the rest of this paper we will consider that gateways implement FCFS discipline and make the comparisons of end-system algorithms without any assumption on the existence of a special congestion control algorithm in intermediate gateways. We would like to get the best performance (efficiency and fair resource sharing) by the use of end-systems mechanisms.

3 The Delay Algorithm

The delay based mechanism we will describe is inspired from early work of [Jai89] and [Hui87]. In this mechanism, sources use packet round trip delays to adjust their window sizes. The objective of the algorithm which we refer to as the Delay Algorithm, or DA is to maintain packet delays at a target level, which we take to be the smallest possible delay (i.e. the delay corresponding to a "no queueing case"), and yet not severely restrict throughput.

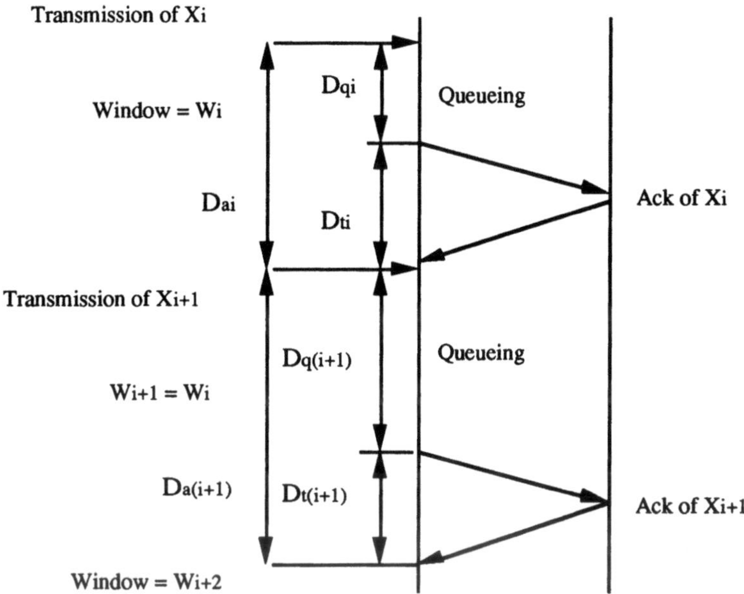

Fig. 3.1 Correspondence between window size and round trip delays

The Delay Algorithm was first proposed in [Hui87] and [Dab91]. It has two phases, like Jacobson's algorithm. At connection set-up and after a packet loss is detected, the window is set to 1 and the "slow-start" phase begins. The window is increased by 1 whenever an acknowledgement is received, until it reaches a threshold value, at which point the congestion avoidance phase takes over. This threshold value is determined by a condition on the *normalized delay gradient* (NDG) as defined in [Jai89]:

$$NDG = (\Delta D/\Delta W)/(D/W)$$

where ΔD is the increase in the round trip delay corresponding to an increase of ΔW of the window. If NDG is smaller than 0.5 the algorithm switches from the exponential phase to congestion avoidance phase (The numerical values for the control parameters were chosen intuitively. Their value should be adapted according to the status of the network.). The 0.5 value corresponds to a "medium" gain in the delay increase.

During the congestion avoidance phase, the principle behind the DA is illustrated on Figure 3.1. We divide time into epochs of length equal to one round trip delay. We refer to the first packet sent in an epoch as a "head of window". Let X_i denote the head of window for epoch i. The source measures the round trip delays of the X_i's. Let D_{ai} denote the round trip delay of packet X_i. D_{ai} can be decomposed into a fixed delay D_{ti} including propagation and transmission delays, and a variable queueing delay D_{qi}. Let us assume for now that the fixed part of the delay does not vary between two successive epochs, i.e.,

$$D_{ti} = D_{t(i+1)} = D_t$$

This implies, for example, that no route or topology change occurs between two epochs (a modified version of the Delay Algorithm which handles such changes is described in section 5). $D_{q(i+1)}$ is the queueing delay caused by the $W_i - 1$ packets sent between X_i and $X_{(i+1)}$. Thus, the delay $D_{a(i+1)}$ represent the time necessary to transmit W_i packets through the network. The same throughput could have been achieved with a window:

$$W = W_i.D_t/D_{a(i+1)}$$

This window corresponds to a "zero queueing" condition.

However, the source does not have a priori knowledge of D_t, the round trip delay in the absence of queueing delays. It is estimated with the smallest round trip delay seen so far, i.e. min (D_{ai}). Thus, at epoch i+1, the source first estimate D_t as

$$D_{t(i+1)} = \min (D_{ti}, D_{a(i+1)})$$

then compute the new window size:

$$
\begin{aligned}
W_{i+2} &= W_i.D_{t(i+1)} /D_{a(i+1)} &\quad &\text{if } Da(i+1) > K.D_{ti} \\
&= W_i &\quad &\text{if } D_{ti} < D_{a(i+1)} < K.D_{ti} \\
&= W_i +1 &\quad &\text{otherwise}
\end{aligned}
$$

That is, we increase the window by 1 when we suppose that no queueing at all occurred. Otherwise, if the delay has "increased substantially" ($Da(i+1) > K.Dti$, with $1.125 < K < 1.25$) the window size is decreased by a multiplicative factor proportional to the increase in the acknowledgement delay. Note that the hypothesis of a constant transmission delay is somewhat antagonistic with the philosophy of connectionless networks, where packets belonging to the same transport connection may follow different routes. We use a factor K in order to filter small fluctuations in the round trip delay. In section 5 we propose a variant of the algorithm that is less sensitive to delay changes over a datagram network.

Indeed, the window W that we have computed here is a maximum value. If for some reason the other side decides to restrain the flow of credits to a value of W_c, e.g. when processing the data takes longer than transmitting them, the transmitting maximum window size will be the minimum of both W and W_c.

The idea of using the delay gradient as a feedback signal was also proposed by Zheng and Crowcroft who described in [WCr92] an algorithm similar to the DA: a delay threshold $D_i = D_p + \beta.Q_{max}.M/\mu$ is defined, where D_p is the round trip propagation delay (i.e. without queueing), Q_{max} is the maximum queue length, M the packet size and μ the bottleneck capacity. D_i can be rewritten as $(1-\beta).D_{min} + \beta.D_{max}$ where $D_{min} = D_p$ and $D_{max} = D_p + Q_{max}.M/\mu$ are respectively the minimum and maximum round trip delays. The β parameter defines the operating point i.e. the ratio of the target queue size to the maximum queue size at the bottleneck. The window adjustment is based on this threshold value: during normal resource probing phase, i.e. when the window size is increased by one every round trip delay, if $D > D_i$ the congestion window is set to 7/8 of the current size. If a packet loss is detected the slow start algorithm is used. Other delay-based mechanisms have been described in [Jai89] and [MSe90]. In section 5 we present a version of the Delay Algorithm allowing to easily adapt to intrinsic network configuration changes and to have improved fairness between competing sources.

4 Simulation Results

In this section, we evaluate the performance of the delay algorithm using simulation. We used a modified version of the REAL simulator [Kes88].

4.1 The Simple Case

We present the simulation results for a simple network topology shown in Figure 4.1: a "stream" source sends continuously packets of the same size (1000 bytes) to a destination. The returned acknowledgements packets are 40 bytes long. Each node has a maximum buffer capacity of 15 packets unless otherwise mentioned. The link between the source and the gateway has a 8 Mbps with a propagation delay of 5 ms. The link between the gateway and the destination have a capacity of 800 Kbps and a propagation delay of 100 ms (the bottleneck link). The total path length is then 221.44 ms and a window of 23 is sufficient to fill the pipe. The TCP connection has a maximum window size of 64 packets (> 23 + 15) in order to allow for a single connection to saturate the gateway buffering capacity. The simulations last for a period of 100 s. Recent measures on the Internet [CDJ91] have shown that most FTP connections last for only a short period of time (90% of the TCP connections exchange less than 10 Kbytes of data). However, recall that we intend to use the Delay Algorithm to support video packet transmission. It is reasonable to expect video conference applications to be active for at least a few minutes.

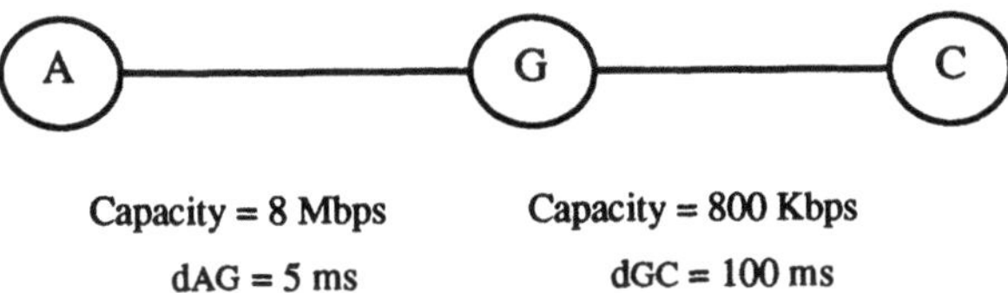

Fig. 4.1 Network configuration with a single user

The results of the simulation are shown in Figure 4.2. The sources implementing Jacobson-Karels algorithm within the REAL simulator are labelled as "JK" sources. A source implementing JK opens the connections with slow start and continues to increase the window until a packet loss occurs. It then enters consecutive cycles of slow start and linear increase phases, the window size oscillating between half the maximum size of 18 and the maximum size of 37 during the linear phase.

The Delay Algorithm follows the same policy as slow start for the connection start-up. However, the exponential increase is stopped and the linear phase is entered at W=13 (corresponding to NDG < 0.5) i.e. before a packet loss occurs.

The window will then stabilize around 22 or 23 which corresponds to the path length with "zero queueing". The simulation also shows that a DA source operates the bottleneck at maximal capacity (800 Kbps) i.e. the source has a throughput of 100 packets per second, while a JK source transmits roughly 90 packets per second. We observe that the delays for the DA source stabilize around 260 ms, while the delays for the JK source vary between 221 ms and 370 ms. Refer to Table 4.1

We next compare the JK and DA algorithms in the case of two users sharing the same bottleneck link and we show how the delay algorithm eliminates the "traffic phase effects" observed in [FJa91].

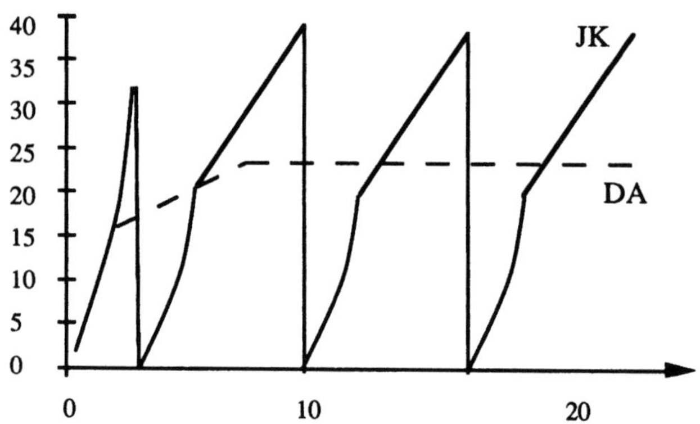

Fig. 4.2 Window size vs time (in s) for both JK and Delay algorithms

Table 4.1 Round trip delays in ms for network topology depicted in Figure 4.1

	Minimum	Average	Maximum
DA source	260	260	260
JK source	221	287	370

4.2 Competition between two sources

It has been noticed in [FJa91] that the JK algorithm may lead to unfair resource allocation between two competing users for some network configurations with FCFS drop tail gateways (When their input queue is full, drop tail gateways drop all incoming packets.). This discriminatory behaviour of the network is a function of two factors: the synchronous transmission of new packets driven by the service of previous packets at the bottleneck gateway, and the relative value of the round trip time of any two sources sharing the bottleneck gateway. In fact, it was shown by simulations [FJa91] that a small change in the round trip time of a particular connection may cause a significant reduction in its effective throughput.

This effect (called the "phase effect") is due to an inherent characteristic of the JK algorithm: users always increase their demand (window size) until a packet loss is detected. In addition, the packet arrivals at the bottleneck gateway are related to packet departures, therefore packet arrivals for a given connection may always precede the arrivals from another connection. When the gateway buffers are full, any increase in the window size will lead a packet loss. Due to the phase effect, the gateway may systematically drop a packet of the same connection at the end of each congestion avoidance phase (the last packet in the gateway queue) resulting in the biased network behaviour shown in simulations.

The Delay algorithm does not present such phase effect related unfairness. The competing sources adjust their window according to feedback information about round trip times and not after a packet loss. This will prevent any systematic drop of packets of a given connection at the bottleneck gateway. This feature of the DA algorithm was verified by simulations, where two "stream" sources share a bottleneck gateway. The network topology is shown in Figure 4.3. The links between the sources A and B and the gateway have respectively 5 ms and d_{BG} propagation delay. For the same packet and acknowledgement sizes (1000 and 40 bytes), the total path length without queueing is $d_1 = 221.44$ ms for ACA and $d_2 = (211.44$ ms $+ d_{BG})$ for BCB. Each source has a maximum window of 32 packets.

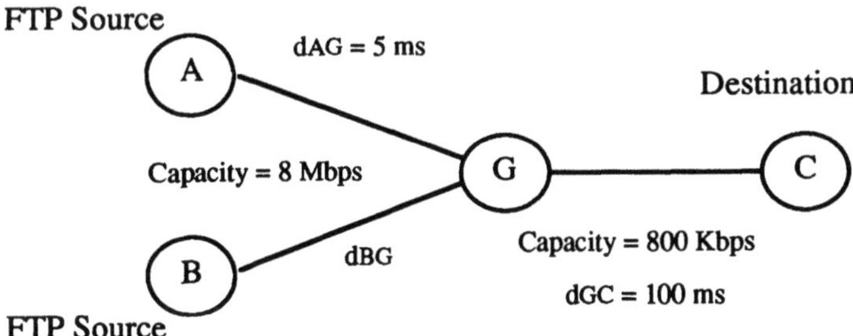

Fig. 4.3 Network configuration with two "stream" sources

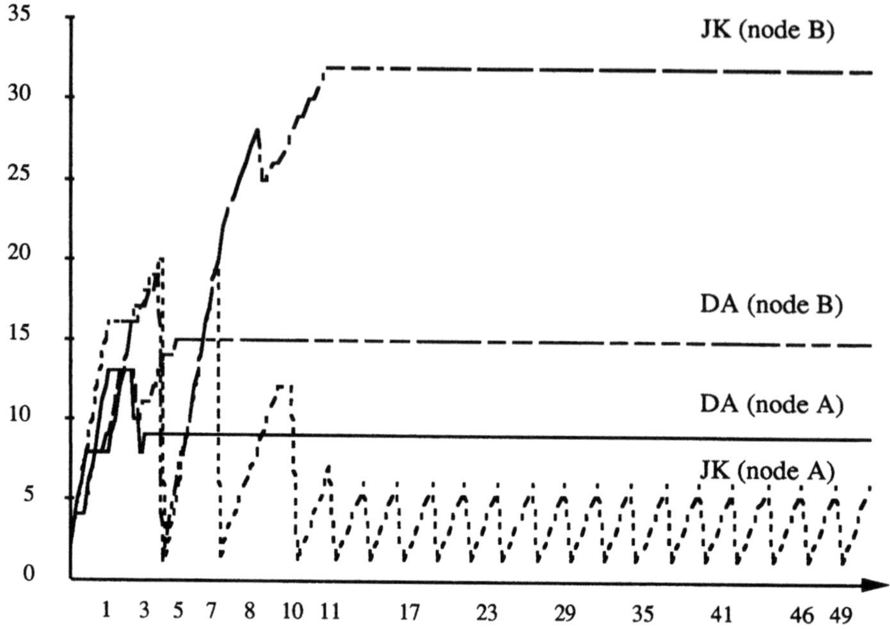

Fig. 4.4 Window size for nodes A and B for both JK and DA algorithms, $d_1/d_2 = 0.962$

The simulation results in Figure 4.4 correspond to $d_{BG} = 9.8$ ms i.e. $d_1/d_2 = 0.962$. Even with such a small difference in the round trip propagation delays for both connections, the simulation shows that when the two sources are implementing the JK algorithm, node B gets a significant part of the available bandwidth. Even if we increase the buffer space at the bottleneck gateway this biased behaviour persists confirming the unfair allocation of resources according to the relative values of the round trip delays. Small variations of the round trip delay may cause node's B throughput to fall to about 10 % of the total throughput.

The results in Figure 4.5 corresponds to a propagation delay between the source B and the gateway: $d_{BG} = 3.5$ ms, i.e. $d_1/d_2 = 1.013$. In fact, the unfairness in resources allocation between the two sources with the JK algorithm is due to the systematic drop of the packets belonging to one of the two connections due to the phase effect. The DA shows improved fairness because there is no systematic packet drop at the bottleneck gateway. Extensive simulations confirmed these results: the DA is not sensitive to small variations of the round trip delay even with drop tail FCFS gateways. However, we might have stable "slightly unfair" sharing of the resources due to the very conservative increase/decrease policy. Refer to Figures 4.4 and 4.5: DA source A has a window size of 9 and 13, while DA source B has a window size of 15 and 10 respectively. In section 5, we describe how to tune the Delay Algorithm in order to eliminate such discriminatory behaviour of the network.

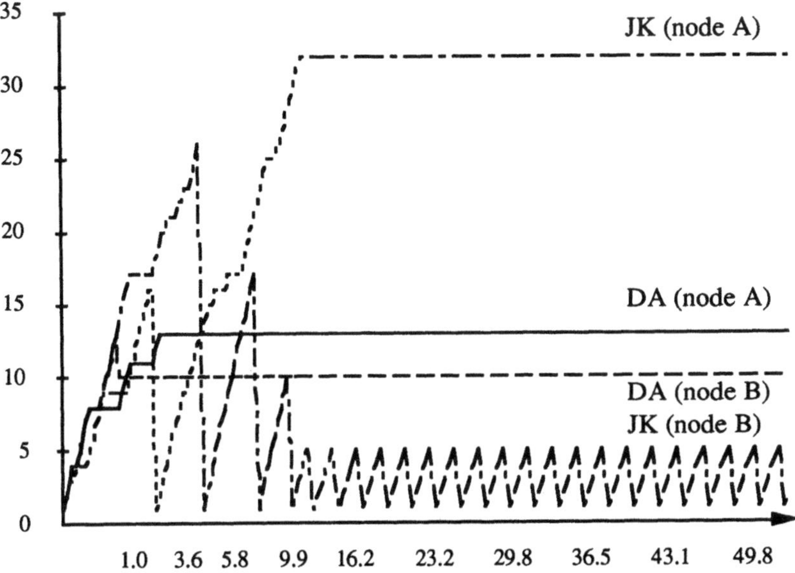

Fig. 4.5 Window size for nodes A and B for both JK and DA algorithms, $d_1/d_2 = 1.013$

4.3 Adapting to changes in network resources

The delay based feedback adjustment of the window size allows to pick the released network resources up rapidly. A load decrease due to the termination of a network connection translates to smaller delays and will be therefore noticed by active sources within a round trip delay period. In Table 4.2 we show the effective throughput (in packets/s) for nodes A and B in a network configuration close to that depicted in Figure 4.3, with a propagation delay $d_{BG} = 5$ ms, but with an additional "background" source (Bg) which transmits to the destination C via G at a constant rate (50 packets/s) for 40 seconds and then stops for 40 seconds. Two simulations for 120 seconds were run with both sources having the same type (JK or DA) in each simulation. These simulations allowed to test the ability of both DA and JK sources to reduce their throughput when a new user join a shared path, and to increase their window when any user stops. Table 4.2 shows that both algorithms update the window size of the sources according to the "residual" resources (the background source is not flow controlled). However, we can make the following remarks:

- the bottleneck link is saturated with the use of the DA sources (2000 packets for both sources in 20 seconds), and operated at 93% of its capacity with the JK sources.
- in the "steady state" i.e. two sources without the background, each JK source systematically has a packet dropped at the bottleneck gateway each 6 seconds. The number of packet dropped increases when the background source joins the network. DA sources, however, reduce their throughput sufficiently early to prevent any

packet loss even when the conditions on the network are changing (e.g. a background source joining or leaving the transmission path).

Table 4.2 Comparison of the dynamic behaviour of both DA and JK algorithms

Time	Source	Type	Sent	Dropped	Type	Sent	Dropped
20	A	DA	976	0	JK	905	3
	B	DA	965	0	JK	883	3
	C	Bg	0	0	Bg	0	0
40	A	DA	996	0	JK	927	3
	B	DA	1004	0	JK	930	3
	C	Bg	0	0	Bg	0	0
60	A	DA	467	0	JK	504	7
	B	DA	534	0	JK	508	7
	C	Bg	999	0	Bg	949	44
80	A	DA	462	0	JK	499	7
	B	DA	538	0	JK	463	7
	C	Bg	1000	0	Bg	991	6
100	A	DA	932	0	JK	901	3
	B	DA	1023	0	JK	882	3
	C	Bg	3	0	Bg	12	0
120	A	DA	957	0	JK	917	3
	B	DA	1043	0	JK	928	3
	C	Bg	0	0	Bg	0	0

5 The Delay Algorithm Revisited

In this section we propose two mechanisms to enhance the flexibility and the performance of the algorithm.

5.1 Raising the Reference Delay

The use of the minimum observed round trip delay as a reference value for the control may have a negative side effect in some cases when the intrinsic propagation delay D_t of the path increases because of route changes e.g. when a failing terrestrial link (d = 20 ms) is replaced by a satellite link (d = 300 ms). This may lead to successive window size reduction as the measured round trip time will be considered too high compared the minimum value. The reference value should then be updated if the network conditions change.

A possible way to solve the problem is to let the reference value to be the minimum of the last M measures of the round trip time. We simulated such dynamic raise of the reference value but we did not obtain satisfactory results: this produced a positive feedback effect, and the "equilibrium point" drifted from the zero queueing at the bottleneck to higher values. This led to frequent packet losses when new users join a loaded link.

Another way to circumvent the problem is to consider that the conditions on the network have changed if the measured delay is too high comparing to the reference delay for MAX-REDUCE consecutive times: it means that we update the reference delay after MAX-REDUCE consecutive reduction of the window size. We tested this policy by simulation: we started the simulation with a network configuration corresponding to Figure 4.1 (with one DA source and $d_{GC} = 50$ ms). At $t = 20$ s the propagation delay of this link is switched to $d_{GC} = 100$ ms. The DA algorithm recovered and the reference delay value was updated after the fourth window reduction (MAX-REDUCE = 4). The results are reported in Figures 5.1 and 5.2.

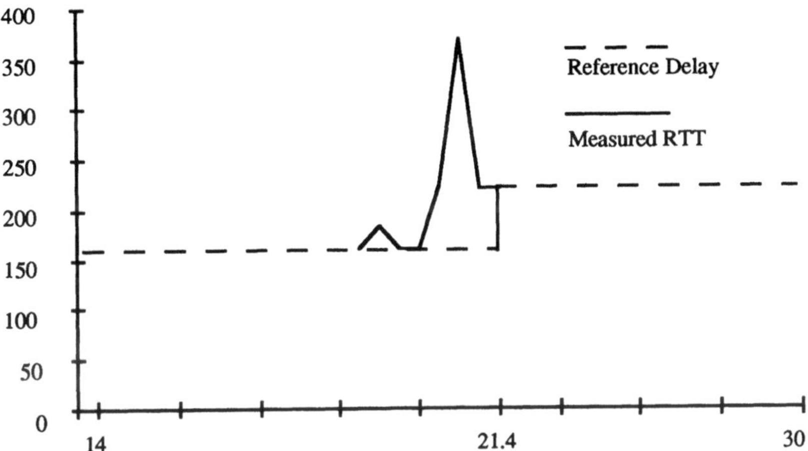

Fig. 5.1 Increase in the reference delay (in milliseconds) after network conditions change

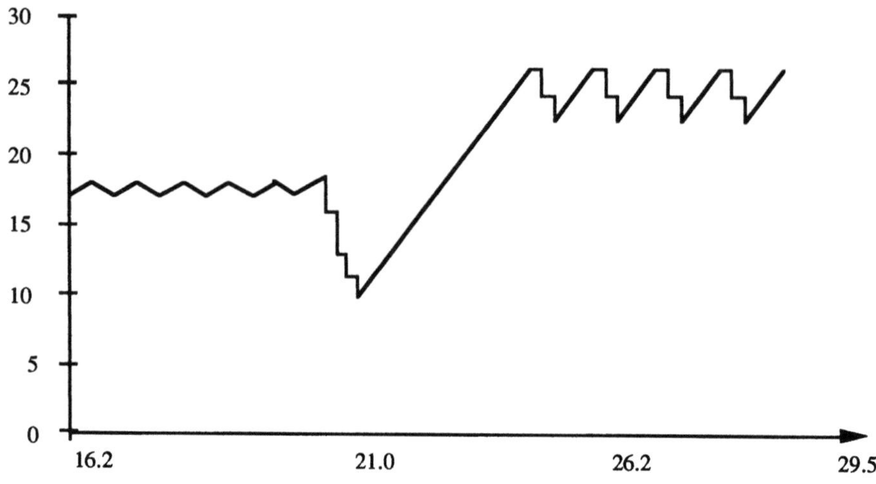

Fig. 5.2 The corresponding window size variation

5.2 A Shade of Aggressivity

The Delay Algorithm uses a "conservative" policy to increase the window: the reference delay value corresponds to "zero queueing" at the bottleneck gateway. Another variant is possible: if the delay increases substantially, the window is reduced to a value that corresponds to a minimal buffer occupation at the gateway and not to "zero queueing". This may be accomplished by adding to the value of congestion window computed according the Delay Algorithm a small increase of 1 or 2 packets corresponding to the minimal buffering at the gateway. This policy avoids a drastic reduction of the window size, even when the demand is increasing. On the other hand, this policy leads to better fairness in the resource sharing: the minimal buffering offset is a more aggressive increase condition which allows to leave a possible unfair stable position. The simulations with a minimal buffering of two packets show that the bandwidth is equally shared by two DA sources even with various small variations of the propagation delay d_{BG}. Figure 5.3 corresponds to the network configuration depicted in Figure 4.3 with a delay $d_{BG} = 9.8$ ms: note the equal share of the bandwidth between the two sources. Figure 4.4 showed that, with the same network configuration, the JK sources received a completely biased bandwidth allocation.

Another optimistic policy is possible for the window size increase: when the delay is stable, the source probes additional resources by increasing its window. If the delay increases "substantially" (i.e. $D_{a(i+1)} > K.D_{ti}$), the window will be decreased after a round trip time. This allows for a more rapid adaptation to changing network state but may possibly lead to the loss of some packets if the buffer capacity of the bottleneck gateway is reduced. The saw tooths in Figure 5.2 are due to this optimization. When the demand and the resources are stable, these oscillations cause delay variations. Thus, a tradeoff should be made between "response time" (aggressivity) and "quality of service in the steady state" (amplitude of the oscillations).

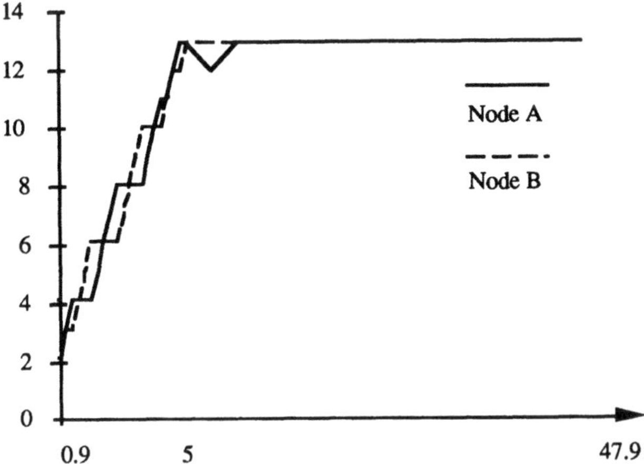

Fig. 5.3 Window size for two DA sources with "Minimal-Buffer = 2"

6 Fairness Issues

The Delay algorithm has also been implemented and tested on the Internet, where hosts implement the standard slow start TCP code. The results of the experiments are detailed in [Dab92]. They show that the delays for the TCP connections follow the same cyclic behaviour as the congestion window. We observe that the average round trip delay is much lower for the DA connection than for the TCP connection. Furthermore, the variance of the delays is also lower for the DA connection. However, the difference between the standard deviations for the DA and the TCP connections is not as important as what simulation results from Section 4 could have lead us to believe (see Table 4.1). Furthermore, we observe almost as many packet losses in the DA and the TCP connections, although we expect packet losses in DA connections to be very infrequent occurrences. These phenomena occur because the experiments were carried out in the Internet, which means that i) traffic from the DA connection competes for network resources with traffic from TCP connections, which does not depend on packet delays, and ii) queueing discipline in the Internet gateways is First-Come-First-Serve (FCFS).

In a FCFS network, queueing delays at a gateway are linked in detail to the arrival patterns of all the conversations in the gateway. Consider for example the situation where a single DA connection shares the resources of a gateway with multiple TCP connections. If any one of the TCP connections sends a large burst of data, then the queueing delays of all other connections increase until the burst has been served. If no packet is lost in the process, the other TCP connection will not react (i.e. will not adjust their window size) to this increase in delays. The DA flow control mechanism will assume that these increased delays mean that the DA connection is causing congestion. However, the DA connection is only affected by congestion caused by other connections. Nevertheless, the flow control mechanism reduces the window size and hence unnecessarily restricts throughput of the DA connection. Thus, we expect that in our experiments, the throughput of a DA connection is lower than the throughput of a TCP connection. This is indeed the case. We observe that the throughput of the DA connection is 30% lower than the throughput of the TCP connection.

We note that this phenomenon would not occur in a so-called rate allocating servers network (RAS network) [KAS91], in which gateways use Fair Queueing or other rate based scheduling disciplines. With FQ discipline, gateways maintain separate queues for packets from each connection. The queues are serviced in a round-robin manner. If a source send packets too quickly, it merely increases the length of its own queue. Thus, the queueing delays experienced by each connection are, to a first approximation, independent of the arrival patterns of other connections. Therefore, if a DA connection observes increasing round trip delays, then this connection is really a source of congestion. As a consequence, we expect the throughput of DA and TCP connections to be equal in a FQ network.

We could not verify the above hypothesis with experiments over the Internet. We used simulations on the simple network configuration shown in Figure 4.3. In this network, two connections send packets over a common path and share a common gateway. One connection uses Jacobson's algorithm, the other the DA algorithm.

Detailed results are presented in [Dab92]. They showed that with the FCFS gateway, the throughput of the TCP connection is almost twice as much as the throughput of the DA connection. Furthermore, we observed that the evolutions of the round trip delays follow the same pattern for both connections. This was expected, since packets from the DA connection will be queued behind packets from the TCP connection. This clearly reflects the inability of FCFS gateways to distinguish between different connections and allocate bandwidth and buffer space independently. The situation is quite different with the FQ gateway. There, both connections evenly share the bottleneck capacity. Furthermore, as expected, the average and the standard deviation of packet delays are smaller for the DA connection.

The above results and the simulation results of Section 4 indicate that the DA algorithm can provide connections with low average delay and low delay variance in networks with rate allocating servers. These results confirmed that fairness with end systems congestion control mechanisms cannot be achieved in the presence of aggressive or malicious users without the arbitration of the intermediate gateways.

7 Conclusion

In this paper we studied some issues related to the congestion control problem. The aim of the study was to find an algorithm suitable to be used within the TPX protocol currently under design within the OSI95 ESPRIT project. We proposed to use a delay based congestion control algorithm that eliminates the periodic packet loss experienced with the slow start algorithm and reduces the round trip delay variation. We compared the simulation results of both slow start and the Delay Algorithm and show that the latter allows to eliminate discriminatory behaviour of the network due to the traffic "phase effects". Fine adjustments of the Delay Algorithm allow to improve its fairness and guarantee its function on a network with dynamic configuration changes (e.g. re-routing after a link failure). However, this study showed that complete fairness cannot be achieved with FCFS gateways: in this case the aggressive policies get higher share of the available bandwidth. Nevertheless, as the Delay Algorithm reduces round trip delays, it could be used for the support of a video conference application in a network where gateways implement a gateway congestion control mechanism such as the Fair Queueing. The delay algorithm is destined to be used as a combined flow/congestion control mechanism for a video conference application currently under study at INRIA [HTu92]. Future research will investigate application level flow control, in which the video conference application uses the network feedback information, namely packet round trip delays, to adjust the coding rate of the video coder and hence the quality of the images sent over the network.

References

[CDJ91]   Caceres R., Danzig P., Jamin S., Mitzel D., **Characteristics of wide-area TCP/IP conversations,** in *Proceedings ACM SIGCOMM '91,* Zurich, Switzerland, September 1991.

[Che86]   Cheriton D. R., **VMTP: a transport protocol for the next generation of communication systems,** in *Proceedings ACM SIGCOMM '86,* Stowe, Vermont, pp. 406-415, August 1986.

[Che89]   Chesson, G., **XTP/PE Design Considerations,** in *Protocols for High Speed Networks,* H. Rudin, R. Williamson, Eds. Elsevier Science Publishers/North-Holland, May 1989.

[CLZ87]   Clark D., Lambert M., Zhang L., **NETBLT: a high throughput transport protocol,** in *Proceedings of the ACM SIGCOMM '87,* Stowe, Vermont, pp. 353-359, August 1987.

[Dab91]   Dabbous, W., *Etude des protocoles de contrôle de transmission à haut débit pour les applications multimédias,* PhD Thesis, Université de Paris-Sud, March 1991.

[DKS89]   Demers A. Keshav S., Shenker S., **Analysis and Simulation of a Fair Queueing Algorithm,** in *Proceedings ACM SIGCOMM '89,* pp. 1-12, September 1989.

[FJa91]   Floyd S., Jacobson V.,**Traffic Phase Effects in Packet-Switched Gateways,** *Computer Communication Review,* Vol. 21, No. 2, pp. 26-42, April 1991.

[GKl80]   Gerla M., Kleinrock L., **Flow control: A comparative survey,** *IEEE Trans. Comm.,* Vol. 28, April 1980.

[Hui87]   Huitema C., **A high speed network connection between NSF and INRIA,** Research note, INRIA Sophia Antipolis, October 1987.

[Jac88]   Jacobson V., **Congestion avoidance and control,** in *Proceedings ACM SIGCOMM '88,* Stanford, California, pp. 314-329, August 1988.

[Jai89]   Jain R., **A Delay-Based Approach for Congestion Avoidance in Interconnected Heterogeneous Computer Networks,** *Computer Communication Review,* Vol. 18, No. 4, pp 56-71, October 1989.

[Kes88]   Keshav S., **The REAL network simulator,** *Technical Report UCB/CSD 88/472,* University of California at Berkeley, December 1988.

[KAS91]   Keshav S., Agrawala A. , Singh S., **Design and analysis of a flow control algorithm for a network of rate allocating servers,** in *Protocols for High-Speed Networks II,* Elsevier Science Publishers/North-Holland, April 1991.

[Man90]   Mankin A., **Random Drop Congestion Control,** *Proceedings ACM SIGCOMM '90,* Philadelphia, Pa., pp. 1-7, September 1990.

[MRa91]   Mankin A., Ramakrishnan K., **Gateway Congestion Control Survey,** *RFC-1254,* August 1991.

[MSe90]   Mitra D., Seery J. B., **Dynamic adaptive windows for high speed data networks: Theory and simulations,** in *Proceedings ACM SIGCOMM '90,* Philadelphia, Pa., pp. 30-40, September 1990.

[OSI95]   ESPRIT Project: High performance OSI protocols with multimedia support on HSLAN's and B-ISDN (OSI95). Technical Annex, September 1990.

[RJa90]   Ramakrishnan K. K., Jain R., **A Binary Feedback Scheme for Congestion Avoidance in Computer Networks,** *ACM Transactions on Computer Systems,* Vol. 8, No. 2, pp. 158-181, May 1990.

[RFS90]   Robinson J., Friedman D.,, Steenstrup M., **Congestion control in BBN packet-switched networks,** *Computer Communication Review,* Vol. 20, No. 1, pp. 30-39, January 1990.

[SZC90]   Shenker S., Zhang L., Clark D., **Some Observations on the Dynamics of a Congestion Control Algorithm,** *Computer Communication Review,* Vol. 20, No. 5, pp. 30-39, October 1990.

[WCr92]   Wang Z., Crowcroft J., **Eliminating Periodic Packet Losses in the 4.3 Tahoe BSD TCP Congestion Control Algorithm,** *CCR,* Volume 22, Number 2, April 1992.

[ZKe91]   Zhang H., Keshav S., **Comparison of Rate-Based Service Disciplines,** in *Proceedings ACM SIGCOMM '91,* Zurich, Switzerland, September 1991.

[Dab92]   Dabbous W., Bolot J., **Study of congestion avoidance mechanisms,** Technical Report, Deliverable INRIA-5, ESPRIT Project OSI95, May 1992.

[HTu92]   Huitema C., Turletti T., **Software Codecs and Workstation Video Conferences,** in *Proceedings of iNET '92,* Kobe, Japan, June 15-18, 1992.

VLSI Support for Transport Protocols

Gregory Votsis and Vangelis Chalkiadakis
INTRACOM S.A., Development Programmes Department, P.O. Box 68, GR-190 02, Peania, Greece

In this paper the design considerations and obstacles for the VLSI implementation of transport protocols are presented, the state of the art of existing implementations are discussed, and finally a new architecture for a high performance transport protocol VLSI implementation is proposed.

This study has been performed in the Framework of A9 Task of OSI95 entitled "Guidelines for Transport Protocols VLSI Implementation".

Keywords : Transport Protocols, VLSI implementation, Protocol Engine PE1000, Network Co-processor

1 Introduction

The advances in communication technology have resulted in the improvement of the transmission bandwidth in the physical layer. FDDI products (100 Mbps) are commercially available, and many research groups investigate Gbps network technologies [WO90] [Par90]. Furthermore, the application environment is changing rapidly due to the introduction of high bandwidth, real time constraints demanding distributed multimedia applications, as well as the penetration of ODP systems requesting short communication delays for interprocess communication.

This new environment imposes:

- the specification of new communication protocols capable to exploit the high bandwidth offered at the physical layer and to provide the appropriate services satisfying requirements stemming from the applications. The transport layer is a key for these issues and several new alternative solutions have been proposed such as XTP [XTP 92], TPX [Dan92], and VMTP [Che86]
- This new environment imposes the careful selection of the protocol's features, the network's interfaces and the method of implementation, to optimise the performance of the protocol.

This paper deals with the second of the above points and investigates the hardware implementation of a transport protocol.

The term "hardware implementation of transport protocol" refers to a couple of main issues: *the VLSI transport protocol implementation* and *the design of the appropriate network interface architecture* accommodating the VLSI chips. It is impossible to dissociate the two above issues, since measurements have shown that protocol processing represents less than 20% of the total overhead and the remaining 80% is devoted to non protocol tasks, such as data movement (copy), buffer and timer management (80/20 principle) [Chs89]. The data movement represents an important part of the non-protocol overhead and that is the most critical one, from the performance point of view. This is because it introduces latency and jitter, and furthermore it wastes CPU cycles and bandwidth of internal busses. The timer management has an impact on the performance [Chs89], but it is directly related to the protocol specification. Therefore, the conception of the appropriate network interface architecture is mandatory to handle both protocol and non-protocol related functions by VLSI circuits.

Today, the software processing of the transport protocol consumes significant parts of both the host CPU processing power and the bandwidth of interfacing buses. Consequently, the implementation of transport protocols in silicon, together with an appropriate network interface design could drastically improve their performance, because the processing of a task using hardware is generally faster than a software execution.

The network interface must handle the available at the physical layer bandwidth in such a manner that it will introduce a minimum delay and jitter in the transmission and reception of TPDUs. For this reason, the processing of the TPDUs must be performed in "real time" and the data movements between different memory locations and the usage of internal buses, associated to the protocol execution, must be minimised.

Furthermore, the VLSI implementation of communication protocols has the following advantages:

- Reduction of the interface's cost, in the case of massive production of VLSI chips.
- It is an alternative way for exploiting of communication tasks' parallelism and their concurrent execution

2 Design Considerations and Obstacles

The objective of the study performed in the framework of OSI95 project was the establishment of guidelines for achieving a compact VLSI implementation of transport protocols [OSI95 Technical Annex].

We can distinguish the remarks on the design considerations and obstacles to the following three main issues:

- Protocol: Related to the complexity, stability and flexibility of the selected transport protocol(s).
- Architecture: Related to the OSI layered approach as well as to the host system accommodating the network interface.
- VLSI implementation: Related to the feasibility and expected integration degree as well as to the potential industrial interest.

By comparing TP4 with XTP (transport part) and TPX protocols, we can see that both of them provide a much richer range of services to the upper layers and that is reflected directly to the variety as well as the complexity of the transport protocol mechanisms. For example, we have seen that the fast connection set-up, the selective retransmission, the variety of proposed connectionless service types and the QoS management increase the implementation's complexity.

The TP4 is a simple protocol, but it has not been designed to be easily implemented either in software or in hardware. Definitions such as the CRC position within a packet, the TPDUs numbering scheme, the TPDU formats, etc., have been designed without any care on simplicity and performance. Nevertheless, we are convinced that the TP4 revision and the specification of a new High Speed Transport Protocol (HSTP) are mainly conducted by the need to support emerging requirements of new services in the HSN environment, such as multicast, expanded QoS and selectable error control, instead of insufficient performance achieved by TP4. By implementing TP4 in hardware or in a parallel platform (e.g. Transputer based [Dio91]) we can provide to the user sufficient performance for some application types like voluminous file transfers.

Nevertheless, the question of which transport protocol should be selected today for VLSI implementation is not easy to be answered, due to the ongoing standardisation activities in ISO (JTC1/SC6). Therefore, the design of VLSI circuits dedicated to specific transport protocols has important industrial risks. On the other hand, for industrial and exploitation constraints, flexibility is a very desirable characteristic. However, this flexibility in the processing of different protocols increases both the VLSI complexity and the development cost.

Based on that, let us now consider the architectural aspects; one of the major obstacles for the hardware implementation of transport protocols - from the architectural point of view - is the decoupling of Network and Transport Layer protocols in the OSI Reference Model. The design of VLSI circuits for the transport layer without consideration of the underlying Network layer protocol is weak from many points of view. Obviously it is impossible to consider carefully the interface between the two layers, as well as to proceed to any implementation optimisation.

The Transfer Layer as it is defined in GAM-T-103 model combines the two layers. This solution minimises redundancies, QoS management, priority and addressing handling as well as interfaces between two layers having inherently a complementary role. The Transfer Layer philosophy adopted in XTP specification, is certainly more appropriate, than the classic one in OSI RM, for VLSI implementation.

Finally, the implementation of communication protocols in VLSI is a very complex task, requiring stability of protocol specification and knowledge of the host system where the interface will be integrated.

We believe that during next years, the need for low cost multimedia stations, having the capability to be connected to High Speed Networks, will create an important market. Today therefore, it seems realistic to specify the appropriate architecture for hardware transport protocol implementation and to undertake the effort for designing a set of fundamental and configurable modules that will be included in one (more) VLSI communication circuit(s), executing one (more) transport(s). This paper is a contribution towards this direction.

3 State of the Art Analysis

Hardware implementation of communication protocols means a hardware system including one or more VLSIs chip(s) that, execute communication protocols of one or more layers, perform non communication protocol tasks (e.g. data movement, timers management) and can interfaced with microprocessors or other chips (e.g. DMA controllers) or standard bus (e.g. VME). Several VLSI manufacturers commercialise products for the physical and MAC layers for all standardised LANs (e.g. FDDI, ETHERNET, Token Bus, token Ring etc.). Today, few experiments and platforms, integrating VLSI chip-sets that implement network an transport layers, have been reported in literature and two of them are reviewed in the next paragraphs.

3.1 Protocol Engine (PE 1000)

Protocol Engine Inc. (PE) has announced, in February 1992, a contract with Hewlett-Packard, for the fabrication of PE chip-set, the PE1000, in CMOS .8 micron technology. The Protocol Engine (PE 1000) [PE91], [Sch90], provides a hardware platform for multiprotocol processing that can handle a variety of them such as, XTP, TCP/IP, UDP, NFS, TP4. However, the PE chipset was designed by considering XTP as his dedicated transfer protocol (transport and network) and XTP was specified in order to be implemented by a PE chipset [Chs89]. The design of PE is a trade-off between flexibility and performance that can be obtained by VLSI chips dedicated to perform protocol related functions and can support different hosts and media.

3.1.1 PE Components - Bus

A Protocol Engine contains the following components:

1. Host Port interfaces to host memory or the host I/O bus and provides the following functions, transmitted frame checksum generation, host RAM control block management and real time packetizing of DMA its intelligent interface.
2. MAC Port (MPORT) that provides an interface to one or more of the same type media. An MPORT provides the following functions, received ISO, TCP, XTP frame check summing, header parsing via programmable finite state machine, error detection, sequence number checking, rate control, real-time address resolution for any syntax and sufficient FIFO capability to support Dbus data bursts. Protocol Engine may have more than one MPORT in order to potentially support different media types.
3. Buffer Controller (BCTL) that controls packet memory storage. The Buffer Control Component provides the following functions, buffer memory management, linked-list buffer queue maintenance, out-of-order packet processing, priority queues for TCP and XTP and data bus (Dbus) arbitration.
4. The optional Control Processor (CP) provides the following functions, connection management, control and data bus command passing and general protocol ACKs and other output messages.
5. Control Memory that provides address resolution and context information storage.
6. Buffer Memory that provides control pointer and data storage.

There are two Protocol Engine configurations. The minimal configuration does not contain a CP circuit. This implies a host processor providing the processing that would be provided by a CP component. The second configuration is the embedded system and unlike the minimal one, it does have a CP component as indicated in Figure 3.1. In either configuration, a specific Protocol Engine may contain multiple HPORTs and multiple MPORTs.

Moreover, inter-component communication is achieved via two busses:

1. The Cbus that handles bursts between control memory, the HPORT(s), the MPORT(s), and a CP (if present).
2. The Dbus that handles bursts between buffer memory, the HPORT(s), the MPORT(s), and a CP (if present)

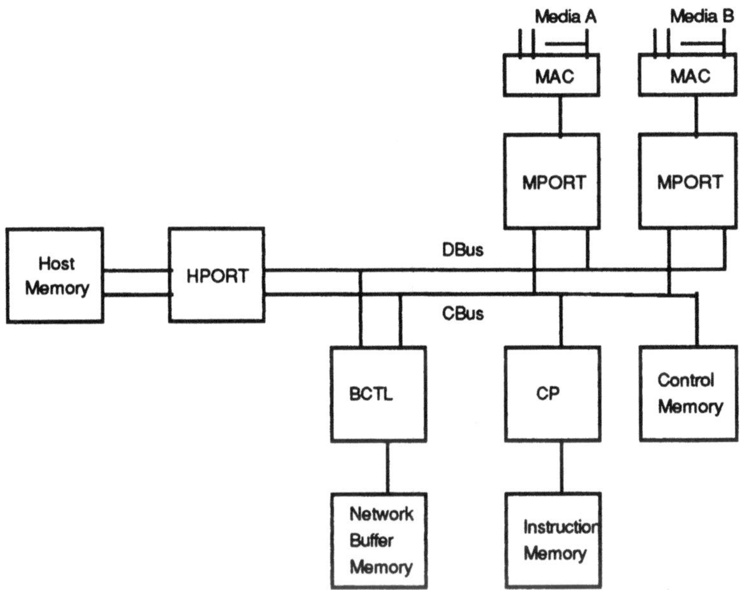

Fig. 3.1 Protocol Engine Embedded System Configuration

3.1.2 Remarks on PE1000 Solution

The PE1000 chip-set offers a multi-transport processing platform. This is certainly outside the initial scope of the design, that it was targeting to XTP, and was decided to facilitate market penetration, and handle any new high performance ISO transport and network protocol. Each chip includes one RISC processor. This selection is explained by the fact that the final XTP specification was not available during design phase of the chips and RISC offers fast context switching as well as more flexibility compared to dedicated hardware structures. The network interface board, based on PE1000 chip-set, represents a trade-off between multiprotocol flexibility and performance, obtained through VLSI chips dedicated to perform protocol related functions. The flexibility achieved by the architecture impacts the complexity of development tools used in implementing the selected transport protocol. While the performance achieved by the XTP platform and other protocols are not thoroughly analysed, preliminary results indicate that, the XTP implementation with PE1000 chip-set, can handle FDDI bandwidth [Sch90].

3.2 Network Co-processor

The concept for a protocol processing system combining a Network Co-processor and a multi-port memory is introduced in [Sid90]. This design targets in the implementation of TP4 OSI protocol. One or two co-processors can collaborate with the same CPU system in order to improve parallelism in the execution of communication functions.

For the identification of functions to be implemented in VLSI, the 80/20 principle is considered. On the basis of that principle, the following functions have been selected to be incorporated in the Network Co - processors: dual port DMA handling a linked list of buffers, checksum calculation, timer queue and variable length buffer allocation. The above functions are executed concurrently and managed by a control block, integrated in the VLSI chip. A single co-processor is sufficient to handle all the above tasks. Nevertheless, if two Co-processors are used, the following partitioning is proposed, to avoid competing bus accesses: the first one undertakes DMA and timer queue management while the other undertakes buffer allocation and checksum calculation.

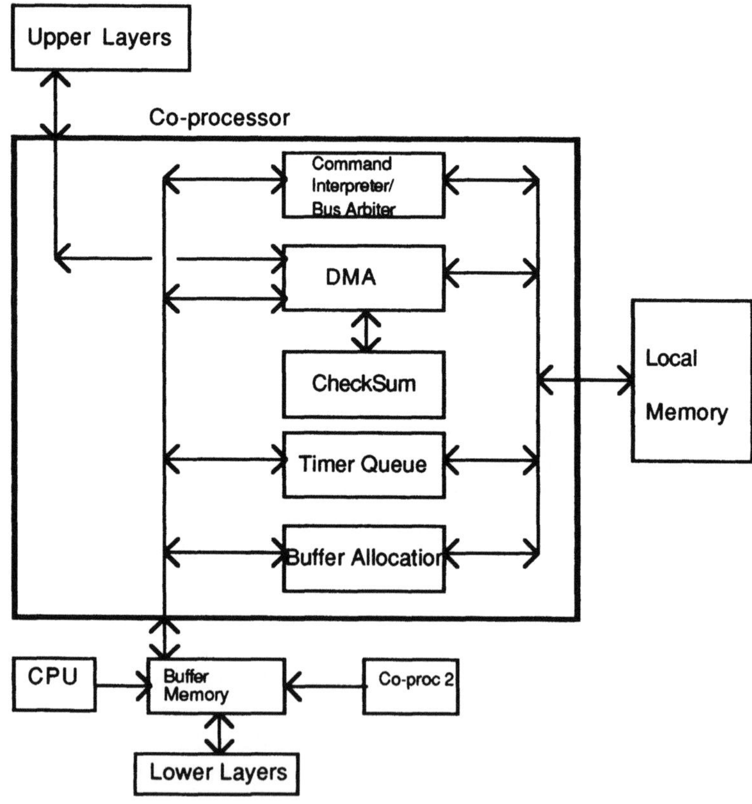

Fig. 3.2 The Network Co-processor Architecture

The utilisation of dual-port RAM memory in the communication interface boards is explored in [DRS87]. In [Sid90] the use of four-port RAM memory is proposed. Four-port RAM can be built up from very fast conventional RAMs associated with a multiplexing logic or by using available today chips. Indisputably, that improves the performance but it is a very costly solution. In the same paper ([Sid90]) the performance issues are not analysed, but a performance gain of 3 times compared to the conventional systems is mentioned.

Figure 3.2 presents the proposed architecture and the internal architecture of the network co-processor.

3.2.1 Remarks on Network Co-processor Solution

The proposed Network Co-processor's architecture, combined with the four-port memory, optimises the performance by enabling concurrent memory access to the memory. The interfacing problems between the Network Co-processor(s) and the conventional microprocessor are not investigated in the paper. A FDDI network usually exceeds 100KTPDUs per second, resulting to a considerable overhead. Furthermore, the implementation of the four-port memory is very costly solution even when using fast SRAMs. The programming of the TP4 in this architecture seems to be a complex task. As far as TP4 is concerned, it is standard and stable but it is certainly not suited for VLSI implementation and it is far from being the ideal candidate for handling distributed multimedia applications.

4 Architecture for Transport Layer Implementation.

The challenge for a high performance architecture implementing a transport protocol is to exploit the available at the physical layer bandwidth and to introduce the minimum transmission/reception delays and jitters caused by the protocol processing. The optimum configuration would allow complete protocol processing within the transmission/reception time of a packet, and this is often referred as real time protocol processing. Note now that the processing of one 128 bytes packet is approximately 1 usec if the bandwidth in the physical layer is 100Mbps. If the bandwidth is 1Gbps the above processing time is further divided by 10!

Therefore, we can conclude that:

- the processing of the transport protocols must be performed faster than at any time in the past
- the data transfer between different memory locations, associated to the protocol execution, must be minimised and
- the transport controller(s) must operate "autonomously" and their interaction with the host should be minimised (e.g. only in transitory phases)

Figure 4.1 illustrates the proposed architecture for transport protocol hardware implementation. It is a data flow architecture, where transport protocol VLSI controllers execute the protocol functionalities (connection management, retransmissions, rate control), process TPDUs on the fly and data for transmission or reception are stored or retrieved directly in or from the host memory. This architecture consists of five major parts:

- The transmission controller
- The reception controller
- The local memory
- The transmission FIFO
- The reception FIFO

Ideally, the implementation of this architecture should be done using one only ASIC of advanced technology (e.g. .5 microns). However, by considering the large number of pins needed and the high complexity, it is more reasonable to split it to several ICs for the very first prototyping step. This will definitely improve the testability of the whole system although it will degrade its performance.

This architecture for transport protocol includes two dedicated VLSI circuits (ASIC) one transmission controller and one reception controller offering a real time protocol processing and addressing the host memory space directly. In this way, data movements between any intermediate storage are avoided. Both the transmission and the reception controller are controlling the data transfers between host memory and FIFO memories.

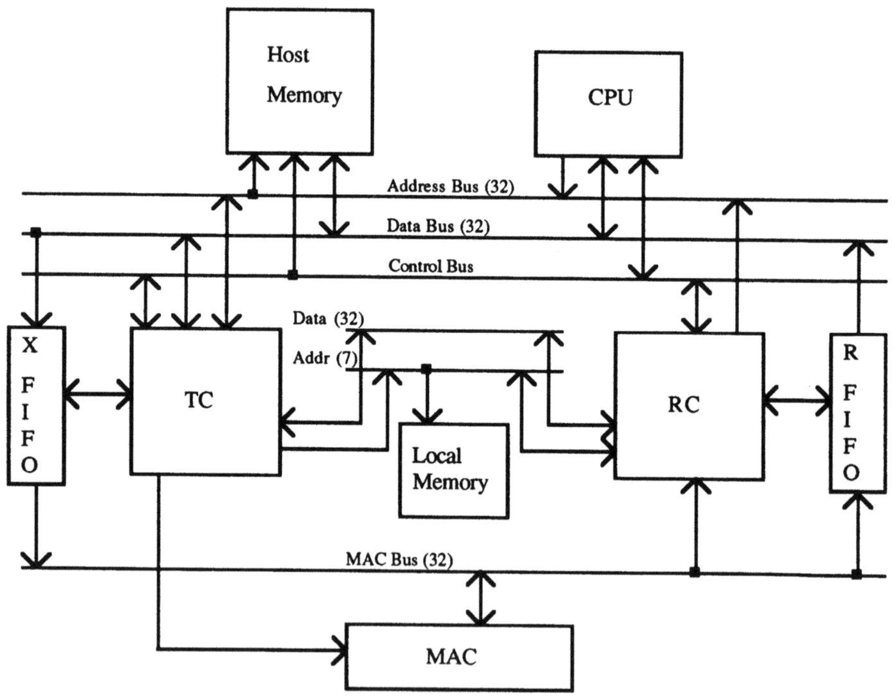

Fig. 4.1 Architecture for hardware transport protocol implementation.

The transmission and reception controllers are interfaced with two small dual ported memories, organised as FIFOs, in order to absorb host bus latencies. These FIFOs can be implemented with commercially available ICs with a typical size of 2Kwords X 9 bits/word and offering the capability of asynchronous read and write operations. Thus, for a FIFO memory of 8Kbytes, four chips can be used.

The host memory is a block of conventional SRAMs shared between the transmission and reception controllers, the host and the peripherals. As far as the transmission and reception space area are concerned, the host memory is organised in four different parts. One part acts as a mailbox between the host and the controllers for exchanging control information, that concerns new connections' establishment and commands. The other three parts are organised as lists of linked descriptors associated to variable size buffers. The host is responsible for organising the host memory and allocating the necessary buffer space, based on the QoS parameters, that are negotiated during the connection establishment phase, as well as the host memory space allocated to the corresponding application. One list is accessed by the transmission controller and holds the data that are to be transmitted, the second list is accessed by the reception controller and holds the received data and the third one is also accessed by the reception controller and holds the received data corresponding to fast connection set up requests (one descriptor is associated to each fast connection request). Figure 4.2 illustrates the host memory organisation. The "Status + Offset" memory word in each descriptor indicates whether the buffer is unused, terminated or currently in use. In the last case, the offset from the beginning of the buffer is provided.

Finally, interrupts activated by the transmission and reception controllers are restricted to the absolute minimum. Only one interrupt is activated by the reception controller when a packet request for a new connection set up (fast or not) is received. The Transmission Controller updates the descriptor, since it gets all the necessary information from the Local Memory and it is responsible for the assembly of the corresponding ACKs. The time consumable polling per packet procedure, that is triggered by interrupts, is unnecessary, because the host can monitor the status of his memory that is allocated to a connection, based on the corresponding QoS parameters.

There are also two other busses, the local data bus and the local address bus, which interconnect the local memory with the transmission and reception controllers, where the contexts and status of active connections are stored. This local memory is a quite complex structure and include a relatively small and fast conventional static RAM, two small CAMs and an intelligent memory controller that accepts a searching by local connection identifiers and returns the context and status of active connection. Figure 4.3 illustrates the local memory organisation.

The underlying protocols are handled by a MAC controller assumed to perform an equivalent set of functions of the FDDI MC chip-set [Mot90]. The problems that are related to the network layer functionality are not examined in this paper.

As it has been already mentioned in section 1, the data movement represents an important part of the protocol processing overhead (80/20 principle). The data flow architecture does not require intermediate storage and manages data directly inside the application area for connection oriented data exchanges. This approach obviously improves the interface performance, because no additional latency and jitter are introduced during the packet movement. It also supports efficiently the multimedia applications that are normally connection oriented and highly sensitive in both the delay and jitters introduced by the network.

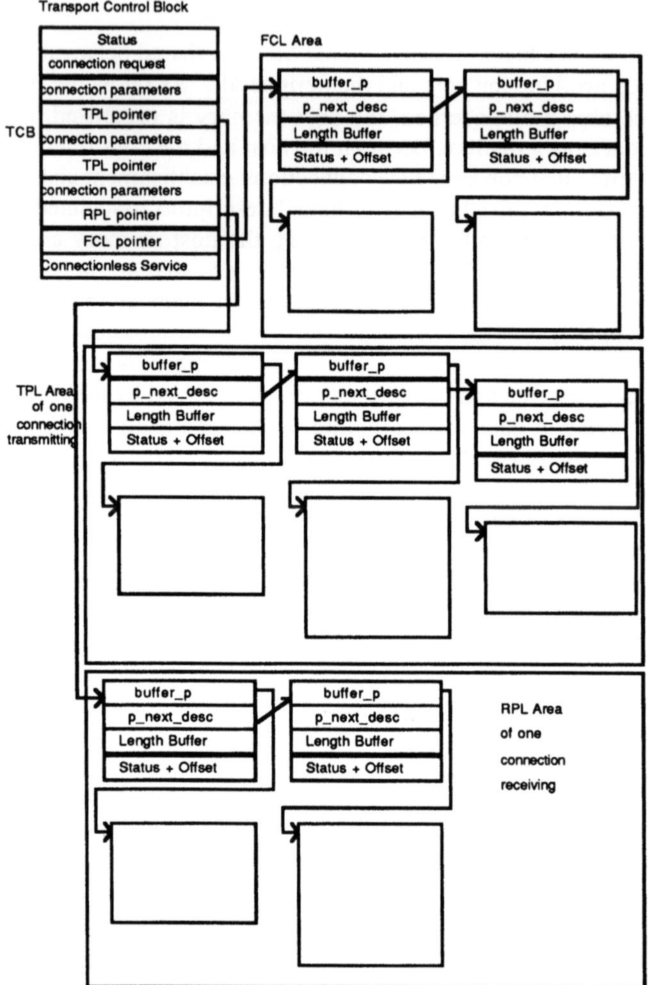

Fig. 4.2 The Host memory organisation

The proposed data flow architecture supports the placement of the CRC at the end of each packet and not at the end of the variable size header of the TPDU as in TP4. Furthermore, the checksum calculation scheme that it has been proposed in XTP is adopted, as it is easier to be implemented in hardware.

Another issue that it must be considered, is the estimation of additional host bus occupation due to retransmissions and reception of damaged packets.

The use of optical fibres in High Speed Networks (HSNs) improves drastically the communications' reliability at the physical layer. Multimedia applications are relatively insensitive to packet loss and bit errors. Furthermore, selective retransmission's strategies are more effective than Go-Back-N ones in HSNs and their introduction in the next generation of standard transport protocols can be certainly assumed.

Since retransmissions will be selective and rare, we afford to read the data, that are to be retransmitted, directly from the host's memory a second time, without noticeably increasing the average load on the host bus. The received packets are stored in the RFIFO and, consequently, the reception controller has the time to discard damaged packets.

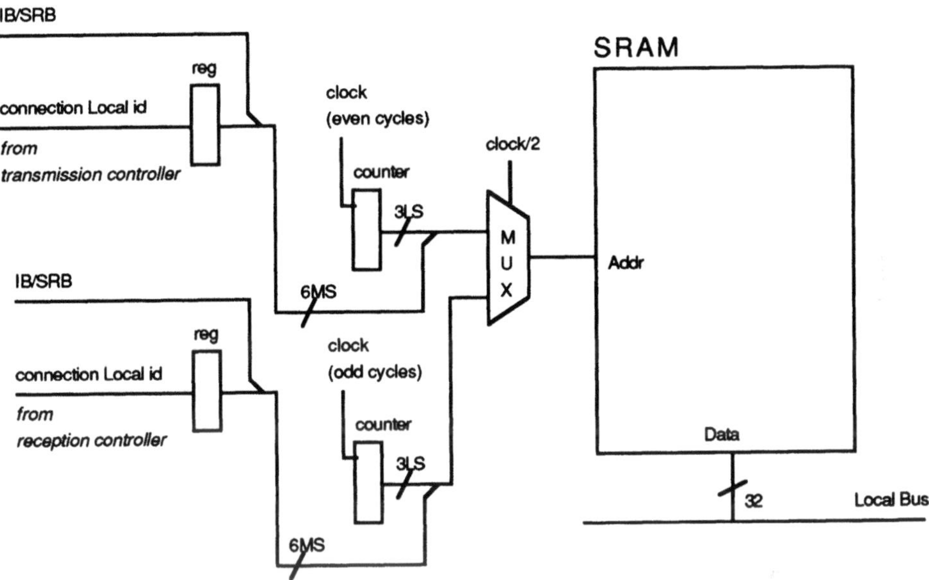

Fig. 4.3   The local memory organisation

5 Conclusions

We believe that the forthcoming High Speed Transport Protocol (HSTP) specifications should be done in such a way that they will ease the VLSI implementation. Therefore, the new protocol should not be a superset of services and protocol mechanisms, but rather a clever selection of services and protocol mechanisms, that are necessary for the forthcoming High Performance Networking environment and support distributed multimedia applications. Furthermore, the TPDU format must be carefully specified in such a way that it will facilitate the VLSI implementation and thus it will be beneficial for the communication performance.

The proposed architecture for a VLSI implementation of a transport protocol aims at offering the following advantages:

- compact, minimum external logic
- TPDU's processing on the fly so that no additional delay or jitter is introduced
- manipulation of connection oriented communication data directly in the application memory area
- easy integration within existing systems

We are convinced that such an architecture is optimal for high performance networking environments.

Today, it seems realistic to undertake the effort for specifying the appropriate network interface architecture and for designing a set of fundamental and configurable modules that they will be included in one ore more communication VLSI circuit (s), executing one or more transport protocol(s) by considering:

- the target host system architecture (CPU host and MAC interfaces),
- the XTP specifications, due to the fact that is "stable" and is supported by many US and some European manufacturers. TPX must also considered because it defines a rich set of services and protocol mechanisms and that may influence the evolution in the area of high performance transport protocols
- the ongoing evolution of transport and network protocols in ISO standardisation committees.

Note: The results of the study performed in the framework of OSI95 ESPRIT project for the task entitled "Transport protocols VLSI implementation" are presented in [Vot92].

References

[Che86]   Cheriton D., **VMTP: a protocol for the next generation of communication systems,** *ACM SIGCOM'86 symposium,* Vermont 1986

[Chs89]   Chesson G., **XTP/PE Design Considerations,** *IFIP Workshop HS LANs,* Zurich 1989

[DRS87]   Dang M., Rolin P., Sponga L., Votsis G., **Some effects of High Speed LANs on the design of MAC Communicating Circuits,** *IFIP Workshop HS LANs,* Aachen 1987

[Dan92]   Danthine A., **A New Transport Protocol for the Broadband Environment,** *IFIP Workshop on Broadband Communication,* 1992

[Dio91]   Diot C., **Architecture pour l'implantation hautes performances des protocoles de communication de niveau transport,** *These INPG,* France, 1991

[Mot90]   Motorola, **M68FDDIADS Hardware User's Manual,** *Preliminary Specification,* 1990

[Par90]   Partridge C., **Internet research steering group workshop on very high speed networks,** *RFC 1152, Workshop report,* April 1990

[PE91]   Protocol Engines, **PE1000 Series: Protocol Engines Chipset,** *Advanced Information,* 26 March 1991

[Sch90]   Schaweder W., **XTP in VLSI,** *Transfer PEI,* Vol. 3No 4, 7-8/90

[Sid90]   Sidenius M., **Hardware Support for Implementation of Transport Layer Protocols,** *IFIP Workshop HSNets,* Palo Alto, 1990
[Vot92]   Votsis G., **Transport protocols VLSI implementation,** *ICOM-2, A9 final deliverable* 11/1992
[XTP92]   **XTP, Protocol Definition,** Rev 3.6, 1/92
[WO90]   Watson G., Ooi S., **What should a Gbit/s network interface look like?,** *IFIP Workshop HSNets,* Palo Alto, 1990

[Si90]   Siekaus M., Hardware Support for Implementation of Transport Layer Protocols, IFIP Workshop HSC'90, Palo Alto, 1990.

[Vo92]   Voila G., Transport protocols VLSI implementation, TENCON'92, 1992.

[XTP91] XTP: Protocol Definition, Rev 3.6, 1992.

[We88]   Watson R., What should a Chars network service like "the"? Note C-IS-83C, 1983.

Time Extension of LOTOS

At the start of the project, it has been anticipated that the more synchronous and isochronous natures of the communication streams related to the new applications would benefit from a time extension of LOTOS. This was confirmed during the formal specification of the OSI95 Transport Service.

An extension of LOTOS, with timed and probabilistic behaviour, has been developed by DIT. This extended LOTOS is upward compatible with LOTOS.

It is classical to assess any formal technique on a test case. For the existing standard FDTs, the test case is the Abracadabra protocol which is a grouping of most of the mechanisms that will be found in any real protocol. Such a test case is useful to allow a fair comparison of the proposed FDTs.

As the same need was emerging for comparing the various proposals to introduce a time extension in LOTOS, ULg designed the "Tick-Tock Protocol". This example will be useful to assess any proposed timed FDT. It consists of a protocol composed of two entities and an underlying service provider whose behaviour is mainly based on various timing constraints such as time-out, isochronous behaviour and rate control.

The first application of the tick-tock case has been the assessment of the time extension of LOTOS-TP. LOTOS-TP was also assessed for its probabilistic behaviour on simplified models of FDDI and of XTP.

It is expected that ISO will soon envisage the possibility of an extension of LOTOS and that LOTOS-TP as well as other proposals for time extension of LOTOS will be considered as basic contributions.

Extended LOTOS*

C. Miguel, L. Vidaller, and A. Fernández
Dpto. Ingeniería de Sistemas Telemáticos (DIT). ETSI Telecomunicación, UPM
Ciudad Universitaria - 28040 Madrid - Spain
e-mail: cmiguel@dit.upm.es

This paper describes an extension of LOTOS with timed and probabilistic behaviours. The resulting language is denoted LOTOS–TP. It has been defined taking into account a set of performance requirements extracted from the analysis of data communication networks, services, and protocols. The timed part extends LOTOS with basically a new operator: "Timed action prefix" which can be combined with other LOTOS operators to specify a wide variety of timed behaviours. The paper defines the strong and weak bisimulation relations taking into account the timed evolution of the specified systems. The main properties of LOTOS remain, being LOTOS–TP upward compatible.
Keywords: LOTOS, timed FDT, equivalences.

1   Introduction

The increasing complexity of communication systems demands the existence of formal specification techniques which support their design. Standardization organizations have reacted to this necessity defining different specification techniques such as the Specification and Description Language (SDL) [CCI85] defined by the Consultative Committee for International Telephone and Telegraph (CCITT), or the Extended State Transition Language (ESTELLE) [ISO88] and the Language Of Temporal Ordering Specification (LOTOS) [ISO89] defined by the International Standards Organization (ISO).

An ideal environment for designing communication systems is shown in Figure 1.1. Such an environment should be based on a machine–readable system specification which could be tested for correctness (verification and validation), and from which the designer can obtain: the performance of the system at early design stages, a prototype or implementation, and a set of test cases used to assure that the implementation conforms to the specification. We have used the term "ideal" because the existing formal specification techniques are not able to completely cope with the needs of this designing environment. This is specially true because the abovementioned standard techniques do not permit the specification of the timing and probabilistic characteristics of systems. Therefore, the performance aspects of systems cannot be specified.

---

*Sections 2 and 3 and parts of sections 1 and 5 of this paper have been published in: "C. Miguel, A. Fernández, L. Vidaller; Extending LOTOS Towards Performance Evaluation; Formal Description Techniques V, M. Diaz, R. Groz (Editors)" and are reprinted here with the kind permission of Elsevier Science Publishers B. V. (North Holland).

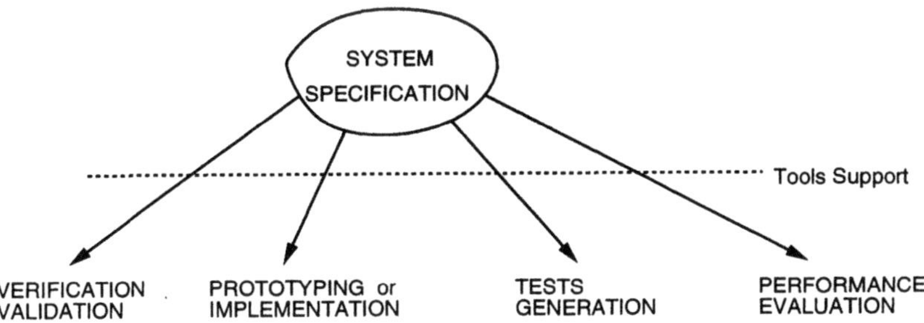

Fig. 1.1 Ideal environment for designing

The performance evaluation is currently made by applying analytical techniques such as queuing networks, or simulation techniques which obtain an approximation to the performance of the system by running (using a computer) a model of the system written in one of the existing, not formalized, simulation languages [Pol89].

Even though formal techniques such as LOTOS are able to model the ordering of system events in time, they cannot express quantitative properties about the instant of time in which such events occur. Taking into account that the behavior of most of the real systems depends on the quantitative passing of time, e.g the use of timer to solve error conditions in data communication systems, the extension of these techniques to cope with the timing requirements of communication systems seems necessary. Such extension is necessary not only if we desire to express some performance figures of the specified systems; but it is also very valuable to model systems because the "time" is a common knowledgement among every component of a system, and it can be used to easily and realistically express dependencies between them.

This chapter proposes some extensions to the formal description technique LOTOS in order to be able to modelize the timing and probabilistic characteristics of systems in addition to the functional ones. Therefore, the "ideal" design environment can be fully built. The resulting formal technique has been termed LOTOS–TP.

LOTOS was specially developed to describe the protocols and services involved in data communication systems. It reached the status of ISO international standard at the end of 1988. The description of the system behaviour in LOTOS is based on a modification of CCS [Mil80, Mil89] that includes elements of the other calculi. LOTOS also permits us the description of data structures and values by using an abstract data type language based on ACT–ONE [HEH83].

The definition of extensions of LOTOS and other formal techniques is currently an active research area which has produced several timed LOTOS proposals as shown in the overviews of the state of the art [NS91, MFV91b]. The final result of these researches should be the standardization of an extended LOTOS which should cope with the specification of timing aspects. Such a technique should be able to model synchronous and asynchronous behaviors, to describe waitings, delays and timers. It should also be able to measure the time elapsed between the occurrence of two system events. To achieve this expressive power is the objective that guided the definition of the timing part of the extended language defined in this chapter.

A characteristic of LOTOS is that it permits us to describe partially concurrent systems by including non–deterministic behaviours. These non-deterministic components are abstractions of the actual ones, and they can be detailed in successive refinements. LOTOS–TP is an enrichment of the above abstraction. It is a formal description technique which is able to characterize the non-determinism in a probabilistic way.

The performance evaluation of systems can be done by applying the defined LOTOS–TP calculus, but also by *simulation*. Simulation means that we are able to obtain the desired results with a given confidence level. The simulation is carried out by observing repeated trials of the system behaviour and applying the well known *Law of the Large Numbers* [Pap89].

This chapter is organized as follows: Section 2 presents a cursory introduction of LOTOS in order to provide a first view to those readers who are not familiar with the formal technique. Section 3 describes the timing part of LOTOS–TP, which is called LOTOS–T. First we make an informal introduction to the language which uses some examples to show its expressive capabilities. Then the formal semantics is defined. Section 4 presents an informal description of the probabilistic component of LOTOS–TP which is called LOTOS–P. The conclusions of this work are presented in Section 5.

2   Cursory Overview of LOTOS

As the introductory section has outlined, LOTOS–TP is an upward compatible extension of LOTOS. So, let's start describing LOTOS before introducing the timed and probabilistic expressions. A complete definition of LOTOS can be found in [ISO89]. This section does not try to be complete or exhaustive. Its objective is to give an informal and simplified idea of the language on which the chapter is based to the readers who are not familiar with LOTOS.

LOTOS has two main components. The first one is an abstract data types (ADT) definition language. It permits us to define data objects and their operations in an equational way. The second one is the description of the dynamic behaviour of the system in terms of the ordering in time of system events. Roughly speaking, LOTOS is based on four operators which are described below. Let's denote events as "a", "b", or "c" and LOTOS expressions as "B" or "C":

Sequence .- The expression "a; B" means that when the system performs event "a", then it will behave as "B" describes. The expression "B >> C" represent the sequential composition of behaviour B and C.

Choice .- The expression "B [] C" represents an alternative between two possible behaviours: "B" or "C".

Parallel .- The expression "B |[a,b,···]| C" means that the behaviours B and C run concurrently, but they have to synchronize in the events "a", "b", ···. A particular case of the parallel is when the behaviours are not forced to synchronize in any event. This case is represented as "B ||| C".

Hiding .- The "hide" operator provides a mechanism to hide the occurrence of events so that they cannot be observed externally; e.g. the expression "hide a in B" makes that the occurrence of event "a" in the behaviour B becomes an unobservable (internal) event. This is a basic and powerful mechanism of abstraction which characterizes almost every technique based on process algebras.

Besides, there are facilities to handle processes and data values such as the passing of values in a synchronization ("a!value" is an event that offers a value, and "a?x:sort" is an event that accepts values of a given sort, e.g. naturals), or guards. Another interesting expression is the choice of values. The expression

```
choice x:nat [] B,
```

where x can be a free variable in B, behaves as the choice of every B expressions resulting from substituting all occurrences of variable x in B for all possible values of sort (type) nat.

3   Timed Behaviours in LOTOS–TP (LOTOS–T)

3.1   LOTOS–T Informal Syntax and Semantics

LOTOS syntax and semantics have been extended in order to cope with the modeling of quantitative time requirements of a system.

Table 3.1 Informal comparison of semantics

Expression	LOTOS informal Semantics	LOTOS–T informal Semantics
a; B	Action "a" occurs before behaviour "B", but at an unspecified instant of time.	Action "a" must occur as soon as possible (ASAP), and after that the system behaves like "B". Time may pass if the environment prevents the occurrence of action "a", but it is always ready to occur at any instant of time. It must occur as soon as the environment permits its occurrence.
exit	The successful termination occurs, but at an unspecified instant.	The successful termination[1] must occur ASAP.
a{t}; B	———	Action "a" occurs *just* when a time "t" has elapsed. If the environment prevents the occurrence of action "a" at such instant of time, action "a" will not occur. Time passing is not blocked in spite of the possible occurrence of a blocking in action "a".
exit{t}	———	The successful termination[1] occurs *just* when time "t" has elapsed.

LOTOS–T extends the LOTOS syntax with two new operators:

- Timed Action Prefix.- *a* {*t*}; *B* where "a" denotes an action, "t" is a value expression, and "B" is a LOTOS–T expression.
- Timed Successful Termination.- *exit* {*t*}   where "t" is a value expression.

Table 3.1 summarizes the informal semantics of the extended technique in comparison with the standard LOTOS one.

As shown in the table, in addition to the new operators (Timed action prefix, and Timed successful termination), the interpretation of the LOTOS action prefix and successful termination have been extended to cope with timing. Standard LOTOS actions occur at unspecified instants, but LOTOS–T actions occur as soon as possible (ASAP); that is, "*now*" if there are no environment constraints. It is important to point out that "a; B" is *NOT* equivalent to "a {0}; B" because the first behaviour can wait until action "a" is possible, but the second one cannot.

It is important to comment that the environment of an action is completely defined when the action is hidden, so that no other constraint can be added to the occurrence of the action. Therefore, we will not be able to determine the time in which an observable action occurs until it is hidden.

Timed operators force actions to occur at a single time instant. This time constraint seems too strong; but if it is combined with the LOTOS operators, we will be able to specify any type of time constraint. For example, the expression (It is supposed that "time" is modelled by natural numbers "nat")

```
choice  t: nat []  [f(t)]  ->  a {t}; B
```

specifies that action "a" will occur as soon as it can, but only at those time instants (t) that make condition "f(t)" be true. New syntactical elements could be easily defined in order to represent the above expression of a general time constraint in a more concise way, but we do not make such extension in this chapter because it is not necessary for the definition of the semantic model.

We will illustrate the expressive power of LOTOS–T through a set of examples:

Delays.- Behaviour expression "`choice t:nat [] [t > 5]-> a {t}; B`"
   specifies that action "a" can only occur after 5 time units have elapsed.
Waits.- ASAP criterion permits the specification of waitings until the environment is
   ready to accept the action occurrence (synchronization). Behaviour

```
        data_req?d:data; Data_transfer_Phase
  []disc_req; Disc_Phase
```

specifies a communication service which is waiting until the environment (upper layer) desires to synchronize in actions "data_req" or "disc_req". If data_req synchronizes, the behaviour expression called "Data_transfer_Phase" handles the data. If disc_req synchronizes, a disconnection phase is started.
Synchronous behaviours.- The following process models a system that generates clock ticks each 3 time units.

---

[1] It is actually a special action prefix "δ; *stop*". Therefore, the same comments as in action prefix are applied.

```
process clock [tick]:noexit:=
   tick {3}; clock [tick]
endproc
```

Time-Outs.- The behaviour

```
      ack; send_next_msg
   [] i {5}; retransmission
```

specifies that if action "ack" has not occurred before 5 time units, the system will evolve to behave as "retransmission" (if "ack" occurs at 5, we will have a non-deterministic choice). Remember that because of maximum progress or ASAP criterion, internal action will occur at 5.
A timer can be specified by the process:

```
process timer [start, time_out]:noexit:=
  start?to:nat; (  time_out {to}; timer[start,time_out]
              [] timer[start, time_out])
endproc
```

It can be started to time out after "to", and it can be also restarted at any time.
Time measurement and handling.- The process:

```
process global_time [current_time](t:nat):noexit:=
  choice t1: nat [] current_time!t+t1 {t1};
                 global_time[current_time](t+t1)
endproc
```

can inform of the global time (time elapsed from its first instantiation: "global_time [current_time] (0)") through "current_time" gate. Parameter "t" has the value of the time elapsed until the last synchronization in gate "current_time", and "t1" the time elapsed since that instant.
If we desire to measure the time between two actions ("send" and "receive"), it can be specified by

```
      send; (choice t: nat [] receive {t}; B)
```

where behaviour "B" knows the transmission delay (value of variable t), and it can use this value to make some calculations, or to determine its future behaviour.

3.2   Time Domain

Discrete event systems, the ones we are interested in, are those whose state changes or system events occur at a countable set of time instants, but time is considered continuous.

Some formal description techniques, such as timed LOTOS [Azc90, Led91], SCCS [Mil83], ATP [NRSV90], TPCCS [Han91] or TPL [HR90], model time in a discrete way; most of them consider that there exists a special action (tick) whose occurrence means the passing of one unit of time. Such discrete time domain seems suitable to model systems whose actions occur synchronously with a single clock. If that condition does not hold, the discrete approximation could lead to unbound errors in the action occurrence time (single error can be accumulative). In this cases the "time unit" (tick) should be properly selected in order to obtain the desired maximum error, but this selection depends on the dynamics of the system.

This kind of problems can be solved if a *dense time domain* is used to model real time. A domain D is a dense domain if for every $t_1, t_2 \in D$ with $t_1 > t_2$, there exists a value

$t \in D$ such that $t_1 > t > t_2$. Rational and real numbers are examples of dense domains.

The time domain used in LOTOS–T should be defined by the designer, and it can be chosen as either discrete or dense domain. One of the requirements imposed on LOTOS–T states that time values have to be handled as normal data values; therefore, LOTOS–T defines the time domain as the set of values of a given data sort (Q(time) where **time** is a LOTOS sort) defined in LOTOS data language. Furthermore, in order to consider a specific *sort* as a time domain it should contain certain semantical elements, and such elements should have a set of properties. These semantical elements and the requirements to be imposed on the specification of a *sort* in order to consider it as a correct specification of a time domain are described below. The following elements have to be defined over $Q(\textbf{time})$:

- A total order relation, that is represented by "$>$".
- An element $0 \in Q(\textbf{time})$ such that for each value $r \in Q(\textbf{time})$ and $r \neq 0$

$$r > 0$$

  This element models the present: *now*.
- A commutative and associative operation "$+ : Q(\textbf{time}), Q(\textbf{time}) \longrightarrow Q(\textbf{time})$" such that

  - for every $r, r_1 \in Q(\textbf{time})$:

$$r > r_1 \iff \exists r' > 0 \text{ such that } (r' + r_1) = r;$$

    that is, time always passes: $r > 0 \implies r + r_1 > r_1$,
    and the time interval between two time instants can always be measured.
  - $0 \in Q(\textbf{time})$ is an identity element of "$+$" such that for all value $r \in Q(\textbf{time})$:

$$r + 0 = r$$

For example, natural numbers as specified in the standard library [ISO89] of LOTOS fulfill the requirements for being a discrete time domain. Also, in appendix A there is a specification of the positive rational numbers which can be used as a dense time domain. Appendix A lists the abstract data type definitions of the natural (nat) and positive rational numbers (p_rational) which make up, with the booleans (logical values), the basic data library used in this chapter.

3.3   LOTOS–T Syntax

LOTOS syntax [ISO89] has been extended with three new elements:

Time domain definition .- Time domain elements have to be defined within the global definitions of the specification with the following syntax:

```
time-domain-definition =
        "timedef" "sort" sort-identifier
           "opns" "+" "for" operation-identifier
                  "0" "for" value-expression
                  ">" "for" operation-identifier
        "enddef"
```

Timed action prefix .- A new action prefix expression is defined in LOTOS–T:

```
timed-action-prefix-expression =
        action-denotation "{" value-expression "}"
      ";" action-prefix-expression.
```

For example: a?x:nat {5}; exit

Timed termination .- A new atomic expression is defined in LOTOS–T:

```
timed-termination = "exit" [exit-parameter-list]
                    "{" value-expression "}"
```

For example: exit {t+2}, where t is a variable.

3.4   LOTOS–T Formal Semantics

Let AS be a canonical algebraic specification as defined in [ISO89]:

$$AS =< S,\ OP,\ E >$$

where S is a set of sort names, OP is a set of operations and E is a set of conditional
equations with respect to the signature $< S,\ OP >$. Its interpretation [ISO89] as
a many-sorted algebra gives us for each sort ($s \in S$) a set of values, Q(s). A term
t of a sort $s \in S$ is interpreted as a value in $Q(s)$, denoted as $[t]$. An operation
$op \in OP$ with arguments of sorts $s_1, \cdots, s_n \in S$, and result of sort s_r is interpreted as
a function: $Q(op) : Q(s_1) \times \cdots \times Q(s_n) \longrightarrow Q(s_r)$. Let DD be the set of all values,
$DD = \bigcup_{s \in S} Q(s)$.

A process definition *pdef* is a 2–tuple $< p, B >$ where p is the process name, and B is
a LOTOS–T behaviour expression. Its formal gates are denoted as $fg(p)$, and its formal
parameters as $fv(p)$.

In order to define the LOTOS–T formal semantics we have assumed a context that
consists of a set of process definitions, PDEFS, and a canonical algebraic specification,
AS.

3.4.1   Semantics of Time Domain

From the time domain definition given in section 3.3 we define the following tuple:

$$\text{TIME} =< \text{time}, t_0, op_+, op_> >$$

where **time** $\in S$ is the sort of the time, t_0 is a ground term of sort **time**, $op_+ \in OP$ applies
two terms of sort **time** over another term of the same sort (op_+ : **time, time** -> **time**),
and $op_> \in OP$ applies two terms of sort **time** over another term of a sort $s \in S$ ($op_>$:
time, time -> s).

As it has been previously mentioned, quantitative time is modelled with four semantical
elements:

- The set of time values Q(**time**).
- A least element $0 = [t_0] \in Q(\text{time})$.
- An internal operation $+ = Q(op_+)$.

- An order relation defined as $\quad \forall v, v' \in Q(\mathsf{time}), v > v' \iff Q(op_>)(v, v') = [true]$.

These elements *must conform* to the semantical requirements stated in section 3.2 in order to model correctly the time domain.

3.4.2   Semantics of LOTOS–T Behaviour Expressions

The semantic model of a LOTOS–T behaviour expression, B, is a timed version of the LOTOS *structured labelled transition system* [ISO89]. We have called it a Timed structured labelled transition system, TTS(B):

$$TTS(B) = < S_t, G \cup \{\delta, i\}, AS, \mathsf{TIME}, T, s_0 >$$

where

- S_t is a set of states.
- G is a set of gate names.
- AS and TIME have been already defined.
- T is a set of timed labelled transitions. $T = \{ \ -a\,t\ \mapsto\ \ | a \in Act, \ t \in Q(\mathsf{time})\}$
  with $-a\,t\ \mapsto\ = \{< B_1, B_2 > \ | \ B_1 \ -a\,t\ \mapsto\ B_2 \ , \ B_1, B_2 \in S_t\}$ and
  Act $= \{i\} \cup \{gv \ | \ g \in G \cup \{\delta\}, \ v \in DD^*\}$.

- The initial state. $s_0 = B$.

The elements S_t and T can be obtained in one of the following ways:

- Defining a set of the axioms and inference rules that include the ASAP criterion, and permit to obtain directly the transition system. Such approach was taken by C. Smith & S. Rudkin [SR88], M. Hennessy & T. Regan [HR90], L. Chen [CAM90], W. Yi [Yi91], and H. Hansson [Han91]. These sets of axioms and inference rules for LOTOS–T can be found in [MFV91c, MFOV93].
- Defining a set of the axioms and inference rules that do not include the ASAP criterion. The transition system, TTS(B), will *contain only the ASAP evolutions* derived from these sets of axioms and inference rules. A similar approach was used by R. Cleaveland and M. Hennessy in [CH88] when they extend CCS with priorities.

We are going to present here the second approach. Table 3.2 shows the set of axioms and inference rules of the LOTOS–T transition derivation system without ASAP criterion where a, b are actions (including the internal action, $a, b \in Act$), B, B_i, B', B_i' are LOTOS-T behaviour expressions, t_t is a value expression, s is a sort name, and A is a list of gate identifiers. The function *name(a)* gives us the gate name associated with action "a".

As in the TIC model [QAF93], a semantical operator " Age " is defined in order to make time pass at the same pace in all the concurrent components of the system. The expression Age (t, B) represents the behaviour "B" when a time "t" has elapsed. Therefore, the actions that "B" may derive before "t" will be impossible, and the actions that "B" may derive after a time $t' \geq t$ will be derived at a time t_1 such that $t_1 + t = t'$.

If we obtain the timed structured labelled transition system directly from the derivatives of the sets of axioms and rules in Table 3.2, the resulting model is very similar to the

Table 3.2 LOTOS-T Transition Derivation System (without ASAP)

Action Prefix:	$a; B \; - \; a \; t \; \rightarrow \; B \qquad \forall t \in Q(\text{time})$
Timed Act. Prefix:	$a \{t_t\}; B \; - \; a \; t \; \rightarrow \; B \qquad t = [t_t]$
Termination:	$exit \; - \; \delta \; t \; \rightarrow \; stop \qquad \forall t \in Q(\text{time})$
Timed Termin.:	$exit\{t_t\} \; - \; \delta \; t \; \rightarrow \; stop \qquad t = [t_t]$
Choice:	$$\dfrac{B_1 \; - \; a \; t \; \rightarrow \; B_1'}{B_1[]B_2 \; - \; a \; t \; \rightarrow \; B_1'} \qquad \dfrac{B_2 \; - \; a \; t \; \rightarrow \; B_2'}{B_1[]B_2 \; - \; a \; t \; \rightarrow \; B_2'}$$
Hiding:	$$\dfrac{B \; - \; a \; t \; \rightarrow \; B'}{\text{hide A in } B \; - \; a \; t \; \rightarrow \; \text{hide A in } B'} \quad \text{if name } (a) \notin A$$ $$\dfrac{B \; - \; a \; t \; \rightarrow \; B'}{\text{hide A in } B \; - \; i \; t \; \rightarrow \; \text{hide A in } B'} \quad \text{if name } (a) \in A$$
Parallel:	$$\dfrac{B_1 \; - \; a \; t \; \rightarrow \; B_1'}{B_1|[A]|B_2 \; - \; a \; t \; \rightarrow \; B_1'|[A]| \text{ Age } (t, B_2)} \quad \text{if name } (a) \notin A \cup \{\delta\}$$ $$\dfrac{B_2 \; - \; a \; t \; \rightarrow \; B_2'}{B_1|[A]|B_2 \; - \; a \; t \; \rightarrow \; \text{Age } (t, B_1)|[A]|B_2'} \quad \text{if name } (a) \notin A \cup \{\delta\}$$ $$\dfrac{B_1 \; - \; a \; t \; \rightarrow \; B_1', \; B_2 \; - \; a \; t \; \rightarrow \; B_2'}{B_1|[A]|B_2 \; - \; a \; t \; \rightarrow \; B_1'|[A]|B_2'} \quad \text{if name } (a) \in A \cup \{\delta\}$$
Disabling:	$$\dfrac{B_1 \; - \; a \; t \; \rightarrow \; B_1'}{B_1[> B_2 \; - \; a \; t \; \rightarrow \; B_1'[> \text{Age } (t, B_2)} \quad \text{if name } (a) \notin A \cup \{\delta\}$$ $$\dfrac{B_1 \; - \; \delta \; t \; \rightarrow \; B_1'}{B_1[> B_2 \; - \; \delta \; t \; \rightarrow \; B_1'}$$ $$\dfrac{B_2 \; - \; a \; t \; \rightarrow \; B_2'}{B_1[> B_2 \; - \; a \; t \; \rightarrow \; B_2'}$$
Enabling:	$$\dfrac{B_1 \; - \; a \; t \; \rightarrow \; B_1'}{B_1 >> B_2 \; - \; a \; t \; \rightarrow \; B_1' >> B_2} \qquad \text{if name } (a) \neq \delta$$ $$\dfrac{B_1 \; - \; \delta \; t \; \rightarrow \; B_1'}{B_1 >> B_2 \; - \; i \; t \; \rightarrow \; B_2}$$
Aging:	$$\dfrac{B_1 \; - \; a \; t \; \rightarrow \; B_1'}{\text{Age } (t_1, B_1) \; - \; a \; t' \; \rightarrow \; B_1'} \quad \text{if } t' + t_1 = t.$$
Summation:	$$\dfrac{[r/x]B \; - \; a \; t \; \rightarrow \; B'}{\text{CHOICE } x : s[]B \; - \; a \; t \; \rightarrow \; B'} \quad \text{if } [r] \in Q(s)$$
Relabeling:	$$\dfrac{B \; - \; a \; t \; \rightarrow \; B'}{(B)[g_1/h_1, \cdots, g_n/h_n] \; - \; a' \; t \; \rightarrow \; (B')[g_1/h_1, \cdots, g_n/h_n]}$$ with $\quad a = a' \qquad\qquad \text{if name(a)} \notin \{h_1, \cdots, h_n\}$ $\qquad\quad a' = g_i v1 \cdots v_m \quad \text{if } a = h_i v1 \cdots v_m.$
Proc. Instantiation:	$$\dfrac{([t_1/x_1, \cdots, t_m/x_m]B_p)[g_1/h_1, \cdots, g_n/h_n] \; - \; a \; t \; \rightarrow \; B'}{p[g_1, \cdots, g_n](t_1, \cdots, t_m) \; - \; a \; t \; \rightarrow \; B'}$$ where $\quad fg(p) =< h_1, \cdots, h_n >, fv(p) =< x_1, \cdots, x_m >, \text{ and}$ $\qquad\quad < p, B_p > \in PDEFS$

one proposed by J. Quemada and A. Fernández in [QF87], and it also would have some anomalous system evolutions. The following examples clarify the causes of the anomalous behaviours and how the ASAP criterion is able to resolve them (natural numbers are used as time domain).

Example 3.1
Suppose that our system has two components that evolve independently; the first one offers an action "a" at instant 3, and the second one offers event "b" at instant 5:

`(a {3}; stop)   ||| (b {5}; stop)`

The possible transitions of this system (applying the derivation system of Table 3.2) are:

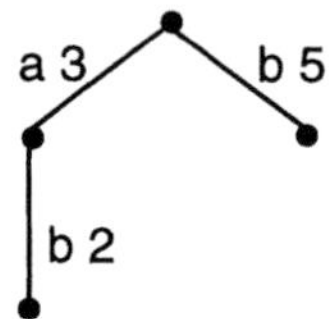

Both evolutions are reasonable because if the environment does not permit action "a" to occur at instant 3, we will be able to observe action "b" at instant 5. Moreover, if action "a" occurs at time 3, action "b" may occur 2 time units later.

Let us consider, now, the same system, but with internal evolutions:

`(i {3}; a {0}; stop)   ||| (i {5}; b {0}; stop)`

The possible transitions of this system (applying derivation system of Table 3.2) will be:

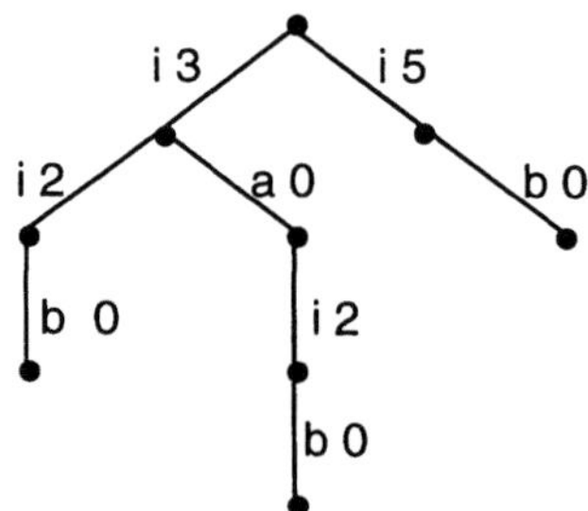

The system evolution through transition "i 5" will lead the system to an anomalous deadlock because action "a" cannot occur even though the environment wants to synchronize on it. This anomalous behaviour can be avoided if we apply the ASAP criterion.

In this case, ASAP means that transition "i 5" is not possible because transition "i 3" *must* occur before, and the environment can always observe action "a" at instant 3 and/or action "b" at 5. After the occurrence of transition "i 3", the choice between "i 2" and "a 0" represents a reasonable system behaviour: The environment can synchronize with action "a" now, but if it does not, time passes (system evolves through "i 2") and action "b" is offered.   □

Example 3.2

Because ASAP criterion means that system evolution can only be decided when it evolves internally, hiding operator has to be analyzed. Hiding makes actions occur because the environment can no longer prevent their occurrence. Suppose that we hide the action "a" in the above systems:

Expression `hide a in (a {3}; stop)  ||| (b {5}; stop)` has a single ASAP evolution: "i 3, b 2", and

```
hide a in (i {3}; a {0}; stop)  ||| (i {5}; b {0}; stop)
```

has a single ASAP evolution: "i 3, i 0, i 2, b 0". Both evolutions correspond to our intuition.

The combination of hiding and choice also produces intuitive results. For example, expression:

```
hide a in (    a {3}; B
          [] b {5}; C)
```

will evolve at instant 3 to behave as "B" because the environment cannot prevent the occurrence of action "a" at instant 3.   □

Definition 3.1 If $B - a\,t \rightarrow B_1$ is a transition from B it is an *ASAP transition* iff for all behaviour expressions B', and $t' \in Q(time)$

$$B - i\,t' \rightarrow B' \;\Rightarrow\; t' \geq t$$

ASAP transitions are represented as $B - a\,t \mapsto B_1$. That is, $B - a\,t \mapsto B_1$ iff no internal transition is possible before t.   □

Definition 3.2 We say that a behaviour expression B is *stable at* $t \in Q(\text{time})$ iff there is no B' such that

$$B - i\,t' \rightarrow B' \;\text{ with } t' < t.$$

We can redefine ASAP transition as follows: If $B - a\,t \rightarrow B_1$ is a transition from B, it is an *ASAP transition* iff B is stable at t.   □

The labelled transition system, TTS(B), will be based on the ASAP transitions which are a subset of the derivatives obtained from the derivation system in Table 3.2 with:

$$S_t = Der^{asap}(B)$$

The set $Der^{asap}(B)$, the set of ASAP derivatives of a behaviour expression B, is the smallest set satisfying:

- $B \in Der^{asap}(B)$; and
- if $B' \in Der^{asap}(B)$ and $B' - a\,t \mapsto B''$ for some "a t", then $B'' \in Der^{asap}(B)$.

As we can check, the transitions removed in the second step must always be removed even when the expression is composed in a context. This fact makes the semantics be compositional, and a full derivation system can be given [MFV91c, MFOV93].

324   C. Miguel et al.

3.5   A Simple System Specification

The following specification shows a simplified communication channel with a fixed propagation delay (1.5 msec.). Messages can be control or information (data). The length (in time) of control messages is 1 msec, and 4 msec for data messages. Data library is listed in appendix A.

```
specification channel[send,rec]:noexit
 library PositiveFracNumber, Boolean, NaturalNumber endlib
 timedef sort   p_rational
         opns   + for + 0 for 00 > for gt
 enddef
 type msgs is PositiveFracNumber
  sorts msg
  opns   control, data :  -> msg
         t_len: msg          -> p_rational
         t_prop:             -> p_rational
  eqns ofsort p_rational
         t_prop = 3/2; t_len(control) = 1/1;
         t_len(data) = 4/1;
 endtype
 behaviour
  channel[send,rec]
 where
  process channel[s,r]:noexit:=
    s?m:msg; i {t_len(m)}; (       (r!m {t_prop}; stop)
                            ||| channel[s,r])
  endproc
endspec
```

The timed action "i {t_len(m)}" models the time required to put the whole message over the channel. After that time, a new message can be accepted for transmission. After t_prop = 1.5 msec the message is offered through gate r. Observe that this channel preserves the ordering of the transmitted messages.

3.6   Strong Bisimulation. ASAP Congruence

Definition 3.3 Let S be the set of closed LOTOS–T expressions in the context of an algebraic specification (AS) and a complete process environment.
A relation $R \subseteq S \times S$ is a *Strong Timed Bisimulation* iff $\forall < B_1, B_2 > \in R$, and for all timed actions "$a\ t$" with $a \in Act, t \in Q(\textbf{time})$ we have

1. $B_1 - a\,t \mapsto B_1' \implies \exists B_2'\ B_2 - a\,t \mapsto B_2'$ and $< B_1', B_2' > \in R.$
2. $B_2 - a\,t \mapsto B_2' \implies \exists B_1'\ B_1 - a\,t \mapsto B_1'$ and $< B_1', B_2' > \in R.$

We say that B_1 and B_2 are strong timed bisimilar or ASAP equivalent, and we will write $B_1 \asymp B_2$, iff there exists some strong timed bisimulation R with $< B_1, B_2 > \in R.$ □

The most interesting property of an equivalence relation is that a behaviour expression can be substituted for an equivalent expression in any context; that is, the relation is a congruence.

Proposition 3.1 ASAP relation is preserved by Parallel Composition, Choice, Hiding, Action Prefix, Disabling, Enabling and Relabeling.   □

The proof of this proposition as well as a set of laws for ASAP congruence can be found in [MFV91c]. These laws include the timed version of the expansion theorems.

3.7   Observational Equivalence Relations

The ASAP relation differentiates between expressions that are equivalent from the observational point of view; for example, the expressions

```
B1 = a{3}; i{2}; b{5}; stop  and  B2 = a{3}; b{7}; stop
```

are not ASAP equivalent, but they cannot be distinguished by an observer. Both of them perform action "a" at 3 and "b" 7 instants of time later. This section defines a weak equivalence relation for LOTOS–T.

3.7.1   Weak Timed Bisimulation Equivalence

A new operational semantics is going to be defined. It considers only the occurrence of the observable actions, the time between them, and the time elapsed since the occurrence of the last of them. The time associated with the internal actions is accumulated in the forthcoming observable action, or constitutes a part of the time elapsed from the execution of the last of them. All this is formalized by the new transition system defined below [QAF93].

Definition 3.4 • $AT = \{a\,t \mid a \in Act,\ t \in Q(\text{time})\}$ is the set of timed actions.
• $OT = \{a\,t \mid a \in Act - \{i\},\ t \in Q(\text{time})\}$ is the set of observable timed actions.
• Let B, B' be LOTOS–T behaviour expressions, and $a_1 t_1, \cdots, a_n t_n \in AT$ with $n > 0$. We say that $B - a_1 t_1, \cdots, a_n t_n \mapsto B'$ is an *extended asap transition* iff $\exists B_0, \cdots, B_n$ behaviours expressions with $B = B_0$ and $B' = B_n$, such that $\forall i \in \{1, \cdots, n\}\ B_{i-1} - a_i t_i \mapsto B_i$.
• Let $B - I_0, a_0 t_0, I_1, a_1 t_1, \cdots, I_{n-1}, a_{n-1} t_{n-1}, I_n \mapsto B'$ be an extended asap transition, where for each $i \in \{0, \cdots, n-1\}\ a_i t_i \in OT$, and each I_i represents a possible empty sequence of internal timed action $it_{i,1}, \cdots, it_{i,n_i}$. Then, for each $j \in \{0, \cdots, n\}$ we define t'_j by

$$t'_j = t_j + \sum_{k=1}^{n_j} t_{j,k} \quad \forall j \in \{0, \cdots, n-1\}; \quad t'_n = \sum_{k=1}^{n_n} t_{n,k}$$

Then, we say that $B = a_0 t'_0, a_1 t'_1, \cdots, a_{n-1} t'_{n-1}, t'_n \Rightarrow B'$ is the *observable timed transition* associated with the given extended asap transition.

□

Definition 3.5 Let S be the set of closed LOTOS–T expressions in the context of an algebraic specification (AS) and a complete process environment.
A relation $R \subseteq S \times S$ is a *Weak Timed Bisimulation* iff $\forall < B_1, B_2 > \in R,\ \forall q \in OT^*$ we have

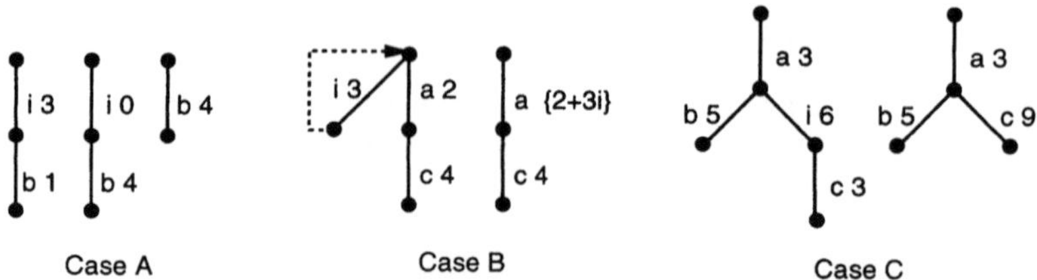

Fig. 3.1 Examples of Weak Timed Bisimilar Behaviours

1. $B_1 = q, t_n \Rightarrow B_1' \implies \exists B_2'$ stable at t such that $B_2 = q, t_n' \Rightarrow B_2'$,
    and $< B_1'$, Age $(t, B_2') >\in R$ with $t + t_n' = t_n$.
2. $B_2 = q, t_n \Rightarrow B_2' \implies \exists B_1'$ stable at t such that $B_1 = q, t_n' \Rightarrow B_1'$,
    and $< B_2'$, Age $(t, B_1') >\in R$ with $t + t_n' = t_n$.

We say that B_1 and B_2 are weakly timed bisimilar, and we will write $B_1 \sim B_2$, iff there exists some weak timed bisimulation R with $< B_1, B_2 >\in R$.   □

The idea is the following: if B_1 performs the timed observable sequence q, and after that time t_n passes resulting B_1', then B_2 must derive an equivalent behaviour after performing the same sequence q, and time t_n has elapsed. It is possible that there exists some behaviour B_2' such that $B_2 = q, t_n' \Rightarrow B_2'$ with $t_n' < t_n$, and no other internal transition from B_2' allows us to represent the passing of time up to t_n (B_2' is stable at $t = t_n - t_n'$). This is the reason why we need to age B_2' so that the resulting behaviour is B_2 after performing q and then t_n has elapsed.

Some interesting examples of weak timed bisimilar behaviours are in Figure 3.1. Case A) shows how the time consumption due to an internal action can be accumulated to the occurrence time of the observable action. In case B) we can see how internal loops can be removed, which in fact already happened in the untimed case. But in the timed model the effect of the passing of time because of the internal transitions is kept in the time instants in which action "a" may occur $\{2 + 3i \mid i \in \text{nat}\}$. This behaviour can be expressed in LOTOS–T as: `choice i:nat [] a{2 + (3*i)}; c{4}; stop.` Case C) shows how internal transitions can be removed, even if some other alternatives exist. This internal transition can be removed because its occurrence time (6) is greater than the occurrence time (5) of the rest of the alternatives.

Proposition 3.2 Weak timed bisimulation is an equivalence relation.   □

The proof of this proposition can be found in [MFV91c].

It is immediate to prove that strong relation is included in the Weak Timed Bisimulation relation: $\asymp \subseteq \sim$.

Proposition 3.3 Weak timed bisimulation is preserved by Parallel Composition, Hiding, Action Prefix, Enabling and Relabeling.   □

The proof of this proposition can be found in [MFV91c].

3.8   Upward Compatibility

The LOTOS–T specifier decides the time domain that he wants to use. As we have seen, natural or rational numbers can be used to model time, but time can be completely abstracted if the specifier defines a time domain with a single element, e.g:

```
timedef sort  no_time
   opns  + for +  0 for zero  > for gt
enddef

type temporal_abstraction is Boolean
  sorts no_time
  opns  zero   :                        -> no_time
        _ + _  : no_time, no_time -> no_time
        _ gt _ : no_time, no_time -> bool
  eqns ofsort no_time zero + zero = zero;
       ofsort bool     zero gt zero = false
endtype
```

Using this time domain, LOTOS–T is identical to Standard LOTOS. This identity means that given a LOTOS expression, the transition systems generated by the semantics of LOTOS is the same as that generated by the semantics of LOTOS–T assuming that time labels (0) are removed from the transitions, and that $Age\,(0, B) = B$. With this time domain, the standard LOTOS equivalence relations are the same as the ones defined for LOTOS–T. In this sense, we can say that LOTOS is a subcalculus of LOTOS–T. We have also seen [MFV93a, MFV91c] that LOTOS properties remain valid in the extended model regardless of the defined time domain.

4   Probabilistic Behaviours in LOTOS–TP (LOTOS–P)

4.1   Probabilistic Behaviours and Experiments.

The definition of a complete system model from which performance evaluation can be obtained requires the definition of random behaviours. These random behaviors can be seen as a probabilistic characterization of non–deterministic ones. Non–deterministic LOTOS behaviours can be expressed by

```
    i; B
[] i; C
```

which represents that system behaves as "C" or "B" with unspecified probability, or by a more general expression:

```
choice t: s [] i; B
```

that models the election of a value of sort "s" (for example, a time value), but with unspecified probability. The proposed technique, LOTOS–TP, will be able to specify the probability of these internal system decisions because it permits the definition of random experiments and the specification of behaviours which depend on them.

As well as the classical probability theory, the probabilistic models are based on the concept of *experiment.* An experiment E consists of a set of possible outcomes. Each experiment outcome has associated a real number which represents the probability of its occurrence when the experiment is performed. The formal definition of experiment can be found in [Pap89].

Previous works on probabilistic techniques apply the experiment concept by defining a n–ario choice operator [GSST90, Tof90, Zic90] and giving its formal semantics:

$$E = \sum_{n=1}^{k} p_i B_i$$

This expression represents an experiment with "k" outcomes, $B_i, 1 \leq i \leq k$, and probabilities p_i. Let us analyse two possible informal interpretations of such expression when it is composed with an environment. A first approach, which is closer to the above referred works, is that the behaviour of the environment can influence (conditioned experiment) the experiment outcomes. The second approach consists in considering that the outcomes of experiment E are B_i with probability p_i independently of the behaviour of the environment. A single example will make clear these different interpretations. Let's consider the following expression:

$$E = \sum_{n=1}^{3} p_i B_i$$

where
$$
\begin{aligned}
p_1 B_1 &= 0.2\ a;\ B_1' \\
p_2 B_2 &= 0.5\ b;\ B_2' \\
p_3 B_3 &= 0.3\ c;\ B_3'
\end{aligned}
$$

It could be interpreted, following the generative model described in [GSST90], as a PTS (probabilistic transition system):

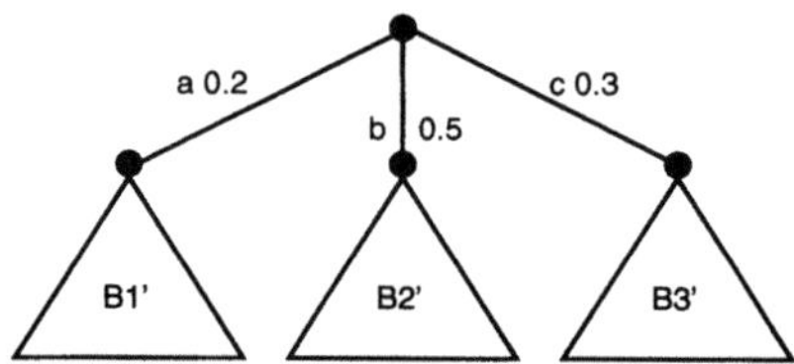

Let's suppose, now, that E has to synchronize with an environment C $(E||C)$ which prevents the occurrence of action "b". With which probability will actions "a" and "c" occur?. Using the first approach, we have the original experiment, but conditioned by the environment:

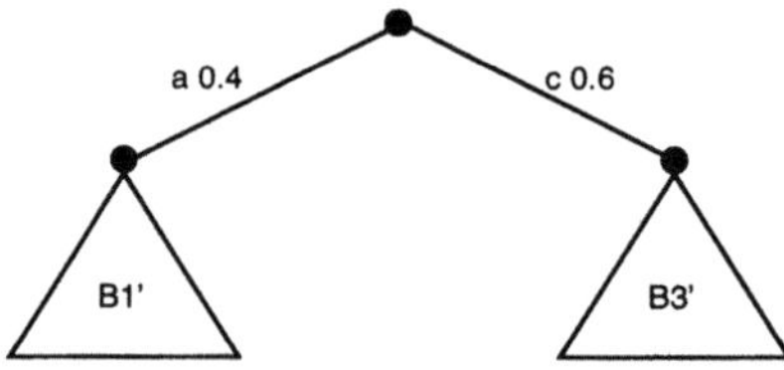

That is, the experiment E has been conditioned to the occurrence of the actions that the environment permits. Using the second approach, the probability of actions "a" and "c" will continue being 0.2, and 0.3 respectively. Behaviour E will be blocked by the environment with probability 0.5 because E will behave as $b; B_2'$ with that probability, and the environment will prevent its occurrence.

The technique proposed in this chapter uses experiments which are independent of the environment. This independence is achieved by modeling the experiment trials as internal (i) actions. These experiments are defined in a more flexible way by using the LOTOS capability of describing sets of data values (sorts). Let us explain the model with a communication example. We want to describe the behaviour of a *Binary Symmetric Channel (BSC)*. A BSC permits the transmission of "bits" between two ends, but each bit can be received with error at the destination end with a given probability (p):

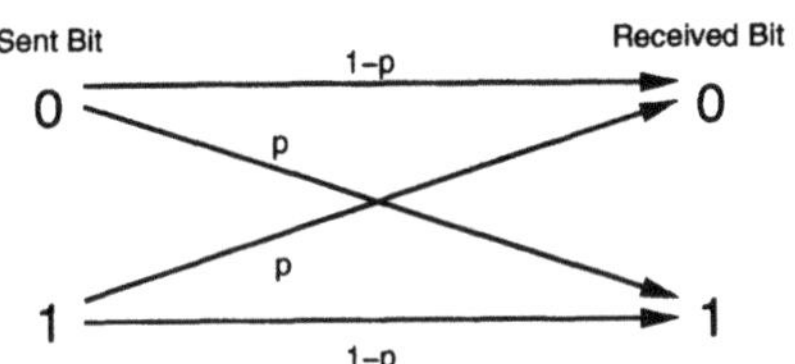

Standard LOTOS allows the specification of an abstract model of the BSC which does not include the probabilistic characterization:

```
process BSC [in,out]: noexit:=
  in?b: bit;
  (  i; out!not(b); BSC [in,out]   (* Error *)        (1)
  [] i; out!b; BSC [in,out])       (* No error *)
endproc
```

We have supposed that there exists a sort "bit" with two different values. This sort is actually a renaming of booleans:

```
  type bits is
    sorts bit
    opns 0,1 : -> bit
         not : bit -> bit
    eqns ........
         not(0) = 1; not(1) = 0;
  endtype
```

The previous process is also a LOTOS–P process, and it will be interpreted in the same way; that is, we do not know the probability of receiving an erroneous bit. Nevertheless, we can detail its behaviour to obtain the complete model of the BSC:

```
process BSC [in,out]: noexit:=
  in?b: bit; Random e: Error in                      (2)
      (  [e]-> out!not(b); BSC [in,out]
      [] [not(e)]-> out!b; BSC [in,out])
endproc
```

where "Error" is an experiment with two possible outcomes

$$\{true, false\}$$

with probabilities $P(true) = p$, and $P(false) = 1 - p$. We have supposed that the booleans are defined and the set of values of sort "bool" are $Q(s) = \{true, false\}$.

A random expression:

Random v: Exp in B

is interpreted as the trial of experiment "Exp" whose outcomes are values of a sort "s". The outcome of the trial will be the value of the variable "v" which can be used in the behaviour "B" (v is a free variable of B).

The *certain behaviour* of BSC will be the behaviour that considers that any outcome may occur. It is the same behaviour as the previous LOTOS expression (1) and also the same as the following one:

```
process BSC [in,out]: noexit:=
   in?b: bit;
      Choice e: bool [] i; ( [e]-> out!not(b); BSC [in,out]
                           [][not(e)]-> out!b; BSC [in,out])
   endproc
```

The certain behaviour can be interpreted as the following transition system:

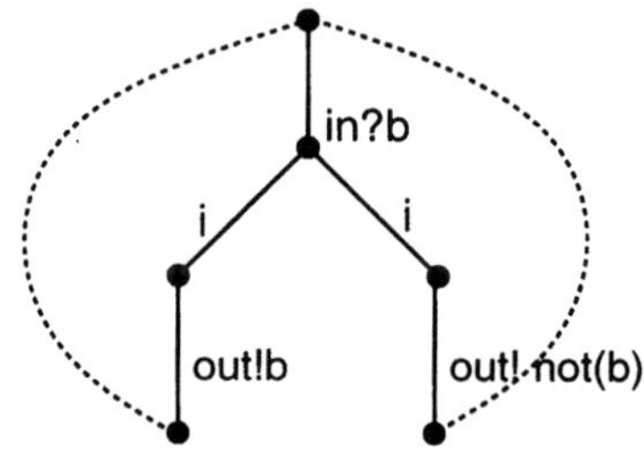

The experiment trial is modeled by the occurrence of an internal event which guarantees the independence of the experiment outcomes with respect to the environment. Observe that *Every property that we can obtain from the certain behaviour will be property of the system with probability 1*. For example, we can conclude from the certain behaviour of the BSC that when a bit is sent to the channel, another bit is received at the other end with probability 1.

The experiment "Error" has to be performed every time a bit is transmitted. Therefore, we can obtain the behaviour of the BSC conditioned to the occurrence of an event of a combined experiment Z which results from performing the experiment "Error" m times. Each performing is independent from the others:

$$Z = \text{Error}^m, \quad m \geq 1$$

For example, the behaviour of the BSC conditioned to the event s = {false true} of the experiment $Z = \text{Error}^2$, represented as BSC/s, is the one shown in figure 4.1 as a transition system.

Other examples of the BSC behaviour conditioned to different events are in Figure 4.2. Each behaviour has associated the probability of the event of Z.

We will see that every property that we can obtain from the behaviour of a system "B" conditioned to an event "s" (B/s) will be property of "B" with probability greater or equal to the probability of the event "s".

The complete LOTOS–P calculus and its application to the evaluation of the system can be found in [MFV93b, MFV91a]. The calculus is based on the existence of k

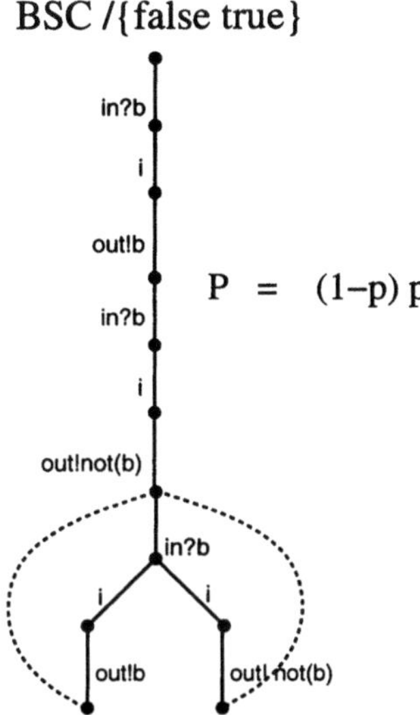

Fig. 4.1 BSC behaviour conditioned to event {false true}

experiments,

$$EX = < E_1, \cdots, E_k >,$$

and the interpretation of behaviour expressions conditioned to the occurrence of events of combined experiments:

$$Z = E_1^{m_1} \times \cdots \times E_k^{m_k}, \quad m_i \in \mathbf{N}, \quad 1 \le i \le k$$

4.2 Randomness and Non-determinism

The main objective in the definition of LOTOS–P is the upward compatibility with standard LOTOS. It means that LOTOS–P probabilistic calculus should be able to handle non–determinism. When we say "non–determinism", we mean that it is not characterized probabilistically. This section sketches the effects of the non–determinism over the LOTOS–P calculus.

Let us use the binary symmetric channel example to explain such effects. The BSC specification is completely characterized probabilistically. Therefore, we can obtain the exact probability of the verification of a given property, e.g:

P(receiving 1 bit without error) = 1-p, and
P(receiving 1 bit with error) = p

Let's suppose now that we have a channel with two noise sources. The first one is known enough to model its effects in a probabilistic way, but the second one is not. The specification of this channel can be:

```
process NoisyChan [in,out]: noexit:=
```

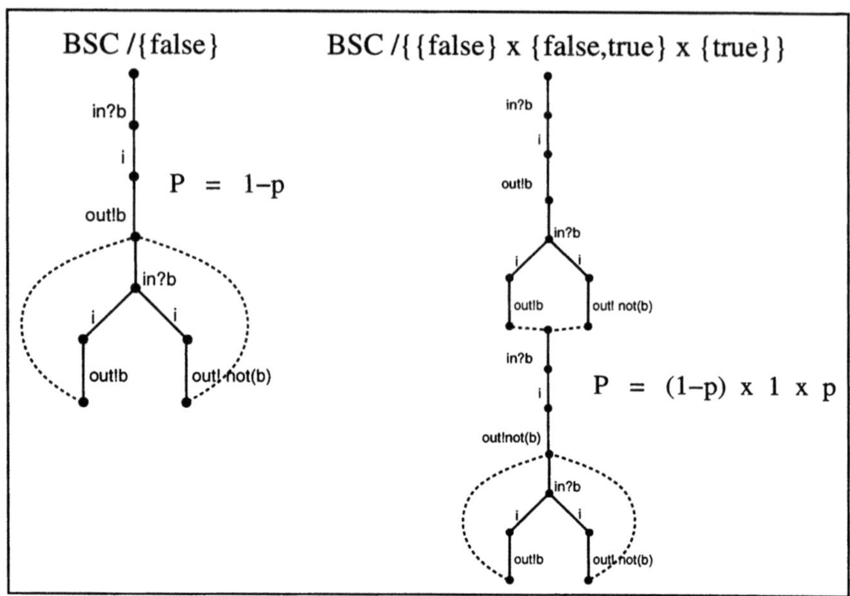

Fig. 4.2 BSC behaviour conditioned to different events

```
in?b: bit;
   Random e: FirstNoise in
        (  [e]-> out!not(b); BSC [in,out]
        [][not(e)]-> (* here it is the second noise *)
             (  i; out!not(b); BSC [in,out]
             [] i; out!b; BSC [in,out]))
endproc
```

Where the experiment "FirstNoise" models the probabilistic effects of the first source of noise which leads to an error with probability "p". The second source can lead to errors, but with unknown probability. From this system behaviour we can only obtain bounds to the probability of the system properties, e.g.:

$$0 \leq P(\text{receiving 1 bit without error}) \leq 1\text{-p, and}$$
$$p \leq P(\text{receiving 1 bit with error}) \leq 1$$

because we cannot assure that the bit is received without error even though the first noise does not produce the error, but we are sure that the bit will be erroneous with at least probability "p".

As the example sketches, the non–determinism causes uncertainty in the probabilistic analysis of a system whose behaviour is "B". This uncertainty is expressed by a range of probabilities:

B satisfies a property e with probability $P \in [p_1, p_2]$, $p_1, p_2 \in \mathbf{R}$, and $p_2 \geq p_1$. If "B" is characterized enough in terms of probability, p_1 will be equal to p_2 and the uncertainty disappears.

The complete formalization of the probabilistic model and the properties verification can be found in [MFV93b, MFV91a].

5   Conclusions

This chapter describes the formal model of LOTOS extended with timed and probabilistic behaviours, and the goals that guide such extensions.

LOTOS–T, the timing part of LOTOS–TP, with its operational semantics is able to model the quantitative passing of time, where the time domain is defined by the specifier. Both discrete and dense time domains can be used.

Asynchronous behaviours can be specified as well as synchronous ones. Synchronous behaviours can be expressed by the new timed action prefix operator which makes actions occur at a single and specific instant of time. Due to the fact that event occurrences do not consume any time, concurrency is modelled as the interleaving of events.

LOTOS–T applies the ASAP (as soon as possible) time criterion to interpret the timed expressions in the same way as Timed CSP [RR88], TPL [HR90], Timed LOTOS [SR88], and Timed CCS models [CAM90, Yi91, Han91]. This criterion is also known by the names of *maximum progress*, minimal delay, or *maximum parallelism for internal actions*. It means that:

If the system can evolve, it does. Therefore, internal actions occur as soon as they can.

LOTOS–T extends LOTOS with basically one new operator: *Timed action prefix* which can be combined with other LOTOS operators. The result of this combination is a very expressive timed language which is able to express a wide variety of time constraint.

Measurement of the time elapsed between two events is also possible. This fact will permit us to measure performance parameters in the system we are specifying, e.g. round trip time. This capability makes the model able to be used as a performance evaluation technique.

Following the equivalence relations defined in standard LOTOS, we have defined two equivalence relation in LOTOS–T: Strong and weak timed bisimulation relations. We have shown that the LOTOS–T strong and weak bisimulation relations preserve the properties of the LOTOS relations.

LOTOS–P, the probabilistic part of LOTOS–TP, is an enrichment of LOTOS that permits us to analyse the probabilistic properties of the specifications in addition to the non–probabilistic ones. From the design point of view, the probabilistic analysis is needed because many of the requirements imposed to the design are harder than "this requirement *must* be satisfied" or "this requirement *may* be satisfied". Many of the actual requirements, e.g. for a data communication system, are given in terms of the probabilities of their fulfillment. In spite of this enrichment, LOTOS–P permits us the specification of non–deterministic behaviours that will produce some uncertainty in the probabilistic analysis. This uncertainty means that perhaps we will not be able to obtain the exact probability of a property, but a range of probabilities.

A key objective in the definition of LOTOS–TP has been the upward compatibility with LOTOS. We have seen that the upward compatibility is achieved because LOTOS is actually a subcalculus of LOTOS–TP.

Another important point related to the usefulness of LOTOS–TP is the possibility of developing tools which support the design and verification of data communication systems. An example of this tool is TOPOSIM [MFOV93] which supports partially LOTOS–TP and permits us to obtain, by simulation, the performance figures of the specified systems.

Appendix A. Data Types Library

DATA TYPES LIBRARY

BOOLEANS

```
type Boolean is
  sorts Bool
  opns
    true, false                :      -> Bool
    not                        : Bool -> Bool
    _and_, _or_, _xor_, _implies_,
    _iff_, _eq_, _ne_          : Bool, Bool -> Bool
  eqns forall x, y : Bool ofsort Bool
      not(true) = false;
      not(false) = true;
      x and true = x;
      x and false = false;
      x or true = true;
      x or false = x;
      x xor y = (x and not(y)) or (y and not(x));
      x implies y = y or not(x);
      x iff y = (x implies y) and (y implies x);
      x eq y = x iff y;
      x ne y = x xor y;
  endtype
```

NATURALS

```
type BasicNaturalNumber is
  sorts Nat
  opns
    0                 :                   -> Nat
    Succ              : Nat               -> Nat
    _+_, _*_, _**_    : Nat, Nat          -> Nat
  eqns forall m, n : Nat ofsort Nat
      m + 0 = m;
      m + Succ(n) = Succ(m) + n;
      m * 0 = 0;
      m * Succ(n) = m + (m * n);
      m ** 0 = Succ(0);
      m ** Succ(n) = m * (m ** n);
endtype

type NaturalNumber is BasicNaturalNumber, Boolean
  opns _eq_, _ne_, _lt_, _le_, _ge_, _gt_ : Nat, Nat -> Bool
  eqns forall m, n : Nat ofsort Bool
      0 eq 0 = true;
      0 eq Succ(m) = false;
      Succ(m) eq 0 = false;
      Succ(m) eq Succ(n) = m eq n;
      m ne n = not(m eq n);
      0 lt 0 = false;
      0 lt Succ(n) = true;
      Succ(n) lt 0 = false;
      Succ(m) lt Succ(n) = m lt n;
      m le n = m lt n or (m eq n);
      m ge n = not(m lt n);
      m gt n = not(m le n);
endtype

type ExtendedNaturals is NaturalNumber
  opns 1,2,3,4,5,6,7,8,9,10: -> Nat
      pred            : Nat -> Nat
      _-_             : Nat, Nat -> Nat
  eqns forall n,n1: Nat
      ofsort Nat
      pred(0) = 0; pred(succ(n)) = n;
      n - 0 = n; n - succ(n1) = pred(n) - n1;
      1 = succ(0); 2 = succ(1); 3 = succ(2); 4 = succ(3); 5 = succ(4);
      6 = succ(5); 7 = succ(6); 8 = succ(7); 9 = succ(8); 10 = succ(9);
endtype
```

Positive RATIONAL NUMBERS

```
type PositiveFracNumber is ExtendedNaturals,
              NaturalNumberWithoutZero, Boolean
  sorts p_rational
  opns 00                :              -> p_rational
    real                 : Nat          -> p_rational
    _/_                  : Nat, Nat0     -> p_rational
    id                   : p_rational    -> p_rational
    _*_, _+_, _-_        : p_rational, p_rational -> p_rational
    _*_, _+_, _-_, _**_  : p_rational, Nat   -> p_rational
    _*_, _+_, _-_        : Nat, p_rational   -> p_rational
    _/_                  : p_rational, Nat0  -> p_rational
    _gt_, _lt_, _ge_     : p_rational, p_rational -> bool
  eqns forall n, n1, n2: Nat,  m1, m2: Nat0, r,r1: p_rational
      ofsort p_rational
      00 = 0/1;
      real(n) = n/1;
      id(r) = r;
      (n1 * N(m2)) = (n2 * N(m1)) => n1/m1 = n2/m2;

      (n1/m1) + (n2/m2) = ((n1*N(m2))+(n2*N(m1)))/(m1*m2);
      (n1/m1) - (n2/m2) = ((n1*N(m2))-(n2*N(m1)))/(m1*m2);
      (n1/m1) * (n2/m2) = (n1*n2)/(m1*m2);

      r * n = r * real(n);    n * r = real(n) * r;
      r + n = r + real(n);    n + r = real(n) + r;
      r - n = r - real(n);    n - r = real(n) - r;
      r ** 0 = 1/1;           r ** succ(n) = r * (r ** n);

      (n1/m1) / m2 = n1/(m1*m2);

      ofsort bool
      (n1/m1) gt (n2/m2) = (n1 * N(m2)) gt (n2 * N(m1));
      (n1/m1) lt (n2/m2) = (n1 * N(m2)) lt (n2 * N(m1));
      r ge r1 = not(r lt r1);
endtype
```

NATURALS Without 0

```
type NaturalNumberWithoutZero is ExtendedNaturals
  sorts Nat0
  opns 1,2,3,4,5,6,7,8,9,10:          -> Nat0
      succ                  : Nat0      -> Nat0
      _+_, _*_,_**_         : Nat0, Nat0 -> Nat0
      N                     : Nat0      -> Nat
  eqns forall m,n: Nat0
      ofsort Nat0
      2 = succ(1); 3 = succ(2); 4 = succ(3); 5 = succ(4);
      6 = succ(5); 7 = succ(6); 8 = succ(7); 9 = succ(8);
      10 = succ(9);
      m + 1 = succ(m);
      m + Succ(n) = Succ(m) + n;
      m * 1 = m;
      m * Succ(n) = m + (m * n);
      m ** 1 = m;
      m ** Succ(n) = m * (m ** n);
      ofsort Nat
      N(1) = 1;
      N(succ(m)) = succ(N(m));
endtype
```

References

[Azc90]   A. Azcorra. **Formal Modeling of Synchronous Systems**. PhD thesis, ETSI Telecomunicación, UPM, MADRID, Ciudad Universitaria s/n, September 1990.

[CAM90]   L. Chen, S. Anderson, and F. Moller. **A timed calculus of communicating systems.** Technical Report ECS-LFCS-89-104, Laboratory for Foundations of Computer Science LFCS, University of Edinburgh, December 1990.

[CCI85]   CCITT. **Specification and Description Language (SDL)**. Red Book Recommendations Z.101 to Z.104, Committee XI, Geneva, 1985.

[CH88]   R. Cleaveland and M. Hennessy. **Priorities in Process Algebras**. In *Proceedings Symposium on Logic in Computer Science*. LICS 88, 1988.

[GSST90]   R. Glabbeek, S.A. Smolka, B. Steffen, and C.M.N. Tofts. **Reactive, Generative, and Stratified Models of Probabilistic Proccesses**. In *Proceedings 5th Annual Symposium on Logic in Computer Science*, Philadelphia, USA, 1990. LICS 90.

[Han91]   H. Hansson. **Modeling Timeouts and Unreliable Media with a Timed Probabilistic Calculus**. In K. Parker and G. Rose, editors, *Fourth International Conference on Formal Description Techniques*, Sydney, November 1991. FORTE 91.

[HEH83]   W. Fey H. Ehrig and H. Hansen. **Act one: An algebraic language with two levels of semantics**. Technical Report Bericht Nr. 83.103, Tech. Universitat Berlin, 1983.

[HR90]   M. Hennessy and R. Regan. **A temporal process algebra**. In J. Quemada, J. Mañas, and E. Vázquez, editors, *Third International Conference on Formal Description Techniques*, Madrid, Spain, November 1990. FORTE 90.

[ISO88]   ISO. **ESTELLE– a formal description technique based on an extended state transition model**. IS 9074, TC97/SC21, 1988.

[ISO89]   ISO. **LOTOS a Formal Description Technique based on the Temporal Ordering of Observational Behaviour**. IS 8807, TC97/SC21, 1989.

[Led91]   G. Leduc. **An upward compatible timed extension to LOTOS**. In K. Parker and G. Rose, editors, *Fourth International Conference on Formal Description Techniques*, Sydney, November 1991. FORTE 91.

[MFOV93]   C. Miguel, A. Fernández, J.M. Ortuño, and L. Vidaller. **A LOTOS based Performance Evaluation Tool**. *Special Issue of "Computer Networks and ISDN Systems" on: TOOLS FOR FDTs*, 25(7):791–813, 1993.

[MFV91a]   C. Miguel, A. Fernández, and L. Vidaller. **LOTOS Extended with Probabilistic Behaviours**. Technical Report OSI95/DIT/B5/7/TR/R/V2, DIT. OSI95 ESPRIT II Project, Nov. 13. 1991.

[MFV91b]   C. Miguel, A. Fernández, and L. Vidaller. **State of the Art on Timed & Probabilistic Models**. Technical Report OSI95/DIT/B5/3/TR/R/V3, DIT. OSI95 ESPRIT II Project, Nov. 13. 1991.

[MFV91c]   C. Miguel, A. Fernández, and L. Vidaller. **Timed LOTOS definition. LOTOS–T**. Technical Report OSI95/DIT/B5/5/TR/R/V3, DIT. OSI95 ESPRIT II Project, Nov. 13. 1991.

[MFV93a]   C. Miguel, A. Fernández, and L. Vidaller. **Extending LOTOS towards Performance Evaluation**. In M. Diaz and R. Groz, editors, *Formal Description Techniques V*, pages 103–118, Amsterdam, 1993. Elsevier Science Publisher (North Holland).

[MFV93b]   C. Miguel, A. Fernández, and L. Vidaller. **LOTOS Extended with Probabilistic Behaviors**. *Formal Aspects of Computing. The international journal of Formal Methods*, 5(3):253–281, 1993.

[Mil80]   R. Milner. **A Calculus of Communicating Systems**. Number 92 in Lecture Notes in Computer Science. Springer Verlag, Berlin, 1980.

[Mil83]   R. Milner. **Calculi for Synchrony and Asynchrony**. *Theoretical Computer Science*, 25:267–310, 1983.

[Mil89]   R. Milner. **Communication and Concurrency**. International Series in Computer Science. Prentice Hall, 1989.

[NRSV90]   X. Nicollin, J.L. Richier, J. Sifakis, and J. Voiron. **ATP: an algebra for timed processes.** In M. Broy and C.B. Jones, editors, *IFIP Working Conference on Programming Concepts and Methods*, The Netherlands, 1990. IFIP, Elsevier Science.

[NS91]   X. Nicollin and J. Sifakis. **An Overview on Timed Process Algebras**. In *CAV'91*, Alborg, Denmark, July 1991.

[Pap89]   A. Papoulis. **Probability, Random Variables, and Stochastic Processes**. McGraw–Hill, 1989.

[Pol89]   L.F. Pollacia. **A survey of discrete event simulation and state–of–the–art discrete event languages**. *Simulation Digest*, Fall. 1989.

[QAF93]   J. Quemada, A. Azcorra, and D. Frutos. **TIC: A TImed Calculus**. *Formal Aspects of Computing. The international journal of Formal Methods*, 5(3):224–252, 1993.

[QF87]   J. Quemada and A. Fernández. **Introduction of quantitative relative time into LOTOS.** In *Workshop on Protocol Specification, Testing and Verification: VII*, Zurich, May 1987. IFIP.

[RR88]   G.M. Reed and A.W. Roscoe. **A timed Model for Communicating Sequential Processes**, volume 58, pages 249–261. North-Holland, 1988.

[SR88]   C. Smith and S. Rudkin. **Time Guards and ASN.1 in LOTOS**. Technical Report Ref. 40, Formal Methods Group, British Telecom, St. Vincen House, Ipswich, UK, 1988.

[Tof90]   C. Tofts. **A Synchronous Calculus of Relative Frequency**, pages 467–480. Number LNCS-458, ISBN 3-540-53048-7 in Lecture Notes in Computer Science. Springer-Verlag, Berlin Heidelberg, New York, 1990.

[Yi91]   W. Yi. **CCS + Time = an Interleaving Model for Real Time Systems**. In *ICALP'91*, 1991.

[Zic90]   J.J. Zic. **Some thoughts on communication system performance evaluation.** In *Proceedings of the Open Distributed Processing Workshop*, Sydney, January 1990.

The Tick-Tock Case Study for the Assessment of Timed FDTs

Luc Léonard[1] Guy Leduc[2] and André Danthine
[1] Research Assistant of the Belgian National Fund for Scientific Research (F.N.R.S.)
[2] Research Associate of the F.N.R.S.
Université de Liège, Institut d'Electricité Montefiore, B 28, B-4000 Liège 1, Belgium
Email: leonard@montefiore.ulg.ac.be

The initial purpose of this paper was to design a case study to assess LOTOS-T [MFV 93] which is a temporal extension of LOTOS developed within the ESPRIT II OSI95 project. However, we think that it can be useful for any proposed timed FDT.

It consists of a protocol composed of two entities and an underlying service provider, whose behaviour is mainly based on various timing constraints such as time-out, isochronism, rate-control, ...

The selection of the mechanisms was guided by the two following characteristics:

- Realism: the selected mechanisms have been inspired by similar and existing protocol mechanisms or service facilities, even if we have tried to (over)simplify them in order to focus the case study on the timing constraints.

- Temporal modelling facilities: this means that the specificity and the variety of the timing constraints are intended to assess whether timed FDTs have enough power and flexibility to tackle a maximum number of aspects of timed behaviours. Of course, we do not pretend to have explored all possible timed behaviours which may exist in protocols. We have simply tried to cover a broad spectrum of them.

Keywords: time, FDT, case study, assessment.

1 Introduction

The main reason for the development of FDTs was to obtain a way to describe services and protocols, in a manner that would be non ambiguous, clear and precise. This required them to be expressive enough, to allow the description of the behaviour of any system. Most of them indeed have the expressive power of a Turing machine. This also required them to be designed so that the description of most of the classical mechanisms could be made clear, with light constructs, and without having to use "tricks". However, initially, most of the FDTs were conceived without integrating explicitly any notion of time. In other words, it was not possible to describe easily the influence of time on the behaviour of a system. The recourse to extremely heavy and complicated artificial mechanisms was needed, leading to specifications of an unacceptable complexity.

At the present time, because of an increasing number of protocols including mechanisms where time plays a non negligible role, the real need to remedy to this problem has already led to the creation of many "timed" FDTs, most of which being either new timed process algebra or quantitative time extensions of well-known asynchronous process algebra. For example: ACP_ρ [BaB 90], $ACP^t_{t\epsilon}$ [Gro 90], ATP (Algebra of Timed Processes) [NRS 90, NiS 91, NSY 91][1], CIRCAL [Mil 85], Estelle [ISO 9074], LOTOS-T [MFV 92], Meije [AuB 84], PADS (Process Algebra for Distributed Systems) [Azc 90], SCCS [Mil 83], SDL [CCITT Z100], TCCS [MoT 90], TIC [QAF 90], TiCCS (Timed CCS) [Wan 90, Wan 91], TiCSP (Timed CSP) [ReR 88, Ree 90], Timed-arc PN [Wal 83], Timed PN [MeF 76], TinLOTOS (Timed-Interaction LOTOS) [BLT 90], TLOTOS [Led 92], Timed LOTOS [LeL 92], U-LOTOS & T-LOTOS [BoL 92], TPCCS [Haj 90, Han 91], TPL (Temporal Process Language) [HeR 91].

As we can see, there exist many proposals, and interesting novelties appear regularly. This situation reflects the diversity of the philosophies envisaged about the way time should be introduced in the FDTs, and the variety of the options explored in order to realize them. An overview and synthesis may be found in [NiS 92], a classification is also proposed in [LeL 93].

But questions thus arise: are all the choices equally valuable? What are the advantages and disadvantages of each one? What are the improvements a given formalism would require? We think that the most instructive way to get pieces of answers to these questions is simply to try to apply these FDTs to a concrete case. The main reason is that this seems to be the only way to determine the difficulties that could arise, and the deficiencies or the needs they will bring to light.

However, until now, most of the researches on timed FDTs did remain within a quite theoretical scope, the most complex example of practical use being often a simple alternating bit protocol (other examples are also treated in [Han 91]). Anyway, applying a timed FDT to the specification of a real system is not the ideal way to assess it. Two main reasons for this:

- It is likely that a real system will confront with only a limited number of problems, among the ones a timed FDT should be able to treat.
- Usually, the majority of the features of a real system are not concerned by timing aspects, and one would waste much time in specifying them.

It thus appeared to us that the definition of a case study, specially designed for the assessment of timed FDTs, would be useful, in particular to the authors of such FDTs. In this perspective, our work of assessment of LOTOS-T [MFV 93], developed within the ESPRIT II OSI95 project, gave us the opportunity to propose the case study we present in this paper. We did not focus it on LOTOS-T, and we think it could be of interest for any proposed timed FDT.

It consists of a protocol that groups together several mechanisms, chosen according to two main criteria:

- Either because they correspond to "classical", often encountered, mechanisms.
- Or because they illustrate the need for some facilities that, according to our knowledge of several formalisms, appear to be among the usual deficiencies we noticed.

---

[1] A new formalism based on ATP but with some different properties (e.g. the possibility of a dense time domain) is presented in [NSY 92]

The well-known time-out or the rate control are examples of the first type mechanisms. Also, the present development of the R.N.I.S. confronts us with the specification of isochronous systems, with discrete variations of the period.

Among the special difficulties, let us point out:

- The problems related to the definition of non-deterministic delays. Such delays are an useful abstraction, for instance to describe a propagation time into a transmission medium.
- The possibility for the temporal data to be treated and manipulated as any other data type. Many mechanisms (for instance congestion avoidance mechanisms) adapt their behaviour according to temporal information they deduce from the observation of their environment.
- The freedom to treat time as a dense or as a non dense domain. Some timed FDTs are restricted to a discrete time domain. Using a discrete time domain requires a careful definition of the grain of time of the specified system, in order to be able to express any timed constraint w.r.t. this basic unit of reference. This turns out to be inconvenient in a practical design, in which the grain of time is tightly bound to the abstraction level of the specification. Therefore, when a process is refined, this may require the selection of a smaller grain of time, and the whole specification needs to be rewritten for consistency. Such problems can be avoided by using a dense time domain, like the rational numbers.
- And, last but not least, as this request usually appears to be both the most fundamental one and the most difficult to tackle: the ability to express some constraints on the occurrence of interactions between concurrent processes. This capability is, in particular, of prime importance to allow for the specification in a structured style (see next chapter).

Although we have tried to keep it realistic, this case study is mainly focused on the timing aspects, and we have lightened it from the details that were not relevant to this concern.

2 Expressiveness and Style

Before describing the case study, we would like to come back to an important point, and bring some more ideas about it: What is a "good" timed FDT? And thereby, what could be a good specification of this case study?

As we already said above, one should not think that the (theoretical) expressive power is the sole criterion to assess the capabilities of an FDT. This is a very naive approach which does not give any interesting results. The reason is that most *non-temporal* FDTs (e.g. Basic LOTOS, CCS, ... but not Petri Nets) have already the maximal expressing power, i.e. they can express Turing machines, and thus are able in particular to express quantitative timed behaviours.

In other words, every timed FDT permits probably to describe the case study we present here. But what we really need is an FDT that provides compact and readable solutions to these problems. The keywords are therefore "flexibility", "facility", "readability" and "modularity" rather than "(theoretical) expressive power".

Another keyword is also often encountered in this context: the *style* of a formal description. In LOTOS, four styles have been proposed [VSS 88], each one presenting its own advantages. Among them, the most useful are usually the constraint-oriented style and the resource-oriented style. The constraint-oriented style (for more information, see [Bri 89]) is the most important during the first stage of the design because it is structured and extensional. That means that it allows the production of abstract, modular and easily extensible specifications, which are fundamental properties at this stage. The resource-oriented style is the one that best allows the structured description of possible implementations. This concept of style could be extended to some other FDTs which have a parallel composition operator à la LOTOS. The idea is that one should be able, when it is possible, to choose the most convenient style in accordance with the aim of the specification. In particular, the structure brought by the styles appears to be of paramount help, if not simply mandatory at all, to master the complexity of large systems .

So, we think that the objective when trying to extend FDTs with time, must not simply be to allow clear and light descriptions of timed behaviours but also to provide the capability to choose the style of them. Trying to describe the case study according to different styles is thus a good test in order to determine whether this requirement is met or not. According to our own experience, the most difficult problems would be faced with the constraint-oriented style.

3 The Tick-Tock Case Study

The Tick-Tock system is a protocol composed of two entities and an underlying service, which we will simply call *sender, receiver* and *service*. Let us recall that although we have tried to keep it realistic, it is lightened from the details that are not relevant to timing aspects.

It will first be defined in a simpler version, limited to its main features, in order to exhibit the essential constraints that need to be described. Then, as options, some more complex mechanisms will be presented, which could be encountered in practice and create new problems.

The three elements of the system will be described mainly as "black boxes", through the different constraints imposed to their observable behaviours, and without any detail on their internal organization. This suggests to try to specify them, as far as possible, in a structured and abstract style (e.g. the constraint-oriented style in LOTOS), in order to reflect at best the separation of timing concerns. If some a priori independent mechanisms cannot be specified independently of each other but must be combined in a single process, this should be considered as a shortcoming or, stated otherwise, as a lack of support of modularity. Such intertwinings would be especially annoying for the description of real (i.e. large) protocols. But, of course, one is invited to check that specifying in the other styles is possible too.

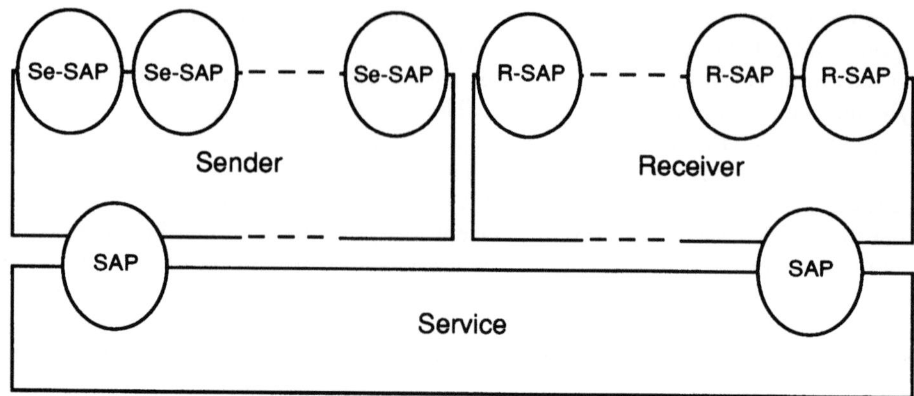

Fig. 3.1 Sketch of the system

The parameters that appear in the definition will never be given a precise value: everyone is free to realize its own implementation. However, we will sometimes point out some obvious rules that should be satisfied to preserve a sensible functioning.

We will use, in the sequel, a referential time unit, simply called "unit". Note however that this "unit" is not an elementary grain of time: the times expressed in the following description could be fractions of a unit.

4 Presentation of the Simple Version of the System

4.1 Service

In order to keep things simple, we have restricted the *service* specification to its interactions with its two users, *sender* and *receiver*, through their respective SAPs. Under these conditions, *service* can be characterized as follows:

Service Primitives: They carry a 48-byte cell as parameter. Primitives are instantaneous and atomic events. Our system is so simplified (the exchanges are always done between the same two SAPs) that no other parameter needs to be specified. In the sequel, these exchanges between the service and its users will simply be referred to as cells instead of primitives.

Isochronism: The given service is isochronous: a cell from *sender* is only accepted at some precise, punctual instants, that follow one another regularly in time, with a given rate (period: π units). An opportunity of emission can be neglected by *sender*. Just one cell can be exchanged at any instant.

Spacing Between the Deliveries: There is always a delay of at least α units between two successive deliveries at a same SAP.

Transmission Delays: A cell is always delivered between τmin and τmax after its emission.

Immediate Acceptation: A cell offered to *receiver* must be immediately accepted by *receiver*. If it is not possible, *service* losses the cell immediately.

Loss Free Transmission: The previous point describes the only way a cell received from *sender* can be lost: no cell is lost during the transit through *service*.

FIFO-Ordering of Cells: The cells arrive in their emission order.

The last two constraints - "Loss free transmission" and "FIFO-ordering of cells" - are not strictly necessary and might seem less realistic. However, they help to avoid unnecessary complications either in the specification of the service or in the definition of other mechanisms that we will meet later.

4.1.1 Comments About *Service*

It is mandatory that a delivered cell be accepted immediately by *receiver* or be lost. With just one constraint (Transmission delays) to determine the instant of arrival of a cell, expressing this urgency would have been easy with most of the formalisms. That is the reason why we have tried to introduce an additional constraint (the minimal delay α between two successive deliveries), that can modify the result of the first one, and thereby makes it impossible to impose urgency unilaterally. So, this problem of expressing urgency will have to be solved as a joint effect of the two constraints, a task for which most of the formalisms are not directly kited out (remember that, with concern of clarity, a structured and abstract style is suggested to specify the system, which supposes a separate representation of the different temporal constraints).

Let us notice that there is no incompatibility between the constraint "transmission delays" and the constraints "spacing between the deliveries" and "FIFO-ordering of cells", as long as the period of admission π is greater than or equal to the minimal delay between two deliveries α.

Besides this difficulty, the modelling of *service* also faces the problem of describing an isochronous system and of an undetermined delay between two bounds.

4.2 Sender

This entity multiplexes on *service* the data received from its n users, with which it communicates through their respective Se-SAP. It is characterized as follows:

Received Data: *Sender* receives from its users primitives that contain 3 parameters :

- a source address (20 bytes)
- a destination address (20 bytes)
- a SDU (we propose, for reasons of simplicity, that the size of each SDU be a multiple of 40 bytes)

Minimal Delay Between Primitives on a Given Se-SAP: On a given Se-SAP, every primitive must be at least spaced out from the previous one by a delay that is proportional to the size of its own SDU (β units per 40 bytes).

Emitted Data: *Sender* emits 48-byte cells on *service*, through their common SAP.

Emission of a Cell: *Sender*, when it desires to emit a cell, waits for an opportunity from *service*. When *service* is ready, the operation occurs immediately. If it has nothing to emit, *sender* just neglects the offers of *service*.

Segmentation of a SDU: Each SDU received by *sender* is emitted on *service* as a sequence of consecutive 48-byte cells. This sequence corresponds to the segmentation of the SDU into 40-byte fragments, completed with a 6-byte header and a 2-byte trailer. The sequence begins with a cell that carries the source and destination addresses.

Identification of the SDU: Each SDU receives an identification number that, from SDU to SDU, is increased of one unit, following the order determined by the instant of occurrence of the primitive. This number is written in the second and third bytes of the header of each cell.

Order of Emission of the Cells Related to Different SDUs: All the cells related to a SDU are emitted consecutively. The order of emission between the cells from different SDUs follows the one determined by the instant of occurrence of the primitives.

Order of Emission of the Cells Related to a Same SDU: The fourth byte and the first six bits of the fifth byte of the header of each cell indicate the rank, in the initial SDU, of the 40-byte fragment carried. These values are set to zero for the first cell of the sequence (the one that carries the source and destination addresses).

Emission of the Cells Related to a SDU: The order of emission of the cells corresponds to the one indicated by the fourth byte and the first six bits of the fifth byte of their header.

Minimal Delay Before the Emission of a Cell: A cell that is ranked x in the order of emission of the cells related to its SDU cannot be emitted before a delay $(x * \upsilon)$ after the occurrence of the primitive.

Types of Cells: The last two bits of the fifth byte of the header of each cell indicate its type as follows:

00 -> the cell carries the source and destination addresses ;
01 -> the cell carries a fragment of a SDU that is not the last one ;
10 -> the cell carries the last fragment of a SDU.

In the case of a 40-byte SDU, a 10-type cell follows directly the 00-type one.

Urgency of the Emission: *Sender* emits its cells as soon as it is allowed to, according to the previous rules.

4.2.1 Comments About *Sender*

Sender presents mainly two potential difficulties. First, one must be able to express an ASAP (As Soon As Possible) constraint between *sender* when it wants to emit and *service* when it agrees to receive a cell. This case is a typical example of the need for such constraints.

Next, one must try to express in a light and clear way, how the delay between two consecutive primitives on a same Se-SAP is related to the size of the SDU carried by the last one. In particular, this supposes the ability to treat time values as data.

4.3 Receiver

This entity de-multiplexes on its users the data received from *sender* through *service*.

Received Data: *Receiver* receives 48-byte cells from *service* through the SAP that joins them.

Frequency of Receptions: *Receiver* always listens to *service*, except during a period of "deafness" of ϕ units after the receipt of a cell.

Prolongation of the Deafness: An undetermined prolongation of the "deafness" may occur, because of internal congestion reasons.

Emitted Data: *Receiver* offers to its users primitives that carry two parameters: a source address and a SDU.

Creation of the Primitives: *Receiver* proposes a primitive for every SDU properly received and reassembled. This means that all the cells of the sequence have been received successively and in the right order. These conditions can be checked thanks to the parameters carried by the cells.

Emission Point of a Primitive: A primitive is proposed on the R-SAP that corresponds to the received destination address.

Discipline of Exit: If more than one SDU happen to wait on a same R-SAP, they will be proposed one by one, according to a FIFO discipline.

Minimal Delay Between Primitives on a R-SAP: On a R-SAP, every primitive is at least spaced out from the previous one by a delay proportional to the size of its SDU (ω units per 40 bytes).

4.3.1 Comments About Receiver

As it is defined here, *receiver* does not present special difficulties. Let us notice that it would be logical for the usual "deafness" delay ϕ to be smaller than the minimal delay α between the delivery of two consecutive cells by *service*.

5 Description of Additional Features

5.1 Adaptation of the Access Period to *Service*

The period between two consecutive interaction offers made by *service* may vary in time. The aim is to adapt the access to *service* to the presumed needs of *sender*, estimated from the use *sender* makes of the actual capacities he has at its disposal.

The mechanism proposed here segments the stream of proposals into consecutive and separated sequences of 10 proposals, at the end of which *service* is allowed to modify the period. Two main rules apply:

- at the end of a sequence, if all the offers have been accepted (10 cells have been emitted), the period is divided by two.
- at the end of 3 consecutive sequences, during which the period has not been updated by the previous or the present rule, the period is multiplied by a coefficient determined thanks to the following table, according to the number of proposals effectively used among the 30. (Multiplying by 1, in the case 26 --> 30, is also considered to be an update.)

Table 5.1 Determination of the multiplying coefficient

Number of proposals used			Coefficient
0	------->	10	2
11	------->	20	3/2
21	------->	25	6/5
26	------->	30	1

Bounds are however imposed on the possibilities of variation of the period, which must always remain between η and ψ units. Remember that initially, the period is supposed to be equal to π units.

5.2 "Crash" of *Service*

At any instant, without any reason, *service* may "crash". All the cells in transit are then lost and *service* needs an unpredictable delay before restarting. It restarts free of any cell and with a period π.

5.3 Rate-Control at the Access to the Service Provided by *Sender*

In section 4.2., a minimal delay between primitives occurring at a Se-SAP has been introduced. It may be interpreted as a kind of rate-control or any other kind of upper limit on the capabilities of *sender*. In this section, two other mechanisms are introduced which do not replace the minimal delay, but add rate-control mechanisms of a different nature. They impose an upper bound to the average rate *sender* accepts. Their effects are described by the next two additional constraints:

Individual Control on Each Se-SAP: Through each Se-SAP, during any period of κ units, the amount of information exchanged in the SDUs must not exceed φ blocks of 40 bytes. In other words, at any instant, a primitive may occur at a given Se-SAP only if the size (expressed in blocks of 40 bytes) of the SDU it carries is smaller than or equal to φ minus the sum of the sizes of the SDUs carried by all the primitives that have occurred at this Se-SAP since κ units of time (or since the start of *sender* if it has been functioning for less than κ units of time).

However, a primitive that carries a SDU with more than φ blocks of 40 bytes (let us say ζ blocks, with $\zeta > \varphi$) can be accepted if it is spaced out from the previous primitive on the Se-SAP (or from the start of *sender*) by a delay that is at least of $((\zeta / \varphi) * \kappa)$ units.

Figure 5.1 illustrates this mechanism, with $\kappa = 4$ and $\varphi = 20^2$.

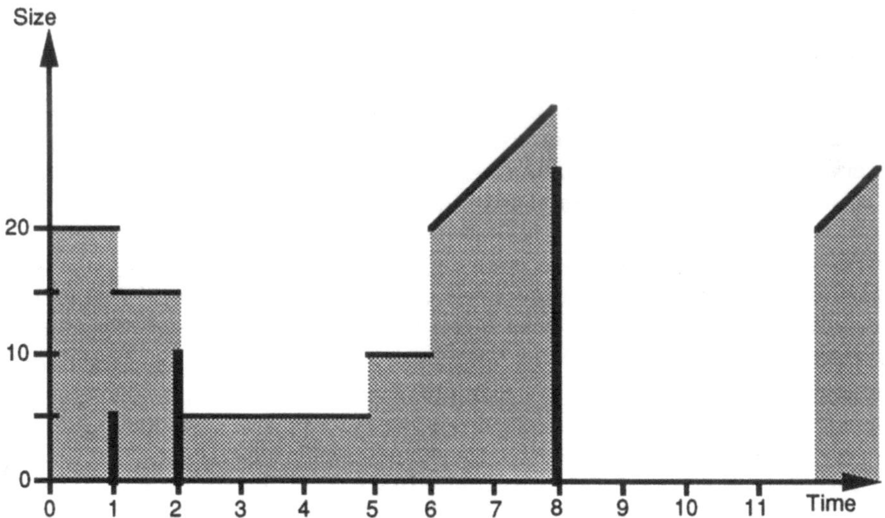

Fig. 5.1 Illustration of the individual control mechanism

---

^{2} Note that κ and φ are not necessarily natural numbers.

A vertical bold line indicates the occurrence of a primitive at a given instant, and the size of its SDU. The horizontal or oblique lines above grey surfaces indicate, at each time, the maximal possible size of SDU that could be accepted. Initially, the maximal size is 20 (blocks of 40 bytes). Primitives carrying a SDU of more than 20 blocks are not yet allowed. At time 1, a SDU of size 5 is accepted and the maximal authorised size decreases of 5 units. At time 2, a SDU of size 10 is accepted and the maximal authorised size decreases of 10 units. At time 5, i.e. 4 time units after its occurrence, the first SDU stops being taken into account in the sum of the sizes, and the maximal authorised size grows of 5 units. At time 6, the same occurs for the second SDU. At this time, i.e. 4 time units after the occurrence of the last primitive, primitives carrying a SDU of more than 20 blocks begin to be accepted. The maximal authorised size grows proportionally to the elapsed time. At time 8, such a primitive occurs. Its effect is that for 4 units of time, i.e. until it stops being taken into account in the sum of the sizes, no primitive at all can be accepted.

Global Control of all the Se-SAPs: Through the whole set of Se-SAPs, during any period of λ units, the amount of information exchanged in the SDUs must not exceed μ blocks of 40 bytes.

However, a primitive that carries a SDU with more than μ blocks of 40 bytes (let us say ζ blocks, with $\zeta > \mu$) can be accepted if it is spaced out from the previous primitive on all the Se-SAPs of *sender* (or from the start of *sender*) by a delay that is at least of $((\zeta / \mu) * \lambda)$ units.

This mechanism is totally similar to the previous one but cares for all the Se-SPAs at a time.

5.4 Rate-Control on the Reception of Cells by *Receiver*

A rate-control mechanism, aimed at preventing the congestion of *receiver*, limits both its capability to accept cells from *service* and to keep SDUs not yet accepted by the users. The constraints "frequency of receptions" and "prolongation of the deafness" of the simplified definition are forgotten and replaced by the following ones.

First Sufficient Condition to Accept a Cell: A cell proposed by *service* is accepted if, at this moment[3], the difference between the number of blocks of 40 bytes already received and the number of blocks of 40 bytes already evacuated (the notion of evacuation will be explained later) by the entity since the beginning of its functioning is smaller than θ. The only bytes taken into account when counting the received or evacuated blocks are the ones from the exchanged SDUs, thus excluding the ones from the headers, the trailers or the source and destination addresses.

Second Sufficient Condition to Accept a Cell: When the first condition alone does not apply, a cell proposed by *service* is however accepted if, and only if, there exists at least one SDU among the ones proposed to the users that has been waiting for more than ρ units to be accepted. The oldest of the SDUs in this situation is immediately

---

[3] Remember that the offers of service are "punctual", they must be accepted immediately or never.

deleted and the next SDU on the same R-SAP (according to the FIFO rule), if it exists, takes its place.

Let us insist on the fact that one deletes an old enough SDU only if and when it is expressly necessary.

Evacuation of Bytes: There are three ways to evacuate some bytes:

- the bytes of a complete SDU are considered evacuated at the instant of occurrence of the primitive that transmits it from *receiver* to its user ;
- the bytes of the already received fragments of a still incomplete SDU are considered evacuated at the moment when one notices that a loss occurred, that definitely prevents a correct reassembling of the SDU (this case is detailed in the next point) ;
- when there is a lack of place to receive a coming cell, we saw that a complete SDU, waiting for delivery, can possibly be deleted if it is waiting for more than ρ units.

Evacuation Policy of the Useless Fragments: The main rule is simple: the already received fragments of a still incomplete SDU (a "last fragment" 10-type cell is not yet arrived) are evacuated if, and when, a cell arrives that is not the expected one, either because it does not carry the right sequence number, or because it carries an identification number different from the current SDU's one. The faulty cell is also thrown away, except if it is a "first fragment" 00-type cell, that announces the arrival of a new SDU. Also, if there is currently no pending SDU, either because the last cell arrived was of 10-type, or after the arrival of an unexpected cell that is not a 00-type one, each new cell arriving that is not of the 00-type will be rejected.

Another rule is added to this main one, that prevents from keeping fragments waiting for the arrival of a new cell, when it is known that, without loss, this cell had to arrive within a certain time. A time-out mechanism is then implemented, that is restarted after the arrival of each new cell and evacuates the pending fragments when it expires. The value of this time-out theoretically depends on the access period to *service* that, as we saw in a previous improvement, may vary in time in a non negligible way. *Receiver* thus tries to estimate this period, and at each restarting, a new time-out value is recalculated, according to the next formula: ($2 * $ *min-delay* $+ 3 *$ *marge*). In this formula, *min-delay* is the smallest delay between two consecutively received cells noticed since ($29 * \eta$) units and *marge* is the difference between the maximal and the minimal possible transmission delays from *sender* to *receiver* (*marge* $= \tau$max $- \tau$min). If some cells are lost, or if their emissions were spaced out by a long period, it might happen that *min-delay* be quite important, or even do not exist, because less than two cells were received for the previous ($29 * \eta$) units. The time-out value must however always remain smaller than ($\psi +$ *marge*) units. It will then be reduced to this value if it happens to be greater, or if *min-delay* does not exist (η and ψ are the previously defined minimal and maximal bounds on the access period to *service*).

5.5 Comments About the Additional Features

As announced, these features present different, and quite complex mechanisms that could be practically encountered. Especially, the rate-control of *receiver*, and the adaptation of the access period to *service* request the ability for the formalism to treat the temporal values as any data type, and the "crash" scenario of *service* poses a problem that, according to us, could sometimes be difficult to solve in a smart way: an interruption that can occur at any time without any reason.

6 Conclusion

The objective of this case study was to propose a large panel of temporal mechanisms that could be faced while specifying real protocols or services and that a temporal FDT should thus be able to describe in a smart way. In comparison to an already existing, real protocol or service, this case study presents, in our opinion, two advantages:

- all its features not directly relevant to the time have been (over)simplified;
- it offers a wider scope of possible aspects of time modelling.

We thus think, and hope, that it could be of interest to anyone who would like to test a temporal FDT in a more concrete situation, as did the *Abracadabra* service and protocol and the *Daemon Game* examples [ISO 10167] for the "classical" FDTs.

References

[AuB 84]   D. Austry, G. Boudol, **Algèbre de Processus et Synchronisation,** *Theoretical Computer Science 30* (1984) 91 - 131 (North-Holland, Amsterdam).

[Azc 90]   A. Azcorra-Saloña, **Formal Modeling of Synchronous Systems,** Ph. D. Thesis, ETSI Telecomunicación, Universidad Politécnica de Madrid, Spain, Nov. 1990.

[BaB 90]   J.C.M. Baeten, J.A. Bergstra, **Real time process algebra,** Rept. No. P8916b, University of Amsterdam, Amsterdam, March 1990.

[BLT 90]   T. Bolognesi, F. Lucidi, S. Trigila, **From Timed Petri Nets to Timed LOTOS,** in: L. Logrippo, R. Probert, H. Ural, eds., *Protocol Specification, Testing and Verification X,* (North-Holland, Amsterdam, 1990) 395-408.

[BoL 92]   T. Bolognesi, F. Lucidi, **LOTOS-like process algebras with urgent or timed inter-actions,** in: K. Parker, G. Rose, eds, *Formal Description Techniques IV,* (North Holland, Amsterdam, 1992), 249-264.

[Bri 89]   Ed Brinksma, **Constraint-oriented specification in a constructive formal description technique,** in: J.W. de Bakker, W.-P. de Roever, G. Rozenberg, eds., *Stepwise Refinement of Distributed Systems, Models, Formalisms, Correctness, LNCS 430* (Springer - Verlag, Berlin Heidelberg New York, 1990) 130-152.

[CCITT Z100]   CCITT-Study group X,**Specification and Description Language (SDL), Z.100,** (Blue Book, Vol. X, Fascicle X.1, 1988).

[Gro 90]   J. F. Groote, **Specification and Verification of Real Time Systems in ACP**, in: L. Logrippo, R. Probert, H. Ural, eds., *Protocol Specification, Testing and Verification X*, (North-Holland, Amsterdam, 1990), 261-274.

[HaJ 90]   H. Hansson, B. Jonsson, **A calculus for communicating systems with time and probabilities**, in: *11th IEEE Real-Time Systems Symposium*, Orlando, Florida, 1990, IEEE Computer Society Press

[Han 91]   H. Hansson, **Time and Probability in Formal Design of Distributed Systems**, Ph. D Thesis, DoCS 91/27, Uppsala University, Dept. of Computer Science, P.O. Box 520, S-75120 Uppsala, Sweden.

[HeR 91]   M. Hennessy, T. Regan, **A temporal process algebra**, in: J. Quemada, J. Mañas, E. Vazquez, eds., *Formal Description Techniques III*, (North-Holland, Amsterdam, 1991) 33-48.

[ISO 8807]   ISO/IEC-JTC1/SC21/WG1/FDT/C, **IPS - OSI - LOTOS, a Formal Description Technique Based on the Temporal Ordering of Observational Behaviour**, IS 8807, February 1989.

[ISO 9074]   ISO/IEC-JTC1/SC21/WG1/FDT/B,**Information Processing Systems - Open Systems Interconnection - Estelle, a formal description technique based on an extended state transition model**; IS 9074, July 1989.

[ISO 10167] ISO/IEC DTR 10167, **Guidelines for the Application of Estelle, LOTOS and SDL**, January 90.

[Led 92]   G. Leduc, **An upward compatible timed extension to LOTOS**, in: K. Parker, G. Rose, eds, *Formal Description Techniques IV* (North Holland, Amsterdam, 1992), 217-232.

[LeL 93]   G. Leduc, L. Léonard, **A timed LOTOS supporting a dense time domain and including new timed operators.** in: M.Diaz, R.Groz eds., *Formal Description Techniques V*, (North-Holland, Amsterdam, 1993), 87-102.

[MeF 76]   P. Merlin and D. Farber. **Recoverability of communication protocols - implications of a theoretical study.** *IEEE Trans. on Computers*, 24(9):1036-1043,1976.

[MFV 93]   C.Miguel, A.Fernandez, L.Vidaller, **Extending LOTOS towards performance evaluation.** in: M.Diaz, R.Groz eds., *Formal Description Techniques V*, (North-Holland, Amsterdam, 1993).

[Mil 83]   A.J.R.G. Milner, **Calculi for Synchrony and Asynchrony**, *Theoretical Computer Science*, Vol. 25, No. 3, July 1983, 267-310 (North-Holland, Amsterdam).

[Mil 85]   G. Milne, **CIRCAL and the Representation of Communication, Concurrency and Time**, *ACM Transactions on Programming Languages and Systems*, Vol. 7, No. 2, April 1985, 270-298.

[MoT 90]   F.Moller, C. Tofts, **A temporal calculus of communicating systems**, in: J.C.M. Baeten, J.W. Klop, eds., CONCUR '90, *Theories of Concurrency: Unification and Extension*, *LNCS 458* (Springer - Verlag, Berlin Heidelberg New York, 1990) 401-415.

[NRS 90]   X. Nicollin, J.-L. Richier, J. Sifakis, J. Voiron, **ATP: An algebra for timed processes**, in: M. Broy, C.B. Jones, eds., *IFIP Working Conference on Programming Concepts and Methods*, Sea of Gallilee, Israel (North-Holland, Amsterdam, 1990).

[NiS 91]   X. Nicollin, J. Sifakis, **The Algebra of Timed Processes ATP: Theory and Application**, Rept. No. RT-C26, Projet Spectre, LGI-IMAG, Nov. 1991.

[NiS 92]   X. Nicollin, J. Sifakis, **An Overview and Synthesis on Timed Process Algebras**, in: K.G. Larsen, A. Skou, eds., *Computer-Aided Verification III*, (LNCS 575, Springer-Verlag, Berlin Heidelberg New York, 1992) 376-398. Also in: LNCS 600.

[NSY 91]   X. Nicollin, J. Sifakis, S. Yovine, **From ATP to Timed Graphs and Hybrid Systems**, in: J.W. de Bakker, C. Huizing, W.P. de Roever, G. Rozenberg, eds., *Real-Time: Theory and Practice* (LNCS 600, Springer-Verlag, Berlin Heidelberg New York, 1992) 549-572.

[NSY 92]   X. Nicollin, J. Sifakis, S. Yovine, **Compiling Real-Time Specifications into Extended Automata**, in *IEEE/TSE Special issue on Real-Time*, Vol 18, n°9

[QAF 90]   J. Quemada, A. Azcorra, D. Frutos, **A timed calculus for LOTOS**, in: S. T. Vuong, ed., *Formal Description Techniques II*, (North-Holland, Amsterdam, 1990) 195-209.

[Ree 90]   G.M.Reed, **A Hierarchy of Domains for Real Time Distributed Computing,** in: M. Main, A. Melton, M. Mislove, D. Schmidt, eds., *Mathematical Foundations of Programming Semantics* (LNCS 442, Springer-Verlag, Berlin Heidelberg New York, 1990) 80-128.

[ReR 88]   G.M.Reed, A.W. Roscoe, **A Timed Model for Communicating Sequential Processes,** *Theoretical Computer Science 58* (1988) 249 - 261 (North-Holland, Amsterdam).

[VSS 88]   C.A. Vissers, G. Scollo, M. van Sinderen, **Architecture and Specification Style in Formal Descriptions of Distributed Systems,** in: S. Aggarwal, K. Sabnani, eds., *Protocol Specification, Testing and Verification, VIII* (North-Holland, Amsterdam, 1988), 189-204.

[Wal 83]   B.Walter, **Timed Petri Nets for Modelling and Analysing Protocols with Real-Time Characteristics,** in: H. Rudin and C.H. West, eds., *Protocol Specification, Testing and Verification III,* (North-Holland, Amsterdam, 1983) 149-160.

[Wan 90]   Y. Wang, **Real-Time Behaviour of Asynchronous Agents,** in: J.C.M. Baeten, J.W. Klop, eds., *CONCUR '90, Theories of Concurrency: Unification and Extension, LNCS 458* (Springer - Verlag, Berlin Heidelberg New York, 1990) 502-520.

[Wan 91]   Y. Wang, **CCS + Time = an Interleaving Model for Real Time Systems,** in: J. Leach Albert, B. Mounier, M. Rodríguez Artalego, eds., *Automata, Languages and Programming 18,* (LNCS 510, Springer-Verlag, Berlin Heidelberg New York, 1991) 217-228.

Assessment of Extended LOTOS

C. Miguel, L. Vidaller, and A. Fernández
Dpto. Ingeniería de Sistemas Telemáticos (DIT). ETSI Telecomunicación, UPM
Ciudad Universitaria - 28040 Madrid - Spain
e–mail: cmiguel@dit.upm.es

This paper describes the assessment of LOTOS extended with timed and probabilistic behaviours, LOTOS–TP, with three case studies which are real or very close to real systems: Tick-Tock case study, FDDI network, and XTP protocol.

Keywords: LOTOS-TP, timed FDT, Probabilistic FDT, Case study.

1   Introduction

Nowadays the research on timed and probabilistic formal description techniques has produced a lot of models which are properly formalized. Nevertheless, their usefulness to specify and describe performance aspects of communication systems has not been assessed.

This chapter presents the assessment of LOTOS extended with timed and probabilistic behaviours, LOTOS–TP [MFV93a, MFV93b], with three case studies which are real or very close to real systems:

- Section 2 makes a detailed description of the LOTOS–T specification of Tick-Tock case study defined in this book.
- Section 3 presents the conclusion obtained from the application of LOTOS–TP to the performance evaluation of a FDDI network.
- Section 4 shows the main results of the XTP protocol specification in LOTOS–T.

2   Tick-Tock Case Study

The Tick-Tock case study is a clear example of the need of formal description techniques which support the description of time constraints in addition to the so–called *functional requirements*. Even though the system is not a real one, its requirements are similar to the actual requirements of real systems. LOTOS–T has been used to formally describe the above system, including the time and functional requirements.

The specification of the case study has been done following a constraint oriented style which is very close to the informal Tick–Tock definition. We say that it is a constraint oriented style because each, or almost each, constraint of the definition has been specified with a LOTOS–T expression. These expressions have been synchronized so that the occurrence of events are the addition of every constraint imposed on them. It is important to point out that the hiding of events is used in LOTOS–T to express that no more constraints can be added on them.

2.1   Introduction to the Case Study

The system consists of a data communication system with three main components: A *sender*, a *receiver* and an underlying *service* which interconnects them. Figure 2.1 shows the structure of the system specification where the boxes represent processes and the arrows mean communication through gates. This structure is valid for both simple and full versions of the system. A shaded box means that the internal structure of the process will be further detailed in another picture.

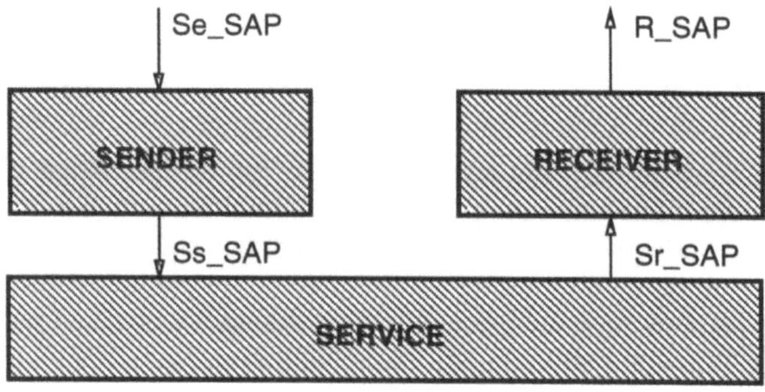

Fig. 2.1 System structure

The sender gets data from the users through the "Se_SAP" gate and uses the "Ss_SAP" gate of the service entity to send the user data to the other communication end.

The receiver gets data from the "Sr_SAP" gate of the service entity and delivers user data through "R_SAP" gate.

The service gets data from the sender and delivers them to the receiver.

Appendix A lists the data library used to specify the system. This library contains the LOTOS specification of the booleans, naturals, positive fractional numbers, and some other data types which will be described later on. Because the system requirements use time values which are fractions of a unit, the fractional numbers have been used as time domain (dense domain). This fact has been expressed in the beginning of the specifications as follows:

```
timedef
   sort p_rational
   opns 0 for 00   + for +   > for gt
enddef
```

The data objects exchanged between the system entities, and between the system and users are three:

Addresses .- Source and destination users are identified by their address. The specification of this data object can be found in the *addresses* type of appendix A.

SDUs .- Data units that source users send to destination users. The specification of this data object can be found in the *SDUs* type of appendix A. These data units have been specified as a sequence of *data fragments* (40 bytes). The operation "add_fragment" adds a fragment to a SDU. The operation "rest_fragment" removes a fragment from the SDU. The operation "first_fragment" returns the first fragment of the SDU. And the operation "num_of_fragments" permits us to know the number of data fragments in a SDU.

Cells .- Data units exchanged between sender and service, and between service and receiver. There are three types of cells:

- Cells carrying the source and destination addresses (*add_cell*).
- Cells carrying a data fragment that is not the last of a SDU (*data_cell*).
- Cells carrying a data fragment that is the last of a SDU (*last_cell*).

The specification of this data object can be found in the *Cells* type of appendix A. Two sorts have been defined: "cell", and "cell_type". The values of the first one are the cells themselves. The sort "cell_type" contains three values, one per each type of cell: *add_cell*, *data_cell*, and *last_cell*. Cells are built depending on their type:

add_cell : SDU identification number + source add. + dest. add.
data_cell : SDU identification number + cell rank into de SDU + data fragment.
last_cell : SDU identification number + cell rank into de SDU + data fragment.

The type of a cell can be obtained with operation "the_cell_type". Each of the components of a cell can be also obtained by applying the operations: "the_SDU_id", "the_cell_rank", "the_src_add", "the_dest_add", and "the_data_fragment".

The representation of theses data objects as a sequence of bytes has not been specified because of simplicity. This simplification does not affect in any sense the assessment of the timed model.

2.2   System's Simple Version

The LOTOS–T specification of the system's simple version can be found in appendix B. The sections below explain the behaviour of each system component following the structure of the requirement of Tick–Tock case study.

2.2.1   Service

Figure 2.2 shows the structure of the service.

We can see that each box in the figure (LOTOS–T process) imposes a new constraint to the service:

Exchanged primitives .- "service" exchanges *cells* with "sender" and "receiver":
$$Ss\_SAP?c{:}cell, \text{ and } Sr\_SAP!c$$
where *c* is of sort *cell*.

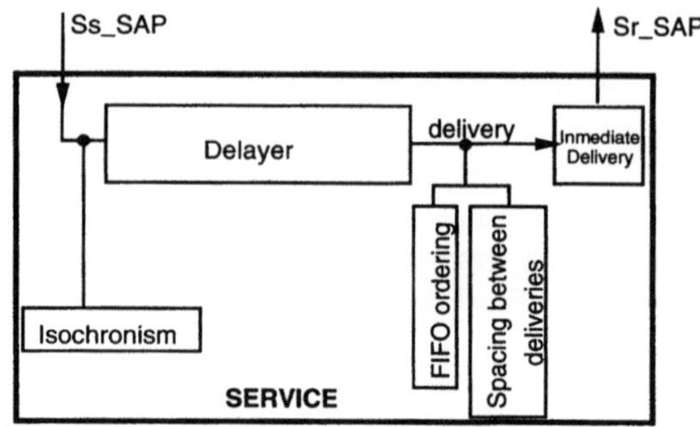

Fig. 2.2 Simple version: Service structure

Isochronism .- This constraint is imposed to the acceptation of cells at some precise, punctual instants, that follow one another regularly in time, with a given rate. It has been specified in the LOTOS–T process called isochronism whose behaviour is:

```
choice n:nat [] Ss_SAP?c:cell {(n+1) * service_period};
               isochronism[Ss_SAP]
```

We can see that cells are accepted only at instants "(n+1) * service_period" where "n+1" is a natural number higher than 0, and "service_period" is a constant which determines the acceptation period (π units). In case the sender neglects the emission of cell at some of these instants, it will have the opportunity in the forthcoming instants. We can see in the specification that just one cell can be exchanged at any instant because when a cell is accepted, the next one can only be accepted at the next service periods (this is the reason of the term n+1).

Transmission delays .- Process "delayer" takes cells from the sender and delays them a non–deterministic time between "transmission_delay_min" (τ_{min}) and "transmission_delay_max" (τ_{max}). The non–deterministic choice is expressed as:

```
choice delay:p_rational [] [delay ge transmission_delay_min]->
                           [delay le transmission_delay_max]-> i;
........................................................
```

After the internal action occurrence, variable "delay" has the non–deterministic time value chosen which is used to determine the delivery time.

Because other constraints on the delivery (fifo ordering and spacing between deliveries) could make the cells not to be able to be delivered just when the chosen delay has elapsed, the delivery time value will be greater or equal to the chosen delay:

```
(choice deliv_time: p_rational [] [deliv_time ge delay] ->
       delivery!n_order!c {deliv_time}; stop)
```

Observe that the deliveries are numbered consecutively according to the order of acceptation of the cell by the service. See Appendix B to clarify the combination of the previous two expressions.

Spacing between the deliveries .- The behaviour of the process:

```
process Spacing_between_deliveries[delivery]:noexit:=
  choice t:p_rational [] [t ge spacing_between_deliveries]->
```

```
delivery?n:nat?c:cell {t};
Spacing_between_deliveries[delivery]
```
```
endproc
```

imposes that there should be a delay of at least "spacing_between_deliveries" (α units) between two successive deliveries.

Immediate acceptation .- When every constraint imposed to the delivery of cells has been fulfilled (delay, FIFO ordering and Spacing between deliveries) the cell must be immediately accepted by receiver, or otherwise the cell is lost: This behaviour is expressed in LOTOS–T by the process "immediate_delivery":

```
delivery?n:Nat?c:Cell;
(  Sr_SAP!c {00}; immediate_delivery[delivery,Sr_SAP]
[] immediate_delivery[delivery,Sr_SAP])
```

We can see that, when service decides to deliver a cell (delivery?n:Nat?c:Cell synchronization), the cell is passed to the receiver if it can receive the cell immediately (Sr_SAP!c {00}). Otherwise, the cell is lost. In any case, new deliveries are accepted.

Loss free transmission .- This characteristic imposes that every cell accepted in gate "Ss_SAP" has to be delivered. The only way in which a cell can be lost has been specified in the above–mentioned process: "immediate_delivery".

FIFO–ordering of cells .- Because the deliveries are numbered consecutively according to the order of acceptation of the cell by the service, the process "FIFO_ordering" imposes cells to be delivered in the order of emission in a very simple way:

```
delivery!n_order?c:cell; FIFO_ordering[delivery](n_order+1)
```

In order to avoid incompatibilities between the time constraints, we have specified that constant "service_period" is greater or equal to "spacing_between_deliveries".

2.2.2 Sender

Figure 2.3 shows the structure of the sender.

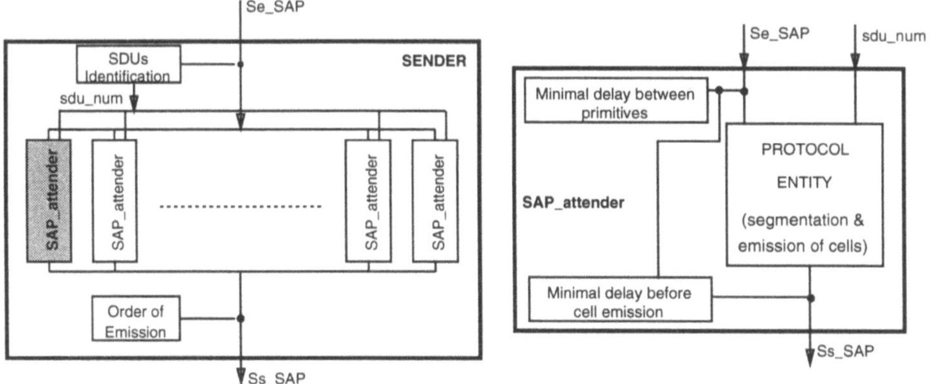

Fig. 2.3 Simple version: Sender structure

For each source user, the sender has a SAP attender which handles the SDUs from

358   C. Miguel et al.

the SAP belonging to that user. The internal structure of such a SAP attender can be
also seen in figure 2.3. We have assumed for the specification that the SAP attender can
accept a new SDU only when all the cells from the previous one have been emitted.

Received data .- Sender receives from its users primitives that contain 3 parameters:

$$Se\_SAP?sa,da:address?d:SDU$$

That is, source and destination addresses, and an SDU.

Minimal delay between primitives on a given Se–SAP .- This constraint has to be applied independently to each user access point. It imposes that every primitive must be
at least spaced out from the previous one by a delay that is "Beta * num_of_fragments
of the SDU". Therefore, this constraint has to be imposed in each "SAP_attender"
process, and it is described in the process:

```
process Minimal_delay_between_primitives[Se_SAP]
                                        (user:nat):noexit:=
  choice t:p_rational []
    Se_SAP!add(user)?da:address?d:SDU
                      [t ge (Beta * num_of_fragments(d))] {t};
    Minimal_delay_between_primitives[Se_SAP](user)
endproc
```

Notice that the constraint is only applied to the primitives from the user with address
"add(user)", and that the delay depends on the number of data fragments of the SDU
received on that synchronization.

Emitted data .- Sender emits cells: *Ss_SAP!c* where c is of sort cell.

Emission of a cell and urgency of the emission .- Sender, when it desires to emit a cell,
emits it as soon as the service is ready to do that because sender and service are
synchronizing on Ss_SAP, and no other process imposes more constraints. This
behaviour is implicit in the semantics of LOTOS–T.

Segmentation of a SDU .- When a SAP attender receives an SDU, it emits the data as
a sequence of cells that contains the data fragments (one per cell) with which the
SDU has been built (see SDUs data type). The first cell of the sequence is a cell that
contains the source and destination addresses (add_cell). This fragmentation is done
in the process called protocol_entity.

Order of emission and emission of the cells related to a same SDU .- The cells related
to a SDU are emitted sequencially, beginning with the cell that contains the addresses.
Each cell is numbered with the rank, in the SDU, of the data fragment that the cell
contains. The emission of the cells related to a same SDU is done in the process
called emission_of_cells:

```
process emission_of_cells[Ss_SAP]
                  (sdu_id,rank:Nat, sa,sd:address, d:SDU):exit:=
  ( [rank eq 0] -> Ss_SAP!add_cell(sdu_id,sa,sd);
    emission_of_cells[Ss_SAP](sdu_id,rank+1,sa,sd,d)
  [][rank gt 0] -> [num_of_fragments(d) gt 1] ->
    Ss_SAP!data_cell(sdu_id,rank,first_fragment(d));
    emission_of_cells[Ss_SAP]
                  (sdu_id,rank+1,sa,sd,rest_fragment(d))
  [][rank gt 0] -> [num_of_fragments(d) eq 1] ->
    Ss_SAP!last_cell(sdu_id,rank,first_fragment(d)); exit)
endproc
```

Identification of a SDU .- There is a process which determines the identification number of the SDU. This number increases one unit each time a primitive (Se_SAP) occurs. This identification number is given to a SAP attender when it receives an SDU. This is done through the gate "sdu_num".

Order of emission of the cells related to different SDUs .-
The process "Order_of_emission" warranties that all the cells related to a SDU are emitted consecutively, and the order of emission between the cells from different SDUs follows the one determined by the instant of occurrence of the primitive (or the SDU identification number):

```
process Order_of_emission [Ss_SAP](n:Nat):noexit:=
  Ss_SAP?c:cell [n = the_SDU_id(c)];
    ( [the_cell_type(c) = last_cell]->
                      Order_of_emission [Ss_SAP](n+1)
    [][the_cell_type(c) = add_cell]->
                      Order_of_emission [Ss_SAP](n)
    [][the_cell_type(c) = data_cell]->
                      Order_of_emission [Ss_SAP](n))
endproc (* Order_of_emission *)
```

Minimal delay before the emission of a cell .-
The process "minimal_delay_before_cell_emission" imposes a minimal delay between a cell emission and the occurrence of the primitive (Se_SAP):

```
process minimal_delay_before_cell_emission[Se_SAP,Ss_SAP]:
                                                 noexit:=
  Se_SAP?sa,da:address?d:SDU; time_constraint[Ss_SAP] (00)
     >> minimal_delay_before_cell_emission[Se_SAP,Ss_SAP]
  where
   process time_constraint[Ss_SAP](t:p_rational):exit:=
    choice t0:p_rational []
     Ss_SAP?c:cell
       [(t+t0) ge (cell_min_delay * the_cell_rank(c))] {t0};
       (   [the_cell_type(c) = add_cell]->
                 time_constraint[Ss_SAP](t+t0)
        [] [the_cell_type(c) = data_cell]->
                 time_constraint[Ss_SAP](t+t0)
        [] [the_cell_type(c) = last_cell]-> exit)
   endproc (* time_constraint *)
endproc (* minimal_delay_before_cell_emission *)
```

Observe that a cell (c) can be emitted only if a time "cell_min_delay * the_cell_rank(c)" has elapsed from the primitive occurrence where "cell_min_delay" (ν) is a constant. The value of (t+t0) is the time elapsed from the primitive occurrence.

2.2.3 Receiver

Figure 2.4 shows the structure of the receiver.

For each destination user, the receiver has an emission point which handles the SDUs to the SAP corresponding to that user. The internal structure of such an emission point can be also seen in figure 2.4.

Received data .- Receiver receives cells from service: *Ss_SAP?c:cell*

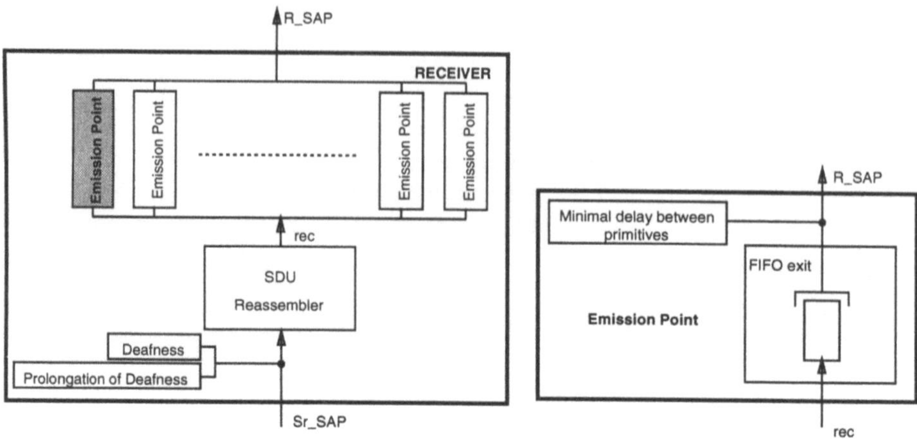

Fig. 2.4 Simple version: Receiver structure

Frequency of receptions .- Process "deafness" specifies that the receiver always listens
to service, except during a period of time after the reception of a cell:

```
process deafness [Sr_SAP]:noexit:=
   Sr_SAP?c:cell; i {deafness}; deafness [Sr_SAP]
endproc (* deafness *)
```

where "deafness" is a constant (ϕ units).

Prolongation of deafness .- Process "prolongation_of_deafness" expresses how an undetermined prolongation of the deafness may occur:

```
process prolongation_of_deafness [Sr_SAP]:noexit:=
   Sr_SAP?c:cell;
     ((choice t:p_rational [] [t ge deafness]-> i;  exit {t})
                        >> prolongation_of_deafness [Sr_SAP])
endproc (* prolongation_of_deafness *)
```

Emission point of a primitive .- Receiver has as many "emission points" as destination
    users, so that a primitive is proposed on the R_SAP that corresponds to the received
    destination address.

Emitted data .- Receiver proposes the following synchronization to its users:
$$R\_SAP!da!sa!d$$
    where "da" is the address of the destination user with which the receiver is synchro-
    nizing, "sa" is the source address, and "d" is an SDU.

Creation of the primitives .- Process "SDU_reassembler" offers (rec) every SDU prop-
    erly received and reassembled to the emission point that corresponds to the received
    destination address.

Discipline of exit .- Primitives on a R_SAP are proposed according to a FIFO discipline.
    Process "FIFO_exit" describes such a behaviour. It uses the type "Queue_of_SDUs"
    that specifies a FIFO queue of SDUs.

Minimal delay between primitives on a R–SAP .-
    Process "Minimal_delay_between_primitives" imposes that every primitive to a user

is at least spaced out from the previous one by a delay "W * num_of_fragments(d)":

```
process Minimal_delay_between_primitives[R_SAP]
                                  (user:nat):noexit:=
  choice t:p_rational []
   R_SAP!add(user)?sa:address?d:SDU
       [t ge (W * num_of_fragments(d))] {t};
         Minimal_delay_between_primitives[R_SAP](user)
endproc (* Minimal_delay_between_primitives *)
```

where "W" is a constant, and "num_of_fragments(d)" is the number of data fragments of the SDU "d".

2.3   System's Full Version. Addition of New Features

New features have to be added to the components of the system's simple version: service, sender and receiver. The full system specification can be found in appendix C. The following sections explain the specification of these new features.

2.3.1   Adaptation of the Access Period to "Service"

This new feature implies to enrich the previous "isochronism" constraint. In fact, the time period between two consecutive interactions in "Ss_SAP" has to be modified according to the cells traffic through this service access point. Figure 2.5 shows how this new feature has been specified. As expected, it only affects the process "isochronism" of the system's simple version. The rest of the components remain unchanged.

The enriched process has been called "adaptative_isochronism" which consists of two main components (see this process in appendix C):

- *Adaptation of period.-* This process updates (if needed) the service period after each sequence of 10 proposals, according to the algorithm fixed in the requirements.
- *Emission control.-* This process imposes the time constraint to the occurrence of cells emissions (action on gate "Ss_SAP"). Its behaviour is the same as the "isochronism" process of the simple version of the system, but in this case the period can be updated by the above mentioned process.

2.3.2   "Crash of Service"

Service may "crash" without any reason (internal) at any instant. Service behaviour is disabled at an undeterministic instant of time:

```
choice crash_time:p_rational [] i;
      <service behaviour> [> i {crash_time}; ....
```

Also, service needs an unpredictable time before restating. After the disabling, an undeterministic delay is inserted before the process is restarted:

```
choice delay_time:p_rational []
       i; exit {delay_time} >> Service [Ss_SAP,Sr_SAP]
```

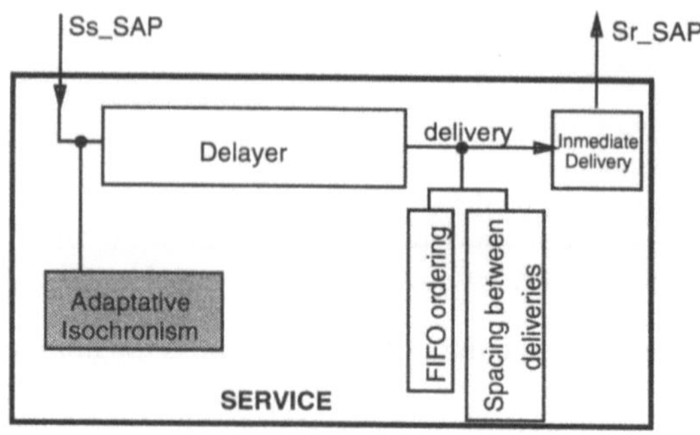

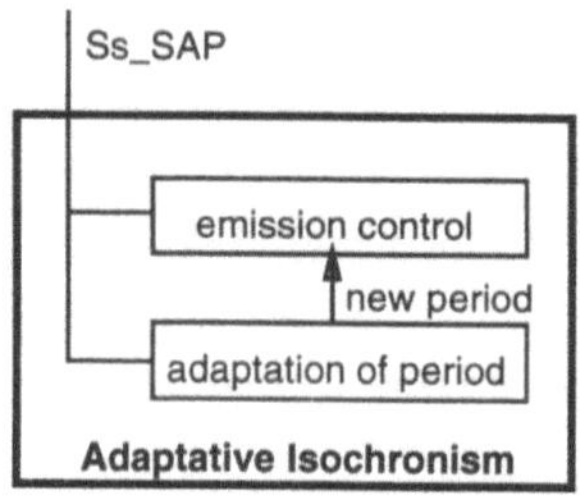

Fig. 2.5 Full version: Service structure

2.3.3   Rate Control at the Access to the Service Provided by "Sender"

Sender is extended with two new features: *Individual rate control on each Se–SAP* and *Global rate control of all the Se–SAPs*. Each of these new requirements are added to the specification as two new processes which constraint the occurrence of "Se–SAP" gate. Figure 2.6 shows how an *individual rate control* process constraints locally the occurrence of each SAP event; and a *global rate control* process constraints globaly the occurrence of SAP event.

It is important to realize that the addition of these new features does not imply any modification of the former simple version. This addition has been done following a constraint oriented approach.

The *global rate control* process behaves in a very simple way. It allows the occurrence of "Se_SAP" event if the amount of information exchanged, during any *period* of λ units, is below a given threshold. If this condition does not hold for a given instant of time (present), the occurrence of "Se_SAP" should be delayed up to the rate is below the threshold. The checking of the condition can be easily done in LOTOS by using a boolean operation:

```
enable_time: p_rational, Nat, prim_history -> bool
```

whose arguments are: (1) The time elapsed since the last "Se_SAP" occurrence. (2) The number of blocks of 40 bytes of the SDU that is going to be passed through the SAP. And, (3) the history of previous "Se_SAP" occurrences from which the rate in a given *period* can be evaluated. This operation has been defined formally in the type

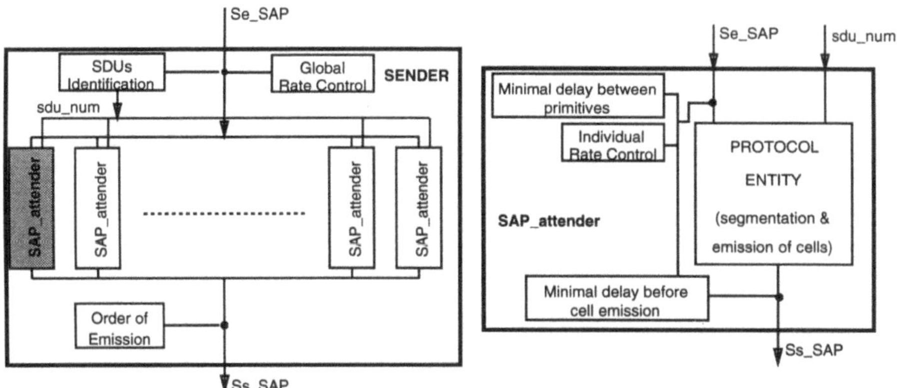

Fig. 2.6 Full version: Sender structure

global_rate of the specification. The resulting process is:

```
process global_rate_control [Se_SAP](h:prim_history):noexit:=
  choice t:p_rational []
   Se_SAP?sa,da:address?d:SDU
     [enable_time(t,num_of_fragments(d),h)] {t};
   global_rate_control [Se_SAP](store(num_of_fragments(d),t,h))
endproc
```

Let us observe that the number of blocks of each SDU and occurrence time of the SAP event, relative to the previous occurrence time, are stored in the process parameter in order to be able to check the global rate condition.

The *individual rate control* process behaves in the same way as the previous global rate control process. The difference is that it is applied locally to each SAP; that is, one per destination user.

2.3.4   Rate Control on the Reception of Cells by "Receiver"

The two requirements about "deafness" of the simplified version of the system have been removed so that the two processes that imposed such constraints have also been removed. In addition, new constraints have been imposed to the acceptation of cells as well as a time-out mechanism that prevents from keeping fragments of an SDU waiting for the arrival of a new cell when such a cell has been lost.

Figure 2.7 shows the new structure of the receiver where the two processes that handle the deafness in the simplified version have disappeared, and two new processes impose the new requirements:

- *Cells acceptation control.* Two sufficient conditions have been imposed to the acceptation of cells. The first depends on the total amount of bytes in the receiver:

$$\text{(bytes received)} - \text{(bytes evacuated)}$$

The bytes received are evaluated through the synchronization in "Sr_SAP" (bytes of the cell). The bytes evacuated are evaluated through the synchronization in "R_SAP"

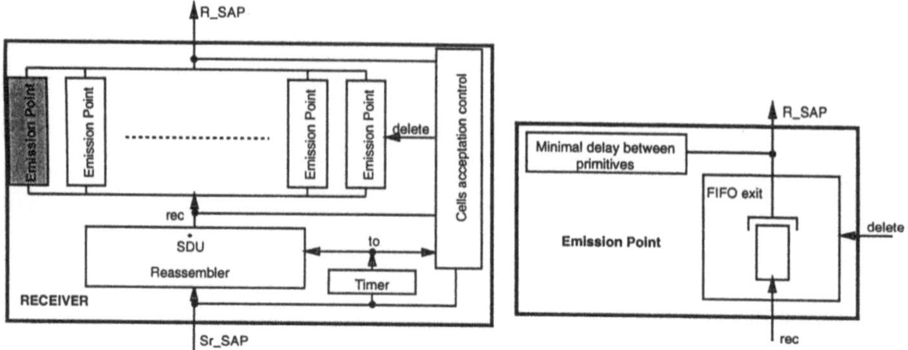

Fig. 2.7 Full version: Receiver structure

(bytes of the SDU) and "to" (bytes of the fragments of the SDU that is being received). If the first condition does not apply, the second sufficient condition is considered. It depends on the age of the SDUs on the emission points (time since the "rec" occurrence). If there exist some SDUs older than ρ units, the incoming cell is accepted and the oldest SDU is immediately deleted (occurrence of "delete" event). The age of the SDUs at each emission point is evaluated in this process because it stores the time at which each SDU was received ("rec" occurrence). When an SDU is delivered, it is not considered in this age evaluation any more.
The behaviour of this process is as follows:

```
process Cells_acceptation_ctrl[R_SAP,Sr_SAP,rec,delete,to]:
 noexit:=
   evacuation_control[R_SAP,Sr_SAP,rec,delete,to]
                              (0,0,0,no_waiting,00)
   where
       ....................  (* type definition *)

  process evacuation_control[R_SAP,Sr_SAP,rec,delete,to]
          (sdu_id,rank,blocks:Nat, w:waiting_times,
                  current_time:p_rational): noexit:=
   choice t:p_rational []
   let current_time:p_rational = current_time +t in
    (  Sr_SAP?c:cell [first_condition_holds(blocks)] {t};
        evacuation_control[R_SAP,Sr_SAP,rec,delete,to]
          (next_cell_SDU_id(c.sdu_id,rank),
           next_cell_rank(c,sdu_id,rank),
           total_blocks(c,sdu_id,rank,blocks),w,current_time)
    [] Sr_SAP?c:cell
        [    not(first_condition_holds(blocks))
         and second_condition_holds(current_time,w)] {t};
        delete!oldest_add(w) {00};
        evacuation_control[R_SAP,Sr_SAP,rec,delete,to]
          (next_cell_SDU_id(c,sdu_id,rank),
           next_cell_rank(c,sdu_id,rank),
           total_blocks(c,sdu_id,rank,blocks)-oldest_blocks(w),
```

```
                  remove_oldest(w),current_time)
    [] R_SAP?da,sa:address?d:SDU {t};
         evacuation_control[R_SAP,Sr_SAP,rec,delete,to]
            (sdu_id,rank,blocks-num_of_fragments(d),
             remove_oldest(da,w),current_time)
    [] rec?da,sa:address?d:SDU {t};
         evacuation_control[R_SAP,Sr_SAP,rec,delete,to]
           (sdu_id,rank,blocks,
            add_wt(current_time,da,num_of_fragments(d),w),
            current_time)
    [] to {t}; evacuation_control[R_SAP,Sr_SAP,rec,delete,to]
                   (sdu_id+1,0,total_b_after_to(rank,blocks),w,
                    current_time)
    )
 endproc (* evacuation_control *)
endproc (* Cells_acceptation_ctrl *)
```

The process "evacuation_control" has the following parameters: the SDU identification, cell rank of the next expected cell, the total number of block (40 bytes) in the
receiver, the age of each SDU at each emission point (waiting times), and the current
time (absolute time elapsed since the process instantiation). A cell is accepted if the
first sufficient condition holds:

```
         Sr_SAP?c:cell [first_condition_holds(blocks)]
```

If the first condition does not apply, but the second does, the cell is accepted:

```
  Sr_SAP?c:cell [    not(first_condition_holds(blocks))
                 and second_condition_holds(current_time,w)]
```

Observe that the first condition only depends on the total amount of data in the
receiver; and that the second one depends on the age of the SDUs. The age of the
SDUs can be evaluated by knowing the current time and the SDUs waiting times
stored in "w".
Each time an event occurs, the current time is updated because the value of "t" is the
amount of time since the last event (Sr_SAP, rec, to, or R_SAP). When these events
occur, the parameters of the process are accordingly updated.

- *Timer.-* Its behavior is:

```
process timer [Sr_SAP,to](h:cells_timing):noexit:=
 choice t: p_rational [] Sr_SAP?c:cell {t};
  (let h:cells_timing = cell(t,h) in
   (choice t:p_rational []
     [time_out(t,h)]-> to {t}; stop) [> timer [Sr_SAP,to](h))
endproc (* timer *)
```

This timer is started/restarted after the arrival of each new cell (Sr_SAP?c:cell). The
occurrence time of each cell acceptation, relative to the previous one, is stored in
the process parameter (h) so that the time–out value can be computed from this
information. The timer expires (it offers "to") when a time "t" has elapsed such that
the "time_out" condition hold. This time–out condition depends on the timing of
previous cell acceptations which is stored in parameter "h". The specification of this
condition has been done as a normal boolean operation.

The time–out event (to) is synchronized with the "SDU reassembler" and the "cells acceptation control" process in order to discard fragments of an SDU, and consider such fragments as evacuated.

The inclusion of the new requirements to the receiver has implied the modification of the behavior of:

- the "emission points" to delete the oldest (first of the queue) SDU when it is required (delete occurrence),
- and the "SDU reassembler to take into account the time–out mechanism (occurrence of "to").

3   FDDI Network

The LOTOS–TP specification and performance evaluation of a FDDI network has been also used to assess the suitability of the language.

The structure of the network is given in figure 3.1.

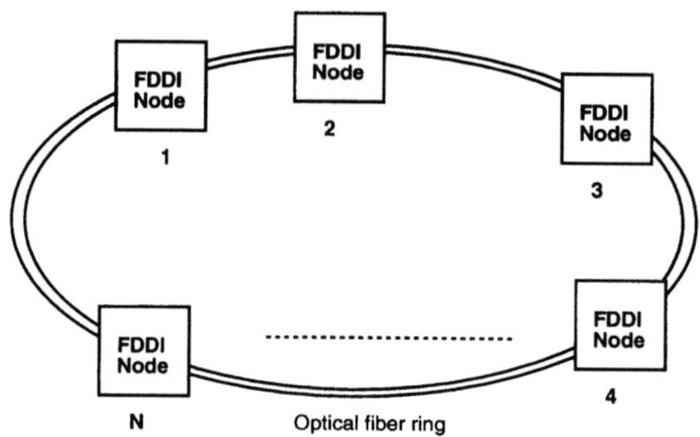

Fig. 3.1 FDDI network structure

Such a network has been modelled and specified in LOTOS–TP following the process architecture shown in figure 3.2.

Process "Channel" includes the specification of delays. Process "FDDI MAC" specifies the behaviour of the Token Rotation Timer (TRT), and the Token Holding Timer (THT) [ISO89]. The process "Traffic generator" generates user data whose rate follows different probabilistic distributions.

The performance evaluation of this system, specified in LOTOS–TP, was performed by using TOPOSIM simulation tool [MFOV93, MO91]. The result of this evaluation was validated against analytical and well known FDDI performance figures [DB88].

As an example of results, figure 3.3 shows the high degree of similarity between the simulation results (points) and the abovementioned analytical model (continuous line). This figure represents the variation of the total throughput of the network versus the number of active stations, supposed that the total delay is fixed (1 msec). T_Opr

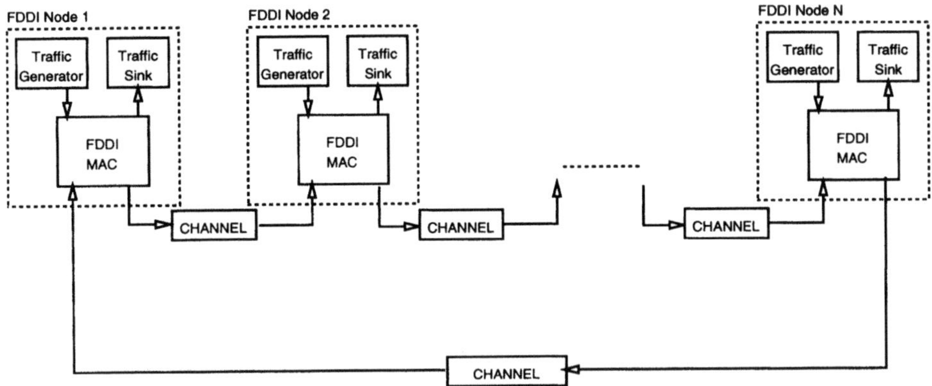

Fig. 3.2 FDDI network model

parameter (Operative Target Token Rotation Time) is also fixed, and its value is 5 msec. In the figure we can see that the total throughput increases as the number of active station increases, with the condition of a fixed total delay.

With this case study we have covered several aspects related to each other, but from which we can obtain conclusions separately. These aspects are: (1) The use of formal techniques LOTOS–TP to specify communication systems. And (2) The performance evaluation tool TOPOSIM which was used to evaluate the performance figures of the network.

The application of LOTOS–TP to this case study has shown that it is a powerful specification technique. It inherits from LOTOS its suitability to design communication systems, and extends the LOTOS capabilities with the possibility of specifying and evaluating the performance figures of the systems. Even though we have used simulation techniques to evaluate the system, the formal semantics of LOTOS–TP provides also facilities for analytical evaluations.

TOPOSIM has proved to be a useful tool. Nevertheless, it is still a prototype tool that has to be improved in several directions: (1) Increasing the efficiency of the simulation. (2) Improving the user interface. And (3) Adding new simulation control mechanisms which permit us to improve the evaluation of the confidence in the results obtained from the simulation.

4   XTP Protocol

The XTP specification that was used to be specified is based on the revision 3.5 of the XTP protocol definition [Pro90], but with some simplifications: (1) Only one active context is supported. (2) There is no priorities, multicast, selective retransmission and routing.

The definition of the protocol includes *timers* (Connection timers, CTIMERs. And,

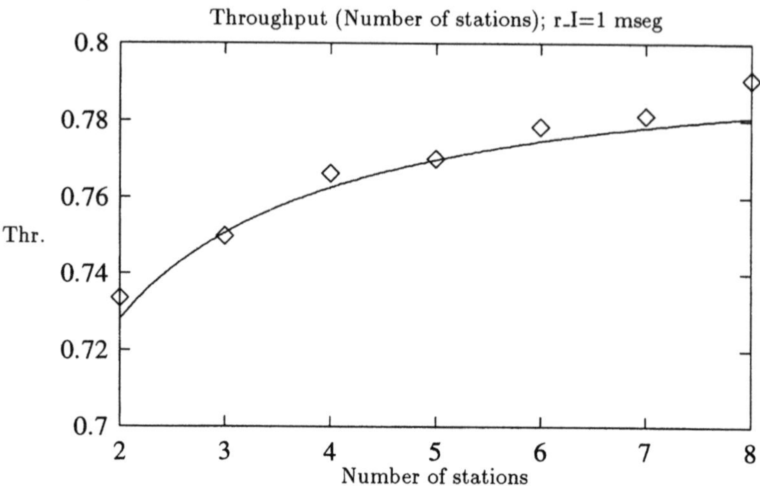

Fig. 3.3 Simulation: Throughput(n); r_I=1 msec; T_Opr=5 msec

Wait timer, WTIMER.). The round trip time (rtt) has also to be measured as the difference between two time instants obtained from a *clock*.

We have shown that the specification of those components in LOTOS–T are simple and close to the informal specification of the protocol. The new timed operators have been needed only to specify the timers and clock. The rest of the specification has been specified as we would have done with LOTOS. In fact, we first specified the protocol in LOTOS by making the timeouts and time reading external events that can be controlled by the tests. After that we specified the timed part in LOTOS-T by hiding the above–mentioned events and specifying the time pass with the new timed operators.

The specification has been done in an implementation oriented style which contrasts with the constraint oriented style used in the specification of the Tick–Tock case study. So we can conclude that LOTOS–T is suitable to specify with different styles as LOTOS is.

5   Conclusions

This chapter presents the application of the extended LOTOS to three "real" case studies. Only the Tick–Tock case study has been described in detail, but it is a very good and representative example of the set of time constraints of real systems.

The main conclusion obtained from the extended LOTOS assessment is that LOTOS–TP inherits from LOTOS its suitability to design communication systems, and extends the LOTOS capabilities with the possibility of specifying and evaluating the performance figures of the systems.

Every time constraint of the case studies has been specified with LOTOS–T. From this specification experience, we can identify some weak and strong points of the defined language. The use of the "choice" LOTOS operator to define time alternatives seems to be very powerful because it permits us to define general time constraint and to measure the actual time occurrence of the events. Such a powerfulness is achieved because an expression

$$\text{choice t:time} \quad [] \quad [f(t)] - > B(t)$$

can be interpreted as:

$$\forall \, t \, \in \, time \text{ such that } f(t) \text{ then } B(t)$$

Nevertheless, it is very verbose. An extensive use of this expressions can generate specifications that are difficult to read. An easy solution to this problem will be the definition of a syntactically simpler expression with the same semantics.

The non-determinism in time is expressed in LOTOS–T with expressions such as

$$\text{choice t:time} \quad [] \quad [f(t)] - > i; \, B(t)$$

which are also verbose and require the occurrence of an internal action. This fact makes that the expression of non-determinism in time should be done with some care, e.g. to avoid that the internal occurrence decides some choice. Nevertheless, the occurrence of the internal event is meaningful because it represents the internal decision of a non-deterministic value.

An important aspect related to the usefulness of LOTOS–TP is the possibility of developing tools which support the design and verification of data communication systems. An example of this tool is TOPOSIM [MFOV93] which supports partially LOTOS–TP and permits us to obtain, by simulation, the performance figures of the specified systems. This tool has been used in one of the application cases described in this chapter. TOPOSIM has not been used to validate the other two case studies because it supports a subset of LOTOS–TP.

Appendix A: Tick-Tock Data Types Library

```
type Boolean is
 sorts Bool
 opns
   true, false                 :             -> Bool
   not                       : Bool      -> Bool
   _and_, _or_, _xor_, _implies_,
    _iff_, _eq_, _ne_          : Bool, Bool -> Bool
 eqns forall x, y : Bool ofsort Bool
      not(true) = false;
      not(false) = true;
      x and true = x;
      x and false = false;
      x or true = true;
      x or false = x;
      x xor y = (x and not(y)) or (y and not(x));
      x implies y = y or not(x);
      x iff y = (x implies y) and (y implies x);
      x eq y = x iff y;
      x ne y = x xor y;
endtype


type BasicNaturalNumber is
 sorts Nat
 opns
   0           :              -> Nat
   Succ        : Nat          -> Nat
   _+_,_*_,_**_ : Nat, Nat     -> Nat
 eqns forall m, n : Nat ofsort Nat
      m + 0 = m;
      m + Succ(n) = Succ(m) + n;
      m * 0 = 0;
      m * Succ(n) = m + (m * n);
      m ** 0 = Succ(0);
      m ** Succ(n) = m * (m ** n);
endtype


type NaturalNumber is BasicNaturalNumber, Boolean
 opns _eq_, _ne_, _lt_, _le_, _ge_, _gt_ : Nat, Nat -> Bool
 eqns forall m, n : Nat ofsort Bool
      0 eq 0 = true;
      0 eq Succ(m) = false;
      Succ(m) eq 0 = false;
      Succ(m) eq Succ(n) = m eq n;
      m ne n = not(m eq n);
      0 lt 0 = false;
      0 lt Succ(n) = true;
      Succ(n) lt 0 = false;
      Succ(m) lt Succ(n) = m lt n;
      m le n = m lt n or (m eq n);
      m ge n = not(m lt n);
      m gt n = not(m le n);
endtype


type ExtendedNaturals is NaturalNumber
 opns 1,2,3,4,5,6,7,8,9,10: -> Nat
   pred          : Nat -> Nat
   _-_           : Nat, Nat -> Nat
 eqns forall n,n1: Nat
    ofsort Nat
      pred(0) = 0; pred(succ(n)) = n;
      n - 0 = n; n - succ(n1) = pred(n) - n1;
      1 = succ(0); 2 = succ(1); 3 = succ(2); 4 = succ(3); 5 = succ(4);
      6 = succ(5); 7 = succ(6); 8 = succ(7); 9 = succ(8); 10 = succ(9);
endtype


type NaturalNumberWithoutZero is ExtendedNaturals
 sorts Nat0
 opns 1,2,3,4,5,6,7,8,9,10:        -> Nat0
      Succ          : Nat0        -> Nat0
      _+_,_*_,_**_  : Nat0, Nat0  -> Nat0
      N             : Nat0        -> Nat
 eqns forall m,n: Nat0
    ofsort Nat0
      2 = Succ(1); 3 = Succ(2); 4 = Succ(3); 5 = Succ(4);
      6 = Succ(5); 7 = Succ(6); 8 = Succ(7); 9 = Succ(8);
      10 = Succ(9);
      m + 1 = Succ(m);
      m + Succ(n) = Succ(m) + n;
      m * 1 = m;
      m * Succ(n) = m + (m * n);
      m ** 1 = m;
      m ** Succ(n) = m * (m ** n);
    ofsort Nat
      N(1) = 1;
      N(Succ(m)) = Succ(N(m));
endtype


type PositiveFracNumber is ExtendedNaturals, NaturalNumberWithoutZero,
     Boolean
 sorts p_rational
 opns 00          :              -> p_rational
      real        : Nat          -> p_rational
      _/_         : Nat, Nat0     -> p_rational
      id          : p_rational    -> p_rational
      _*_,_+_,_-_ , _-_    : p_rational, p_rational -> p_rational
      _*_,_+_,_-_,_-_  **  : p_rational, Nat      -> p_rational
      _*_,_+_,_-_          : Nat, p_rational      -> p_rational
      _/_         : p_rational, Nat0       -> p_rational
      _gt_,_ge_,_lt_,_le_: p_rational, p_rational -> bool
 eqns forall n, n1, n2: Nat, m1, m2: Nat0, r,r1: p_rational
    ofsort p_rational
      00 = 0/1;
      real(n) = n/1;
      id(r) = r;
      (n1 * N(m2)) = (n2 * N(m1)) => n1/m1 = n2/m2;

      (n1/m1) + (n2/m2) = ((n1*N(m2))+(n2*N(m1)))/(m1*m2);
      (n1/m1) - (n2/m2) = ((n1*N(m2))-(n2*N(m1)))/(m1*m2);
      (n1/m1) * (n2/m2) = (n1*n2)/(m1*m2);
      r * n = r * real(n);   n * r = real(n) * r;
      r + n = r + real(n);   n + r = real(n) + r;
      r - n = r - real(n);   n - r = real(n) - r;
      r ** 0 = 1/1;          r ** succ(n) = r * (r ** n);

      (n1/m1) / m2 = n1/(m1*m2);

    ofsort bool
      (n1/m1) gt (n2/m2) = (n1 * N(m2)) gt (n2 * N(m1));
      (n1/m1) lt (n2/m2) = (n1 * N(m2)) lt (n2 * N(m1));
      r ge r1 = not(r lt r1);
      r le r1 = not(r gt r1);
endtype

type addresses is NaturalNumber
 sorts address
 opns add: Nat -> address
    _ eq _: address, address -> bool
 eqns forall n1,n2: nat
    ofsort bool
      add(n1) eq add(n2) = n1 eq n2;
endtype

type SDUs is ExtendedNaturals
 sorts SDU, data_fragment
 opns fragment: Nat -> data_fragment

      null_SDU: -> SDU
      add_fragment: data_fragment, SDU -> SDU
      first_fragment: SDU -> data_fragment
      rest_fragment: SDU -> SDU
      num_of_fragments: SDU -> Nat
 eqns forall d1,d2:data_fragment, s:sdu
    ofsort SDU
      rest_fragment(add_fragment(d1,null_SDU)) = null_SDU;
      rest_fragment(add_fragment(d2,add_fragment(d1,s))) =
                  add_fragment(d2,rest_fragment(add_fragment(d1,s)));
    ofsort data_fragment
      first_fragment(add_fragment(d1,null_SDU)) = d1;
      first_fragment(add_fragment(d2,add_fragment(d1,s))) =
                  first_fragment(add_fragment(d1,s));
    ofsort Nat
      num_of_fragments(null_SDU) = 0;
      num_of_fragments(add_fragment(d1,s)) = 1 + num_of_fragments(s);
endtype

type Cells is SDUs, Addresses
 sorts cell, cell_type
 opns add_cell, data_cell, last_cell: -> cell_type
      add_cell: nat, address, address -> cell
      data_cell,last_cell: nat, nat, data_fragment -> cell
          (* First nat.-  The identification of the SDU
             Second nat.- Rank, in the SDU, of the 40-byte fragment
                  carried *)
      (* SELECTORS: *)
      the_cell_type: cell -> cell_type
      the_SDU_id, the_cell_rank: cell -> nat
      the_src_add, the_dest_add: cell -> address
      the_data_fragment: cell -> data_fragment
 eqns forall d:data_fragment, ct:cell_type, ns,nc: nat, a1,a2:address
    ofsort cell_type
      the_cell_type(add_cell(ns,a1,a2)) = add_cell;
      the_cell_type(data_cell(ns,nc,d)) = data_cell;
      the_cell_type(last_cell(ns,nc,d)) = last_cell;
    ofsort nat
      the_SDU_id(add_cell(ns,a1,a2)) = ns;
      the_SDU_id(data_cell(ns,nc,d)) = ns;
      the_SDU_id(last_cell(ns,nc,d)) = ns;
      the_cell_rank(add_cell(ns,a1,a2)) = 0;
      the_cell_rank(data_cell(ns,nc,d)) = nc;
      the_cell_rank(last_cell(ns,nc,d)) = nc;
    ofsort address
      the_src_add(add_cell(ns,a1,a2)) = a1;
      the_dest_add(add_cell(ns,a1,a2)) = a2;
      the_src_add(data_cell(ns,nc,d)) = add(0); (* It has no sense *)
      the_dest_add(data_cell(ns,nc,d)) = add(0); (* It has no sense *)
      the_src_add(last_cell(ns,nc,d)) = add(0); (* It has no sense *)
      the_dest_add(last_cell(ns,nc,d)) = add(0); (* It has no sense *)
    ofsort data_fragment
      the_data_fragment(data_cell(ns,nc,d)) = d;
      the_data_fragment(last_cell(ns,nc,d)) = d;
      the_data_fragment(add_cell(ns,a1,a2)) = fragment(0); (* It has no sense *)
endtype
```

Appendix B: Tick-Tock, Simple Version

```
specification system[Se_SAP,R_SAP](n_users:nat):noexit

  timedef
    sort p_rational
    opns 0 for 00  + for +  > for gt
  enddef

  library Boolean, ExtendedNaturals, PositiveFracNumber,
          addresses, SDUs, Cells endlib

behaviour
  hide Ss_SAP, Sr_SAP in
    (Sender [Se_SAP,Ss_SAP](n_users) ||| Receiver [R_SAP,Sr_SAP](n_users))
  |[Ss_SAP,Sr_SAP]|
    Service [Ss_SAP,Sr_SAP]
where
  process Sender [Se_SAP,Ss_SAP](users:nat):noexit:=
  hide sdu_num in
    SAP_attenders [Se_SAP,Ss_SAP,sdu_num](users)
  || ( SDU_identification [Se_SAP,sdu_num](0)
    ||| Order_of_emission [Ss_SAP](0))
  where
  type SenderParameters is PositiveFracNumber
    opns cell_min_delay, Beta : -> p_rational
    eqns ofsort p_rational
      cell_min_delay = ?????;
      Beta = ?????;
  endtype

  process SDU_identification [Se_SAP,sdu_num](sdu_id:Nat):noexit:=
    Se_SAP?sa,da:address?d:SDU; sdu_num!sdu_id;
                  SDU_identification [Se_SAP,sdu_num](sdu_id + 1)
  endproc (* SDU_identification *)

  process Order_of_emission [Ss_SAP](n:Nat):noexit:=
    Ss_SAP?c:cell [n = the_SDU_id(c)];
    ( [the_cell_type(c) = last_cell] -> Order_of_emission [Ss_SAP](n + 1)
    [] [the_cell_type(c) = add_cell] -> Order_of_emission [Ss_SAP](n)
    [] [the_cell_type(c) = data_cell] -> Order_of_emission [Ss_SAP](n))
  endproc (* Order_of_emission *)

  process SAP_attenders [Se_SAP,Ss_SAP,sdu_num](users:Nat):noexit:=
    [users gt 0] -> ( SAP_attender [Se_SAP,Ss_SAP,sdu_num](users)
                  ||| SAP_attenders [Se_SAP,Ss_SAP,sdu_num](users – 1))
  where
  process SAP_attender [Se_SAP,Ss_SAP,sdu_num](user:Nat):noexit:=
      protocol_entity [Se_SAP,Ss_SAP,sdu_num](user)
  |[Se_SAP,Ss_SAP]|
      ( Minimal_delay_between_primitives[Se_SAP](user)
      |[Se_SAP]| minimal_delay_before_cell_emission[Se_SAP,Ss_SAP])
  endproc (* SAP_attender *)

  process protocol_entity[Se_SAP,Ss_SAP,sdu_num](user:Nat):noexit:=
    Se_SAP!add(user)?da:address?d:SDU; sdu_num?sdu_id:Nat;
    emission_of_cells[Ss_SAP] (sdu_id,0,add(user),da,d)
      >> protocol_entity[Se_SAP,Ss_SAP,sdu_num](user)
  where
  process emission_of_cells[Ss_SAP]
                  (sdu_id,rank:Nat, sa,sd:address, d:SDU):exit:=
    ( [rank eq 0] -> Ss_SAP!add_cell(sdu_id,sa,sd);
      emission_of_cells[Ss_SAP](sdu_id,rank+1,sa,sd,d)
    [][rank gt 0] -> [num_of_fragments(d) gt 1] ->
      Ss_SAP!data_cell(sdu_id,rank,first_fragment(d));
      emission_of_cells[Ss_SAP](sdu_id,rank+1,sa,sd,rest_fragment(d))
    [][rank gt 0] -> [num_of_fragments(d) eq 1] ->
      Ss_SAP!last_cell(sdu_id,rank,first_fragment(d)); exit)
  endproc
  endproc

  process Minimal_delay_between_primitives[Se_SAP](user:Nat):noexit:=
    choice t:p_rational []
    Se_SAP!add(user)?da:address?d:SDU [t ge (Beta * num_of_fragments(d))] {t};
        Minimal_delay_between_primitives[Se_SAP](user)
  endproc (* Minimal_delay_between_primitives *)

  process minimal_delay_before_cell_emission[Se_SAP,Ss_SAP]:noexit:=
    Se_SAP?sa,da:address?d:SDU; time_constraint[Ss_SAP] (00)
      >> minimal_delay_before_cell_emission[Se_SAP,Ss_SAP]
  where
  process time_constraint[Ss_SAP](t:p_rational):exit:=
    choice t0:p_rational []
    Ss_SAP?c:cell [(t+t0) ge (cell_min_delay * the_cell_rank(c))] {t0};
    ( [the_cell_type(c) = add_cell] -> time_constraint[Ss_SAP](t+t0)
    [] [the_cell_type(c) = data_cell] -> time_constraint[Ss_SAP](t+t0)
    [] [the_cell_type(c) = last_cell] -> exit)
  endproc (* time_constraint *)
  endproc (* minimal_delay_before_cell_emission *)
  endproc (* SAP_attenders *)
  endproc (* Sender *)


  process Receiver [R_SAP,Sr_SAP](users:nat):noexit:=
  hide rec in
    Emission_points [R_SAP,rec](users)
  |[rec]|
    (      SDU_reassembler [Sr_SAP,rec](0,0,add(0),add(0),null_SDU)
    |[Sr_SAP]| deafness [Sr_SAP]
    |[Sr_SAP]| prolongation_of_deafness [Sr_SAP])
  where
  type ReceiverParameters is PositiveFracNumber
    opns W, deafness: -> p_rational
    eqns ofsort p_rational
      W = ?????;
      deafness = ?????;
  endtype

  process Emission_points [R_SAP,rec](users:Nat):noexit:=
    [users gt 0] -> ( Emission_point [R_SAP,rec](users)
            ||| Emission_points [R_SAP,rec](users – 1))
  where
  process Emission_point [R_SAP,rec](user:Nat):noexit:=
          FIFO_exit [R_SAP,rec](user,no_SDUs)
    |[R_SAP]| Minimal_delay_between_primitives[R_SAP](user)
  where
  type Queue_of_SDUs is SDUs, Addresses
    sorts SDUs_queue
    opns no_SDUs: -> SDUs_queue
      add : SDU, address, SDUs_queue -> SDUs_queue
      first_sdu : SDUs_queue -> SDU
      first_add : SDUs_queue -> address
      rest : SDUs_queue -> SDUs_queue
      length: SDUs_queue -> nat
    eqns forall s1,s2: SDU, a1,a2: address, q:SDUs_queue
      ofsort SDU
      first_sdu(add(s1,a1,no_SDUs)) = s1;
      first_sdu(add(s1,a1,add(s2,a2,q))) = first_sdu(add(s2,a2,q));
      ofsort address
      first_add(add(s1,a1,no_SDUs)) = a1;
      first_add(add(s1,a1,add(s2,a2,q))) = first_add(add(s2,a2,q));
      ofsort SDUs_queue
      rest(no_SDUs) = no_SDUs;
      rest(add(s1,a1,no_SDUs)) = no_SDUs;
      rest(add(s1,a1,add(s2,a2,q))) = add(s1,a1,rest(add(s2,a2,q)));
      ofsort nat
      length(no_SDUs) = 0;
      length(add(s1,a1,q)) = succ(length(q));
  endtype

  process FIFO_exit [R_SAP,rec](user:Nat,q:SDUs_queue):noexit:=
    rec!add(user)?sa:address?s:SDU; FIFO_exit [R_SAP,rec](user,add(s,sa,q))
  [] [length(q) gt 0] -> R_SAP!add(user)!first_add(q)!first_sdu(q);
                  FIFO_exit [R_SAP,rec](user,rest(q))
  endproc (* FIFO_exit *)

  process Minimal_delay_between_primitives[R_SAP](user:nat):noexit:=
    choice t:p_rational []
    R_SAP!add(user)?sa:address?d:SDU [t ge (W * num_of_fragments(d))] {t};
        Minimal_delay_between_primitives[R_SAP](user)
  endproc (* Minimal_delay_between_primitives *)
  endproc (* Emission_point *)
  endproc (* Emission_points *)

  process SDU_reassembler [Sr_SAP,rec]
          (sdu_id,rank:nat,sa,da:address,s:SDU): exit:=
    Sr_SAP?c:cell;
    ( [(the_SDU_id(c) eq sdu_id) and (the_cell_rank(c) eq rank)] ->
      ( [the_cell_type(c) = add_cell] ->
            SDU_reassembler [Sr_SAP,rec]
              (sdu_id,rank+1,the_src_add(c),the_dest_add(c),s)
      [] [the_cell_type(c) = data_cell]->
            SDU_reassembler [Sr_SAP,rec]
              (sdu_id,rank+1,sa,da,add_fragment(the_data_fragment(c),s))
      [] [the_cell_type(c) = last_cell] ->
            rec!da!sa!add_fragment(the_data_fragment(c),s);
            SDU_reassembler [Sr_SAP,rec] (sdu_id+1,0,sa,da,null_SDU)
      )
    [] [not((the_SDU_id(c) eq sdu_id) and (the_cell_rank(c) eq rank))] ->
      ( [the_cell_type(c) = add_cell]->
          SDU_reassembler [Sr_SAP,rec]
                  (the_SDU_id(c),(1 of Nat),
                  the_src_add(c),the_dest_add(c),null_SDU)
      [] [the_cell_type(c) = data_cell]->
          SDU_reassembler [Sr_SAP,rec](0,0,add(0),add(0),null_SDU)
      [] [the_cell_type(c) = last_cell] ->
          SDU_reassembler [Sr_SAP,rec](0,0,add(0),add(0),null_SDU)
      )
    )
  endproc (* SDU_reassembler *)
```

```
process deafness [Sr_SAP]:noexit:=
  Sr_SAP?c:cell; i {deafness}; deafness [Sr_SAP]
endproc (* deafness *)

process prolongation_of_deafness [Sr_SAP]:noexit:=
  Sr_SAP?c:cell;
  ((choice t:p_rational [] [t ge deafness]-> i;  exit {t})
                     >> prolongation_of_deafness [Sr_SAP])
  endproc (* prolongation_of_deafness *)
endproc (* Receiver *)

process Service [Ss_SAP,Sr_SAP]:noexit:=
 hide delivery in
      (        delayer[Ss_SAP,delivery](0)
     |[delivery]| immediate_delivery[delivery,Sr_SAP])
  |[Ss_SAP,delivery]|
      (        isochronism [Ss_SAP]
     |||       Spacing_between_deliveries[delivery]
     |[delivery]| FIFO_ordering[delivery](0))
 where
 type ServiceParameters is PositiveFracNumber
   opns spacing_between_deliveries, service_period,
      transmission_delay_min, transmission_delay_max: -> p_rational
    eqns ofsort p_rational
       spacing_between_deliveries = ?????;
       service_period = spacing_between_deliveries +  ?????;
        (* service_period >= spacing_between_deliveries *)
       transmission_delay_min = ?????;
       transmission_delay_max = transmission_delay_min + ?????;
        (* transmission_delay_max >= transmission_delay_min *)
 endtype

process delayer [Ss_SAP,delivery](n_order:nat):noexit:=
 Ss_SAP?c:cell; (   delay [delivery](n_order,c)
          ||| delayer [Ss_SAP,delivery](n_order+1))
 where
  process delay [delivery](n_order:nat,c:cell):noexit:=
    choice delay:p_rational [] [delay ge transmission_delay_min]->
               [delay le transmission_delay_max]-> i;
      (choice deliv_time: p_rational [] [deliv_time ge delay] ->
         delivery!n_order!c {deliv_time}; stop)
    endproc (* delay *)
 endproc (* delayer *)

process immediate_delivery[delivery,Sr_SAP]:noexit:=
 delivery?n:Nat?c:Cell;
  ( Sr_SAP!c {00}; immediate_delivery[delivery,Sr_SAP]
  [] immediate_delivery[delivery,Sr_SAP])
 endproc (* immediate_delivery *)

process isochronism [Ss_SAP]:noexit:=
 choice n:nat [] Ss_SAP?c:cell {(n+1) * service_period};
                       isochronism[Ss_SAP]
 endproc

process Spacing_between_deliveries[delivery]: noexit:=
 choice t:p_rational [] [t ge spacing_between_deliveries]->
      delivery?n:nat?c:cell {t}; Spacing_between_deliveries[delivery]
 endproc

process FIFO_ordering[delivery](n_order:nat):noexit:=
  delivery!n_order?c:cell; FIFO_ordering[delivery](n_order+1)
 endproc
endproc (* Service *)
endspec
```

Appendix C: Tick-Tock, Full Version

```
specification system[Se_SAP,R_SAP](n_users:nat):noexit

  timedef
    sort p_rational
    opns 0 for 00  + for +  > for gt
  enddef

  library Boolean, ExtendedNaturals, PositiveFracNumber,
          addresses, SDUs, Cells endlib

behaviour
  hide Ss_SAP, Sr_SAP in
    (Sender [Se_SAP,Ss_SAP](n_users) ||| Receiver [R_SAP,Sr_SAP](n_users))
  |[Ss_SAP,Sr_SAP]|
    Service [Ss_SAP,Sr_SAP]
where
type Parameters is PositiveFracNumber
  opns max_period, min_period,
    transmission_delay_min, transmission_delay_max: -> p_rational
  eqns ofsort p_rational
    min_period = ????;
    max_period = min_period + ????;
      (* max_period >= min_period *)
    transmission_delay_min = ????;
    transmission_delay_max = transmission_delay_min + ????;
      (* transmission_delay_max >= transmission_delay_min *)
endtype

process Sender [Se_SAP,Ss_SAP](users:nat):noexit:=
hide sdu_num in
    SAP_attenders [Se_SAP,Ss_SAP,sdu_num](users)
  || (    SDU_identification [Se_SAP,sdu_num](0)
       |[Se_SAP]| global_rate_control [Se_SAP] (no_prim)
       |||    Order_of_emission [Ss_SAP](0))
where
type SenderParameters is PositiveFracNumber
  opns cell_min_delay, Beta : -> p_rational
  eqns ofsort p_rational
    cell_min_delay = ????;
    Beta = ????;
endtype

process SDU_identification [Se_SAP,sdu_num](sdu_id:Nat):noexit:=
  Se_SAP?sa,da:address?d:SDU; sdu_num!sdu_id;
                    SDU_identification [Se_SAP,sdu_num](sdu_id + 1)
endproc (* SDU_identification *)

process Order_of_emission [Ss_SAP](n:Nat):noexit:=
Ss_SAP?c:cell [n = the_SDU_id(c)];
  ( [the_cell_type(c) = last_cell] -> Order_of_emission [Ss_SAP](n + 1)
   [] [the_cell_type(c) = add_cell] -> Order_of_emission [Ss_SAP](n)
   [] [the_cell_type(c) = data_cell] -> Order_of_emission [Ss_SAP](n))
endproc (* Order_of_emission *)

type primitive_history is PositiveFracNumber
  sorts prim_history
  opns no_prim: -> prim_history
    store: Nat, p_rational, prim_history -> prim_history
    num_of_blocks_in_a_period: p_rational, prim_history -> nat
  eqns forall t,t1:p_rational, n,n1:Nat, h: prim_history
    ofsort Nat
    num_of_blocks_in_a_period(t,no_prim) = 0;
    t ge t1  => num_of_blocks_in_a_period(t,store(n1,t1,h)) =
                      n1 + num_of_blocks_in_a_period(t-t1,h);
    t lt t1  => num_of_blocks_in_a_period(t,store(n1,t1,h)) = n1;
endtype

process global_rate_control [Se_SAP](h:prim_history):noexit:=
choice t:p_rational []
Se_SAP?sa,da:address?d:SDU [enable_time(t,num_of_fragments(d),h)] {t};
    global_rate_control [Se_SAP](store(num_of_fragments(d),t,h))
where
type global_rate is primitive_history
  opns enable_time: p_rational, Nat, prim_history -> bool
    period: -> p_rational
    max_blocks: -> Nat0
  eqns forall t:p_rational, n:Nat, h: prim_history
    ofsort bool
    (n gt N(max_blocks)) =>
        enable_time(t,n,h) = t ge ((n/max_blocks)*period);
    (n le N(max_blocks)) and (t ge period) => enable_time(t,n,h) = true;
    (n le N(max_blocks)) and (t lt period) =>
        enable_time(t,n,h) =
          num_of_blocks_in_a_period(period-t,h) le (N(max_blocks) - n);
    ofsort Nat0
    max_blocks = ????;
endtype
endproc (* global_rate_control *)
```

```
process SAP_attenders [Se_SAP,Ss_SAP,sdu_num](users:Nat):noexit:=
  [users gt 0] -> (  SAP_attender [Se_SAP,Ss_SAP,sdu_num](users)
           ||| SAP_attenders [Se_SAP,Ss_SAP,sdu_num](users - 1))
where
 process SAP_attender [Se_SAP,Ss_SAP,sdu_num](user:Nat):noexit:=
   protocol_entity [Se_SAP,Ss_SAP,sdu_num](user)
 |[Se_SAP,Ss_SAP]|
   (        Minimal_delay_between_primitives[Se_SAP](user)
    |[Se_SAP]| individual_rate_control[Se_SAP](user,no_prim)
    |[Se_SAP]| minimal_delay_before_cell_emission[Se_SAP,Ss_SAP])
 endproc (* SAP_attender *)

 process protocol_entity[Se_SAP,Ss_SAP,sdu_num](user:Nat):noexit:=
   Se_SAP!add(user)?da:address?d:SDU; sdu_num?sdu_id:Nat;
     emission_of_cells[Ss_SAP] (sdu_id,0,add(user),da,d)
       >> protocol_entity[Se_SAP,Ss_SAP,sdu_num](user)
   where
   process emission_of_cells[Ss_SAP]
                 (sdu_id,rank:Nat, sa,sd:address, d:SDU):exit:=
   ( [rank eq 0] -> Ss_SAP!add_cell(sdu_id,sa,sd);
       emission_of_cells[Ss_SAP](sdu_id,rank+1,sa,sd,d)
   [][rank gt 0] -> [num_of_fragments(d) gt 1] ->
       Ss_SAP!data_cell(sdu_id,rank,first_fragment(d));
       emission_of_cells[Ss_SAP](sdu_id,rank+1,sa,sd,rest_fragment(d))
   [][rank gt 0] -> [num_of_fragments(d) eq 1] ->
       Ss_SAP!last_cell(sdu_id,rank,first_fragment(d)); exit)
   endproc
 endproc

 process Minimal_delay_between_primitives[Se_SAP](user:Nat):noexit:=
   choice t:p_rational []
   Se_SAP!add(user)?da:address?d:SDU [t ge (Beta * num_of_fragments(d))] {t};
     Minimal_delay_between_primitives[Se_SAP](user)
 endproc (* Minimal_delay_between_primitives *)

 process minimal_delay_before_cell_emission[Se_SAP,Ss_SAP]:noexit:=
   Se_SAP?sa,da:address?d:SDU; time_constraint[Ss_SAP] (00)
     >> minimal_delay_before_cell_emission[Se_SAP,Ss_SAP]
   where
   process time_constraint[Ss_SAP](t:p_rational):exit:=
   choice t0:p_rational []
   Ss_SAP?c:cell [(t+t0) ge (cell_min_delay * the_cell_rank(c))] {t0};
   ( [the_cell_type(c) = add_cell] -> time_constraint[Ss_SAP](t+t0)
   [] [the_cell_type(c) = data_cell]-> time_constraint[Ss_SAP](t+t0)
   [] [the_cell_type(c) = last_cell] -> exit)
   endproc (* time_constraint *)
 endproc (* minimal_delay_before_cell_emission *)

 process individual_rate_control [Se_SAP](user:Nat, h:prim_history):noexit:=
 choice t:p_rational []
 Se_SAP!add(user)?da:address?d:SDU [enable_time(t,num_of_fragments(d),h)] {t};
   individual_rate_control [Se_SAP](user,store(num_of_fragments(d),t,h))
 where
 type individual_rate is primitive_history
   opns enable_time: p_rational, Nat, prim_history -> bool
     period: -> p_rational
     max_blocks: -> Nat0
   eqns forall t:p_rational, n:Nat, h: prim_history
     ofsort bool
     (n gt N(max_blocks)) =>
         enable_time(t,n,h) = t ge ((n/max_blocks)*period);
     (n le N(max_blocks)) and (t ge period) =>
         enable_time(t,n,h) = true;
     (n le N(max_blocks)) and (t lt period) =>
         enable_time(t,n,h) =
           num_of_blocks_in_a_period(period-t,h) le (N(max_blocks) - n);
     ofsort Nat0
     max_blocks = ????;
 endtype
 endproc (* individual_rate_control *)
endproc (* SAP_attenders *)
endproc (* Sender *)

process Receiver [R_SAP,Sr_SAP](users:nat):noexit:=
hide rec,delete,to in
  (     Emission_points [R_SAP,rec,delete](users)
  |[rec]| SDU_reassembler [Sr_SAP,rec,to](0,0,add(0),add(0),null_SDU))
  ||
  (        Cells_acceptation_ctrl[R_SAP,Sr_SAP,rec,delete,to]
  |[Sr_SAP,to]| timer [Sr_SAP,to] (no_cell))
where
type ReceiverParameters is PositiveFracNumber
  opns W, max_waiting: -> p_rational
  eqns ofsort p_rational
    W = ????;
    max_waiting = ????;
endtype
```

```
process Emission_points [R_SAP,rec,delete](users:Nat):noexit:=
[users gt 0] -> ( Emission_point [R_SAP,rec,delete](users)
            ||| Emission_points [R_SAP,rec,delete](users - 1))
where
process Emission_point [R_SAP,rec,delete](user:Nat):noexit:=
        FIFO_exit [R_SAP,rec,delete](user,no_SDUs)
|[R_SAP]| Minimal_delay_between_primitives[R_SAP](user)
where
 type Queue_of_SDUs is SDUs, Addresses
  sorts SDUs_queue
  opns no_SDUs: -> SDUs_queue
    add : SDU, address, SDUs_queue -> SDUs_queue
    first_sdu : SDUs_queue -> SDU
    first_add : SDUs_queue -> address
    rest : SDUs_queue -> SDUs_queue
    length: SDUs_queue -> nat
  eqns forall s1,s2: SDU, a1,a2: address, q:SDUs_queue
    ofsort SDU
    first_sdu(add(s1,a1,no_SDUs)) = s1;
    first_sdu(add(s1,a1,add(s2,a2,q))) = first_sdu(add(s2,a2,q));
    ofsort address
    first_add(add(s1,a1,no_SDUs)) = a1;
    first_add(add(s1,a1,add(s2,a2,q))) = first_add(add(s2,a2,q));
    ofsort SDUs_queue
    rest(no_SDUs) = no_SDUs;
    rest(add(s1,a1,no_SDUs)) = no_SDUs;
    rest(add(s1,a1,add(s2,a2,q))) = add(s1,a1,rest(add(s2,a2,q)));
    ofsort nat
    length(no_SDUs) = 0;
    length(add(s1,a1,q)) = succ(length(q));
 endtype

 process FIFO_exit [R_SAP,rec,delete](user:Nat,q:SDUs_queue):noexit:=
    rec!add(user)?sa:address?s:SDU;
                FIFO_exit [R_SAP,rec,delete](user,add(s,sa,q))
  [] [length(q) gt 0] -> R_SAP!add(user)!first_add(q)!first_sdu(q);
                FIFO_exit [R_SAP,rec,delete](user,rest(q))
  [] delete!add(user); FIFO_exit [R_SAP,rec,delete](user,rest(q))
  endproc (* FIFO_exit *)

 process Minimal_delay_between_primitives[R_SAP](user:nat):noexit:=
  choice t:p_rational []
  R_SAP!add(user)?sa:address?d:SDU [t ge (W * num_of_fragments(d))] {t};
            Minimal_delay_between_primitives[R_SAP](user)
  endproc (* Minimal_delay_between_primitives *)
 endproc (* Emission_point *)
endproc (* Emission_points *)

process SDU_reassembler [Sr_SAP,rec,to]
            (sdu_id,rank:nat,sa,da:address,s:SDU):noexit:=
 Sr_SAP?c:cell;
   ( [(the_SDU_id(c) eq sdu_id) and (the_cell_rank(c) eq rank)] ->
     ( [the_cell_type(c) = add_cell] ->
            SDU_reassembler [Sr_SAP,rec,to]
            (sdu_id,rank+1,the_src_add(c),the_dest_add(c),s)
     [] [the_cell_type(c) = data_cell]->
            SDU_reassembler [Sr_SAP,rec,to]
            (sdu_id,rank+1,sa,da,add_fragment(the_data_fragment(c),s))
     [] [the_cell_type(c) = last_cell] ->
         rec!da!sa!add_fragment(the_data_fragment(c),s);
         SDU_reassembler [Sr_SAP,rec,to] (sdu_id+1,0,sa,da,null_SDU)
     )
   [] [not((the_SDU_id(c) eq sdu_id) and (the_cell_rank(c) eq rank))] ->
     ( [the_cell_type(c) = add_cell]->
            SDU_reassembler [Sr_SAP,rec,to]
                (the_SDU_id(c),(1 of Nat),
                the_src_add(c),the_dest_add(c),null_SDU)
     [] [the_cell_type(c) = data_cell]->
            SDU_reassembler [Sr_SAP,rec,to](0,0,add(0),add(0),null_SDU)
     [] [the_cell_type(c) = last_cell] ->
            SDU_reassembler [Sr_SAP,rec,to](0,0,add(0),add(0),null_SDU)
     )
   )
  [] to; SDU_reassembler [Sr_SAP,rec,to](0,0,add(0),add(0),null_SDU)
 endproc (* SDU_reassembler *)

process Cells_acceptation_ctrl[R_SAP,Sr_SAP,rec,delete,to]:noexit:=
    evacuation_control[R_SAP,Sr_SAP,rec,delete,to](0,0,0,no_waiting,00)
 where
  type evacuation_control is ReceiverParameters, Cells
  sorts waiting_times
  opns no_waiting: -> waiting_times
    add_wt: p_rational, address, Nat, waiting_times -> waiting_times
    oldest_add: waiting_times -> address
    oldest_blocks: waiting_times -> Nat
    remove_oldest: waiting_times -> waiting_times
    remove_oldest: address, waiting_times -> waiting_times
    is_waiting: address, waiting_times -> bool
    first_condition_holds: Nat -> bool
    second_condition_holds: p_rational, waiting_times -> bool
    next_cell_SDU_id, next_cell_rank: cell, Nat, Nat -> Nat
    total_blocks: cell, Nat, Nat, Nat -> Nat
    total_b_after_to: Nat, Nat -> Nat
    max_blocks: -> Nat
  eqns forall a,a1,a2: address, p,p1,t: p_rational, h: waiting_times,
        si1,si2,r1,r2,n,n1: Nat, d: data_fragment
    ofsort Address
    oldest_add(no_waiting) = add(0);
    oldest_add(add_wt(p,a,n,no_waiting)) = a;
    oldest_add(add_wt(p,a,n,add_wt(p1,a1,n1,h))) =
                        oldest_add(add_wt(p1,a1,n1,h));
    ofsort Nat
    oldest_blocks(no_waiting) = 0;
    oldest_blocks(add_wt(p,a,n,no_waiting)) = n;
    oldest_blocks(add_wt(p,a,n,add_wt(p1,a1,n1,h))) =
                        oldest_blocks(add_wt(p1,a1,n1,h));
    next_cell_SDU_id(add_cell(si1,a1,a2),si2,r2) = si1;
    next_cell_SDU_id(data_cell(si1,r1,d),si2,r2) = si1;
    next_cell_SDU_id(last_cell(si1,r1,d),si2,r2) = si1+1;
    next_cell_rank(add_cell(si1,a1,a2),si2,r2) = 1;
    next_cell_rank(data_cell(si1,r1,d),si1,r1) = r1+1;
    (si1 ne si2) or (r1 ne r2) =>
        next_cell_rank(data_cell(si1,r1,d),si1,r1) = 0;
    next_cell_rank(last_cell(si1,r1,d),si2,r2) = 0;
    total_blocks(add_cell(si1,a1,a2),si2,0,n) = n;
    (r2 gt 0) => total_blocks(add_cell(si1,a1,a2),si2,r2,n) = n-(r2-1);
    total_blocks(data_cell(si1,r1,d),si1,r1,n) = n+1;
    (si1 ne si2) or (r1 ne r2) =>
        total_blocks(data_cell(si1,r1,d),si2,r2,n) = n-(r2-1);
    total_blocks(last_cell(si1,r1,d),si1,r1,n) = n+1;
    (si1 ne si2) or (r1 ne r2) =>
        total_blocks(last_cell(si1,r1,d),si2,r2,n) = n-(r2-1);
    total_b_after_to(0,n) = n;
    r1 gt 0 => total_b_after_to(r1,n) = n - (r1-1);
    max_blocks = ????;
    ofsort waiting_times
    remove_oldest(no_waiting) = no_waiting;
    remove_oldest(add_wt(p,a,n,no_waiting)) = no_waiting;
    remove_oldest(add_wt(p,a,n,add_wt(p1,a1,n1,h))) =
            add_wt(p,a,n,remove_oldest(add_wt(p1,a1,n1,h)));
    remove_oldest(a,no_waiting) = no_waiting;
    is_waiting(a,h) => remove_oldest(a,add_wt(p,a1,n,h)) =
            add_wt(p,a1,n,remove_oldest(a,h));
    not(is_waiting(a,h)) => remove_oldest(a,add_wt(p,a,n,h)) = h;
    not(is_waiting(a,h)) => remove_oldest(a,h) = h;
    ofsort bool
    is_waiting(a,no_waiting) = false;
    is_waiting(a,add_wt(p,a,n,h)) = true;
    not(a eq a1)  => is_waiting(a,add_wt(p,a1,n,h)) = is_waiting(a,h);
    first_condition_holds(n) = n lt max_blocks;
    second_condition_holds(t,no_waiting) = false;
    second_condition_holds(t,add_wt(p,a,n,no_waiting)) =
                        (t - p) gt max_waiting;
    second_condition_holds(t,add_wt(p,a,n,add_wt(p1,a1,n1,h))) =
            second_condition_holds(t,add_wt(p1,a1,n1,h));
 endtype

 process evacuation_control[R_SAP,Sr_SAP,rec,delete,to]
  (sdu_id,rank,blocks:Nat, w:waiting_times, current_time:p_rational):
                                        noexit:=
 choice t:p_rational []
 let current_time:p_rational = current_time +t in
 ( Sr_SAP?c:cell [first_condition_holds(blocks)] {t};
    evacuation_control[R_SAP,Sr_SAP,rec,delete,to]
        (next_cell_SDU_id(c,sdu_id,rank),next_cell_rank(c,sdu_id,rank),
        total_blocks(c,sdu_id,rank,blocks),w,current_time)
 [] Sr_SAP?c:cell [   not(first_condition_holds(blocks))
            and second_condition_holds(current_time,w)] {t};
    delete!oldest_add(w) {00};
    evacuation_control[R_SAP,Sr_SAP,rec,delete,to]
        (next_cell_SDU_id(c,sdu_id,rank),next_cell_rank(c,sdu_id,rank),
        total_blocks(c,sdu_id,rank,blocks)-oldest_blocks(w),
        remove_oldest(w),current_time)
 [] R_SAP?da,sa:address?d:SDU {t};
    evacuation_control[R_SAP,Sr_SAP,rec,delete,to]
        (sdu_id,rank,blocks-num_of_fragments(d),remove_oldest(da,w),
        current_time)
 [] rec?da,sa:address?d:SDU {t};
    evacuation_control[R_SAP,Sr_SAP,rec,delete,to]
        (sdu_id,rank,blocks,add_wt(current_time,da,num_of_fragments(d),w),
        current_time)
 [] to {t}; evacuation_control[R_SAP,Sr_SAP,rec,delete,to]
        (sdu_id+1,0,total_b_after_to(rank,blocks),w,current_time)
 )
 endproc (* evacuation_control *)
endproc (* Cells_acceptation_ctrl *)
```

```
type Cells_timing_information is Parameters
 sorts cells_timing
 opns no_cell: -> cells_timing
    cell: p_rational, cells_timing -> cells_timing

    time_out: p_rational, cells_timing -> bool

    exist_min_delay: p_rational, cells_timing -> bool
    num_of_cells_in_period: p_rational, cells_timing -> Nat
    min_delay: p_rational, cells_timing -> p_rational
    min_delay_aux: p_rational, nat, cells_timing -> p_rational
    marge: -> p_rational
    29: -> Nat
 eqns forall t,t1,m: p_rational, h: cells_timing, n: Nat
    ofsort bool
    not(exist_min_delay(t,h)) => time_out(t,h) = (t ge (max_period+marge));
    exist_min_delay(t,h) and
    (((2*min_delay(t,h)) + (3 * marge)) gt (max_period+marge)) =>
                        time_out(t,h) = (t ge (max_period+marge));
    exist_min_delay(t,h) and
    (((2*min_delay(t,h)) + (3 * marge)) le (max_period+marge)) =>
            time_out(t,h) = t ge ((2*min_delay(t,h)) + (3 * marge));
    t gt (29*min_period)  => exist_min_delay(t,h) = false;
    t le (29*min_period)  =>
        exist_min_delay(t,h) =
                    num_of_cells_in_period((29*min_period)-t,h) gt 1;
    ofsort p_rational
    marge = transmission_delay_max - transmission_delay_min;
    exist_min_delay(t,h) =>
      min_delay(t,h) =
        min_delay_aux(00,num_of_cells_in_period((29*min_period)-t,h),h);
    not(exist_min_delay(t,h)) => min_delay(t,h) = max_period;
                            (* It does not matter *)
    (n gt 1) and (m le t) =>
        min_delay_aux(m,n,cell(t,h)) = min_delay_aux(m,n-1,h);
    (n gt 1) and (m gt t) =>
        min_delay_aux(m,n,cell(t,h)) = min_delay_aux(t,n-1,h);
    (n le 1) => min_delay_aux(m,n,h) = m;
    ofsort Nat
    num_of_cells_in_period(t,no_cell) = 0;
    t ge t1 => num_of_cells_in_period(t,cell(t1,h)) =
                    1 + num_of_cells_in_period(t-t1,h);
    t lt t1 => num_of_cells_in_period(t,cell(t1,h)) = 1;
    29  = (2 * 10) + 9;
endtype

process timer [Sr_SAP,to](h:cells_timing):noexit:=
 choice t: p_rational [] Sr_SAP?c:cell {t};
 (let h:cells_timing = cell(t,h) in
  (choice t:p_rational []
    [time_out(t,h)]-> to {t}; stop) [> timer [Sr_SAP,to](h))
endproc (* timer *)
endproc (* Receiver *)

process Service [Ss_SAP,Sr_SAP]:noexit:=
choice crash_time:p_rational [] i;(* The time of crash is non-deterministic *)
 (hide delivery in
    (          delayer[Ss_SAP,delivery](0)
    |[delivery]| immediate_delivery[delivery,Sr_SAP])
  |[Ss_SAP,delivery]|
    (          adaptative_isochronism [Ss_SAP]
    |||     Spacing_between_deliveries[delivery]
    |[delivery]| FIFO_ordering[delivery](0))
 ) [> i {crash_time};
    (choice delay_time:p_rational []
     i; exit {delay_time} >> Service [Ss_SAP,Sr_SAP])
where
type ServiceParameters is PositiveFracNumber
 opns spacing_between_deliveries, initial_service_period : -> p_rational
 eqns ofsort p_rational
    spacing_between_deliveries = ????;
    initial_service_period = spacing_between_deliveries + ????;
        (* initial_service_period >= spacing_between_deliveries *)
endtype

process delayer [Ss_SAP,delivery](n_order:nat):noexit:=
 Ss_SAP?c:cell; (   delay [delivery](n_order,c)
        ||| delayer [Ss_SAP,delivery](n_order+1))
 where
 process delay [delivery](n_order:nat,c:cell):noexit:=
  choice delay:p_rational [] [delay ge transmission_delay_min]->
                [delay le transmission_delay_max]-> i;
    (choice deliv_time: p_rational [] [deliv_time ge delay] ->
      delivery!n_order!c {deliv_time}; stop)
 endproc (* delay *)
endproc (* delayer *)


process immediate_delivery[delivery,Sr_SAP]:noexit:=
 delivery?n:Nat?c:Cell;
  ( Sr_SAP!c {00}; immediate_delivery[delivery,Sr_SAP]
  [] immediate_delivery[delivery,Sr_SAP])
endproc (* immediate_delivery *)

process adaptative_isochronism [Ss_SAP]:noexit:=
 hide new_period in
    emission_ctrl [Ss_SAP,new_period] (initial_service_period)
  || adaptation_of_period [Ss_SAP,new_period] (initial_service_period)
where
 process emission_ctrl [Ss_SAP,new_period](service_period:p_rational):noexit:=
    (choice n:nat [] Ss_SAP?c:cell {(n+1) * service_period};
            emission_ctrl[Ss_SAP,new_period] (service_period))
 [] new_period?service_period:p_rational;
            emission_ctrl[Ss_SAP,new_period] (service_period)
 endproc (* emission_ctrl *)

 process adaptation_of_period [Ss_SAP,new_period]
                (service_period:p_rational):noexit:=
  emissions_in_a_sequence[Ss_SAP,new_period](service_period,00,0)
  >> accept service_period:p_rational, e1:Nat in
    emissions_in_a_sequence[Ss_SAP,new_period](service_period,00,0)
    >> accept service_period:p_rational, e2:Nat in
     emissions_in_a_sequence[Ss_SAP,new_period](service_period,00,0)
     >> accept service_period:p_rational, e3:Nat in
      ( [(e1 eq 10) or (e2 eq 10) or (e3 eq 10)]->
         adaptation_of_period [Ss_SAP,new_period] (service_period)
      [][not((e1 eq 10) or (e2 eq 10) or (e3 eq 10))]->
         (let prop_used:Nat = e1 + e2 + e3 in
          new_period!updated_period(prop_used,service_period);
          adaptation_of_period [Ss_SAP,new_period]
                (updated_period(prop_used,service_period)))
      )
 where

 type adaptation_coefficients is Parameters
  opns 20,25,30: -> Nat
    updated_period: Nat, p_rational -> p_rational
  eqns forall n:nat, p:p_rational
    ofsort p_rational
    (n le 10) and ((p*2) le max_period) => updated_period(n,p) = p*2;
    (n le 10) and ((p*2) gt max_period) =>
                    updated_period(n,p) = max_period;
    (n gt 10) and (n le 20) and ((p*(3/2)) le max_period) =>
                    updated_period(n,p) = p*(3/2);
    (n gt 10) and (n le 20) and ((p*(3/2)) gt max_period) =>
                    updated_period(n,p) = max_period;
    (n gt 20) and (n le 25) and ((p*(6/5)) le max_period) =>
                    updated_period(n,p) = p*(6/5);
    (n gt 10) and (n le 25) and ((p*(6/5)) gt max_period) =>
                    updated_period(n,p) = max_period;
    (n gt 25) => updated_period(n,p) = p;
    ofsort Nat
    20 = 2 * 10; 25 = (2 * 10) + 5; 30 = 3 * 10;
 endtype

 process emissions_in_a_sequence[Ss_SAP,new_period]
        (service_period:p_rational,seq_time:p_rational,e:Nat):
                            exit(p_rational,Nat):=
  choice t:p_rational []
  ( [t gt ((10 * service_period) - seq_time)]->
    ( [e eq 10] -> [(service_period/2) ge min_period]->
            new_period!service_period/2 {t}; exit(service_period/2,e)
    [] [e eq 10] -> [(service_period/2) lt min_period]->
            new_period!min_period {t}; exit(min_period,e)
    [] [e lt 10] -> exit(service_period,e))
  [] Ss_SAP?c:cell {t}; emissions_in_a_sequence[Ss_SAP,new_period]
                (service_period,seq_time+t,e+1))
  endproc (* emissions_in_a_sequence *)
 endproc (* adaptation_of_period *)
endproc (* adaptative_isochronism *)

process Spacing_between_deliveries[delivery]: noexit:=
 choice t:p_rational [] [t ge spacing_between_deliveries]->
        delivery?n:nat?c:cell {t}; Spacing_between_deliveries[delivery]
endproc

process FIFO_ordering[delivery](n_order:nat):noexit:=
 delivery!n_order?c:cell; FIFO_ordering[delivery](n_order+1)
endproc
endproc (* Service *)
endspec
```

References

[DB88]   Doug Dykeman and Werner Bux. **Analysis and Tuning of the FDDI Media Access Protocol**. *IEEE Journal on Selected Areas in Communications*, 6(6):997–1010, July 1988.

[ISO89]   ISO. **Fiber Distributed Data Interface (FDDI) – Token Ring Media Access Control (MAC)**. IS 9314–2, ISO, 1989.

[MFOV93]   C. Miguel, A. Fernández, J.M. Ortuño, and L. Vidaller. **A LOTOS based Performance Evaluation Tool**. *Special Issue of "Computer Networks and ISDN Systems" on: TOOLS FOR FDTs*, 25(7):791–813, 1993.

[MFV93a]   C. Miguel, A. Fernández, and L. Vidaller. **Extending LOTOS towards Performance Evaluation**. In M. Diaz and R. Groz, editors, *Formal Description Techniques V*, pages 103–118, Amsterdam, 1993. Elsevier Science Publisher (North Holland).

[MFV93b]   C. Miguel, A. Fernández, and L. Vidaller. **LOTOS Extended with Probabilistic Behaviors**. *Formal Aspects of Computing. The international journal of Formal Methods*, 5(3):253–281, 1993.

[MO91]   C. Miguel and J.M. Ortuño. **TOPOSIM Tool. User's Guide**. Technical report, Department of Telematic Systems Engineering (DIT)., ETSI Telecomunicacion, UPM. Ciudad Universitaria – 28040 Madrid, Sept. 12. 1991.

[Pro90]   Protocol Engines Inc. **XTP Protocol Definition. Revision 3.5**. Technical Report 90-120, Protocol Engines Inc, 10 September 1990.

Standardisation

The involvement of OSI95 in the standardisation activities started at the very beginning of the project and aimed first at creating new work items in ISO and in ETSI.

The OSI95 proposed new work item was accepted by ISO in July 1991 and confirmed by a three months ballot with three countries casting a negative vote.

Several meetings were necessary to define the content of the ECFF project (Enhanced Communication Function and Facilities) and in the meantime, several groups (including OSI95) submitted proposals for transport service types and transport protocols, the US delegation submitting diverging documents.

It is in February 1993, at the ISO meeting in London, that important technical discussions took place and that generic facilities such as the Enhanced QoS and several facilities for the transport service were well identified and discussed. This London meeting showed a more positive view for the progress of the ECFF work and a new project on an Enhanced Transport Service is very likely to be favourably considered soon.

The work in ETSI, where the new work item proposed by OSI95 has also been accepted, faced the difficulty that the transport layer sits above the B-ISDN/ AAL which are still under development. The ETSI/NA5 delegates appreciate the approach followed by OSI95 but intend to stay at the network level leaving the transport to "End-terminal protocol" group which presently has less expertise in the B-ISDN.

It is clear that the creation of new work items in ISO and ETSI was clearly a success criterion for the OSI95 project.

To complete these standardisation activities, it is interesting to indicate that ORL activity in the ATM Forum has contributed to the adoption of AAL Type 5 (SEAL) and to the allocation of an essential bit in the header of the ATM cell.

OSI95 Contributions to ISO/IEC and ETSI

Guy Leduc [1,2], André Danthine [2] and Helmut Leopold [3]
[1] Research Associate of the National Fund for Scientific Research (Belgium)
[2] Université de Liège, Institut d'Electricité Montefiore, B 28, B-4000 Liège 1, Belgium
[3] Alcatel Austria Forschungszentrum, Ruthnergasse 1-7, A - 1210 Vienna, Austria

This paper summarizes the activities that were carried out within the standardization bodies ISO/IEC and ETSI during the two-year period of the OSI95 project. However it focuses mainly on the OSI95 contributions.

Keywords: OSI95, standardization, ISO, ETSI

1 Activities within ISO/IEC JTC1/SC6

The Sub-Committee 6 (SC6) of the Joint Technical Committee 1 (JTC1) of ISO/IEC is responsible for the four lower layers of the OSI Reference Model (RM) of ISO ([ISO 7498]). It is composed of several Working Groups (WG). In the sequel we will essentially report our contributions within the WG4 (responsible for the Transport Layer of the OSI RM) and within the newly created ECFF activity. ECFF stands for "Enhanced Communications Functions and Facilities for OSI lower layers". It is a joint activity of several WGs of SC6 dealing with new services and protocols for the lower layers. Its creation during the two-year period of OSI95 results from the acceptation by SC6 of a New Project (NP) on this topic. This will be further detailed in the sequel.

The University of Liège (ULg) attended four SC6 meetings: Berlin (July 1991), Paris (February 1992), San Diego (July 1992) and London (February 1993). The University of Lancaster attended the Paris and London meetings, and Bull, the Paris meeting.

1.1 The Berlin Meeting

At the WG4 meeting in Berlin, ULg presented a contribution ([ISO 4N695]) in which the need for new high speed transport service and protocol was emphasized. This was substantiated by referring to the OSI95 project whose main objectives were presented.

This presentation generated a lot of interest in WG4. Among them, the US representatives tried to see how far our proposal was different from their proposal of 8 NWIs that they were going to present ([ISO 6N6618]) and whose titles are listed hereafter:

- Enhanced Transport Mechanisms Guidelines,
- Additional Data link Layer Service Functions,
- Additional Data link Layer Protocol Functions and Procedures

- Additional Network Layer Service Functions,
- Additional Network Layer Protocol Functions and Procedures
- Additional Transport Layer Service Functions,
- Additional Transport Layer Protocol Functions and Procedures
- Group NSAP addressing

At the plenary session of SC6 two NP (New Project) proposals were finally accepted and were proposed to JTC1 for a three-month ballot:

- Enhanced transport mechanism guidelines ([ISO 1N1515])
- Group NSAP addressing (for multicast operation) ([ISO 1N1514])

The first one was aligned with the OSI95 proposal.

These two NPs were accepted, so that the first success criterion of OSI95 was achieved. Note however that three countries had cast a negative vote on these proposals: UK, Canada and Germany.

1.2 The Paris Meeting

This is during this meeting that the ECFF term was defined as the new title of the first accepted NP.

During the resolution of comments on the two accepted NPs, it already appeared clearly that some countries (mainly Canada and Japan) were against any new design of OSI service or protocol unless they were justified by clear application requirements and were "compatible" with existing OSI services and protocols. It was proposed that the base text of the ECFF NP reflect a top-down approach which favours a clear identification of requirements coming from the applications prior to any proposal for the addition of new facilities or functions in the existing standards. More precisely, the following 4-step procedure was accepted:

1. Identification of application requirements that have implications for the OSI lower layer services and protocols
2. The examination of existing OSI lower layer services and protocols to determine if the requirements identified in (1) can be met by existing or pending OSI standards
3. In those cases in which requirements cannot be met by existing or pending OSI standards, the consideration of proposals for modification/extension of existing OSI services and protocols
4. In those cases in which neither of the approaches outlined in (2) and (3) is sufficient to satisfy identified requirements, the consideration of proposals for new services and/or protocol.

The OSI95 contribution from Lancaster ([ISO P19]) dealing with application requirements was adapted and included in the draft base text of the ECFF NP ([ISO 6N7068]). From then on, this text was also referred to as the "ECFF guidelines" document.

The US and ULg (via the Belgian National Body: IBN) also input contributions on the transport layer ([ISO 6N7070, ISO 6N7312]). In [ISO 6N7312] we gave some general considerations on the design of new transport service and protocol.

1.3 The San Diego Meeting

The first draft of the ECFF guidelines document [ISO 6N7309] was highly criticized. A new structure was set up with several editors responsible for specific parts of the document. ULg was appointed editor for the section on Quality of Service (QoS).

Besides the guidelines, several contributions focused on specific lower layers.

As regards the transport layer, the US National Body (ANSI) proposed HSTS ([ISO SD4.026]), based on XTP [PEI XTP], where several transport service types are defined. ANSI also submitted another transport service definition ([ISO 6N7445]) which was a multipeer extension of the Connection-mode Transport Service ([ISO 8072]). The corresponding two protocols, HSTP and TP5, were submitted as experts contributions ([ISO 6N7429]). HSTP is based on the transport sub-protocol of XTP. TP5 is a multipeer version of TP4. The US Delegation was obviously divided into these two groups. The second group, say the TP5 group, seemed to limit the objective of the new transport service and protocol to the multipeer issue and tries to foster a simple enhancement of TP4. On the other hand, the first group, say the XTP group, was against that approach because they feared that another important ECFF issue, viz. the performance, would not be taken into account.

ULg submitted, via the Belgian National Body (IBN), a second OSI95 contribution ([ISO 6N7323]) in which several types of transport services were presented informally and in LOTOS. Finally, ULg submitted a third document ([ISO 6N7759]) which focused on one of these service types: the Connection-mode Transport Service of OSI95. An important contribution of this paper was dealing with enhanced QoS semantics and negotiation.

A liaison was also established between SC21 and SC6 on QoS ([ISO 6N7788]). SC21 started an NP on QoS, and SC6 will be in a position to contribute to this project in the context of ECFF.

1.4 The London Meeting

The second draft of the ECFF guidelines document ([ISO 6N7788]) was again criticized by several National Bodies that found it mainly too superficial and also inadequate for preparing specific new project proposals (NPs) on ECFF. It was decided at the end of the meeting not to request a new version of this guidelines document.

On the other hand, for the first time since the beginning of the ECFF activity, important technical discussions took place on all the possible enhancements of the transport service: QoS, multipeer, fast connect, graceful release, out-of-band, request-response service and acknowledged connectionless-mode service. We summarize the main contributions hereafter.

1.4.1 QoS

The main contribution was from ULg on QoS enhancement of OSI95 ([ISO 4L19]). It generated many discussions and was considered as very relevant for the future work on ECFF. It was decided that SC6 will circulate this document for comments to its members (Annex A of [ISO 6N7989]) and to SC21 ([ISO 6N8010]) via the already existing liaison on QoS. SC21 has indeed started working on a QoS Framework to be considered in a future version of the OSI Reference Model. Our OSI95 contributions might be included in the draft text of this framework.

1.4.2 Multipeer

The main contributions on multipeer were from US experts and French experts. There are two proposals from US experts: the multicast stream of HSTS ([ISO 6/4N806]) and another more ambitious multicast connection-mode service with an Active Group Integrity (AGI) concept, as discussed in [ISO 6N7445]. The French ETS [ISO 6N7883] also includes multicast, but goes beyond the classical $1 \rightarrow N$ connection, and considers also other multipeer connections like $N \rightarrow 1$ or $N \rightarrow N$. The US also proposed amendments to the connectionless-mode transport service and protocol ([6/4N807, 6/4N808]).

1.4.3 Fast Connect

The main contributions were from US experts and CCITT. Some US experts support a fast connect facility as described in HSTS ([ISO 6/4N806]). CCITT has expressed the need for improving the OSI efficiency ([ISO 6N7856]) and has also considered the fast connect as a way to do so. However CCITT would prefer a simpler solution which consists in allowing a larger data field in the CONNECT.request primitives in all OSI layers.

1.4.4 Graceful Release

The main contribution was from OSI95 by ULg ([ISO 6/4N821, ISO 6/4N822]). We argued in favour of this facility because it enhances the reliability of the connection and also improves the efficiency of the connection release. It was decided that SC6 will circulate to its members excerpts from these documents for comments (Annex F of [ISO 6N7989]).

1.4.5 Out-of-band

The main contribution was also from OSI95 by ULg ([ISO 6/4N821, ISO 6/4N822]). The term out-of-band, as defined therein, was criticized by some experts who found it completely misleading. According to them, what we have defined is in fact an in-band facility because the TSDUs which are transferred that way are associated with a connection. An out-of-band TSDU transfer, they say, is by definition completely independent of any connection. If so, then other Transport Service types may be used for this purpose without having to define such an 'out-of-band' facility. After the CCITT representative pointed out that the terms 'in-band' and 'out-of-band signalling' have always been used with confusion, the convener proposed to give to this new facility a temporary neutral name: the ABC_service. It was decided that SC6 will circulate to its members excerpts from these documents for comments (Annex E of [ISO 6N7989]).

1.4.6 Request-Response Service

The main contributions were from US experts and from OSI95 by ULg. US experts proposed a request-response service type in their HSTS ([ISO 6/4N806]), and ULg proposed another one in OSI95 ([ISO 6/4N821]). They are roughly the same but in OSI95, we do not use transaction_ids in primitive parameters: association endpoints are better suited for this purpose. It was decided that SC6 will circulate to its members excerpts from these documents for comments (Annex D of [ISO 6N7989]).

1.4.7 Acknowledged Connectionless-mode Service

The main contributions were from OSI95 by ULg ([ISO 6/4N821]) and French experts in ETS ([ISO 6N7883]). The main discussion was on the difference between the semantics of the confirmation primitive in this service and the confirmation primitive in the request-response service. Having defined both services in OSI95, we explained the difference between these semantics, and we argued that both services may be useful. It was decided that SC6 will circulate to its members excerpts from these documents for comments (Annex B of [ISO 6N7989]).

1.4.8 Other Discussions

In addition to ULg's contributions, Lancaster presented the results of their evaluation of the OSI95 transport service ([6/4N823]).

The following NPs on transport layer enhancements were proposed, but rejected during the meeting since unanimity was required:

- Three variants of an NP on the connection-mode transport service: the first one proposed an amendment to ISO 8072 to add new service facilities; the second one proposed a completely different standard; and the third one proposed an amendment to ISO 8072 to add only multicast.
- One NP on the connection-mode transport protocol.
- Two NPs on multicast extensions of the connectionless transport service and protocol.

An NP on a new transport protocol was considered premature. And some National Bodies (e.g. Japan, Germany, UK) stated that issuing NPs on the transport service was also premature, and that more technical discussions were needed.

It is not excluded however that a National Body will make another NP proposal for ballot before the next SC6 meeting in Seoul (October 93).

1.5 Conclusion

The London Meeting has clearly shown a change of attitude of delegations which, in the past meetings, were very reluctant to see the progression of the ECFF work.

It seems that most of the delegations that were attending the London meeting are ready to move to a new project on at least an Enhanced Transport Service Definition.

If such a new project is approved before the Seoul meeting it seems reasonable to believe that many sessions in Seoul may be devoted to in-depth discussions of the various facilities that have been already tackled in London.

The problem of an Enhanced Transport Protocol seems to be less mature, many delegations believing that it would be better to work for a while on the Transport Service Definition before beginning to work on the Transport Protocol.

It is worth mentioning that the WG1 of SC21, being responsible for the maintenance of the OSI RM, also plays an important role in two matters: the QoS framework and the multipeer communications. As regards QoS the SC21 framework might benefit from our work. It is too early to draw conclusions on this liaison. As regards multipeer SC21 might re-activate its project.

1.6 Other ISO/IEC Activities

The work on the LOTOS methodology and its related specification of the ISO transport protocol TP4 ([ISO 8073]) in LOTOS also led ULg to send 10 new defect reports on ISO 8073. These defects were found as a result of a development of a LOTOS description of TP4 in the context of OSI95.

2 Activities within ETSI/NA5

The ETSI Technical Committee (TC) ETSI/NA is responsible for Network Aspects (NA) in general, and the Sub-Technical Committee (STC) ETSI/NA5 particularly takes account of Broadband aspects; i.e. it is responsible for the standardization of B-ISDN within Europe. Included in the scope of the latter is ATM technology and its related Physical Layer, and the ATM Adaptation Layer (AAL) protocols which sit on top of the ATM Layer.

The aim of participation of OSI95 within ETSI/NA5 was the introduction of a new work item on "high performance transport protocols" within B-ISDN. The STC ETSI/NA5 was selected as this is the group within ETSI where most of the expertise of the new evolving area of broadband communication is located. However, first of all it was necessary to clarify the assignment of responsibility for such work within ETSI.

To reach this goal it was necessary to consider the following important points: (i) the need for clarification of the relationship between the OSI Protocol Reference Model (PRM) and the B-ISDN/PRM; (ii) the necessity for layers (protocols) above the B-ISDN/AAL; and (iii) the requirements to be taken into account, when specifying the B-ISDN/AAL interface to the layers above the AAL.

The OSI95 contribution at the The Hague meeting in April 1991 ([ETSI 91/4]) identified that there is a need for layers above the B-ISDN/AAL, and provided the rationale for higher layers protocols within B-ISDN. Contribution at the Karlsruhe meeting in February 1991 ([ETSI 91/119]) shows the relationship between the AAL type 4 service and the OSI Data Link service and asks for studies on high bit rate protocols up to and including OSI Layer 4.

From the discussion stimulated by these contributions, a conclusion was reached on the need to check the relationship of the B-ISDN/PRM and the OSI/PRM on the one hand and to study the implications of higher layer protocols in the AAL on the other hand. Therefore it was proposed to introduce a new study item ([ETSI 91/172]) in NA5 (ETSI/NA5 meeting in Karlsruhe, February 1991, joint meeting AAL/PRS).

It is important to note that other members had also identified the need to develop protocols above the AAL and thus have supported the intentions raised by OSI95. This need was conformed by the Working Party ETSI/NA5/AAL. As a conclusion of the discussion around this issue, the terms of reference for the study point "Specification of the AAL interface to Higher Layers" ([ETSI 91/147-7]) was worked out at the NA5 meeting in the The Hague meeting (April 1991, joint meeting AAL/PRS).

Due to the fact that neither an ETSI working party nor a CCITT SG XVIII Working Party had already been charged to do this work (ETSI/NA5 Meeting held in Corfu, October 1991), a Liaison Statement to NA ([ETSI 91/135-2]) and a CCITT contribution ([ETSI 91/135-3]) for the CCITT SG XVIII meeting in Melbourne, 1991, was produced on the issue of "Future work on the layers above the AAL" (ETSI/NA5 meeting held in Corfu, October 1991).

Lastly it should be noted that, besides this activity, stimulated mainly by OSI95, proposals to commence study on new "high performance AAL types" to support data transfer were made (e.g. by British Telecom and France Telecom). However, no agreement on this subject has been reached so far.

2.1 Main Achievements within ETSI

A decision was made (ETSI/NA5 meeting in Montpellier, January 1992) that the Technical Committee ETSI/TE (Terminal Equipment) should be responsible for the standardization activity on transport and higher layers. However, since the expertise on B-ISDN is currently under ETS/NA5 it was decided that the latter group should start with this activity. As soon as ETSI/TE is able to take over the work, they will consider this activity. Within NA5, the AAL Working Party is dealing with this aspect.

Thus the main goal, to create a new work item which will deal with "high performance protocols above the B-ISDN/AAL" was fulfilled, and the responsible working group identified.

2.2 Current Situation on the "Offered Service of B-ISDN AAL for Connection Oriented (CO) Data Transport"

Although the basic goal, to identify a group within ETSI responsible for the specification of protocols above the B-ISDN/AAL, was achieved, there is still a very important open issue. Up to now, there is no clear understanding on the kind of service (in the OSI sense) the B-ISDN/AAL for data communication (i.e. U-plane of AAL-3/4 and 5) should offer to the higher layers, in order to allow existing protocols to be used.

ETSI/NA5 proposes to define the service in such a way that a well known OSI lower layer service is offered. Candidate services are, for example, Connection-oriented Data Link service (CCITT recommendation X.212), Connection-oriented Network Service (CCITT recommendation X.213) and Connection-oriented Transport Service TP0 (CCITT recommendation X.214). ETSI/NA5 is inviting the ETSI/STC's NA1, NA2 and TE to comment on this suggestion (ETSI/NA5 meeting in Berne, March 1993).

At the ETSI/NA5 meeting in Chester (May 1993), a discussion on the U-plane service to be provided at the AAL-SAP reflected a clear preference to provide the OSI Transport Service as defined in CCITT recommendation X.214.

However, as already mentioned above, no clear decision was made on this subject and further study is required. ETSI/NA5 is now awaiting technical contributions on this issue to fulfil the mandate of this new working field on "high performance protocol above the B-ISDN/AAL".

2.3 Interest of ETSI in the OSI95 Approach

The ETSI/NA5 delegates appreciate the approach followed by OSI95. They found it very important to identify first of all the communication requirements of future applications to influence the protocol specifications at an very early point in time. Furthermore, the idea to initiate forthcoming work on protocols above the AAL was welcomed.

However, the main problem within ETSI/NA5 was an organisational one. On the one hand, since the ETSI/NA5 group (as originally constituted) is only responsible for network issues, they did not want to extend their scope to include "end-terminal protocols". On the other hand, there is no other group where such an extensive knowledge on broadband communication is available.

References

[ETSI 91/4]	H. Leopold, R. Peschi, **B-ISDN and Higher Layer Protocols**, ETSI/NA5, TD 91/4, 12-03-1991, (OSI95/ELIN/A8/02/TN/P/V1).
[ETSI 91/119]	H. Leopold, R. Peschi, **New Work Item : "High Speed Transport Protocols"**, ETSI/NA5, TD 91/119, 14-02-1991 (OSI95/ELIN/A8/01/TN/P/V2).
[ETSI 91/135-2]	Source: ETSI/NA5/AAL drafting group, **Future work on the layers above the AAL - Liaison Statement to ETSI/NA**, ETSI/NA5, TD 91/135-Annex 2, Corfu, 14-18 October 1991, (OSI95/ELIN/A8/03/TN/P/V1).
[ETSI 91/135-3]	Source: ETSI/NA5/AAL drafting group, **Future work on the layers above the AAL - Contribution to CCITT SG XVIII**, ETSI/NA5, TD 91/135-Annex 3, Corfu, 14-18 October 1991 (OSI95/ELIN/A8/03/TN/P/V1).
[ETSI 91/147-7]	Source: ETSI/NA5/AAL-PRS drafting group, **Terms of Reference for the study point: Specification of the AAL interface to Higher Layers**, ETSI/NA5, TD 91/147-Annex 7, The Hague, 15-19 April 1991 (OSI95/ELIN/A8/02/TN/P/V1).
[ETSI 91/172]	Source: ETSI/NA5/AAL-PRS drafting group, **Proposal for a new work item: Reliable high speed data transfer in B-ISDN**, ETSI/NA5, TD 91/172, Karlsruhe, 18-22 Feb. 1991, and TD 91/173-Annex 4 (OSI95/ELIN/A8/01/TN/P/V2).

[ISO 7498]   Source: ISO TC97/SC16/WG1, **Information Processing Systems - Open Systems Interconnection - Basic Reference Model**, IS 7498, 1984.

[ISO 8072]   ISO-TC97/SC6/WG4, **Information Processing Systems - Open Systems Interconnection - Transport Service Definition**, IS 8072, 1986-06-15.

[ISO 8073]   ISO/IEC JTC1/SC6/WG4, **Information Processing Systems - Open Systems Interconnection - Connection oriented transport protocol specification (3nd edition)**, IS 8073, 1992-12-15.

[ISO 1N1514]   Source: ISO/IEC JTC1, **Proposal for a New Work Item: Group NSAP addressing**, ISO/IEC JTC1/N1514, 1991-08-16.

[ISO 1N1515]   Source: ISO/IEC JTC1, **Proposal for a New Work Item: Enhanced transport mechanisms**, ISO/IEC JTC1/N1515, 1991-08-16.

[ISO 6N6618]   Source: USA, **US Contribution - Eight New Work Item Proposals (NPs)**, ISO/IEC JTC1/SC6/N6618, 1991-05-15.

[ISO 6N7068]   Source: USA, **Draft proposed text for guidelines for enhanced transport mechanisms**, ISO/IEC JTC1/SC6/N7068, 1991-11-28.

[ISO 6N7070]   Source: USA, **Proposed working draft for high-speed transport service definition**, ISO/IEC JTC1/SC6/N7070, 1991-11-28.

[ISO 6N7309]   Source: USA, **Draft guidelines for ECFF**, ISO/IEC JTC1/SC6/N7309, 1992-05-11.

[ISO 6N7312]   Source: Belgium, **Belgian National Body Contribution - Issues Surrounding the Specification of High-Speed Transport Service and Protocol**, ISO/IEC JTC1/SC6/N7312, 11-05-1992, 46 p. (OSI95/ULg/A/15/TR/P/V2, January 1992)

[ISO 6N7323]   Source: Belgium, **Belgian National Body Contribution - Four Types of Enhanced Transport Services and their LOTOS Specifications**, ISO/IEC JTC1/SC6/N7323, 13-05-1992, 58 p. (OSI95/ULg/A/22/TR/P, May 1992)

[ISO 6N7429]   Source: USA, **Contribution on New Transport Protocol Development**, ISO/IEC JTC1/SC6/N7429, 1992-06-30.

[ISO 6N7445]   Source: USA, **USA Contribution on Multipeer Data Transmission Transport Service**, ISO/IEC JTC1/SC6/N7445, 1992-07-13.

[ISO 6N7548]   Source: ISO/IEC JTC1/SC6, **Liaison to SC21 on Enhanced Quality-of-Service (QoS)**, ISO/IEC JTC1/SC6/N7548 Rev.

[ISO 6N7759]   Source: Belgium, **Belgian National Body Contribution - The Enhanced Connection-mode Transport Service of OSI95**, ISO/IEC JTC1/SC6/N7759, 1992-10-14, also in: OSI95/ULg/A/24/TR/P, June 1992

[ISO 6N7788]   Source: Project editor, **Second WD of a type 3 TR - Guidelines for Enhanced Transport Mechanisms**, ISO/IEC JTC1/SC6/N7788, 1992-12-15.

[ISO 6N7856]   Source: CCITT SG VII, **Status of Work on Efficiency of OSI Protocol**, ISO/IEC JTC1/SC6/N7856, 1993-01-11.

[ISO 6N7883]   Source: France (not an AFNOR position), **Specification of ETS, the enhanced transport service**, ISO/IEC JTC1/SC6/N7883, 1993-01-13.

[ISO 6N7989]   Source: ISO/IEC JTC1/SC6, **Request for Comments on "Issues on the Enhanced Communications Functions and Facilities"**, ISO/IEC JTC1/SC6/N7989, 1993-03-17.

[ISO 6N8010]   Source: ISO/IEC JTC1/SC6, **Liaison Statement to ISO/IEC JTC1/SC21 on Qualities of Service (QoS) Work**, ISO/IEC JTC1/SC6/N8010, 1993-03-17.

[ISO 6/4N695]   Source: Belgian Expert, **High Speed Transport Protocol and Service**, ISO/IEC JTC1/SC6/WG4/N695, 1991-05-14.

[ISO 6/4N806]   Source: US expert, **Proposed Draft Text for a High Speed Transport Service (HSTS) Specification**, ISO/IEC JTC1/SC6/WG4/N806, 1992-12-19.

[ISO 6/4N807]   Source: USA, **Proposed Draft Text for an Amendment to ISO/IEC 8602 (Protocol for Providing the Connectionless-mode Transport Service) Covering Multicast Extensions**, ISO/IEC JTC1/SC6/WG4/N807, 1992-12-19.

[ISO 6/4N808]   Source: USA, **Proposed Draft Text for an Amendment to ISO/IEC 8072 (Transport Service Definition) Covering Multicast Extensions to the Connectionless Transport Service**, ISO/IEC JTC1/SC6/WG4/N808, 1992-12-19.

[ISO 6/4N821]   Source: Belgian Expert, **The OSI95 Enhanced Transport Services**, ISO/IEC JTC1/SC6/WG4/N821, 1993-04-19, also in ISO/IEC JTC1/SC6/WG4/4L20, 1993-01-27.

[ISO 6/4N822]   Source: Belgian Expert, **Enhanced Transport Service Definition**, ISO/IEC JTC1/SC6/WG4/N822, 1993-04-19, also in ISO/IEC JTC1/SC6/WG4/4L21, 1993-02-05.

[ISO 6/4N823]   Source: UK Expert, **The OSI95 Transport Service and the New Environment: a contribution to the Discussion on the Enhanced Communications Functions and Facilities**, ISO/IEC JTC1/SC6/WG4/N823, 1993-04-19, also in OSI95/LANC/A8/03/TR/P/V1.

[ISO P19]   Source: UK Expert, **A Continuous Media Orchestration Service**. ISO/IEC JTC1/SC6/Paris-19, Feb. 1992

[ISO SD4.026]   Source: USA, **High-Speed Transport Service Definition (HSTS)**, ISO/IEC JTC1/SC6/WG4/SD4.026, July 1992.

[ISO 4L19]   Source: Belgian Expert, **An Enhancement of the QoS Concept**, ISO/IEC JTC1/SC6/WG4/4L19, 1993-01-26.

[PEI XTP]   Protocol Engines Inc., **XTP Protocol Definition - Revision 3.6**, PEI 92-10, 1900 State Street, Santa Barbara, CA 93101, Jan. 1992.

Publications and Available Reports from OSI95

This paper is the complete list of the publications and available reports which have been produced in the framework of the project OSI95.

Publications of OSI95

B. Alkhechi, Y. Souissi, *Modélisation et analyse de performances du protocole XTP dans un environement FDDI*, in: CFIP 93, Hermes, 1993

Y. Baguette, A. Danthine, *Comparison of TP4, TCP and XTP - Part 1: Connection Management Mechanisms,* in: European Transactions on Telecommunications, vol. 3, N. 5, Sept.-Oct. 1992, 523-533

Y. Baguette, A. Danthine, *Comparison of TP4, TCP and XTP - Part 2: Data Transfer Mechanisms,* in: European Transactions on Telecommunications, vol. 3, N. 6, Nov.-Dec. 1992, 625-636

Y. Baguette, L. Léonard, G. Leduc, A. Danthine, *The OSI95 Connection-Mode Transport Service,* in: The OSI95 Transport Service with Multimedia Support on HSLAN's and B-ISDN, A. Danthine, ed., Springer Verlag, 181-198

Y. Baguette, G. Leduc, *The Connection Release Facilities in the OSI95 Transport Service,* in: The OSI95 Transport Service with Multimedia Support on HSLAN's and B-ISDN, A. Danthine, ed., Springer Verlag, 199-211

Y. Baguette, *The OSI95 Connectionless-Mode Transport Services,* in: The OSI95 Transport Service with Multimedia Support on HSLAN's and B-ISDN, A. Danthine, ed., Springer Verlag, 212-223

J. Boerjan, R. Peschi, *B-ISDN Services and their Relation with the OSI Protocol Reference Model,* in: The OSI95 Transport Service with Multimedia Support on HSLAN's and B-ISDN, A. Danthine, ed., Springer Verlag, 247-259

S. Budkowski, B. Alkhechi, M.L. Benalycherif, P. Dembinski, M. Gardie, E. Lallet, J.-P. Mouchel La Fosse, Y. Souissi, *Formal Specification, Validation and Performance Evaluation of the XTP Protocol,* in : A. Danthine, G. Leduc, P. Wolper, eds, Protocol Specification, Testing and Verification XIII, Liège, May 1993, Elsevier Science Publishers (North Holland), Amsterdam, 1993, 191-206

S. Budkowski, B. Alkhechi, M.L. Benalycherif, P. Dembinski, M. Gardie, E. Lallet, J.P. Mouchel La Fosse, Y. Souissi, *Modelling and Analysing the Xpress Transfer Protocol (XTP),* in: The OSI95 Transport Service with Multimedia Support on HSLAN's and B-ISDN, A. Danthine, ed., Springer Verlag, 20-41

A. Campbell, G. Coulson, F. Garcia, D. Hutchison, *A Continuous Media Transport and Orchestration Service,* in: ACM SIGCOMM'92, Baltimore, USA, October 1992, 99-110

A. Campbell, G. Coulson, F. Garcia, D. Hutchison, *Orchestration Services for Distributed Multimedia Synchronisation,* in :A. Danthine, O. Spaniol, eds., High Performance Networking IV, Liège, December 1992, Elsevier Science Publishers (North Holland), Amsterdam, 1993, 153-168

A. Campbell, G. Coulson, F. Garcia, D. Hutchison, H. Leopold, *Integrated Quality of Service for Multimedia Communications,* in : IEEE INFOCOM'93, San Francisco, April 1993, 732-739

A. Campbell, G. Coulson, F. Garcia, D. Hutchison, *Resource Management in Multimedia Communication Stacks,* in : 4th IEE Conference on Telecommunications, Manchester, April 1993, IEE, London, Conference Publication Number 371, 287-295

A. Campbell, G. Coulson, F. García, D. Hutchison, *Orchestration Services for Distributed Multimedia Synchronisation ,* in: The OSI95 Transport Service with Multimedia Support on HSLAN's and B-ISDN, A. Danthine, ed., Springer Verlag, 82-100

A. Campbell, G. Coulson, F. García, D. Hutchison, *Integrated Quality of Service for Multimedia Communications,* in: The OSI95 Transport Service with Multimedia Support on HSLAN's and B-ISDN, A. Danthine, ed., Springer Verlag, 101-122

G. Coulson, F. Garcia, D. Hutchison, D. Shepherd, *Protocol Support for Distributed Multimedia Applications,* in:R.G. Herrtwich, ed., Network and Operating System Support for Digital Audio and Video, Springer-Verlag, 1992, 45-56

W. Dabbous, *Analysis of a Delay Based Congestion Avoidance Algorithm,* in: A. Danthine, O. Spaniol, eds., High Performance Networking IV, Liège, December 1992, Elsevier Science Publishers (North-Holland), Amsterdam, 1993, 283-298

W. Dabbous, *Analysis of a Delay Based Congestion Avoidance Algorithm,* in: The OSI95 Transport Service with Multimedia Support on HSLAN's and B-ISDN, A. Danthine, ed., Springer Verlag, 280-296

A. Danthine, *A New Transport Protocol for the Broadband Environment,* in: A. Casaca, ed., Broadband Communications, Estoril, January 1992, C-4, Elsevier Science Publishers (North-Holland), Amsterdam, 1992, 337-360

A. Danthine, *Esprit Project OSI95 - New Transport Services for High-Speed Networking,* in: Computer Networks and ISDN Systems 25 (4-5), Elsevier Science Publishers (North-Holland), Amsterdam, Nov. 1992, 384-399.

A. Danthine, Y. Baguette, G. Leduc, L. Léonard, *The OSI95 Connection-mode Transport Service - The Enhanced QoS,* in: A. Danthine, O. Spaniol, eds., High Performance Networking IV, Liège, December 1992, Elsevier Science Publishers (North-Holland), Amsterdam, 1993, 235-252

A. Danthine, O. Bonaventure, G. Leduc, *The QoS Enhancements in OSI95,* in: The OSI95 Transport Service with Multimedia Support on HSLAN's and B-ISDN, A. Danthine, ed., Springer Verlag, 125-150

A. Danthine, *The Networking Environment of the Nineties and the Need for New Standards,* in: The OSI95 Transport Service with Multimedia Support on HSLAN's and B-ISDN, A. Danthine, ed., Springer Verlag, 1-12

P. Dembinski, *Queueing Network Model for Estelle,* in: M. Diaz, R. Groz, eds, Formal Description Techniques V, Perros Guirec, October 1992, C-10, Elsevier Science Publishers (North Holland), Amsterdam, 1993, 73-86

M. De Prycker, R. Peschi, T. Van Landegem, *B-ISDN and the OSI Protocol Reference Model,* in: O. Spaniol, A. Danthine, eds, High Speed Networking III, Berlin, March 1991, Elsevier Science Publishers (North Holland), Amsterdam, 1993, 39-57

A. Fernandez, C. Miguel, L. Vidaller, *Development of Satellite Communication Networks based on LOTOS,* in : R.J. Linn, M.U. Uyar, eds, Protocol Specification, Testing and Verification XII, Lake Buena Vista, June 1992, C-8, Elsevier SciencePublishers (North Holland), Amsterdam, 1992, 179-192

D.J. Greaves, K. Zielinski, *Preliminary performance results for the CBN half-duplex VME stations (VIS),* in : G. Pujolle, R. Puigjaner, eds., IFIP WG 6.4 Workshop on Local Communication Systems, Palma, Spain, Elsevier Science Publishers (North Holland) 1992, 391-406

D.J. Greaves, D. McAuley, *Private ATM Networks,* in : B. Pehrson, P. Gunningberg, S. Pink, eds., Protocols for High Speed Networks III, Stockholm, May 1992, C-9, Elsevier Science Publishers (North Holland)1992, 171-181

D.J. Greaves, D. McAuley, *ATM Network Services for Workstations,* in: The OSI95 Transport Service with Multimedia Support on HSLAN's and B-ISDN, A. Danthine, ed., Springer Verlag, 260-278

E. Lallemand, G. Leduc, *A LOTOS Data Facility Compiler (DAFY) ,* in: K. Parker, G. Rose, eds, Formal Description Techniques IV, Sydney, November 1991, C-2, Elsevier Science Publishers (North Holland), Amsterdam, 1992, 313-327.

E. Lallet, Ch.-A. Lebrun, J.-F. Martin, S. Budkowski, *Un outil de génération automatique de l'environnement d'exécution de spécifications Estelle,* in : O. Rafiq, ed., Colloque Francophone sur l'Ingéniérie des Protocoles (CFIP'91), Hermes, 1991, 277-286

E. Lallet, P. Dembinski, J.P. Mouchel La Fosse, B. Alkhechi, M. Gardie, O. Catrina, S. Budkowski, *Validation of the Xpress Transfer Protocol (XTP),* in: The OSI95 Transport Service with Multimedia Support on HSLAN's and B-ISDN, A. Danthine, ed., Springer Verlag, 42-62

G. Leduc, A. Danthine, *On the Provision of a Fast Connect Facility in a Connection-Mode Transport Service,* in: The OSI95 Transport Service with Multimedia Support on HSLAN's and B-ISDN, A. Danthine, ed., Springer Verlag, 225-238

G. Leduc, A. Danthine, H. Leopold, *OSI95 Contributions to ISO/IEC and ETSI,* in: The OSI95 Transport Service with Multimedia Support on HSLAN's and B-ISDN, A. Danthine, ed., Springer Verlag, 378-387

G. Leduc, L. Léonard, *A Timed LOTOS supporting a dense time domain and including new timed operators,* in: M. Diaz, R. Groz, eds, Formal Description Techniques V, Perros Guirec, October 1992, C-10, Elsevier Science Publishers (North Holland), Amsterdam, 1993, 87-102

G. Leduc, *A Method for Applying LOTOS at an Early Design Stage and its Application to the ISO Transport Protocol,* in: The OSI95 Transport Service with Multimedia Support on HSLAN's and B-ISDN, A. Danthine, ed., Springer Verlag, 151-180

L. Léonard, G. Leduc, A. Danthine, *The Tick-Tock Case Study for the Assessment of Timed FDTs,* in: The OSI95 Transport Service with Multimedia Support on HSLAN's and B-ISDN, A. Danthine, ed., Springer Verlag, 338-352

L. Léonard, *The LOTOS Specification of the Enhanced Transport Service,* in: The OSI95 Transport Service with Multimedia Support on HSLAN's and B-ISDN, A. Danthine, ed., Springer Verlag, 239-244

H. Leopold, A. Campbell, N. Singer, *Towards an Integrated Quality of Service Architecture (QOS-A) for Distributed Multimedia Communications,* in : A. Danthine, O Spaniol, eds, High Performance Networking IV, Liège, December 1992, Elsevier Science Publishers (North Holland), Amsterdam, 1993, 169-182

H. Leopold, A. Campbell, N. Singer, *Will BISDN Services Meet the Needs of Distributed Mutlimedia Communications ?,* in : 4th IEE Conference on Telecommunications, Manchester, April 1993, IEE, London, Conference Publication Number 371, 139-145

H. Leopold, G. Coulson, K. Frimpong-Ansah, D. Hutchinson, N. Singer, *The evolving relationship between OSI and ODP in the new communications environment,* in: 2nd International Conference on Broadband Islands, Athen, Greece, June 14-16, 1993

H. Leopold, A. Campbell, D. Hutchison, N. Singer, *Distributed Multimedia Communication System Requirements* , in: The OSI95 Transport Service with Multimedia Support on HSLAN's and B-ISDN, A. Danthine, ed., Springer Verlag, 64-81

C. Miguel, A. Fernandez, L. Vidaller, *Extending LOTOS towards Performance Evaluation,* in : M. Diaz, R. Groz, eds, Formal Description Techniques V, Perros Guirec, October 1992, C-10, Elsevier Science Publishers (North Holland), Amsterdam, 103-117

C. Miguel, A. Fernandez, L. Vidaller, *LOTOS Extended with Probabilistic Behaviors,* to appear in : C. Jones, J. Cooke, eds, Formal Aspects of Computing (The International Journal of Formal Methods), Springer International, London, 1993

C. Miguel, A. Fernandez, J.M. Ortuno, L. Vidaller, *A LOTOS based Performance Evaluation Tool,* in : Computer Networks and ISDN Systems, Special issue on Tools for FDTs, Elsevier Science Publishers (North Holland), Amsterdam, 1993, N°25, 791-814

C. Miguel, L. Vidaller, A. Fernández, *Extended LOTOS,* in: The OSI95 Transport Service with Multimedia Support on HSLAN's and B-ISDN, A. Danthine, ed., Springer Verlag, 312-337

C. Miguel, L. Vidaller, A. Fernández, *Assessment of Extended LOTOS,* in: The OSI95 Transport Service with Multimedia Support on HSLAN's and B-ISDN, A. Danthine, ed., Springer Verlag, 353-376

G. Votsis, V. Chalkiadakis, *VLSI Support for Transport Protocols* , in: The OSI95 Transport Service with Multimedia Support on HSLAN's and B-ISDN, A. Danthine, ed., Springer Verlag, 297-310

Reports Registered in ETSI

ETSI/NA5/AAL-PRS drafting group (ed. H. Leopold), *Proposal for a new work item: Reliable high speed data transfer in B-ISDN, ETSI/NA5, TD 91/172 and TD 91/173-Annex 4, Karlsruhe, 18-22 February 1991, 1 p. (CP/OSI95/91/10-2), in OSI95/ELIN/A8/01/TN/P/V2, 6-05-1991, 14 p. (CP/OSI95/91/14)*

ETSI/NA5/AAL-PRS drafting group (ed. H. Leopold), *Terms of Reference for the study point : Specification of the AAL interface to Higher Layers, ETSI/NA5, TD 91/147-Annex 7, The Hague, 15-19 April 1991, 1 p. (CP/OSI95/91/16-2), in OSI95 report : OSI95/ELIN/A8/02/TN/P/V1, 24-04-1991, 6 p. (CP/OSI95/91/18)*

ETSI/NA5/AAL drafting group (ed. H. Leopold), *Future work on the layers above the AAL - Liaison Statement to ETSI/NA, ETSI/NA5, TD 91/135-Annex 2, Corfu, 14-18 October 1991, 1 p. (CP/OSI95/91/61-2), in OSI95/ELIN/A8/03/TN/P/V1, 22-10-1991, 5 p. (CP/OSI95/91/61)*

ETSI/NA5/AAL drafting group (ed. H. Leopold), *Future work on the layers above the AAL - Contribution to CCITT SG XVIII, ETSI/NA5, TD 91/135-Annex 3, Corfu, 14-18 October 1991, 1 p. (CP/OSI95/91/61-3), in OSI95/ELIN/A8/03/TN/P/V1, 22-10-1991, 5 p.*

H. Leopold, R. Peschi, *New Work Item : "High Speed Transport Protocols"* , ETSI/NA5, TD 91/119, Karlsruhe, 18-22 Febr. 1991, 14-02-1991, 8 p. (CP/OSI95/91/10) in OSI95/ELIN/A8/01/TN/P/V2, 6-05-1991, 14 p. (CP/OSI95/91/14)

H. Leopold, R. Peschi, *B-ISDN and Higher Layer Protocols*, ETSI/NA5, TD 91/4, The Hague, 15-19 April 1991, 12-03-1991, 2 p. (CP/OSI95/91/16), in OSI95/ELIN/A8/02/TN/P/V1, 24-4-1991, 6 p. (CP/OSI95/91/18)

Reports Registered in ISO

Y. Baguette, L. Léonard, G. Leduc, A. Danthine, *Belgian National Body Contribution - Four Types of Enhanced Transport Services and their LOTOS Specifications*, ISO/IEC JTC1/SC6/N7323, 13-05-1992, 58 p. (OSI95/ULg/A/22/TR/P, May 1992, SART92/12/05)

Y. Baguette, L. Léonard, G. Leduc, A. Danthine, *The OSI95 Enhanced Transport Services*, ISO/IEC JTC1/SC6/WG4 N821, 19-04-1993, 79 p. (R2060/ULg/CIO/DS/P/002/b1, January 1993, SART 93/01/15)

A. Campbell, G. Coulson, F. Garcia, D. Hutchison, H. Leopold, N. Singer, *A Continuous Media Transport and Orchestration Service : a Contribution to the Discussion on the New ISO Work Item*, ISO/IEC JTC1/SC6/Paris 19, 10-1-92, BSI/IST 6/-/2/718, BSI/IST/21/-/5/62, 14 p. (OSI95/LANC/A8/01/TR/P/V1, January 1992, MPG-92-31)

A. Campbell, G. Coulson, F. Garcia, D. Hutchison, H. Leopold, N. Singer, *Key Issues in Distributed Multimedia Communications : a Contribution to Discussion on the New ISO work Item*, ISO/IEC JTC1/SC6/San Diego 8,10-4-92, BSI/IST 6/-/2/738, BSI/IST/21/-/1/5/60, 23 p. (OSI95/LANC/A8/02/TR/P/V1, April 1992, MPG-92-30)

A. Campbell, G. Coulson, F. Garcia, D. Hutchison, H. Leopold, N. Singer, *A Suggested QOS Architecture for Multimedia Communications*, ISO/IEC JTC1/SC21/WGI/N1201, 15-9-92, BSI/IST 21/-/1/5/738, 15 p. (OSI95/LANC/A8/03/TR/P/V1, September 1992, MPG-92-37)

A. Danthine, Y. Baguette, G. Leduc, *Belgian National Body Contribution - Issues Surrounding the Specification of High-Speed Transport Service and Protocol*, ISO/IEC JTC1/SC6/N7312, 11-05-1992, 46 p. (OSI95/ULg/A/15/TR/P/V2, January 1992, SART 92/04/05)

A. Danthine, Y. Baguette, G. Leduc, L. Léonard, *Belgian National Body Contribution - The Enhanced Connection-mode Transport Service of OSI95*, ISO/IEC JTC1/SC6/N7759, 14-10-1992, 16 p. (OSI95/ULg/A/24/TR/P, June 1992, SART92/14/05)

A. Danthine, Y. Baguette, L. Léonard, G. Leduc, *An Enhancement of the QoS Concept*, in ISO/IEC JTC1/SC6/N8010 (Liaison Statement to ISO/IEC JTC 1/SC 21 on Qualities of Service (Qos)), 17-03-93, 16 p.(OSI95/ULg/A/28/TR/P, January 1993, SART 93/02/05)

C. Miguel, A. Fernandez, L. Vidaller, *Contribution to the Enhancements of LOTOS*, in : ISO/IEC JTC 1/SC21/WG1 N1180 (OSI95/DIT/B5/5/TR/P/V3, May 1992, 59 p.)

Reports Available from Alcatel Austria Forschungszentrum (Previously Alcatel Austria ELIN Forschungszentrum)

Contact :   Helmut Leopold, Email : Helmut.Leopold@rcvie.co.at
Alcatel Austria Forschungszentrum
Ruthnergasse 1-7, A - 1210 Vienna, Austria

H. Leopold, G. Blair, A. Campbell, G. Coulson, P. Dark, F. Garcia, D. Hutchison, N. Singer, N.Williams, *Distributed Multimedia Communication System Requirements,* OSI95/Deliverable ELIN-1/P/V4, 21-07-1992, 142 p. (3BY 00082 0000 UPZZA ED.04)

N. Singer, H. Leopold, G. Blair, A. Campbell, G. Coulson, K. Frimpong-Ansah, F. Garcia, D. Hutchison, *OSI/ULA - ODP Architecture Mapping,* OSI95/Deliverable ELIN-3/P/V1, 9-12-1992, 94 p. (3BY 00097 0003 UPZZA ED.01)

N. Singer, H. Leopold, J. Bernadat, *A Study of Distributed Processing Projects,* OSI95/ELIN/D2/05/TR/P/V3, 17-06-1992, 110 p. (3BY 00081 0000 UPZZA ED.03)

H. Leopold, *Overview of the B-ISDN AAL Protocols and Services,* OSI95/ELIN/D3/04/TN/P/V4, 3-12-1992, 13 p. (3BY 00097 0013 UPZZA ED.02)

Reports Available from Alcatel Bell

Contact: J. Boerjan, R. Peschi (Tel. : +32 3 240 40 11)
           Alcatel Bell, Research Centre - RC, F. Wellesplein 1
           B - 2018 Antwerpen, Belgium

J. Boerjan, R. Peschi, *ATM LLC Assessment;* OSI95/Deliverable Bell-1 Version 2/P, 30-10-92, 77 p., Esprit Project OSI95

Reports Available from DIT

Contact :   Prof. L. Vidaller, Email : lvidaller@dit.upm.es
            Departamento de Ingenieria de Sistemas Telematicos (DIT),
            ETSI Telecomunicacion, UPM,
            Ciudad Universitaria, E - 28040 Madrid, Spain

C. Miguel, L. Vidaller, A.Fernandez, *Timed LOTOS definition. LOTOS-T,* OSI95/DIT/B5/5/TR/P/V3, 31-1-1992, 55 p.

C. Miguel, L. Vidaller, A. Fernandez, *LOTOS Extended with Probabilistic, Behaviours,* OSI95/DIT/B5/7/TR/P/V2, 31-1-1992, 32 p.

C. Miguel, L. Vidaller, A. Fernandez, *State of the Art on Timed & Probabilistic, Models,* OSI95/DIT/B5/3/TR/P/V3, 13-11-1991, 32 p.

C. Miguel, L. Vidaller, A. Fernandez, *Extended LOTOS Definition,* OSI95/DIT/B5/8/TR/P/V1 (OSI95/Deliverable DIT-1/R/V1), 31-1-1991, 149 p.

C. Miguel, L. Vidaller, A. Fernandez, *LOTOS-TP assessment,* OSI95/Deliverable DIT-3/P/V1, 5-11-1992, 151 p.

C. Miguel, J.M. Ortuno, *TOPOSIM Tool. User's Guide,* Department of Telematic Systems Engineering (DIT), Internal report, 12-9-1991, 24 p.

Reports Available from INT

Contact :   Professor Stanislaw Budkowski, Email : stan@int-evry.fr
Institut National des Telecommunications (INT),
Systems and Networks Department
F - 91011 Evry Cedex, France

P. Dembinski, *Using QNAP2 to evaluate performance of systems specified in Estelle,*
OSI95/INT/A6/01/TR/P/V3, April 1991, 45p.

E. Lallet, S. Budkowski, *Universal Test Drivers Generator,* OSI95/INT/A3/01/TR/P/V2,
April 1992, 21 p.

B. Alkhechi, M.L. Benalycherif, S. Budkowski, P. Dembinski, M. Gardie, E. Lallet,
J.-P. Mouchel La Fosse, Y. Souissi, *Formal specification, validation and
performance evaluation of the XTP protocol,* OSI95/Deliverable INT-BULL/P/V3,
December 1992, 152p.

Reports Available from Lancaster University

Contact :   Professor David Hutchison, Email: dh@comp.lancs.ac.uk
Computing Department, School of Engineering, Computing and
Mathematical Sciences
Lancaster University, Lancaster LA1 4YR, United Kingdom

G. Blair, D. Hutchison, D. Shepherd, H. Leopold, *State of the Art in Distributed
Multimedia Computing,* OSI95/Deliverable LANC-1/P, MPG-91-20, June 1991, 74 p.

P. Dark, D. Hutchison, D. Shepherd, *Remote Procedure Calls and Distributed
Multimedia Systems,* OSI95/Deliverable LANC-2/P, MPG-91-26, October 1991, 81 p.

J. Berrocal, A. Campbell, G. Coulson, F. Garcia, D. Hutchison, A. Linares,
Assessment of the Synchronisation Functions, OSI95/LANC/C2/05/TR/P/V1, MPG-92-
33, January 1992, 68 p.

J. Boerjan, A. Campbell, G. Coulson, F. Garcia, D. Hutchison, H. Leopold, N. Singer,
The OSI95 Transport Service and the new Environment,
OSI95/Deliverable LANC-3/P, MPG-92-38, December 1992, 107 p.

Reports Available from ORL

Contact :   David J. Greaves, Email : djg@cam-orl.co.uk
Olivetti Research Limited
Old Addenbrookes Site, 24a Trumpington Street
Cambridge CB2 1QA, United Kingdom
Documents are in postscript in /pub/atm on ftp.cam-orl.co.uk

D.J. Greaves, *Analysis of the ORL LLC in an ATM-like environment,*
   OSI95/Deliverable ORL1/P/V3, May 91, 17 p.
D.J. Greaves, *An MSNA segmentation and reassembly ASIC,*
   OSI95/ORL/B3/05/TN/P, May 91, 10 p.
D.J. Greaves, *Specification of a Workstation ATM interface ASIC,*
   OSI95/ORL/B3/07/TN/P, Jul 91, 10 p.
D.J. Greaves, L.J. French, D. Milway, D. McAuley, *Specification of an High
   Performance Workstation ATM Interface,* OSI95/ORL/Deliverable-2/V3, Oct 92, 37 p.
D.J. Greaves, D. McAuley, L.J. French, E. Hyder, *Protocol and Interface for ATM
   LANs',* OSI95/ORL/B3/14/TN/P/V2, Jan 93, 13 p.

Reports Available from ULg

Contact: Professor André Danthine - Email : danthine@vm1.ulg.ac.be
           Université de Liège, Institut d'Electricité Montefiore B28,
           B-4000 Liège, Belgium.

Y. Baguette, *Questions about the XTP Protocol Definition - Revision 3.5 and
   comments on some proposed XTP changes,* OSI95/ULg/A/05/TN, 27-02-1991, 25 p.
   (SANT 91/02/05)
Y. Baguette, A. Danthine, *Comparison of transport protocol mechanisms,*
   OSI95/Deliverable ULg-1/P, 30-04-1991, 62 p. (SART 91/02/05)
Y. Baguette, *Comments and questions about the XTP Protocol Definition - Revision
   3.5 plus the addendum 1a to this Revision 3.5,*OSI95/ULg/A/12/TN, 22-05-1991, 24 p.
   (SANT 91/08/05)
Y. Baguette, A. Danthine, *Will XTP Fit between LLC Type 1 and ISO Connection-
   Oriented Transport Service ???,* OSI95/Deliverable ULg-1bis/P, 28-10-1991, 71 p.
   (SART91/09/05)
A. Danthine, *High Speed Transport Protocol: Do we need a new OSI Standard ???*
   OSI95/ULg/A/20/TR/P, 31-03-1992, 14 p. (SART 92/09/05)
A. Danthine, Y. Baguette, G. Leduc, *Questionnaire about Multimedia Requirements,*
   OSI95/ULg/A/21/TN/P, 30-04-1992, 5 p.
A. Danthine, G. Leduc, *Report on the ISO/IEC JTC1/SC6 meeting in London, 8-12
   February 1993,* OSI95/ULg/A/29/TR/P, 6 p. (SART 93/05/15)
G. Leduc, *A methodology for the design of large LOTOS specifications and its
   application to ISO 8073,* OSI95/Deliverable ULg-3/P/V3, 11-09-1992, 89 p. (SART 92/19/05)
G. Leduc, *Standardisation activities of ULg in OSI95,* OSI95/ULg/A/26/TR/P, 11-12-1992,
   10 p. (SART 92/24/05)
Y. Baguette, L. Léonard, G. Leduc, A. Danthine, O. Bonaventure, *OSI95 Enhanced
   Transport Facilities and Functions,* OSI95/Deliverable ULg-A/P, 11-12-1992, 277 p.
   (SART 92/25/05)

Postscript

In 1992, the proposal of pursuing the goals of OSI95 up to the end of 1995 was rejected by the OBS sector of ESPRIT. It was the time when the aims of the ESPRIT programme were redirected towards short term developments of products and when a project with more than 15 percent of the budget allocated to universities had almost no chance to be accepted.

Fortunately, the end of the support of ESPRIT for OSI95 was not the end of the progress of the ideas conveyed in the project. Many partners of OSI95 are pursuing the work under other umbrellas.

The development of the OSI95 transport service and protocol is going on in the RACE project 2060 CIO (Coordination, Implementation and Operation of Multimedia Teleservices).

The end of 1992 saw the effective start of the COST project 237 "Multimedia Telecommunications Services". This project aims at the specification of the communication-related service requirements necessary to allow the distribution of those advanced multimedia applications which have been developed in national and pan-european projects but are still mainly standalone. It also intends to do the assessment of suitable networks, service elements and QoS for the implementation of the communication-related service requirements of these multimedia applications. Several OSI95 partners are involved in this COST project.

The Lancaster University has launched the Group Communication Support project with the support of the Science and Engineering Research Council of Great Britain.

In January 1993, a new ESPRIT experimental project, EUCALYPTUS, was launched. It involves cooperation between Europe and North America. The University of Grenoble (France) is the main contractor. The partners are the Universities of Liège (Belgium), Montreal (Canada) and Ottawa (Canada). The goal of this project is to integrate a toolset able to support the use of LOTOS for the development of large specifications. The results of this project will contribute to the final success of the OSI95 goals by allowing the use of the tools for more steps in the protocol development process.

Appendix : LOTOS Specification of the OSI95 Transport Service

This specification has been verified with regard to its syntax and its static semantics, by the use of the LITE tool, a product from the LOTOSPHERE toolset. Nevertheless, it has neither been tested nor completely simulated yet, which would require, with the existing tools and for a specification of such a size, a lot of time. Just some sub-processes have been partially simulated by means of the SMILE simulator of LITE. (*

1 Scope

We present, in this chapter the LOTOS definition of the OSI 95 enhanced transport service, using the formal description technique LOTOS, which is defined in ISO 8807.

This LOTOS specification is organised as follows: point 2 provides some general information about the specification, point 3 presents the data types used in it, and especially the QoS parameters, point 4 is very small, it gives the general structure of the specification, points 5, 6, 7 & 8 respectively describe the behaviour of the PP-ECOTS, of the PP-ERCLTS, of the PP-EACLTS and of the PP-EUCLTS.

Informal explanations precede the formal LOTOS definitions to which they refer, separated from them by a line identical to the first line below. Separation of formal definitions from subsequent informal explanations is indicated by a line identical to the second line below.

NOTE - This convention is consistent with the rules on delimitation of comments defined for LOTOS in ISO 8807.

2 Presentation of the formal description

This LOTOS specification has been built up from the description of the OSI transport service, defined in ISO 8072, with the numerous necessary additions and modifications. However, some conventions and comments are kept the same.

The whole service boundary is formally represented by the single gate t. The event structure at t depends on the type of service it is related to.

For the PP-ECOTS, it consists of a triple of values, resp. of sorts TAddress, TCEI, TSP. The first value identifies the TSAP where the interaction occurs. The second value identifies the TCEP, within that TSAP, where the interaction occurs. The third value is the Transport Service Primitive (TSP) executed in the interaction. As several out-of-band data transfers can occur on a same connection on overlapping periods of time, a way to distinguish them is necessary, so that each request can be related to its corresponding confirmation or rejection indication. This is done by an association identifier, TAEI, that is joined, when necessary, to the classical Connection End-Point Identifier, represented here by a value of sort IBTCEI. The structure of the TCEI is thus, depending on the primitive that passes through it, just an IBTCEI value, or a pair IBTCEI, TAEI.

For the PP-ERCLTS and the PP-EACLTS, the event structure at t is a triple of values, resp. of sorts TAddress, TAEI, TSP. No connection is set up with these modes, but again, as for the out-of-band transfers, it is necessary to establish an association between primitives relative to a same transfer, what explains the use of TAEI.

For the PP-EUCLTS, the event structure at t is just a pair TAddress, TSP.

Full account is taken of the multiplicity aspects of the service.

The behaviour of a never terminating service provider is described.

The specification is not parameterized.

A constraint-oriented specification style is adopted since it is the most appropriate with respect to the definitive character of a service standard. The description is aimed at describing modalities of behaviour of the service provider in terms of the history of executed TSPs only, i.e. with no reference to nor suggestion of any internal structure of the provider itself.

The data types referred to in the library construct are imported from the LOTOS library of data types.

The following convention is adopted: the term 'Request' refers both to request and to response primitives, and the term 'Indication' refers both to indication and to confirm service primitives.   *)

```
specification EnhancedTransportService [t] : noexit

library Set, Element, OctetString, Octet, NatRepresentations,
NaturalNumber, Boolean, FBoolean, DecNatRepr
endlib
(*
```

3 Interface data types

3.1 General

According to the representation of interactions at the TS boundary (see 2), the interface data types consist of three main definitions that respectively construct the sorts TAddress (see 3.2), TCEI and TAEI (see 3.3), and TSP (see 3.4). The parameters of the primitives are defined in 3.5, except the quality of service TSP parameter, introduced in 3.6. Auxiliary definitions are presented in 3.8.

3.2 Transport address

No structure of transport addresses is defined by ISO 8072. The following definition
uses the definition of GeneralIdentifier (see 3.8) and allows to represent an infinite
number of transport addresses.   *)

```
type TransportAddress
is GeneralIdentifier renamedby
sortnames   TAddress                for Identifier
opnnames    SomeTAddress            for SomeIdentifier
            AnotherTAddress         for AnotherIdentifier
endtype (* TransportAddress *)
(*
```

3.3 TC and TA endpoint identifiers

The first two definitions below allow to represent an infinite number of values of sorts
IBTCEI and TAEI. The third definition defines the sort TCEI, according to the structure explained in clause 2, and the last two present the TConnection and TAssociation
endpoint identifiers, respectively a pair TAddress x TCEI, and a pair TAddress x
TAEI. (see 3.8 for the generic definitions Pair and GeneralIdentifier).   *)

```
type InBandTCEndpointIdentifier
is GeneralIdentifier renamedby
sortnames   IBTCEI          for Identifier
opnnames    SomeIBTCEI      for SomeIdentifier
            AnotherIBTCEI   for AnotherIdentifier
endtype (* InBandTCEndpointIdentifier *)

type TAEndpointIdentifier
is GeneralIdentifier renamedby
sortnames   TAEI        for Identifier
opnnames    SomeTAEI    for SomeIdentifier
            AnotherTAEI     for AnotherIdentifier
endtype (* TAEndpointIdentifier *)

type TCEndpointIdentifier
is InBandTCEndpointIdentifier, TAEndpointIdentifier
sorts   TCEI
opns
    InBId       : IBTCEI          -> TCEI
    OOBId       : IBTCEI, TAEI       -> TCEI
    IsInBTCEI   : TCEI           -> Bool
    IsOOBTCEI   : TCEI           -> Bool
    _eq_        : TCEI, TCEI     -> Bool
    _ne_        : TCEI, TCEI     -> Bool
    InBPart     : TCEI           -> IBTCEI
eqns
forall ibtc,ibtc1:IBTCEI, ta,ta1:TAEI, tc,tc1:TCEI
ofsort Bool
IsInBTCEI(InBId(ibtc))          = true    ;
IsInBTCEI(OOBId(ibtc,ta))           = false   ;
IsOOBTCEI(InBId(ibtc))          = false   ;
IsOOBTCEI(OOBId(ibtc,ta))           = true    ;
```

```
InBId(ibtc) eq OOBId(ibtc1,ta)      = false    ;
OOBId(ibtc,ta) eq InBId(ibtc1)      = false    ;
InBId(ibtc) eq InBId(ibtc1)         = ibtc eq ibtc1   ;
OOBId(ibtc,ta) eq OOBId(ibtc1,ta1)  = (ibtc eq ibtc1)
                                       and (ta eq ta1)    ;
tc ne tc1                           = not(tc eq tc1)  ;
ofsort IBTCEI
InBPart(InBId(ibtc))      = ibtc ;
InBPart(OOBId(ibtc,ta))   = ibtc ;
endtype (* TCEndpointIdentifier *)


type TCEIdentification
is Pair actualizedby TransportAddress,
TCEndpointIdentifier using
sortnames  TAddress    for  Element
           TCEI        for  OtherElement
           Bool        for  FBool
           TCId        for  Pair
opnnames   TId              for  Pair
           TA          for  First
           TCEI        for  Second
endtype (* TCEIdentification *)


type TAEIdentification
is Pair actualizedby TransportAddress,
TAEndpointIdentifier using
sortnames  TAddress    for  Element
           TAEI        for  OtherElement
           Bool        for  FBool
           TAId        for  Pair
opnnames   TId              for  Pair
           TA          for  First
           TAEI        for  Second
endtype (* TAEIdentification *)
(*
```

3.4 ETS primitives

3.4.1 General.

The TSP data type is presented starting with the basic construction of values of sort
TSP (see below). The functions that yield TSP values are referred to as "TSP
constructors". This definition imports the definitions that relate to TSP parameters
(see 3.5 and 3.6).

NOTE - In some TSPs the Userdata parameter is an OctetString that has a fixed
bound. For technical convenience, this requirement is formally represented by a
process constraint (See 5.2.1 and 5.2.4) instead of a type constraint.

In 3.4.2 a classification of TSPs is defined, that enables both to enrich the basic construction with further functions in a simple way (see 3.4.3) and to conservatively
extend the data type in the formal description of the transport protocol.   *)

```
type BasicTSP
is TransportAddress, TPlex,TCQuality, TACKQuality, TRRQuality,
TCLQuality, TCTSDUSize,OctetString, TErrorSignal, TREJReason,
TDISReason, TDISDir, TQualityReport,
TRejectRenegociationReason,
TransDelReport
sorts TSP
opns
    TCONreq
    : TAddress, TAddress, TPlex, TCQOS1, TSS1, OctetString
    -> TSP
    TCONind
    : TAddress, TAddress, TPlex, TCQOS1, TSS1, OctetString
    -> TSP
    TCONresp
    : TAddress, TPlex, TCQOS2, TSS2, OctetString -> TSP
    TCONconf
    : TAddress, TPlex, TCQOS2, TSS2, OctetString -> TSP
    TDTreq
    : OctetString -> TSP
    TDTind
    : OctetString, TErSig -> TSP
    TDISreq
    : TDD, OctetString -> TSP
    TDISind
    : TDD, TDISReason, OctetString -> TSP
    TRELreq
    : OctetString -> TSP
    TRELind
    : OctetString, TErSig -> TSP
    TRELconf
    : -> TSP
    TREPind
    : TQR -> TSP
    TRENreq
    : TCQOS1, TSS1 -> TSP
    TRENind
    : TCQOS1, TSS1 -> TSP
    TRENresp
    : TCQOS2, TSS2 -> TSP
    TRENconf
    : TCQOS2, TSS2 -> TSP
    TNEWQOSind
    : -> TSP
    TREJRENreq
    : -> TSP
    TREJRENind
    : TRejRenReason -> TSP
    TOOBACKreq, TOOBACKind
    : TACKQOS, OctetString -> TSP
    TOOBACKconf
    : TDR -> TSP
    TOOBREJACKind
    : TRejReason -> TSP
    TREQRESPreq, TREQRESPind
    : TAddress, TAddress, TRRQOS, OctetString -> TSP
    TREQRESPresp, TREQRESPconf
    : TAddress, OctetString -> TSP
    TREJREQRESPind
    : TRejReason -> TSP
```

```
        TACKDTreq ,TACKDTind
        : TAddress, TAddress, TACKQOS, OctetString -> TSP
        TACKDTconf
        : TAddress, TDR -> TSP
        TREJACKDTind
        : TRejReason -> TSP
        TUDTreq, TUDTind
        : TAddress, TAddress, TCLQOS, OctetString -> TSP
endtype (* BasicTSP *)
(*
```

3.4.2 Classification of the ETS primitives

3.4.2.1 Basic classification

A basic classification of TSPs is defined by means of TSPSubsort, which consists of a set of constants, each denoting a TSP name, and of classification tests: IsRequest, IsIndication (these two tests reflect the convention introduced in clause 2), IsInBand and IsOutOfBand that determine whether a primitive is related to the classical 'in-band' connection mode or peculiar to the out-of-band transfers, IsConOr and IsConLess that determine whether a primitive is related to the connection mode (either in- or out-of-band) or to one of the three connectionless modes.

The type TSPBasicClassifiers is a functional enrichment of the basic construction in 3.4.1, where

- the Subsort function yields the TSP name;
- Boolean functions on TSPs, termed "TSP (subsort) recognisers", are defined according to the basic classification introduced by TSPSubsort.

NOTE - The auxiliary function h that maps TSP names to natural numbers is defined in order to simplify the specification of the Boolean operations of equality on TSP names (as well as on TSPs, in 3.4.3.3).   *)

```
type TSPSubsort
is NaturalNumber
sorts TSPSubsort
opns
    TCONNECTrequest, TCONNECTindication, TCONNECTresponse,
    TCONNECTconfirm, TDATArequest, TDATAindication,
    TDISCONNrequest, TDISCONNindication,TREPORTindication,
    TRELEASErequest, TRELEASEindication, TRELEASEconfirm,
    TRENEGOTIATErequest, TRENEGOTIATEindication,
    TRENEGOTIATEresponse, TRENEGOTIATEconfirm,
    TNEWQOSindication, TREJECTRENEGOTIATErequest,
    TREJECTRENEGOTIATEindication,
    TOUTOFBANDACKDATArequest, TOUTOFBANDACKDATAindication,
    TOUTOFBANDACKDATAconfirm,
TOUTOFBANDREJECTACKDATAindication,
    TREQUESTRESPONSErequest, TREQUESTRESPONSEindication,
    TREQUESTRESPONSEresponse, TREQUESTRESPONSEconfirm,
    TREJECTREQUESTRESPONSEindication,
    TACKNOWLEDGEDDATArequest, TACKNOWLEDGEDDATAindication,
    TACKNOWLEDGEDDATAconfirm,
TREJECTACKNOWLEDGEDDATAindication,
    TUDATArequest, TUDATAindication
```

```
                         : -> TSPSubsort
   h                     : TSPSubsort -> Nat
   Even, Odd             : Nat -> Bool
   IsRequest, IsIndication, IsInBand, IsOutOfBand, IsConOr,
   IsConLess             : TSPSubsort -> Bool
   _eq_, _ne_            : TSPSubsort, TSPSubsort -> Bool
eqns
forall s,s1:TSPSubsort, n:Nat
ofsort Nat
h(TCONNECTrequest)               = 0        ;
h(TCONNECTindication)            = Succ(h(TCONNECTrequest))
   ;
h(TCONNECTresponse)              = Succ(h(TCONNECTindication))
   ;
h(TCONNECTconfirm)               = Succ(h(TCONNECTresponse))
   ;
h(TDATArequest)                  = Succ(h(TCONNECTconfirm))
   ;
h(TDATAindication)               = Succ(h(TDATArequest))
   ;
h(TDISCONNrequest)               = Succ(h(TDATAindication))
   ;
h(TDISCONNindication)            = Succ(h(TDISCONNrequest))
   ;
h(TREPORTindication)             = Succ(Succ(h(TDISCONNindication)))
   ;
h(TRELEASErequest)               = Succ(h(TREPORTindication))
   ;
h(TRELEASEindication)            = Succ(h(TRELEASErequest))
   ;
h(TRELEASEconfirm)               = Succ(Succ(h(TRELEASEindication)))
   ;
h(TRENEGOTIATErequest)           = Succ(h(TRELEASEconfirm))
   ;
h(TRENEGOTIATEindication)           = Succ(h(TRENEGOTIATErequest))
   ;
h(TRENEGOTIATEresponse)          = Succ(h(TRENEGOTIATEindication))
   ;
h(TRENEGOTIATEconfirm)           = Succ(h(TRENEGOTIATEresponse))
   ;
h(TNEWQOSindication)             = Succ(Succ(h(TRENEGOTIATEconfirm)))
   ;
h(TREJECTRENEGOTIATErequest)        = Succ(h(TNEWQOSindication))
   ;
h(TREJECTRENEGOTIATEindication)        =
   Succ(h(TREJECTRENEGOTIATErequest))                 ;
h(TOUTOFBANDACKDATArequest)         =
   Succ(h(TREJECTRENEGOTIATEindication))              ;
h(TOUTOFBANDACKDATAindication)         =
   Succ(h(TOUTOFBANDACKDATArequest))                  ;
h(TOUTOFBANDACKDATAconfirm)         =
   Succ(Succ(h(TOUTOFBANDACKDATAindication)))   ;
h(TOUTOFBANDREJECTACKDATAindication) =
   Succ(Succ(h(TOUTOFBANDACKDATAconfirm)))      ;
h(TREQUESTRESPONSErequest)          =
   Succ(h(TOUTOFBANDREJECTACKDATAindication))   ;
h(TREQUESTRESPONSEindication)             =
   Succ(h(TREQUESTRESPONSErequest))             ;
```

```
h(TREQUESTRESPONSEresponse)              =
    Succ(h(TREQUESTRESPONSEindication))              ;
h(TREQUESTRESPONSEconfirm)               =
    Succ(h(TREQUESTRESPONSEresponse))                ;
h(TREJECTREQUESTRESPONSEindication)  =
    Succ(Succ(h(TREQUESTRESPONSEconfirm)))           ;
h(TACKNOWLEDGEDDATArequest)              =
    Succ(h(TREJECTREQUESTRESPONSEindication))    ;
h(TACKNOWLEDGEDDATAindication)           =
    Succ(h(TACKNOWLEDGEDDATArequest))            ;
h(TACKNOWLEDGEDDATAconfirm)              =
    Succ(Succ(h(TACKNOWLEDGEDDATAindication)))    ;
h(TREJECTACKNOWLEDGEDDATAindication)      =
    Succ(Succ(h(TACKNOWLEDGEDDATAconfirm)))       ;
h(TUDATArequest)                         =
    Succ(h(TREJECTACKNOWLEDGEDDATAindication))   ;
h(TUDATAindication)     = Succ(h(TUDATArequest))   ;

ofsort Bool
Even(0)                 = true           ;
Even(Succ(0))              = false          ;
Even(Succ(Succ(n)))     = Even(n)        ;
Odd(n)                  = not(Even(n))      ;
IsRequest(s)            = Even(h(s))    ;
IsIndication(s)         = Odd(h(s))       ;
IsInBand(s)             = h(s) lt h(TOUTOFBANDACKDATArequest)
    ;
IsOutOfBand(s)             = not(IsInBand(s))
                          and (h(s) lt h(TREQUESTRESPONSErequest))
    ;
IsConOr(s)              = IsInBand(s) or IsOutOfBand(s)
    ;
IsConLess(s)            = not(IsConOr(s)) ;
s eq s1                 = h(s) eq h(s1)   ;
s ne s1                 = not(s eq s1)    ;
endtype (* TSPSubsort *)

type TSPBasicClassifiers
is BasicTSP, TSPSubsort
opns
    Subsort : TSP -> TSPSubsort
    IsTCON, IsTCON1, IsTCON2, IsTDT, IsTDIS, IsTREL, IsTREN,
    IsTREN1, IsTREN2, IsTOOBACK, IsTOOBACK1,IsTOOBACK2,
    IsTREQRESP, IsTREQRESP1, IsTREQRESP2, IsTACKDT, IsTACKDT1,
    IsTACKDT2, IsTUDT, IsTCONreq, IsTCONind, IsTCONresp,
    IsTCONconf,IsTDTreq, IsTDTind, IsTDISreq, IsTDISind,
    IsTREPind, IsTRELreq, IsTRELind, IsTRELconf,IsTRENreq,
    IsTRENind, IsTRENresp, IsTRENconf,IsTNEWQOSind,
    IsTREJRENreq, IsTREJRENind,IsTOOBACKreq, IsTOOBACKind,
    IsTOOBACKconf, IsTOOBREJACKind, IsTREQRESPreq,
    IsTREQRESPind, IsTREQRESPresp, IsTREQRESPconf,
    IsTREJREQRESPind, IsTACKDTreq, IsTACKDTind, IsTACKDTconf,
    IsTREJACKDTind, IsTUDTreq, IsTUDTind, IsTReq, IsTInd,
  IsInB,
    IsOOB, IsTConO, IsTConL, IsTClREJ          : TSP -> Bool
eqns
forall a,a1,a2:TAddress, pl:TPlex, qc1:TCQOS1, qc2:TCQOS2,
qa:TACKQOS, qr:TRRQOS, ql:TCLQOS, ts1:TSS1, ts2:TSS2,
d:OctetString, es:TErSig, td:TDD, r:TDISReason, re:TRejReason,
tqr:TQR, rrr:TRejRenReason, tdr:TDR, t:TSP
```

```
ofsort TSPSubsort
Subsort(TCONreq(a1,a2,p1,qc1,ts1,d))      = TCONNECTrequest
   ;
Subsort(TCONind(a1,a2,p1,qc1,ts1,d))      = TCONNECTindication
   ;
Subsort(TCONresp(a,p1,qc2,ts2,d))      = TCONNECTresponse
   ;
Subsort(TCONconf(a,p1,qc2,ts2,d))      = TCONNECTconfirm       ;
Subsort(TDTreq(d))               = TDATArequest           ;
Subsort(TDTind(d,es))            = TDATAindication        ;
Subsort(TDISreq(td,d))           = TDISCONNrequest        ;
Subsort(TDISind(td,r,d))           = TDISCONNindication       ;
Subsort(TREPind(tqr))            = TREPORTindication      ;
Subsort(TRELreq(d))              = TRELEASErequest        ;
Subsort(TRELind(d,es))           = TRELEASEindication     ;
Subsort(TRELconf)                = TRELEASEconfirm        ;
Subsort(TRENreq(qc1,ts1))        = TRENEGOTIATErequest    ;
Subsort(TRENind(qc1,ts1))        = TRENEGOTIATEindication ;
Subsort(TRENresp(qc2,ts2))       = TRENEGOTIATEresponse   ;
Subsort(TRENconf(qc2,ts2))       = TRENEGOTIATEconfirm    ;
Subsort(TNEWQOSind)              = TNEWQOSindication      ;
Subsort(TREJRENreq)              = TREJECTRENEGOTIATErequest
   ;
Subsort(TREJRENind(rrr))            =
TREJECTRENEGOTIATEindication ;
Subsort(TOOBACKreq(qa,d))        = TOUTOFBANDACKDATArequest
   ;
Subsort(TOOBACKind(qa,d))        = TOUTOFBANDACKDATAindication
   ;
Subsort(TOOBACKconf(tdr))        = TOUTOFBANDACKDATAconfirm
   ;
Subsort(TOOBREJACKind(re))     =
TOUTOFBANDREJECTACKDATAindication     ;
Subsort(TREQRESPreq(a1,a2,qr,d)) = TREQUESTRESPONSErequest
   ;
Subsort(TREQRESPind(a1,a2,qr,d)) = TREQUESTRESPONSEindication
   ;
Subsort(TREQRESPresp(a,d))       = TREQUESTRESPONSEresponse
   ;
Subsort(TREQRESPconf(a,d))       = TREQUESTRESPONSEconfirm
   ;
Subsort(TREJREQRESPind(re))        =
TREJECTREQUESTRESPONSEindication      ;
Subsort(TACKDTreq(a1,a2,qa,d))   = TACKNOWLEDGEDDATArequest
   ;
Subsort(TACKDTind(a1,a2,qa,d))   = TACKNOWLEDGEDDATAindication
   ;
Subsort(TACKDTconf(a,tdr))       = TACKNOWLEDGEDDATAconfirm
   ;
Subsort(TREJACKDTind(re))          =
TREJECTACKNOWLEDGEDDATAindication      ;
Subsort(TUDTreq(a1,a2,ql,d))         = TUDATArequest
   ;
Subsort(TUDTind(a1,a2,ql,d))         = TUDATAindication
      ;
ofsort Bool
IsTCON(t)        = IsTCON1(t) or IsTCON2(t)
   ;
```

```
IsTCON1(t)          = IsTCONreq(t) or IsTCONind(t)
  ;
IsTCON2(t)          = IsTCONresp(t) or IsTCONconf(t)
  ;
IsTDT(t)            = IsTDTreq(t) or IsTDTind(t)
  ;
IsTDIS(t)           = IsTDISreq(t) or IsTDISind(t)
  ;
IsTREL(t)           = IsTRELreq(t) or IsTRELind(t) or IsTRELconf(t)
  ;
IsTREN(t)           = IsTREN1(t) or IsTREN2(t)
                    or IsTNEWQOSind(t) or IsTREJRENind(t)
  ;
IsTREN1(t)          = IsTRENreq(t) or IsTRENind(t)
  ;
IsTREN2(t)          = IsTRENresp(t) or IsTRENconf(t)
  ;
IsTOOBACK(t)        = IsTOOBACK1(t) or IsTOOBACK2(t)
  ;
IsTOOBACK1(t)         = IsTOOBACKreq(t) or IsTOOBACKind(t)
        ;
IsTOOBACK2(t)         = IsTOOBACKconf(t) or IsTOOBREJACKind(t)
        ;
IsTREQRESP(t)       = IsTREQRESP1(t) or IsTREQRESP2(t)
                    or IsTREJREQRESPind(t)
  ;
IsTREQRESP1(t)        = IsTREQRESPreq(t) or IsTREQRESPind(t)
        ;
IsTREQRESP2(t)        = IsTREQRESPresp(t) or IsTREQRESPconf(t)
          ;
IsTACKDT(t)         = IsTACKDT1(t) or IsTACKDT2(t)
  ;
IsTACKDT1(t)        = IsTACKDTreq(t) or IsTACKDTind(t)
  ;
IsTACKDT2(t)        = IsTACKDTconf(t) or IsTREJACKDTind(t)
  ;
IsTUDT(t)           = IsTUDTreq(t) or IsTUDTind(t)
  ;
IsTCONreq(t)     = Subsort(t) eq TCONNECTrequest          ;
IsTCONind(t)     = Subsort(t) eq TCONNECTindication       ;
IsTCONresp(t)      = Subsort(t) eq TCONNECTresponse        ;
IsTCONconf(t)      = Subsort(t) eq TCONNECTconfirm         ;
IsTDTreq(t)      = Subsort(t) eq TDATArequest             ;
IsTDTind(t)      = Subsort(t) eq TDATAindication          ;
IsTDISreq(t)     = Subsort(t) eq TDISCONNrequest          ;
IsTDISind(t)     = Subsort(t) eq TDISCONNindication       ;
IsTREPind(t)     = Subsort(t) eq TREPORTindication        ;
IsTRELreq(t)     = Subsort(t) eq TRELEASErequest          ;
IsTRELind(t)     = Subsort(t) eq TRELEASEindication       ;
IsTRELconf(t)      = Subsort(t) eq TRELEASEconfirm         ;
IsTRENreq(t)     = Subsort(t) eq TRENEGOTIATErequest   ;
IsTRENind(t)     = Subsort(t) eq TRENEGOTIATEindication   ;
IsTRENresp(t)      = Subsort(t) eq TRENEGOTIATEresponse
  ;
IsTRENconf(t)            = Subsort(t) eq TRENEGOTIATEconfirm  ;
IsTNEWQOSind(t)  = Subsort(t) eq TNEWQOSindication       ;
IsTREJRENreq(t)  = Subsort(t) eq TREJECTRENEGOTIATErequest
    ;
```

```
IsTREJRENind(t)      = Subsort(t) eq
TREJECTRENEGOTIATEindication ;
IsTOOBACKreq(t)      = Subsort(t) eq TOUTOFBANDACKDATArequest
       ;
IsTOOBACKind(t)      = Subsort(t) eq TOUTOFBANDACKDATAindication
       ;
IsTOOBACKconf(t)     = Subsort(t) eq TOUTOFBANDACKDATAconfirm
       ;
IsTOOBREJACKind(t)       = Subsort(t) eq
                         TOUTOFBANDREJECTACKDATAindication
       ;
IsTREQRESPreq(t)     = Subsort(t) eq TREQUESTRESPONSErequest
       ;
IsTREQRESPind(t)     = Subsort(t) eq TREQUESTRESPONSEindication
       ;
IsTREQRESPresp(t)    = Subsort(t) eq TREQUESTRESPONSEresponse
       ;
IsTREQRESPconf(t)    = Subsort(t) eq TREQUESTRESPONSEconfirm
       ;
IsTREJREQRESPind(t)      = Subsort(t) eq
                         TREJECTREQUESTRESPONSEindication
       ;
IsTACKDTreq(t)           = Subsort(t) eq TACKNOWLEDGEDDATArequest
       ;
IsTACKDTind(t)           = Subsort(t) eq
TACKNOWLEDGEDDATAindication   ;
IsTACKDTconf(t)      = Subsort(t) eq TACKNOWLEDGEDDATAconfirm
       ;
IsTREJACKDTind(t)    = Subsort(t) eq
                         TREJECTACKNOWLEDGEDDATAindication
       ;
IsTUDTreq(t)             = Subsort(t) eq TUDATArequest
       ;
IsTUDTind(t)             = Subsort(t) eq TUDATAindication
       ;

IsTReq(t)        = IsRequest(Subsort(t))                    ;
IsTInd(t)        = IsIndication(Subsort(t))                 ;
IsInB(t)         = IsInBand(Subsort(t))                     ;
IsOOB(t)         = IsOutOfBand(Subsort(t))                  ;
IsTConO(t)       = IsConOr(Subsort(t))                      ;
IsTConL(t)       = IsConLess(Subsort(t))                    ;
IsTClREJ(t)      = IsTREJREQRESPind(t) or IsTREJACKDTind(t)  ;
endtype (* TSPBasicClassifiers *)
(*
```

3.4.3 Functions on the ETS primitives

3.4.3.1 General

In 3.4.3.2 the construction presented in 3.4.2 is enriched with functions that allow to determine the value of individual parameters of TSPs. These functions are partially defined, i.e. not for all the values of sort TSP. Using such functions eases and lightens the specification but can be dangerous. In our specification, some guards and predicates could be undefined. To remedy to this problem, we have to introduce the operations "andthen" and "orelse", defined in the type "ExtendedBoolean".

In 3.4.3.3 Boolean equality is added to this construction.

In 3.4.3.4 further functional enrichments are presented, that are useful to represent negotiation (see 5.2.3.2) and provider's non-determinacy (see 5.3.2.2.3).

3.4.3.2 Parameter selectors   *)

```
type ExtendedBoolean is Boolean
opns
    _andthen_   : Bool, Bool -> Bool
    _orelse_    : Bool, Bool -> Bool
eqns
forall x:Bool

ofsort Bool
false andthen x   = false    ;
true andthen x        = x          ;
false orelse x        = x          ;
true orelse x     = true     ;

endtype (* ExtendedBoolean *)


type TSPParameterSelectors
is TSPBasicClassifiers, ExtendedBoolean
opns
    CalledOf, CallingOf, RespondingOf : TSP -> TAddress
    PlexOf            : TSP -> TPlex
    TQOS1Of           : TSP -> TCQOS1
    TQOS2Of           : TSP -> TCQOS2
    TQOSAOf           : TSP -> TACKQOS
    TQOSROf           : TSP -> TRRQOS
    TQOSLOf           : TSP -> TCLQOS
    TTSDU1SizeOf      : TSP -> TSS1
    TTSDU2SizeOf      : TSP -> TSS2
    ErSigOf           : TSP -> TErSig
    DisDirOf          : TSP -> TDD
    DISReasonOf       : TSP -> TDISReason
    REJReasonOf       : TSP -> TRejReason
    QualReportOf      : TSP -> TQR
    REJRenReasonOf    : TSP -> TRejRenReason
    TDReportOf        : TSP -> TDR
    UserData          : TSP -> OctetString
eqns
forall a,a1,a2:TAddress, pl:TPlex, qc1:TCQOS1, qc2:TCQOS2,
ts1:TSS1, ts2:TSS2, qa:TACKQOS, qr:TRRQOS,
ql:TCLQOS, d:OctetString, es:TErSig, td:TDD, r:TDISReason,
re:TRejReason, tqr:TQR, rrr:TRejRenReason, tdr:TDR, t:TSP

ofsort TAddress
CalledOf(TCONreq(a1, a2, pl, qc1, ts1, d))     = a1   ;
CalledOf(TCONind(a1, a2, pl, qc1, ts1, d))     = a1   ;
CalledOf(TREQRESPreq(a1, a2, qr, d))           = a1   ;
CalledOf(TREQRESPind(a1, a2, qr, d))           = a1   ;
CalledOf(TACKDTreq(a1, a2, qa, d))             = a1   ;
CalledOf(TACKDTind(a1, a2, qa, d))             = a1   ;
CalledOf(TUDTreq(a1, a2, ql, d))               = a1   ;
CalledOf(TUDTind(a1, a2, ql, d))               = a1   ;

CallingOf(TCONreq(a1, a2, pl, qc1, ts1, d))     = a2   ;
CallingOf(TCONind(a1, a2, pl, qc1, ts1, d))     = a2   ;
CallingOf(TREQRESPreq(a1, a2, qr, d))           = a2   ;
CallingOf(TREQRESPind(a1, a2, qr, d))           = a2   ;
CallingOf(TACKDTreq(a1, a2, qa, d))             = a2   ;
```

```
CallingOf(TACKDTind(a1, a2, qa, d))                         = a2     ;
CallingOf(TUDTreq(a1, a2, ql, d))                   = a2     ;
CallingOf(TUDTind(a1, a2, ql, d))                   = a2     ;
RespondingOf(TCONresp(a, pl, qc2, ts2, d))          = a        ;
RespondingOf(TCONconf(a, pl, qc2, ts2, d))          = a        ;
RespondingOf(TREQRESPresp(a, d))                    = a        ;
RespondingOf(TREQRESPconf(a, d))                    = a        ;
RespondingOf(TACKDTconf(a, tdr))                    = a        ;
ofsort TPlex
PlexOf(TCONreq(a1, a2, pl, qc1, ts1, d))            = pl     ;
PlexOf(TCONind(a1, a2, pl, qc1, ts1, d))            = pl     ;
PlexOf(TCONresp(a, pl, qc2, ts2, d))                = pl     ;
PlexOf(TCONconf(a, pl, qc2, ts2, d))                = pl     ;
ofsort TCQOS1
TQOS1Of(TCONreq(a1, a2, pl, qc1, ts1, d))           = qc1    ;
TQOS1Of(TCONind(a1, a2, pl, qc1, ts1, d))           = qc1    ;
TQOS1Of(TRENreq(qc1, ts1))                          = qc1    ;
TQOS1Of(TRENind(qc1, ts1))                          = qc1    ;
ofsort TCQOS2
TQOS2Of(TCONresp(a, pl, qc2, ts2, d))               = qc2    ;
TQOS2Of(TCONconf(a, pl, qc2, ts2, d))               = qc2    ;
TQOS2Of(TRENresp(qc2, ts2))                         = qc2    ;
TQOS2Of(TRENconf(qc2, ts2))                         = qc2    ;
ofsort TACKQOS
TQOSAOf(TOOBACKreq(qa, d))                          = qa     ;
TQOSAOf(TOOBACKind(qa, d))                          = qa     ;
TQOSAOf(TACKDTreq(a1, a2, qa, d))                   = qa     ;
TQOSAOf(TACKDTind(a1, a2, qa, d))                   = qa     ;
ofsort TRRQOS
TQOSROf(TREQRESPreq(a1, a2, qr, d))                 = qr     ;
TQOSROf(TREQRESPind(a1, a2, qr, d))                   = qr    ;
ofsort TCLQOS
TQOSLOf(TUDTreq(a1, a2, ql, d))                     = ql     ;
TQOSLOf(TUDTind(a1, a2, ql, d))                     = ql     ;
ofsort TSS1
TTSDU1SizeOf(TCONreq(a1, a2, pl, qc1, ts1, d))        = ts1  ;
TTSDU1SizeOf(TCONind(a1, a2, pl, qc1, ts1, d))        = ts1  ;
TTSDU1SizeOf(TRENreq(qc1, ts1))                     = ts1  ;
TTSDU1SizeOf(TRENind(qc1, ts1))                     = ts1  ;
ofsort TSS2
TTSDU2SizeOf(TCONresp(a, pl, qc2, ts2, d))      = ts2  ;
TTSDU2SizeOf(TCONconf(a, pl, qc2, ts2, d))      = ts2  ;
TTSDU2SizeOf(TRENresp(qc2, ts2))                = ts2  ;
TTSDU2SizeOf(TRENconf(qc2, ts2))                = ts2  ;
ofsort TErSig
ErSigOf(TDTind(d, es))                          = es       ;
ErSigOf(TRELind(d, es))                         = es       ;
ofsort TDD
DISDirOf(TDISreq(td, d))                          = td        ;
DISDirOf(TDISind(td, r, d))                     = td       ;
ofsort TDISReason
DISReasonOf(TDISind(td, r, d))                  = r        ;
ofsort TRejReason
REJReasonOf(TOOBREJACKind(re))                  = re       ;
REJReasonOf(TREJREQRESPind(re))                 = re       ;
```

```
REJReasonOf(TREJACKDTind(re))                    = re          ;
ofsort TQR
QualReportOf(TREPind(tqr))                        = tqr         ;

ofsort TRejRenReason
REJRenReasonOf(TREJRENind(rrr))                   = rrr         ;

ofsort TDR
TDReportOf(TOOBACKconf(tdr))                        = tdr         ;
TDReportOf(TACKDTconf(a, tdr))                    = tdr         ;

ofsort OctetString
UserData(TCONreq(a1, a2, pl, qc1, ts1, d))       = d            ;
UserData(TCONind(a1, a2, pl, qc1, ts1, d))       = d            ;
UserData(TCONresp(a, pl, qc2, ts2, d))           = d            ;
UserData(TCONconf(a, pl, qc2, ts2, d))           = d            ;
UserData(TDTreq(d))                              = d            ;
UserData(TDTind(d, es))                          = d            ;
UserData(TDISreq(td, d))                             = d            ;
UserData(TDISind(td, r, d))                      = d            ;
UserData(TRELreq(d))                             = d            ;
UserData(TRELind(d, es))                             = d            ;
UserData(TOOBACKreq(qa, d))                      = d            ;
UserData(TOOBACKind(qa, d))                      = d            ;
UserData(TREQRESPreq(a1, a2, qr, d))                 = d            ;
UserData(TREQRESPind(a1, a2, qr, d))                 = d            ;
UserData(TREQRESPresp(a, d))                     = d            ;
UserData(TREQRESPconf(a, d))                     = d            ;
UserData(TACKDTreq(a1, a2, qa, d))               = d            ;
UserData(TACKDTind(a1, a2, qa, d))               = d            ;
UserData(TUDTreq(a1, a2, ql, d))                 = d            ;
UserData(TUDTind(a1, a2, ql, d))                 = d            ;
( IsTRELconf(t) or IsTREN(t) or IsTOOBACK2(t) or IsTREPind(t)
or IsTACKDT2(t) or IsTREJREQRESPind(t)) => UserData(t) = <>
    ;
endtype (* TSPParameterSelectors *)
(*
```

3.4.3.3 Equality of ETS primitives

Boolean equality on TSPs is defined as the conjunction of TSP name equality (see
3.4.2.1) and pairwise equality of TSP parameters.   *)

```
type TSPEquality
is TSPParameterSelectors
opns
    _eq_, _ne_ : TSP, TSP -> Bool
eqns
forall a,a1,a2,a3:TAddress, pl,pl1:TPlex, qc1,qc11:TCQOS1,
qc2,qc21:TCQOS2, ts1,ts11:TSS1, ts2,ts21:TSS2, qa,qa1:TACKQOS,
qr,qr1:TRRQOS, ql,ql1:TCLQOS, es,es1:TErSig, d,d1:OctetString,
td,td1:TDD, r,r1:TDISReason, re,re1:TRejReason, tqr,tqr1:TQR,
rrr,rrr1:TRejRenReason, tdr,tdr1:TDR, t,t1:TSP
ofsort Bool
    Subsort(t) ne Subsort(t1) =>
t eq t1 = false   ;
TCONreq(a,a2,pl,qc1,ts1,d) eq TCONreq(a1,a3,pl1,qc11,ts11,d1)
    =
    (a eq a1) and (a2 eq a3) and (pl eq pl1) and (qc1 eq qc11)
    and (ts1 eq ts11) and (d eq d1)                 ;
```

```
TCONind(a,a2,pl,qc1,ts1,d) eq TCONind(a1,a3,pl1,qc11,ts11,d1)
    =
    (a eq a1) and (a2 eq a3) and (pl eq pl1) and (qc1 eq qc11)
    and (ts1 eq ts11) and (d eq d1)                    ;
TCONresp(a,pl,qc2,ts2,d) eq TCONresp(a1,pl1,qc21,ts21,d1)
    =
    (a eq a1) and (pl eq pl1) and (qc2 eq qc21)
    and (ts2 eq ts21) and (d eq d1)                    ;
TCONconf(a,pl,qc2,ts2,d) eq TCONconf(a1,pl1,qc21,ts21,d1)
    =
    (a eq a1) and (pl eq pl1) and (qc2 eq qc21)
    and (ts2 eq ts21) and (d eq d1)                    ;
TDTreq(d) eq TDTreq(d1)                = d eq d1        ;
TDTind(d,es) eq TDTind(d1,es1)    = (d eq d1) and (es eq es1)
    ;
TDISreq(td,d) eq TDISreq(td1,d1) = (td eq td1) and (d eq d1)
    ;
TDISind(td,r,d) eq TDISind(td1,r1,d1) = (td eq td1)
                                    and (r eq r1) and (d eq d1)
    ;
TREPind(tqr) eq TREPind(tqr1)        = tqr eq tqr1
    ;
TRELreq(d) eq TRELreq(d1)            = d eq d1
    ;
TRELind(d,es) eq TRELind(d1,es1) = (d eq d1) and (es eq es1)
    ;
TRELconf eq TRELconf                    = true
    ;
TRENreq(qc1,ts1) eq TRENreq(qc11,ts11)    = (qc1 eq qc11)
                                        and (ts1 eq ts11)
    ;
TRENind(qc1,ts1) eq TRENind(qc11,ts11)    = (qc1 eq qc11)
                                        and (ts1 eq ts11)
    ;
TRENresp(qc2,ts2) eq TRENresp(qc21,ts21) = (qc2 eq qc21)
                                        and (ts2 eq ts21)
    ;
TRENconf(qc2,ts2) eq TRENconf(qc21,ts21)    = (qc2 eq qc21)
                                        and (ts2 eq ts21)
    ;
TNEWQOSind eq TNEWQOSind                    = true
    ;
TREJRENreq eq TREJRENreq                    = true
    ;
TREJRENind(rrr) eq TREJRENind(rrr1)        = rrr eq rrr1
    ;
TOOBACKreq(qa,d) eq TOOBACKreq(qa1,d1)    = (qa eq qa1)
                                        and (d eq d1)
    ;
TOOBACKind(qa,d) eq TOOBACKind(qa1,d1)    = (qa eq qa1)
                                        and (d eq d1)
    ;
TOOBACKconf(tdr) eq TOOBACKconf(tdr1)    = tdr eq tdr1
    ;
TOOBREJACKind(re) eq TOOBREJACKind(re1)        = re eq re1
    ;
TREQRESPreq(a,a2,qr,d) eq TREQRESPreq(a1,a3,qr1,d1)    = (a eq
a1)
            and (a2 eq a3) and (qr eq qr1) and (d eq d1)
    ;
```

```
TREQRESPind(a,a2,qr,d) eq TREQRESPind(a1,a3,qr1,d1)     = (a eq
a1)
                   and (a2 eq a3) and (qr eq qr1) and (d eq d1)
     ;
TREQRESPresp(a,d) eq TREQRESPresp(a1,d1)        = (a eq a1)
                                                  and (d eq d1)

     ;
TREQRESPconf(a,d) eq TREQRESPconf(a1,d1)        = (a eq a1)
                                                  and (d eq d1)

     ;
TREJREQRESPind(re) eq TREJREQRESPind(re1)       = re eq re1
     ;
TACKDTreq(a,a2,qa,d) eq TACKDTreq(a1,a3,qa1,d1)     = (a eq a1)
                   and (a2 eq a3) and (qa eq qa1) and (d eq
d1)      ;
TACKDTind(a,a2,qa,d) eq TACKDTind(a1,a3,qa1,d1)     = (a eq a1)
                   and (a2 eq a3) and (qa eq qa1) and (d eq
d1)      ;
TACKDTconf(a,tdr) eq TACKDTconf(a1,tdr1)        = (a eq a1)
                                                  and (tdr eq tdr1)

     ;
TREJACKDTind(re) eq TREJACKDTind(re1)           = re eq re1
     ;
TUDTreq(a,a2,ql,d) eq TUDTreq(a1,a3,ql1,d1)      = (a eq a1)
                   and (a2 eq a3) and (ql eq ql1) and (d eq
d1)      ;
TUDTind(a,a2,ql,d) eq TUDTind(a1,a3,ql1,d1)      = (a eq a1)
                   and (a2 eq a3) and (ql eq ql1) and (d eq
d1)      ;
endtype (* TSPEquality *)
(*
```

3.4.3.4 Other functions on the ETS primitives

The function NeedsOOBTCEI determines whether a primitive requires a TCEI made of a pair (INBTCEI, TAEI) (see 2). This is the case for all the primitives peculiar to the connection-mode out-of-band transfer, except TOUTOFBANDACKDATAindication that has not to be related to another primitive occurring on the same TCEP.

The function ProviderGeneratedInd characterises TSPs that are solely generated by the service provider. This function is made use of in the description of possible nondeterminacy of the service provider (see 5.3.2.2.3). For reasons coming from the structure of the specification, it does not include the TREJECTRENEGOTIATEindication, that is treated by a special process (see 5.3.2.3).

IsValidTCONreq (resp. IsValidTRENreq, IsValidTOOBACKreq, IsValidTACKDTreq) specifies various constraints that the parameters of a TCONNECTrequest (resp. a TRENEGOTIATErequest, a TOUTOFBANDACK-DATArequest, a TACKNOWLEDGEDDATA request) primitive must respect (for IsTCQOS1, IsBi: see 3.6.1.6, IsTSS1, IsBi: 3.7, IsTACKQOS: 3.6.2).

IsTDISIn, IsTDISOut, IsTDISInOut allow to differentiate the TDISCONNECT primitives, according to the directions of transmission they close.

IsTREPindInp, IsTREPindOut are used to determine the direction of transmission the quality report primitive is related to (see 3.5 for IsTDR,RDir).

IsOutputOf relates execution of a TSP at each endpoint of a TC to previous execution of a corresponding primitive at the other endpoint of the same TC (see 5.3.2.2.3).

Also this function represents negotiation requirements, as relating to possible non-determinacy of the service provider. Let us remark that no constraint is expressed here for the transmission of TDATArequest and TRELEASErequest. For technical reasons, these constraints are defined by a separate process (see 5.3.2.5)

IsValidTCON2For represents TS negotiation requirements that apply locally to each TC endpoint (see 5.2.3.2).   *)

```
type TransportServicePrimitive
is TSPEquality
opns
    NeedsOOBTCEI                     : TSP -> Bool
    ProviderGeneratedInd            : TSP -> Bool
    IsValidTCONreq                    : TSP -> Bool
    IsValidTRENreq                    : TSP -> Bool
    IsValidTOOBACKreq               : TSP -> Bool
    IsValidTACKDTreq                : TSP -> Bool
    IsTDISIn,IsTDISOut,IsTDISInOut    : TSP -> Bool
    IsTREPindInp, IsTREPindOut        : TSP -> Bool
    _IsOutputOf_, _IsValidTCON2For_  : TSP, TSP -> Bool
eqns
forall a,a1,a2,a3:TAddress, pl,pl1:TPlex, qc1,qc11:TCQOS1,
qc2,qc21:TCQOS2, ts1,ts11:TSS1, ts2,ts21:TSS2, qa,qa1:TACKQOS,
qr,qr1:TRRQOS, ql,ql1:TCLQOS, es,es1:TErSig, d,d1:OctetString,
td,td1:TDD, r,r1:TDISReason, re,re1:TRejReason, tqr,tqr1:TQR,
rrr:TRejRenReason, tdr:TDR, t,t1:TSP
ofsort Bool
NeedsOOBTCEI(t)                   = IsOOB(t) and not(IsTOOBACKind(t))
    ;
ProviderGeneratedInd(t)   = IsTREPind(t) or (IsTDISInOut(t)
    and IsTInd(t) andthen (User ne DISReasonOf(t))
    and (UserData(t) eq <>)) ;
IsValidTCONreq(TCONreq(a1,a2,pl,qc1,ts1,d))      = IsTCQOS1(qc1)
    and IsTSS1(ts1)
    and((pl eq both) implies (IsBi(qc1) and IsBi(ts1)))
    and((pl eq him) implies (IsInp(qc1) and IsInp(ts1)))
    and ((pl eq me) implies (IsOut(qc1) and IsOut(ts1)))
    ;
    not(IsTCONreq(t)) =>
IsValidTCONreq(t)                        = false
    ;
IsValidTRENreq(TRENreq(qc1,ts1)) = IsTCQOS1(qc1)
                                   and IsTSS1(ts1)
    ;
    not(IsTRENreq(t)) =>
IsValidTRENreq(t)                          = false
    ;
IsValidTOOBACKreq(TOOBACKreq(qa,d))             = IsTACKQOS(qa)
    ;
    not(IsTOOBACKreq(t)) =>
IsValidTOOBACKreq(t)                            = false
    ;
IsValidTACKDTreq(TACKDTreq(a1,a2,qa,d))        = IsTACKQOS(qa)
    ;
    not(IsTACKDTreq(t)) =>
IsValidTACKDTreq(t)          = false
    ;
IsTDISIn(t)            = IsTDIS(t) andthen (ins eq DISDirOf(t))
    ;
```

```
IsTDISOut(t)          = IsTDIS(t) andthen (out eq DISDirOf(t))
        ;
IsTDISInOut(t)        = IsTDIS(t) andthen (inout eq DISDirOf(t))
        ;
IsTREPindInp(t)       = IsTREPind(t)
                      andthen (RDir(QualReportOf(t)) eq Input)
        ;
IsTREPindOut(t)       = IsTREPind(t)
                      andthen (RDir(QualReportOf(t)) eq Output)
     ;
TCONind(a1,a3,pl1,qc11,ts11,d1) IsOutputOf
TCONreq(a,a2,pl,qc1,ts1,d)    = (a1 eq a) and (a3 eq a2)
     and (pl1 eq trad(pl)) and (qc11 IsValidIndFor qc1)
     and (ts11 IsValidIndFor ts1) and (d1 eq d)
     ;
TCONconf(a1,pl1,qc21,ts21,d1) IsOutputOf
TCONresp(a,pl,qc2,ts2,d)      = (a1 eq a) and (pl1 eq trad(pl))
     and (qc21 eq qc2) and (ts21 eq ts2) and (d1 eq d)
     ;
TDTind(d1,es) IsOutputOf TDTreq(d)              = true
     ;
TDISind(td1,r,d1) IsOutputOf TDISreq(td,d)  = (td1 eq trad(td))
                                    and (r eq User) and (d1 eq d)
     ;
TRELind(d1,es) IsOutputOf TRELreq(d)            = true
     ;
TRELconf IsOutputOf TRELind(d,es)               = true
     ;
TRENind(qc11,ts11) IsOutputOf TRENreq(qc1,ts1)      =
     (qc11 IsValidRenIndFor qc1) and (ts11 IsValidRenIndFor
ts1)     ;
TRENconf(qc21,ts21) IsOutputOf TRENresp(qc2,ts2)    = (qc21 eq
qc2)
                                        and (ts21 eq ts2)
     ;
TNEWQOSind IsOutputOf TRENconf(qc21,ts2)        = true
     ;
TREJRENind(rrr) IsOutputOf TREJRENreq           = rrr eq User
     ;
TOOBACKind(qa1,d1) IsOutputOf TOOBACKreq(qa,d)      = (qa1 eq
qa)
                                        and (d1 eq d)
     ;
TOOBACKconf(tdr) IsOutputOf TOOBACKind(qa,d)    = IsTDR(tdr)
                      and ((tdr ne NoRep) implies
                      (ThldVal(tdr) eq Thld(TACKQOSTrDel(qa))))
     ;
TREQRESPind(a1,a3,qr1,d1) IsOutputOf TREQRESPreq(a,a2,qr,d)
     =
     (a1 eq a) and (a3 eq a2) and (qr1 eq qr) and (d1 eq d)
     ;
TREQRESPconf(a1,d1) IsOutputOf TREQRESPresp(a,d)    = (a1 eq a)
                                        and (d1 eq d)
     ;
TACKDTind(a1,a3,qa1,d1) IsOutputOf TACKDTreq(a,a2,qa,d)    =
     (a1 eq a) and (a3 eq a2) and (qa1 eq qa) and (d1 eq d)
     ;
```

```
TACKDTconf(a1,tdr) IsOutputOf TACKDTind(a,a2,qa,d)      = (a1 eq
a)
                    and IsTDR(tdr) and ((tdr ne NoRep) implies
                    (ThldVal(tdr) eq Thld(TACKQOSTrDel(qa))))
    ;
TUDTind(a1,a3,ql1,d1) IsOutputOf TUDTreq(a,a2,ql,d)      = (a1 eq
a)
              and (a3 eq a2) and (ql1 eq ql) and (d1 eq d)
        ;
    not((IsTCONind(t1) and IsTCONreq(t)) or (IsTCONconf(t1) and
    IsTCONresp(t)) or (IsTDTind(t1) and IsTDTreq(t))
    or (IsTDISind(t1) and IsTDISreq(t)) or (IsTRELind(t1) and
    IsTRELreq(t)) or (IsTRELconf(t1) and IsTRELind(t))
    or (IsTRENind(t1) and IsTRENreq(t)) or (IsTRENconf(t1) and
    IsTRENind(t)) or (IsTNEWQOSind(t1) and IsTRENconf(t))
    or (IsTREJRENind(t1) and IsTREJRENreq(t)) or
    (IsTOOBACKind(t1) and IsTOOBACKreq(t))
    or (IsTOOBACKconf(t1) and IsTOOBACKind(t)) or
    (IsTREQRESPind(t1) and IsTREQRESPreq(t))
    or (IsTREQRESPconf(t1) and IsTREQRESPresp(t)) or
    (IsTACKDTind(t1) and IsTACKDTreq(t))
    or (IsTACKDTconf(t1) and IsTACKDTind(t))
    or (IsTUDTind(t1) and IsTUDTreq(t))) =>
t1 IsOutputOf t = false   ;
TCONconf(a1,pl1,qc2,ts2,d1) IsValidTCON2For
TCONreq(a,a2,pl,qc1,ts1,d)   = (a1 eq a) and (pl1 eq pl)
    and (qc2 IsValidCon2For qc1) and (ts2 IsValidCon2For ts1)
    ;
TCONresp(a1,pl1,qc2,ts2,d1) IsValidTCON2For
TCONind(a,a2,pl,qc1,ts1,d)   = (a1 eq a) and (pl1 eq pl)
    and (qc2 IsValidCon2For qc1) and (ts2 IsValidCon2For ts1)
    ;
TRELconf IsValidTCON2For TRELreq(d)      = true
    ;
TRENconf(qc2,ts2) IsValidTCON2For TRENreq(qc1,ts1)      =
    (qc2 IsValidRenCon2For qc1) and (ts2 IsValidRenCon2For
ts1)   ;
TRENresp(qc2,ts2) IsValidTCON2For TRENind(qc1,ts1)       =
    (qc2 IsValidRenCon2For qc1) and (ts2 IsValidRenCon2For
ts1)   ;
TOOBACKconf(tdr) IsValidTCON2For TOOBACKreq(qa,d) =
IsTDR(tdr) ;
TREQRESPconf(a1,d) IsValidTCON2For
TREQRESPreq(a,a2,qr,d)   = a1 eq a        ;
TREQRESPresp(a1,d) IsValidTCON2For
TREQRESPind(a,a2,qr,d)   = a1 eq a        ;
TACKDTconf(a1,tdr) IsValidTCON2For
TACKDTreq(a,a2,qa,d)  = (a1 eq a) and IsTDR(tdr)      ;
    not((IsTCONconf(t1) and IsTCONreq(t)) or (IsTCONresp(t1)
and
    IsTCONind(t)) or (IsTRELconf(t1) and IsTRELreq(t))
    or (IsTOOBACKconf(t1) and IsTOOBACKreq(t))
    or (IsTRENconf(t1) and IsTRENreq(t))
    or (IsTRENresp(t1) and IsTRENind(t))
    or (IsTREQRESPconf(t1) and IsTREQRESPreq(t))
    or (IsTREQRESPresp(t1) and IsTREQRESPind(t))
    or (IsTACKDTconf(t1) and IsTACKDTreq(t))) =>
t1 IsValidTCON2For t = false   ;
```

```
endtype (* TransportServicePrimitive *)
(*
```

3.5 Parameters of the ETS primitives

Exception made for the qualities of service treated in 3.6, and for the TCTSDU size treated in 3.7, the various parameters of the primitives are defined here.

TPlex is used to specify, when establishing a connection, the directions of transfer that are required to be opened.

TErrorSignal signals whether corruption and/or losses occurred on the data delivered by a TDATAindication or a TRELEASEindication. It specifies the first and the last octet where damages (either corruption or loss) were observed. These two values are set to zero when the data is transmitted correctly.

TQualityReport is used by primitive TREPORTindication to indicate which quality parameter has been observed under the negotiated threshold level (TRepCau) and which direction of transfer is concerned (TRepDir). TRepCau recalls the threshold value, in order to allow the distinction among reports relative to former or new qualities after a re-negotiation, and also, for the TransitDelay and the Jitter, it gives the actual value observed that caused the report.

TDISDir indicates the transfer direction closed by a TDISCONNECT primitive. ins (resp. out) means that, from the point of view of the entity where the TDISCONNECT primitive is emitted, the direction of reception (resp. emission) is closed. inout signifies the closure of the connection.

TDISReason defines the reason of a disconnect indication : either because of a user request, or because a quality of service was observed under the compulsory level, or because of another reason from the provider. When for quality of service reasons, similar indications are given as for the report indication.

TREJReason just represents the various (unspecified) reasons the service provider could give to reject an out-of-band communication.

TRejectRenegotiationReason indicates whether a renegotiation was rejected by the service provider or by the called entity.

TransDelReport is used by TOOBACKconf and TACKDTconf to signal to the emitter of an out-of-band transmission whether or not the transmission delay happened to be higher than the required threshold value. In case of excess, it recalls the threshold value and indicates the actual one.   *)

```
type TPlex
is NaturalNumber
sorts TPlex
opns
    both    : -> TPlex
    him     : -> TPlex
    me      : -> TPlex
    h       : TPlex              -> Nat
    _eq_    : TPlex, TPlex      -> Bool
    _ne_    : TPlex, TPlex -> Bool
    trad    : TPlex             -> TPlex
eqns
forall p1,p11:TPlex
```

```
ofsort Nat
h(both)         = 0                 ;
h(him)          = Succ(0)           ;
h(me)           = Succ(Succ(0))     ;
ofsort Bool
p1 eq p11       = h(p1) eq h(p11) ;
p1 ne p11       = not(p1 eq p11)    ;
ofsort TPlex
trad(both)      = both ;
trad(him)       = me   ;
trad(me)        = him  ;
endtype(* TPlex *)


type TErrorSignal is NatRepresentations, Signal
sorts   TErSig
opns
    ErSig          : Nat, Nat, Sign, Sign     -> TErSig
    FirDam         : TErSig               -> Nat
    LastDam        : TErSig               -> Nat
    CSigOf         : TErSig               -> Sign
    LSigOf         : TErSig               -> Sign
    _eq_           : TErSig, TErSig   -> Bool
eqns
forall n1,n2,n3,n4:Nat,  s1,s2,s3,s4:Sign
ofsort Nat
FirDam(ErSig(n1,n2,s1,s2))    = n1    ;
LastDam(ErSig(n1,n2,s1,s2))   = n2    ;
ofsort Sign
CSigOf(ErSig(n1,n2,s1,s2))    = s1    ;
LSigOf(ErSig(n1,n2,s1,s2))    = s2    ;
ofsort Bool
ErSig(n1,n2,s1,s2) eq ErSig(n3,n4,s3,s4) = (n1 eq n3)
    and (n2 eq n4) and (s1 eq s3) and (s2 eq s4)     ;
endtype (* TErrorSignal *)


type Signal is Doublet renamedby
sortnames   Sign    for Doublet
opnnames    Sig     for TheOne
            NoSig   for TheOther
endtype (* Signal *)


type TQualityReport is Pair actualizedby TReportCause,
TReportDirection
using
sortnames   TQR     for Pair
            TRepCau     for Element
            TRepDir     for OtherElement
            Bool        for FBool
opnnames    TQR         for Pair
            RCau    for First
            RDir    for Second
endtype (* TQualityReport *)


type TReportCause is TQOSValue
sorts   TRepCau
opns
```

```
    Throughput              : QVal              -> TRepCau
    Jitter                  : QVal, QVal        -> TRepCau
    TransitDelay            : QVal, QVal        -> TRepCau
    CorruptErrorRate        : QVal              -> TRepCau
    LossesErrorRate         : QVal              -> TRepCau
    Cau                         : TRepCau          -> Nat
    ThldVal                 : TRepCau           -> QVal
    ActualVal               : TRepCau           -> QVal
    IsCauThr                : TRepCau           -> Bool
    IsCauJit                : TRepCau           -> Bool
    IsCauTrD                : TRepCau           -> Bool
    IsCauCER                : TRepCau           -> Bool
    IsCauLER                : TRepCau           -> Bool
    _eq_                    : TRepCau, TRepCau       -> Bool
    _ne_                    : TRepCau, TRepCau       -> Bool
eqns
forall v,v1,v2,v3:QVal, trc,trc1:TRepCau

ofsort Nat
Cau(Throughput(v))           = 0      ;
Cau(Jitter(v,v1))            = succ(0)  ;
Cau(TransitDelay(v,v1))      = succ(succ(0))     ;
Cau(CorruptErrorRate(v))        = succ(succ(succ(0))) ;
Cau(LossesErrorRate(v))      = succ(succ(succ(succ(0))))    ;

ofsort QVal
ThldVal(Throughput(v))          = v      ;
ThldVal(Jitter(v,v1))           = v      ;
ThldVal(TransitDelay(v,v1))     = v      ;
ThldVal(CorruptErrorRate(v))    = v      ;
ThldVal(LossesErrorRate(v))     = v      ;
ActualVal(Jitter(v,v1))         = v1     ;
ActualVal(TransitDelay(v,v1))      = v1     ;

ofsort Bool
IsCauThr(trc)      = Cau(trc) eq 0                     ;
IsCauJit(trc)      = Cau(trc) eq succ(0)              ;
IsCauTrD(trc)      = Cau(trc) eq succ(succ(0))             ;
IsCauCER(trc)      = Cau(trc) eq succ(succ(succ(0)))          ;
IsCauLER(trc)      = Cau(trc) eq succ(succ(succ(succ(0)))) ;
    not(Cau(trc) eq Cau(trc1)) =>
trc eq trc1        = false                    ;
Throughput(v) eq Throughput(v1)  = v eq v1       ;
Jitter(v,v1) eq Jitter(v2,v3)        = (v eq v2) and (v1 eq
v3)      ;
TransitDelay(v,v1) eq TransitDelay(v2,v3)    = (v eq v2)
                                    and (v1 eq v3)
       ;
CorruptErrorRate(v) eq CorruptErrorRate(v1)      = v eq v1
       ;
LossesErrorRate(v) eq LossesErrorRate(v1)      = v eq v1
    ;
trc ne trc1        = not(trc eq trc1)
    ;
endtype (* TReportCause *)

type TQOSValue is NatRepresentations
sorts QVal
opns
    All       :              -> QVal
    QVal1   : Nat         -> QVal
```

```
    QVal2    : Nat, Nat      -> QVal
    _eq_     : QVal, QVal     -> Bool
    _ne_     : QVal, QVal     -> Bool
eqns
forall n,n1,n2,n3:Nat, qv,qv1:QVal
ofsort Bool
All eq All                          = true        ;
All eq QVal1(n)                     = false       ;
QVal1(n) eq All                     = false       ;
All eq QVal2(n1,n2)                 = false       ;
QVal2(n,n1) eq All                  = false       ;
QVal1(n) eq QVal2(n1,n2)            = false       ;
QVal2(n,n1) eq QVal1(n2)            = false       ;
QVal1(n) eq QVal1(n1)               = n eq n1     ;
QVal2(n,n1) eq QVal2(n2,n3)         = (n eq n2) and (n1 eq n3)     ;
qv ne qv1                           = not(qv eq qv1)   ;
endtype (* TQOSValue *)


type TReportDirection is Doublet renamedby
sortnames   TRepDir      for Doublet
opnnames    Input        for TheOne
            Output       for TheOther
endtype (* TReportDirection *)


type TDISDir is Boolean
sorts   TDD
opns
    ins          : -> TDD
    out          : -> TDD
    inout    : -> TDD
    trad     : TDD           -> TDD
    _eq_     : TDD, TDD       -> Bool
    _ne_     : TDD, TDD       -> Bool
eqns
Forall u,v:TDD
ofsort TDD
trad(ins)       = out       ;
trad(out)       = ins       ;
trad(inout)     = inout     ;
ofsort Bool
u eq u          = true      ;
ins eq out      = false     ;
ins eq inout    = false     ;
out eq ins      = false     ;
out eq inout    = false     ;
inout eq ins    = false     ;
inout eq out    = false     ;
u ne v          = not(u eq v)   ;
endtype (* TDISDir *)


type TDISReason is TDisCause, TDisDirection
sorts   TDISReason
opns
    User         : -> TDISReason
    ProvNoQual   : -> TDISReason
    ProvQual     : TDisCau, TDisDir      -> TDISReason
    DCau         : TDISReason            -> TDisCau
```

```
    DDir          : TDISReason                   -> TDisDir
    IsQualCause   : TDISReason                   -> Bool
    _eq_          : TDISReason, TDISReason -> Bool
    _ne_          : TDISReason, TDISReason    -> Bool
eqns
forall dca,dca1:TDisCau, ddi,ddi1:TDisDir, tdr,tdr1:TDISReason
ofsort TDisCau
DCau(ProvQual(dca,ddi))   = dca          ;
ofsort TDisDir
DDir(ProvQual(dca,ddi))   = ddi          ;
ofsort Bool
IsQualCause(User)            = false    ;
IsQualCause(ProvNoQual)      = false    ;
IsQualCause(ProvQual(dca,ddi))    = true ;
User eq User                         = true  ;
User eq ProvNoQual                   = false    ;
User eq ProvQual(dca,ddi)              = false    ;
ProvNoQual eq User                   = false    ;
ProvNoQual eq ProvNoQual             = true  ;
ProvNoQual eq ProvQual(dca,ddi)   = false    ;
ProvQual(dca,ddi) eq User            = false    ;
ProvQual(dca,ddi) eq ProvNoQual   = false    ;
ProvQual(dca,ddi) eq ProvQual(dca1,ddi1) = (dca eq dca1)
                                    and (ddi eq ddi1) ;
tdr ne tdr1    = not (tdr eq tdr1)    ;
endtype (* TDISReason *)


type TDisCause is TReportCause renamedby
sortnames  TDisCau    for TRepCau
endtype (* TDisCause *)


type TDisDirection is TReportDirection renamedby
sortnames  TDisDir    for TRepDir
endtype (* TDisDirection *)


type TREJReason is Boolean
sorts   TRejReason
opns
    SomeReasons   : -> TRejReason
    _eq_          : TRejReason, TRejReason -> Bool
eqns
ofsort Bool
SomeReasons eq SomeReasons = true    ;
endtype (* TREJReason *)


type TRejectRenegociationReason is Doublet renamedby
sortnames  TRejRenReason  for Doublet
opnnames   User           for TheOne
           Provider        for TheOther
endtype (* TRejectRenegociationReason *)


type TransDelReport is Boolean, NatRepresentations
sorts   TDR
opns
    NoRep       :            -> TDR
    Rep         : Nat, Nat -> TDR
```

```
      ThldVal     : TDR        -> Nat
      ActualVal   : TDR        -> Nat
      IsTDR       : TDR        -> Bool
      _eq_        : TDR, TDR -> Bool
      _ne_        : TDR, TDR -> Bool
eqns
forall n,n1,n2,n3:Nat, tdr,tdr1:TDR

ofsort Nat
ThldVal(Rep(n,n1))     = n          ;
ActualVal(Rep(n,n1))   = n1    ;

ofsort Bool
IsTDR(NoRep)           = true    ;
IsTDR(Rep(n,n1))       = n le n1  ;
NoRep eq NoRep             = true      ;
NoRep eq Rep(n,n1)        = false                       ;
Rep(n,n1) eq NoRep        = false                       ;
Rep(n,n1) eq Rep(n2,n3)   = (n eq n2) and (n1 eq n3)   ;
tdr ne tdr1            = not(tdr eq tdr1)                    ;

endtype (* TransDelReport *)
(*
```

3.6 QoS parameter

In 3.6.1 is defined the quality of service parameter for the PP-ECOTS in-band
transmissions, in 3.6.2 for the PP-EACLTS, in 3.6.3 for the PP-ERCLTS and in 3.6.4
for the PP-EUCLTS. The quality parameter for the PP-ECOTS out-of-band
transmissions is the same as the one for the PP-EACLTS, as both are based on the
same mechanism.

3.6.1 QoS parameter for the PP-ECOTS in-band transmissions

The definition of all the quality parameters is first given. They are then combined into
a quality of service structure. This structure can get three forms, depending on the
directions of transmission opened i.e. from the point of view of the entity that sets the
values: only the direction of emission, only the direction of reception or both. For the
global QoS structure two sorts are defined, that traduce the two phases of the
negotiation of the compulsory values: in the first two connect or re-negotiate
primitives (request and indication), opportunities are left for negotiations, but within
specified bounds. For the last two primitives (response and confirmation) the parameters are definitively fixed. This duality is present in the definition of the
concerned parameters i.e. the throughput, the transit delay, the transit delay jitter and
the residual error rate. The traffic type indicator is a non negotiated parameter that
each entity states for its own direction of emission. The error control parameter is
used to determine the error management, that is negotiated between the two entities.
But each entity also has the opportunity, if losses are admitted for the data it receives,
to specify independently a dummy octet that should replace the missing ones. The
TCProtection and TCPriority parameters are the same as the ones defined in ISO
8072.

NOTE - The values specified for the compulsory QoS are always maximal (or minimal) allowed values, i.e. a disconnection occurs, or an indication is emitted only when these values are exceeded, not when they are just reached.

3.6.1.1 Throughput

The first parameter defined is the throughput. Throughput1 and Throughput2 correspond to the two phases explained above. MakeThr1FreeTresh and MakeThr1DeltaTresh traduce two possible options in order to fix the proposed and the maximal admitted threshold values: either directly, or by the way of a 'delta' parameter that states a gap between the (maximal adm.) threshold and the (maximal adm.) minimum.

Tests (of the form IsThr...) are defined in order to express constraints that the various values composing the throughput must respect.

IsValidIndFor, introduced in the type Throughput, specifies the link between the throughput value appearing in a request primitive and the corresponding value in the indication.

IsValidCon2For does the same from the indication to the corresponding response.
    *)

```
type Throughput1
is NatRepresentations
sorts   Thr1
opns
    MakeThr1FreeTresh        : Nat, Nat, Nat, Nat, Nat      -> Thr1
    MakeThr1DeltaTresh       : Nat, Nat, Nat, Nat        -> Thr1
    Min              : Thr1      -> Nat
    MaxMin           : Thr1      -> Nat
    Thld             : Thr1      -> Nat
    MaxThld          : Thr1      -> Nat
    Max                : Thr1       -> Nat
    _eq_               : Thr1, Thr1   -> Bool
    _ne_               : Thr1, Thr1   -> Bool
    IsThr1FreeTresh    : Thr1      -> Bool
    IsThr1DeltaTresh   : Thr1      -> Bool
    IsThr1             : Thr1      -> Bool
eqns
forall n1,n2,n3,n4,n5,n6,n7,n8,n9,n10:Nat, rt,rt1:Thr1
ofsort Nat
Min(MakeThr1FreeTresh(n1,n2,n3,n4,n5))        = n1        ;
Min(MakeThr1DeltaTresh(n1,n2,n3,n4))          = n1        ;
MaxMin(MakeThr1FreeTresh(n1,n2,n3,n4,n5))     = n2        ;
MaxMin(MakeThr1DeltaTresh(n1,n2,n3,n4))       = n2        ;
Thld(MakeThr1FreeTresh(n1,n2,n3,n4,n5))         = n3         ;
Thld(MakeThr1DeltaTresh(n1,n2,n3,n4))         = n1 + n3   ;
MaxThld(MakeThr1FreeTresh(n1,n2,n3,n4,n5))    = n4        ;
MaxThld(MakeThr1DeltaTresh(n1,n2,n3,n4))      = n2 + n3   ;
Max(MakeThr1FreeTresh(n1,n2,n3,n4,n5))        = n5        ;
Max(MakeThr1DeltaTresh(n1,n2,n3,n4))          = n4        ;

ofsort Bool
MakeThr1FreeTresh(n1,n2,n3,n4,n5) eq
MakeThr1FreeTresh(n6,n7,n8,n9,n10)   = (n1 eq n6) and (n2 eq n7)
    and (n3 eq n8) and (n4 eq n9) and (n5 eq n10)            ;
```

```
MakeThr1DeltaTresh(n1,n2,n3,n4) eq
MakeThr1DeltaTresh(n5,n6,n7,n8)        = (n1 eq n5) and (n2 eq
n6)
     and (n3 eq n7) and (n4 eq n8)
      ;
MakeThr1FreeTresh(n1,n2,n3,n4,n5) eq
MakeThr1DeltaTresh(n6,n7,n8,n9)        = false                 ;
MakeThr1DeltaTresh(n1,n2,n3,n4) eq
MakeThr1FreeTresh(n5,n6,n7,n8,n9)      = false                 ;
rt ne rt1 = not(rt eq rt1)                                     ;
IsThr1FreeTresh(MakeThr1FreeTresh(n1,n2,n3,n4,n5)) = (n1 le
n2)
     and (n3 le n4) and (n1 le n3) and (n2 le n4) and (n4 le
n5)      ;
IsThr1FreeTresh(MakeThr1DeltaTresh(n1,n2,n3,n4))    = false
      ;
IsThr1DeltaTresh(MakeThr1DeltaTresh(n1,n2,n3,n4))   = (n1 le
n2)
     and ((n2 + n3) le n4)
      ;
IsThr1DeltaTresh(MakeThr1FreeTresh(n1,n2,n3,n4,n5))    = false
      ;
IsThr1(rt) = IsThr1FreeTresh(rt) or IsThr1DeltaTresh(rt)
      ;
endtype (* Throughput1 *)

type Throughput2
is NatRepresentations
sorts   Thr2
opns
    MakeThr2      : Nat, Nat, Nat    -> Thr2
    Min              : Thr2       -> Nat
    Thld          : Thr2       -> Nat
    Max              : Thr2       -> Nat
    _eq_          : Thr2, Thr2   -> Bool
    _ne_          : Thr2, Thr2   -> Bool
    IsThr2        : Thr2           -> Bool
eqns
forall n1,n2,n3,n4,n5,n6:Nat, ct,ct1:Thr2
ofsort Nat
Min(MakeThr2(n1,n2,n3))   = n1    ;
Thld(MakeThr2(n1,n2,n3))     = n2    ;
Max(MakeThr2(n1,n2,n3))   = n3    ;
ofsort Bool
MakeThr2(n1,n2,n3) eq MakeThr2(n4,n5,n6)     = (n1 eq n4)
    and (n2 eq n5) and (n3 eq n6)     ;
ct ne ct1  = not(ct eq ct1)         ;
IsThr2(MakeThr2(n1,n2,n3)) = (n1 le n2) and (n2 le n3) ;
endtype (* Throughput2 *)

type Throughput
is Throughput1, Throughput2
opns
    _IsValidIndFor_    : Thr1, Thr1   -> Bool
    _IsValidCon2For_   : Thr2, Thr1   -> Bool
eqns
forall n1,n2,n3,n4,n5,n6,n7,n8,n9,n10:Nat,
     ct1,ct11:Thr1, ct2:Thr2
```

```
ofsort Bool
    not((IsThr1FreeTresh(ct1) and IsThr1FreeTresh(ct11))
    or (IsThr1DeltaTresh(ct1) and IsThr1DeltaTresh(ct11))) =>
ct1 IsValidIndFor ct11    = false      ;
    (IsThr1FreeTresh(ct1) and IsThr1FreeTresh(ct11)) or
    (IsThr1DeltaTresh(ct1) and IsThr1DeltaTresh(ct11)) =>
ct11 IsValidIndFor ct1    = (Min(ct11) eq Min(ct1))
                        and (MaxMin(ct11) le MaxMin(ct1))
                        and (Thld(ct11) eq Thld(ct1))
                        and (MaxThld(ct11) le MaxThld(ct1))
                        and (Max(ct11) le Max(ct1))   ;
    not(IsThr2(ct2) and IsThr1(ct1)) =>
ct2 IsValidCon2For ct1    = false      ;
    IsThr2(ct2) and IsThr1(ct1) =>
ct2 IsValidCon2For ct1    = (Min(ct2) ge Min(ct1))
    and (Min(ct2) le MaxMin(ct1)) and (Thld(ct2) ge Thld(ct1))
    and (Thld(ct2) le MaxThld(ct1)) and (Max(ct2) le Max(ct1))
    ;
endtype (* Throughput *)
(*
```

3.6.1.2 Traffic type indicator

By the way of TrafficTypeInd, an entity indicates to the service provider the type of
traffic (i.e. synchronous or not) it will have to treat. Due to the lack of a timed
LOTOS the influence of this parameters was not traduced in a constraint.   *)

```
type TrafficTypeInd is Doublet renamedby
sortnames  TTI              for Doublet
opnnames   Sync          for TheOne
           AnyTraf       for TheOther
endtype (* TrafficTypeInd *)
(*
```

3.6.1.3 Transit delay and transit delay jitter

The types Delay1 and Delay2 and Delay are generic. By renaming, they are used to
define the TransitDelay and the DelayJitter parameters. Their structure is similar to
the one of the Throughput, except that here, better qualities correspond to smaller
values, and that no Min, equivalent to the Max of the throughput is defined. The type
NatSubs defines the operator mn (minus) on two natural numbers. It will be used in
the Delay types.   *)

```
type NatSubs
is NatRepresentations
opns
    _mn_     : Nat,Nat   -> Nat
eqns
forall n,n1:Nat
ofsort Nat
0 mn succ(n)           = 0               ;
n mn 0                 = n               ;
succ(n) mn succ(n1)    = n mn n1  ;
endtype (* NatSubs *)

type Delay1
```

```
is NatSubs
sorts   Del1
opns
    MakeDel1FreeTresh       : Nat, Nat, Nat, Nat  -> Del1
    MakeDel1DeltaTresh      : Nat, Nat, Nat       -> Del1
    Max                     : Del1      -> Nat
    MinMax          : Del1      -> Nat
    Thld            : Del1      -> Nat
    MinThld         : Del1      -> Nat
    _eq_            : Del1, Del1    -> Bool
    _ne_            : Del1, Del1    -> Bool
    IsDel1FreeTresh     : Del1      -> Bool
    IsDel1DeltaTresh    : Del1      -> Bool
    IsDel1          : Del1      -> Bool
eqns
forall n1,n2,n3,n4,n5,n6,n7,n8:Nat, rt,rt1:Del1
ofsort Nat
Max(MakeDel1FreeTresh(n1,n2,n3,n4))        = n1           ;
Max(MakeDel1DeltaTresh(n1,n2,n3))          = n1           ;
MinMax(MakeDel1FreeTresh(n1,n2,n3,n4))     = n2           ;
MinMax(MakeDel1DeltaTresh(n1,n2,n3))       = n2           ;
Thld(MakeDel1FreeTresh(n1,n2,n3,n4))       = n3           ;
Thld(MakeDel1DeltaTresh(n1,n2,n3))         = (n1 mn n3)   ;
MinThld(MakeDel1FreeTresh(n1,n2,n3,n4))      = n4             ;
MinThld(MakeDel1DeltaTresh(n1,n2,n3))      = (n2 mn n3)   ;
ofsort Bool
MakeDel1FreeTresh(n1,n2,n3,n4) eq
MakeDel1FreeTresh(n5,n6,n7,n8)    = (n1 eq n5) and (n2 eq n6)
and (n3 eq n7) and (n4 eq n8)     ;
MakeDel1DeltaTresh(n1,n2,n3) eq MakeDel1DeltaTresh(n4,n5,n6)
    =
                     (n1 eq n4) and (n2 eq n5) and (n3 eq n6)
    ;
MakeDel1FreeTresh(n1,n2,n3,n4) eq
MakeDel1DeltaTresh(n5,n6,n7) =
                                                      false
    ;
MakeDel1DeltaTresh(n1,n2,n3) eq
MakeDel1FreeTresh(n4,n5,n6,n7)    =
                     false  ;
rt ne rt1 = not(rt eq rt1)          ;
IsDel1FreeTresh(MakeDel1FreeTresh(n1,n2,n3,n4))    = (n1 ge
n2)
    and (n3 ge n4) and (n1 ge n3) and (n2 ge n4) and (n4 gt 0)
    ;
IsDel1FreeTresh(MakeDel1DeltaTresh(n1,n2,n3))   = false
    ;
IsDel1DeltaTresh(MakeDel1DeltaTresh(n1,n2,n3))     = (n1 ge
n2)
                                          and (n2 gt n3)
    ;
IsDel1DeltaTresh(MakeDel1FreeTresh(n1,n2,n3,n4)) = false
    ;
IsDel1(rt) = IsDel1FreeTresh(rt) or IsDel1DeltaTresh(rt)
    ;
endtype (* Delay1 *)

type Delay2
is NatRepresentations
```

```
sorts   Del2
opns
    MakeDel2     : Nat, Nat      -> Del2
    Max             : Del2          -> Nat
    Thld        : Del2         -> Nat
    _eq_        : Del2, Del2   -> Bool
    _ne_        : Del2, Del2   -> Bool
    IsDel2      : Del2         -> Bool
eqns
forall n1,n2,n3,n4:Nat, ct,ct1:Del2

ofsort Nat
Max(MakeDel2(n1,n2))  = n1    ;
Thld(MakeDel2(n1,n2)) = n2    ;

ofsort Bool
MakeDel2(n1,n2) eq MakeDel2(n3,n4) = (n1 eq n3) and (n2 eq
n4)     ;
ct ne ct1                          = not(ct eq ct1)
    ;
IsDel2(MakeDel2(n1,n2))            = (n1 ge n2) and (n2 gt 0)
    ;
endtype (* Delay2 *)


type Delay
is Delay1, Delay2
opns
    _IsValidIndFor_    : Del1, Del1   -> Bool
    _IsValidCon2For_   : Del2, Del1   -> Bool
eqns
forall n1,n2,n3,n4,n5,n6,n7,n8,n9,n10:Nat,
       ct1,ct11:Del1, ct2:Del2

ofsort Bool
    not((IsDel1FreeTresh(ct1) and IsDel1FreeTresh(ct11)) or
    (IsDel1DeltaTresh(ct1) and IsDel1DeltaTresh(ct11))) =>
ct1 IsValidIndFor ct11    = false    ;
    (IsDel1FreeTresh(ct1) and IsDel1FreeTresh(ct11)) or
    (IsDel1DeltaTresh(ct1) and IsDel1DeltaTresh(ct11)) =>
ct11 IsValidIndFor ct1    = (Thld(ct11) eq Thld(ct1))
                           and (MinThld(ct11) ge MinThld(ct1))
                           and (Max(ct11) eq Max(ct1))
                           and (MinMax(ct11) ge MinMax(ct1))
    ;
    not(IsDel2(ct2) and IsDel1(ct1)) =>
ct2 IsValidCon2For ct1    = false    ;
    IsDel2(ct2) and IsDel1(ct1) =>
ct2 IsValidCon2For ct1    = (Max(ct2) le Max(ct1))
                           and (Max(ct2) ge MinMax(ct1))
                           and (Thld(ct2) le Thld(ct1))
                           and (Thld(ct2) ge MinThld(ct1))  ;

endtype (* Delay *)


type TransitDelay is Delay renamedby
sortnames   TrD1     for Del1
            TrD2     for Del2
opnnames    MakeTrD1FreeTresh     for MakeDel1FreeTresh
            MakeTrD1DeltaTresh    for MakeDel1DeltaTresh
            IsTrD1FreeTresh       for IsDel1FreeTresh
            IsTrD1DeltaTresh      for IsDel1DeltaTresh
            IsTrD1                for IsDel1
```

```
            MakeTrD2                   for MakeDel2
            IsTrD2                     for IsDel2
endtype (* TransitDelay *)

type Jitter is Delay renamedby
sortnames   Jit1     for Del1
            Jit2     for Del2
opnnames    MakeJit1FreeTresh          for MakeDel1FreeTresh
            MakeJit1DeltaTresh         for MakeDel1DeltaTresh
            IsJit1FreeTresh            for IsDel1FreeTresh
            IsJit1DeltaTresh           for IsDel1DeltaTresh
            IsJit1                     for IsDel1
            MakeJit2                   for MakeDel2
            IsJit2                     for IsDel2
endtype (* Jitter *)
(*
```

3.6.1.4 Error control and residual error rate

The opportunity is given to the communicating entities to negotiate the kind of error control they require on each direction of transmission. The most demanding point of vue is retained. It must be specified whether corruption of the transmitted data is forbidden, allowed but signalled or allowed but not signalled, and whether the loss of data is not admitted or admitted but signalled. All combinations of these two items are allowed. As an information parameter, an entity may also (but is not obliged to) indicate to the service provider a dummy octet by which the missing ones, in case of loss, must be replaced. If no dummy octet is specified, a lost octet can be replaced by any other octet. In case of the loss of all the octets composing a TSDU, there is no replacement, the UserData parameter of the TDATA indication or of the TRELEA-SEindication remains void.

The error control parameter is tightly bound to the residual error rate one (see 3.6.1.4.1). This comes from the fact that the need to specify an error rate depends on the choice made for the error control. Separate error rates are defined for the corruption and for the losses. For the corruption, no error rate needs to be determined if the control required is: "No Corruption", what is obvious, or "Corruption allowed and not signalled", an option used to deactivate the error control mechanism. For the losses, the choice "No Losses" does not require any error rate parameter.

3.6.1.4.1 Residual error rate

The definition of the residual error rate is similar to the one of the preceding parameters, except that the elements whose Cartesian product gives the parameter are no more simple naturals, but pairs of such numbers. These pairs, of sort ERD, are defined in the type ErrorRateDefinition. The mechanism they describe is quite simple: a maximal number of errors is given that may occur in a sequence whose size is also defined. More detailed explanations are given in the comments of TCCorruptionResErrorRateController and TCLossesResErrorRateController (see 5.3.2.9), the processes that specify these constraints.

A special relation of order is defined among the elements of sort ERD ('le' is to be understood here as: more restrictive): a value is less or equal to another one if on a taller sequence it does not allow more errors or if on a shorter sequence, its ratio: errors number on sequence size, is smaller.   *)

```
type ErrorRateDefinition is NatRepresentations
sorts   ERD
opns
    MakeERD              : Nat, Nat -> ERD
    MaxErNumber    : ERD   -> Nat
    SequenceSize   : ERD   -> Nat
    IsERD               : ERD   -> Bool
    _eq_         : ERD, ERD -> Bool
    _ne_         : ERD, ERD -> Bool
    _le_         : ERD, ERD -> Bool
    _ge_         : ERD, ERD -> Bool
eqns
forall n1,n2,n3,n4:Nat, re,re1:ERD

ofsort Nat
MaxErNumber(MakeERD(n1,n2))        = n1    ;
SequenceSize(MakeERD(n1,n2))       = n2    ;

ofsort Bool
IsERD(MakeERD(n1,n2))              = n1 le n2
    ;
MakeERD(n1,n2) eq MakeERD(n3,n4) = (n1 eq n3) and (n2 eq n4)
    ;
re ne re1                         = not(re eq re1)
    ;
MakeERD(n3,n4) le MakeERD(n1,n2) = ((n4 ge n2) and (n3 le n1))
                    or ((n4 le n2) and ((n3 * n2) le (n1 *
n4)))   ;
MakeERD(n3,n4) ge MakeERD(n1,n2) = MakeERD(n1,n2) le
                                  MakeERD(n3,n4)
    ;

endtype (* ErrorRateDefinition *)

type ResidualErrorRate1 is ErrorRateDefinition
sorts   RER1
opns
    MakeRER1    : ERD, ERD, ERD, ERD  -> RER1
    Max              : RER1              -> ERD
    MinMax      : RER1              -> ERD
    Thld        : RER1              -> ERD
    MinThld     : RER1              -> ERD
    _eq_        : RER1, RER1       -> Bool
    _ne_        : RER1, RER1       -> Bool
    IsRER1      : RER1             -> Bool
eqns
forall n1,n2,n3,n4,n5,n6,n7,n8:ERD, re,re1:RER1

ofsort ERD
Max(MakeRER1(n1,n2,n3,n4))         = n1    ;
MinMax(MakeRER1(n1,n2,n3,n4))      = n2    ;
Thld(MakeRER1(n1,n2,n3,n4))        = n3    ;
MinThld(MakeRER1(n1,n2,n3,n4))     = n4    ;

ofsort Bool
MakeRER1(n1,n2,n3,n4) eq MakeRER1(n5,n6,n7,n8) = (n1 eq n5)
    and (n2 eq n6) and (n3 eq n7) and (n4 eq n8)     ;
re ne re1 = not(re eq re1)    ;
IsRER1(MakeRER1(n1,n2,n3,n4))      = IsERD(n1) and IsERD(n2)
                    and IsERD(n3) and IsERD(n4) and (n1 ge n2)
                    and (n3 ge n4) and (n1 ge n3) and (n2 ge n4)
                    and (SequenceSize(n1) eq SequenceSize(n2))
```

```
                          and (SequenceSize(n1) eq SequenceSize(n3))
                          and (SequenceSize(n1) eq SequenceSize(n4))
      ;
endtype (* ResidualErrorRate1 *)

type ResidualErrorRate2 is ErrorRateDefinition
sorts   RER2
opns
    MakeRER2      : ERD, ERD      -> RER2
    Max           : RER2          -> ERD
    Thld          : RER2          -> ERD
    _eq_          : RER2, RER2    -> Bool
    _ne_          : RER2, RER2    -> Bool
    IsRER2        : RER2          -> Bool
eqns
forall n1,n2,n3,n4:ERD, re,re1:RER2

ofsort ERD
Max(MakeRER2(n1,n2))  = n1    ;
Thld(MakeRER2(n1,n2)) = n2    ;
ofsort Bool
MakeRER2(n1,n2) eq MakeRER2(n3,n4)   = (n1 eq n3)
                                       and (n2 eq n4)
    ;
re ne re1                      = not(re eq re1)
    ;
IsRER2(MakeRER2(n1,n2))   = IsERD(n1) and IsERD(n2)
    and (n1 ge n2) and (SequenceSize(n1) eq SequenceSize(n2))
    ;
endtype (* ResidualErrorRate2 *)

type ResidualErrorRate is
ResidualErrorRate1,ResidualErrorRate2
opns
    _IsValidCon2For_   : RER2, RER1 -> Bool
eqns
forall re,re1:RER1, re2:RER2
ofsort Bool
    (IsRER2(re2) and IsRER1(re1))     =>
re2 IsValidCon2For re1 = (Max(re2) le Max(re1))
    and (Max(re2) ge MinMax(re1)) and (Thld(re2) le Thld(re1))
    and (Thld(re2) ge MinThld(re1))  ;
    not(IsRER2(re2) and IsRER1(re1)) =>
re2 IsValidCon2For re1 = false   ;
endtype (* ResidualErrorRate *)
(*
```

3.6.1.4.2 Error control

The structures of the types Corruption and LosD correspond to the explanations given
in 3.6.1.4. Various constructors are defined, relative to the different options, and
taking the required parameters. Let us notice that for the error controls which imply a
residual error rate, two operators must be defined, one for each sort of RER, according
to the phase of the re-negotiation they correspond to. Also, one can remark the
difference between the operators LosSigD and LosSig. They both allow the
occurrence of losses, but with or without specifying a dummy octet in replacement.

Various Boolean testers are also defined, whose role is double. They are necessary to determine the kind of error control a parameter corresponds to, and they allow to express some constraints such parameters must meet in order to be sensible.

The error control is defined as a pair of sorts Cor, LosD.

As the service provider does not take part to the negotiation, no modification will occur from a request to its indication, and the only operator defined in ErrorControl is: _IsValidCon2For_.   *)

```
type Corruption is NaturalNumber, ResidualErrorRate,
ExtendedBoolean
sorts   Cor
opns
     NoCor         :            -> Cor
     CorSig1       : RER1 -> Cor
     CorSig2       : RER2 -> Cor
     CorNotSig     :            -> Cor
     RER1Of        : Cor   -> RER1
     RER2Of        : Cor   -> RER2
     h             : Cor   -> Nat
     IsCorNotSig      : Cor      -> Bool
     IsCorSig         : Cor      -> Bool
     IsCorSig1        : Cor      -> Bool
     IsCorSig2        : Cor      -> Bool
     IsCor            : Cor      -> Bool
     IsNoCor          : Cor      -> Bool
     IsCorrup1        : Cor      -> Bool
     IsCorrup2        : Cor      -> Bool
     _eq_             : Cor, Cor -> Bool
     _ne_             : Cor, Cor -> Bool
     _ge_             : Cor, Cor -> Bool
     _IsValidCon2For_   : Cor, Cor       -> Bool

eqns
forall re1,re11:RER1, re2,re21:RER2, co,co1:Cor
ofsort RER1
RER1Of(CorSig1(re1))  = re1  ;

ofsort RER2
RER2Of(CorSig2(re2))  = re2  ;

ofsort Nat
h(CorNotSig)              = 0                   ;
h(CorSig1(re1))           = succ(0)  ;
h(CorSig2(re2))           = succ(0)  ;
h(NoCor)                  = succ(succ(0))      ;

ofsort Bool
IsCorNotSig(co)           = h(co) eq 0          ;
IsCorSig(co)              = IsCorSig1(co) or IsCorSig2(co) ;
IsCorSig1(CorSig1(re1))   = IsRER1(re1)                 ;
IsCorSig1(CorSig2(re2))   = false                      ;
   not(h(co) eq succ(0)) =>
IsCorSig1(co)             = false                      ;
IsCorSig2(CorSig2(re2))   = IsRER2(re2)                 ;
IsCorSig2(CorSig1(re1))   = false                      ;
   not(h(co) eq succ(0)) =>
IsCorSig2(co)             = false                      ;
IsCor(co)                 = IsCorNotSig(co) or IsCorSig(co)   ;
IsNoCor(co)               = h(co) eq succ(succ(0))       ;
```

```
IsCorrup1(co)            = IsCorNotSig(co) or IsCorSig1(co)
                          or IsNoCor(co)                        ;
IsCorrup2(co)            = IsCorNotSig(co) or IsCorSig2(co)
                          or IsNoCor(co)                        ;
    co = co1 =>
co eq co1       = true      ;
    h(co) ne h(co1) =>
co eq co1       = false     ;
    re1 ne re11 =>
CorSig1(re1) eq CorSig1(re11)    = false              ;
    re2 ne re21 =>
CorSig2(re2) eq CorSig2(re21)    = false              ;
CorSig1(re1) eq CorSig2(re2) = false            ;
CorSig2(re2) eq CorSig1(re1) = false            ;
co ne co1               = h(co) ne h(co1)  ;
co ge co1               = h(co) ge h(co1)  ;
co IsValidCon2For co1 = IsCorrup1(co1) and IsCorrup2(co)
    and      (h(co) ge h(co1)) and (not(IsCorSig1(co1) and
    IsCorSig2(co)) orelse (RER2Of(co) IsValidCon2For
    RER1Of(co1)))   ;
endtype (* Corruption *)


type LosDum is NaturalNumber, Octet, ResidualErrorRate,
ExtendedBoolean
sorts   LosD
opns
    NoLos        :               -> LosD
    LosSigNoD1 : RER1        -> LosD
    LosSigNoD2 : RER2        -> LosD
    LosSigD1    : Octet, RER1      -> LosD
    LosSigD2    : Octet, RER2      -> LosD
    RER1Of      : LosD            -> RER1
    RER2Of      : LosD            -> RER2
    DumOf       : LosD            -> Octet
    h           : LosD        -> Nat
    IsLSigNoD  : LosD        -> Bool
    IsLSigNoD1 : LosD        -> Bool
    IsLSigNoD2 : LosD        -> Bool
    IsLSigD    : LosD        -> Bool
    IsLSigD1   : LosD        -> Bool
    IsLSigD2   : LosD        -> Bool
    IsLSig     : LosD        -> Bool
    IsNoLos    : LosD        -> Bool
    IsD            : LosD        -> Bool
    IsNoD      : LosD        -> Bool
    IsLosD1    : LosD        -> Bool
    IsLosD2    : LosD        -> Bool
    _eq_        : LosD, LosD  -> Bool
    _ne_        : LosD, LosD  -> Bool
    _ge_        : LosD, LosD  -> Bool
    _IsValidCon2For_   : LosD, LosD -> Bool
eqns
forall re1,re11:RER1, re2,re21:RER2, oc,oc1:Octet, lo,lo1:LosD
ofsort RER1
RER1Of(LosSigNoD1(re1))   = re1  ;
RER1Of(LosSigD1(oc,re1))  = re1  ;
ofsort RER2
RER2Of(LosSigNoD2(re2))   = re2  ;
RER2Of(LosSigD2(oc,re2))  = re2  ;
```

```
ofsort Octet
DumOf(LosSigD1(oc,re1))   = oc    ;
DumOf(LosSigD2(oc,re2))   = oc    ;
ofsort Nat
h(LosSigNoD1(re1))     = 0     ;
h(LosSigNoD2(re2))     = 0     ;
h(LosSigD1(oc,re1))    = succ(0)           ;
h(LosSigD2(oc,re2))    = succ(0)           ;
h(NoLos)               = succ(succ(0))     ;
ofsort Bool
IsLSigNoD(lo)   = IsLSigNoD1(lo) or IsLSigNoD2(lo)   ;
IsLSigNoD1(LosSigNoD1(re1))   = IsRER1(re1)        ;
IsLSigNoD1(LosSigNoD2(re2))   = false             ;
    h(lo) ne 0 =>
IsLSigNoD1(lo)                      = false             ;
IsLSigNoD2(LosSigNoD2(re2))   = IsRER2(re2)        ;
IsLSigNoD2(LosSigNoD1(re1))   = false             ;
    h(lo) ne 0 =>
IsLSigNoD2(lo)                      = false             ;
IsLSigD(lo)                    = IsLSigD1(lo) or IsLSigD2(lo)
    ;
IsLSigD1(LosSigD1(oc,re1))    = IsRER1(re1)        ;
IsLSigD1(LosSigD2(oc,re2))    = false             ;
    h(lo) ne succ(0)   =>
IsLSigD1(lo)                        = false             ;
IsLSigD2(LosSigD2(oc,re2))    = IsRER2(re2)        ;
IsLSigD2(LosSigD1(oc,re1))    = false             ;
    h(lo) ne succ(0)   =>
IsLSigD2(lo)     = false                           ;
IsLSig(lo)       = IsLSigNoD(lo) or IsLSigD(lo)    ;
IsNoLos(lo)      = h(lo) eq succ(succ(0))          ;
IsD(lo)          = IsLSigD(lo)                     ;
IsNoD(lo)        = IsLSigNoD(lo) or IsNoLos(lo)    ;
IsLosD1(lo)      = IsLSigNoD1(lo) or IsLSigD1(lo)
                     or IsNoLos(lo)        ;
IsLosD2(lo)      = IsLSigNoD2(lo) or IsLSigD2(lo)
                     or IsNoLos(lo)        ;
    lo = lo1 =>
lo eq lo1  = true      ;
    h(lo) ne h(lo1) =>
lo eq lo1  = false     ;
    re1 ne re11 =>
LosSigNoD1(re1) eq LosSigNoD1(re11) = false     ;
    re2 ne re21 =>
LosSigNoD2(re2) eq LosSigNoD2(re21) = false     ;
LosSigNoD1(re1) eq LosSigNoD2(re2)  = false     ;
LosSigNoD2(re2) eq LosSigNoD1(re1)  = false     ;
    re1 ne re11 =>
LosSigD1(oc,re1) eq LosSigD1(oc1,re11)  = false     ;
    re2 ne re21 =>
LosSigD2(oc,re2) eq LosSigD2(oc1,re21)  = false     ;
    oc ne oc1  =>
LosSigD1(oc,re1) eq LosSigD1(oc1,re11)  = false     ;
    oc ne oc1  =>
LosSigD2(oc,re2) eq LosSigD2(oc1,re21)  = false     ;
LosSigD1(oc,re1) eq LosSigD2(oc1,re2)   = false     ;
LosSigD2(oc,re2) eq LosSigD1(oc1,re1)   = false     ;
lo ne lo1  = h(lo) ne h(lo1) ;
```

```
lo ge lo1   = (h(lo) eq succ(succ(0))) or not(h(lo1) eq
                                        succ(succ(0)))
      ;
lo IsValidCon2For lo1 = IsLosD1(lo1) and IsLosD2(lo)
     and not(IsLSigD1(lo1) and IsLSigD2(lo))
     and (IsLSig(lo1) or IsNoLos(lo))
     and (not(IsLSig(lo1) and IsLSig(lo))
     orelse (RER2Of(lo) IsValidCon2For RER1Of(lo1)));
endtype (* LosDum *)


type BasicErrorControl is Pair actualizedby Corruption, LosDum
using
sortnames   ErCtrl  for Pair
            Cor         for Element
            LosD    for OtherElement
            Bool    for FBool
opnnames    MakeErCtrl for Pair
            Cor             for First
            LosD            for Second
endtype (* BasicErrorControl *)


type ErrorControl is BasicErrorControl
opns
    IsErCtrl1           : ErCtrl -> Bool
    IsErCtrl2           : ErCtrl -> Bool
    _IsValidCon2For_    : ErCtrl,ErCtrl -> Bool
eqns
forall tec,tec1:ErCtrl

ofsort Bool
IsErCtrl1(tec)          = IsCorrup1(Cor(tec)) and
IsLosD1(LosD(tec))      ;
IsErCtrl2(tec)          = IsCorrup2(Cor(tec)) and
IsLosD2(LosD(tec))      ;
tec IsValidCon2For tec1   = IsErCtrl1(tec1) and IsErCtrl2(tec)
                    and (Cor(tec) IsValidCon2For Cor(tec1))
                    and (LosD(tec) IsValidCon2For LosD(tec1))
    ;
endtype (* ErrorControl *)
(*
```

3.6.1.5 TC priority and TC protection

The TCPriority and TCProtection parameters are the ones defined in ISO 8072

By a renaming of the natural numbers, priority levels are represented as a totally ordered set. ISO 8072 specifies that the number of priority levels is limited. This applies as well to the implementations of the NaturalNumber data type.

An ordered set of four constants represents the protection options defined in clause 10.9 of ISO 8072. The ordering is specified by means of an auxiliary mapping h to the natural numbers.   *)

```
type TCPriority is NaturalNumber renamedby
sortnames       TCPriority      for Nat
opnnames        Lowest          for 0
                Higher          for Succ
endtype (* TCPriority *)
```

```
type TCProtection is NaturalNumber
sorts   TCProtection
opns
    NoProtection, Monitoring, Manipulation, FullProtection
                            :           -> TCProtection
    h                       : TCProtection    -> Nat
    _eq_,_ne_,_le_,_lt_,_ge_,_gt_    : TCProtection,
TCProtection -> Bool
eqns
forall pt,pt1:TCProtection

ofsort Nat
h(NoProtection)    = 0                              ;
h(Monitoring)         = Succ(h(NoProtection))   ;
h(Manipulation)    = Succ(h(Monitoring))     ;
h(FullProtection)  = Succ(h(Manipulation))   ;

ofsort Bool
pt eq pt1   = h(pt) eq h(pt1) ;
pt ne pt1   = not(pt eq pt1)  ;
pt le pt1   = h(pt) le h(pt1) ;
pt lt pt1   = h(pt) lt h(pt1) ;
pt ge pt1   = h(pt) ge h(pt1) ;
pt gt pt1   = h(pt) gt h(pt1) ;
endtype (* TCProtection *)

type ThroughputElement is Element renamedby
sortnames ThroughputEl    for Element

endtype (* ThroughputElement *)

type TransitDelayElement is Element renamedby
sortnames TransitDelayEl for Element

endtype (* TransitDelayElement *)

type JitterElement is Element renamedby
sortnames JitterEl     for Element

endtype (* JitterElement *)
(*
```

3.6.1.6 QoS parameter

FormalTCQuality is a generic type. As already explained, two sorts of TCQualities are defined. Their only difference comes from the sort of their parameters with compulsory values. These are thus defined as elements, that are actualised in BasicTCQuality1 and BasicTCQuality2. TCQuality1 and TCQuality2 add the definition of a test operator that expresses some conditions on the parameters of the QoS. (see previous points for these conditions). In particular, let us point out that an entity never specifies a dummy octet for its direction of emission (condition IsNoD(...)).

Three constructors are given, in order to adapt the structure of the TCQOS expressed in a primitive to the concerned directions of transmission. Selectors are also defined, and tests to determine the structure of a TCQOS.   *)

```
type FormalTCQuality
is ThroughputElement, TrafficTypeInd, TransitDelayElement,
JitterElement, ErrorControl, TCPriority, TCProtection
sorts   TCQOS
opns
```

```
    TCQOSBi       : ThroughputEl, TTI, TransitDelayEl, JitterEl,
                    ErCtrl, ThroughputEl, TransitDelayEl, JitterEl,
                    ErCtrl, TCPriority, TCProtection            ->
TCQOS
    TCQOSOut      : ThroughputEl, TTI, TransitDelayEl, JitterEl,
                    ErCtrl, TCPriority, TCProtection        -> TCQOS
    TCQOSInp      : ThroughputEl, TransitDelayEl, JitterEl,
ErCtrl,
                    TCPriority, TCProtection                 -> TCQOS
    FirThr        : TCQOS       -> ThroughputEl
    TTI            : TCQOS       -> TTI
    FirTrD        : TCQOS       -> TransitDelayEl
    FirJit        : TCQOS       -> JitterEl
    FirErC        : TCQOS       -> ErCtrl
    SecThr        : TCQOS       -> ThroughputEl
    SecTrD        : TCQOS       -> TransitDelayEl
    SecJit        : TCQOS       -> JitterEl
    SecErC        : TCQOS       -> ErCtrl
    Pri            : TCQOS       -> TCPriority
    Pro            : TCQOS       -> TCProtection
    IsBi          : TCQOS       -> Bool
    IsOut         : TCQOS       -> Bool
    IsInp         : TCQOS       -> Bool
    _eq_          : TCQOS, TCQOS       -> Bool
    _ne_          : TCQOS, TCQOS       -> Bool
eqns
forall th,th1,th2,th3:ThroughputEl, ti,ti1:TTI,
td,td1,td2,td3:TransitDelayEl, ji,ji1,ji2,ji3:JitterEl,
ec,ec1,ec2,ec3:ErCtrl, pr,pr1:TCPriority, po,po1:TCProtection,
q,q1:TCQOS
ofsort ThroughputEl
FirThr(TCQOSBi(th,ti,td,ji,ec,th1,td1,ji1,ec1,pr,po))  = th
  ;
FirThr(TCQOSOut(th,ti,td,ji,ec,pr,po))                 = th
  ;
FirThr(TCQOSInp(th,td,ji,ec,pr,po))                    =
th  ;
SecThr(TCQOSBi(th,ti,td,ji,ec,th1,td1,ji1,ec1,pr,po))  = th1
  ;
ofsort TTI
TTI(TCQOSBi(th,ti,td,ji,ec,th1,td1,ji1,ec1,pr,po))         = ti
  ;
TTI(TCQOSOut(th,ti,td,ji,ec,pr,po))                        =
ti  ;
ofsort TransitDelayEl
FirTrD(TCQOSBi(th,ti,td,ji,ec,th1,td1,ji1,ec1,pr,po))  = td
  ;
FirTrD(TCQOSOut(th,ti,td,ji,ec,pr,po))                 = td
  ;
FirTrD(TCQOSInp(th,td,ji,ec,pr,po))                    =
td  ;
SecTrD(TCQOSBi(th,ti,td,ji,ec,th1,td1,ji1,ec1,pr,po))  = td1
  ;
ofsort JitterEl
FirJit(TCQOSBi(th,ti,td,ji,ec,th1,td1,ji1,ec1,pr,po))  = ji
  ;
FirJit(TCQOSOut(th,ti,td,ji,ec,pr,po))                 = ji
  ;
```

```
FirJit(TCQOSInp(th,td,ji,ec,pr,po))                        =
ji ;
SecJit(TCQOSBi(th,ti,td,ji,ec,th1,td1,ji1,ec1,pr,po))  = ji1
    ;
ofsort ErCtrl
FirErC(TCQOSBi(th,ti,td,ji,ec,th1,td1,ji1,ec1,pr,po))  = ec
    ;
FirErC(TCQOSOut(th,ti,td,ji,ec,pr,po))                 = ec
    ;
FirErC(TCQOSInp(th,td,ji,ec,pr,po))                        =
ec ;
SecErC(TCQOSBi(th,ti,td,ji,ec,th1,td1,ji1,ec1,pr,po))  = ec1
    ;
ofsort TCPriority
Pri(TCQOSBi(th,ti,td,ji,ec,th1,td1,ji1,ec1,pr,po))         = pr
    ;
Pri(TCQOSOut(th,ti,td,ji,ec,pr,po))                        =
pr ;
Pri(TCQOSInp(th,td,ji,ec,pr,po))                       = pr
    ;
ofsort TCProtection
Pro(TCQOSBi(th,ti,td,ji,ec,th1,td1,ji1,ec1,pr,po))         = po
    ;
Pro(TCQOSOut(th,ti,td,ji,ec,pr,po))                        =
po ;
Pro(TCQOSInp(th,td,ji,ec,pr,po))                       = po
    ;
ofsort Bool
IsBi(TCQOSBi(th,ti,td,ji,ec,th1,td1,ji1,ec1,pr,po))    = true
    ;
IsBi(TCQOSOut(th,ti,td,ji,ec,pr,po))                   = false;
IsBi(TCQOSInp(th,td,ji,ec,pr,po))                      = false;
IsOut(TCQOSOut(th,ti,td,ji,ec,pr,po))                  = true
    ;
IsOut(TCQOSBi(th,ti,td,ji,ec,th1,td1,ji1,ec1,pr,po))   = false;
IsOut(TCQOSInp(th,td,ji,ec,pr,po))                     = false;
IsInp(TCQOSInp(th,td,ji,ec,pr,po))                     = true
    ;
IsInp(TCQOSBi(th,ti,td,ji,ec,th1,td1,ji1,ec1,pr,po))   = false;
IsInp(TCQOSOut(th,ti,td,ji,ec,pr,po))                  = false;
    not((IsBi(q) and IsBi(q1)) or (IsOut(q) and IsOut(q1))
    or (IsInp(q) and IsInp(q1))) =>
q eq q1 = false    ;
        th eq th1, ti eq ti1, td eq td1, ji eq ji1, ec eq ec1,
        th2 eq th3, td2 eq td3, ji2 eq ji3, ec2 eq ec3,
        pr eq pr1, po eq po1 =>
    TCQOSBi(th,ti,td,ji,ec,th2,td2,ji2,ec2,pr,po) eq
    TCQOSBi(th1,ti1,td1,ji1,ec1,th3,td3,ji3,ec3,pr1,po1) = true
        ;
        th eq th1, ti eq ti1, td eq td1, ji eq ji1, ec eq ec1,
        pr eq pr1, po eq po1 =>
    TCQOSOut(th,ti,td,ji,ec,pr,po) eq
    TCQOSOut(th1,ti1,td1,ji1,ec1,pr1,po1) = true        ;
        th eq th1, td eq td1, ji eq ji1, ec eq ec1, pr eq pr1,
        po eq po1 =>
    TCQOSInp(th,td,ji,ec,pr,po) eq
    TCQOSInp(th1,td1,ji1,ec1,pr1,po1) = true            ;
    FirThr(q) ne FirThr(q1) =>   q eq q1 = false    ;
```

```
FirTrD(q) ne FirTrD(q1) =>   q eq q1 = false    ;
FirJit(q) ne FirJit(q1) =>   q eq q1 = false    ;
    (FirErC(q) ne FirErC(q1)) or (Pri(q) ne Pri(q1))
    or (Pro(q) ne Pro(q1)) =>
q eq q1      = false      ;
    th2 ne th3 =>
TCQOSBi(th,ti,td,ji,ec,th2,td2,ji2,ec2,pr,po) eq
TCQOSBi(th1,ti1,td1,ji1,ec1,th3,td3,ji3,ec3,pr1,po1) = false
    ;
    td2 ne td3 =>
TCQOSBi(th,ti,td,ji,ec,th2,td2,ji2,ec2,pr,po) eq
TCQOSBi(th1,ti1,td1,ji1,ec1,th3,td3,ji3,ec3,pr1,po1) = false
    ;
    ji2 ne ji3 =>
TCQOSBi(th,ti,td,ji,ec,th2,td2,ji2,ec2,pr,po) eq
TCQOSBi(th1,ti1,td1,ji1,ec1,th3,td3,ji3,ec3,pr1,po1) = false
    ;
    ec2 ne ec3 =>
TCQOSBi(th,ti,td,ji,ec,th2,td2,ji2,ec2,pr,po) eq
TCQOSBi(th1,ti1,td1,ji1,ec1,th3,td3,ji3,ec3,pr1,po1) = false
    ;
    ti ne ti1 =>
TCQOSBi(th,ti,td,ji,ec,th2,td2,ji2,ec2,pr,po) eq
TCQOSBi(th1,ti1,td1,ji1,ec1,th3,td3,ji3,ec3,pr1,po1) = false
    ;
    ti ne ti1 =>
TCQOSOut(th,ti,td,ji,ec,pr,po) eq
TCQOSOut(th1,ti1,td1,ji1,ec1,pr1,po1) = false      ;
q ne q1      = not(q eq q1)      ;
endtype (* FormalTCQuality *)

type BasicTCQuality1
is FormalTCQuality actualizedby Throughput, TransitDelay,
Jitter
using
sortnames   TCQOS1  for  TCQOS
            Thr1     for  ThroughputEl
            TrD1     for  TransitDelayEl
            Jit1     for  JitterEl
            Bool     for  FBool
opnnames    TCQOS1Bi    for  TCQOSBi
            TCQOS1Out   for  TCQOSOut
            TCQOS1Inp   for  TCQOSInp

endtype (* BasicTCQuality1 *)

type TCQuality1 is BasicTCQuality1
opns    IsTCQOS1    : TCQOS1    -> Bool
eqns
forall th,th1:Thr1, ti:TTI, td,td1:TrD1, ji,ji1:Jit1,
ec,ec1:ErCtrl, pr:TCPriority, po:TCProtection

ofsort Bool
IsTCQOS1(TCQOS1Bi(th,ti,td,ji,ec,th1,td1,ji1,ec1,pr,po))  =
    IsThr1(th) and IsThr1(th1) and IsTrD1(td) and IsTrD1(td1)
    and IsJit1(ji) and IsJit1(ji1) and IsErCtrl1(ec)
    and IsErCtrl1(ec1) and IsNoD(LosD(ec1))              ;
IsTCQOS1(TCQOS1Out(th,ti,td,ji,ec,pr,po))   = IsThr1(th)
    and IsTrD1(td) and IsJit1(ji) and IsErCtrl1(ec)
    and IsNoD(LosD(ec))                                  ;
```

```
IsTCQOS1(TCQOS1Inp(th,td,ji,ec,pr,po))   = IsThr1(th)
    and IsTrD1(td) and IsJit1(ji) and IsErCtrl1(ec)
    ;
endtype (* TCQuality1 *)

type BasicTCQuality2
is FormalTCQuality actualizedby Throughput, TransitDelay,
Jitter
using
sortnames   TCQOS2  for  TCQOS
            Thr2    for  ThroughputEl
            TrD2    for  TransitDelayEl
            Jit2    for  JitterEl
            Bool    for  FBool
opnnames    TCQOS2Bi    for  TCQOSBi
            TCQOS2Out   for  TCQOSOut
            TCQOS2Inp   for  TCQOSInp
endtype (* BasicTCQuality2 *)

type TCQuality2 is BasicTCQuality2
opns
    IsTCQOS2   : TCQOS2    -> Bool
eqns
forall th,th1:Thr2, ti:TTI, td,td1:TrD2, ji,ji1:Jit2,
ec,ec1:ErCtrl, pr:TCPriority, po:TCProtection
ofsort Bool
IsTCQOS2(TCQOS2Bi(th,ti,td,ji,ec,th1,td1,ji1,ec1,pr,po)) =
    IsThr2(th) and IsThr2(th1) and IsTrD2(td) and IsTrD2(td1)
    and IsJit2(ji) and IsJit2(ji1) and IsErCtrl2(ec)
    and IsErCtrl2(ec1)
    ;
IsTCQOS2(TCQOS2Out(th,ti,td,ji,ec,pr,po))   =
    IsThr2(th) and IsTrD2(td) and IsJit2(ji) and IsErCtrl2(ec)
    ;
IsTCQOS2(TCQOS2Inp(th,td,ji,ec,pr,po))      =
    IsThr2(th) and IsTrD2(td) and IsJit2(ji) and IsErCtrl2(ec)
    ;
endtype (* TCQuality2 *)
(*
```

TCQuality defines operators that transpose to the whole TCQOS structure the constraints already expressed for the parameters. Different operators are defined for the connection opening and for the re-negotiation. This comes from the fact that a direction of transmission can be closed during a re-negotiation, what is marked in the structure of the TCQOS. A 'Bi' TCQOS specified in a TRENEGOTIATErequest can thus appear as a 'Inp' or 'Out' TCQOS in the TRENEGOTIATEindication because in the mean time, a direction has been closed.   *)

```
type TCQuality
is TCQuality1, TCQuality2
opns
    _IsValidIndFor_           : TCQOS1, TCQOS1   -> Bool
    _IsValidCon2For_          : TCQOS2, TCQOS1   -> Bool
    _IsValidRenIndFor_        : TCQOS1, TCQOS1   -> Bool
    _IsValidRenCon2For_       : TCQOS2, TCQOS1   -> Bool
eqns
```

```
forall th,th1,th2,th3:Thr1, th4,th5:Thr2, ti,ti1:TTI,
td,td1,td2,td3:TrD1, td4,td5:TrD2, ji,ji1,ji2,ji3:Jit1,
ji4,ji5:Jit2, ec,ec1,ec2,ec3:ErCtrl, pr,pr1:TCPriority,
po,po1:TCProtection, q,q1:TCQOS1, q2:TCQOS2
ofsort Bool
TCQOS1Bi(th1,ti1,td1,ji1,ec1,th3,td3,ji3,ec3,pr1,po1)
IsValidIndFor TCQOS1Bi(th,ti,td,ji,ec,th2,td2,ji2,ec2,pr,po)
    =
    (th1 IsValidIndFor th) and (ti1 eq ti)
    and (td1 IsValidIndFor td) and (ji1 IsValidIndFor ji)
    and (ec1 eq ec) and (th3 IsValidIndFor th2)
    and (td3 IsValidIndFor td2) and (ji3 IsValidIndFor ji2)
    and (ec3 eq ec2) and (pr1 le pr) and (po1 eq po)
    ;
TCQOS1Out(th1,ti1,td1,ji1,ec1,pr1,po1) IsValidIndFor
TCQOS1Out(th,ti,td,ji,ec,pr,po)           =
    (th1 IsValidIndFor th) and (ti1 eq ti)
    and (td1 IsValidIndFor td) and (ji1 IsValidIndFor ji)
    and (ec1 eq ec) and (pr1 le pr) and (po1 eq po)
    ;
TCQOS1Inp(th1,td1,ji1,ec1,pr1,po1) IsValidIndFor
TCQOS1Inp(th,td,ji,ec,pr,po)               =
    (th1 IsValidIndFor th) and (td1 IsValidIndFor td)
    and (ji1 IsValidIndFor ji) and (ec1 eq ec) and (pr1 le pr)
    and (po1 eq po)          ;
    not((IsBi(q) and IsBi(q1)) or (IsOut(q) and IsOut(q1))
    or (IsInp(q) and IsInp(q1))) =>
q IsValidIndFor q1           = false
    ;
    IsBi(q), IsBi(q1) =>
q IsValidRenIndFor q1        = q IsValidIndFor q1
    ;
    IsInp(q), IsInp(q1) =>
q IsValidRenIndFor q1        = q IsValidIndFor q1
    ;
    IsOut(q), IsOut(q1) =>
q IsValidRenIndFor q1        = q IsValidIndFor q1
    ;
TCQOS1Inp(th1,td1,ji1,ec1,pr1,po1) IsValidRenIndFor
TCQOS1Bi(th,ti,td,ji,ec,th2,td2,ji2,ec2,pr,po)        =
    (th1 IsValidIndFor th2) and (td1 IsValidIndFor td2)
    and (ji1 IsValidIndFor ji2) and (ec1 eq ec2) and (pr1 le
pr)
    and (po1 eq po)
    ;
TCQOS1Out(th1,ti1,td1,ji1,ec1,pr1,po1) IsValidRenIndFor
TCQOS1Bi(th,ti,td,ji,ec,th2,td2,ji2,ec2,pr,po)        =
    (th1 IsValidIndFor th) and (ti1 eq ti)
    and (td1 IsValidIndFor td) and (ji1 IsValidIndFor ji)
    and (ec1 eq ec) and (pr1 le pr) and (po1 eq po)
    ;
    (IsBi(q) and not(IsBi(q1))) or (IsInp(q) and IsOut(q1))
    or (IsOut(q) and IsInp(q1)) =>
q IsValidRenIndFor q1            = false      ;
TCQOS2Bi(th4,ti1,td4,ji4,ec1,th5,td5,ji5,ec3,pr1,po1)
IsValidCon2For TCQOS1Bi(th,ti,td,ji,ec,th2,td2,ji2,ec2,pr,po)
    =
    (th4 IsValidCon2For th2) and (td4 IsValidCon2For td2)
    and (ji4 IsValidCon2For ji2) and (ec1 IsValidCon2For ec2)
```

```
    and (th5 IsValidCon2For th) and (td5 IsValidCon2For td)
    and (ji5 IsValidCon2For ji) and (ec3 IsValidCon2For ec)
    and (pr1 le pr) and (po1 le po)
    ;
TCQOS2Inp(th4,td4,ji4,ec1,pr1,po1) IsValidCon2For
TCQOS1Out(th,ti,td,ji,ec,pr,po) = (th4 IsValidCon2For th)
    and (td4 IsValidCon2For td) and (ji4 IsValidCon2For ji)
    and (ec1 IsValidCon2For ec) and (pr1 le pr) and (po1 le
po)    ;
TCQOS2Out(th4,ti,td4,ji4,ec1,pr1,po1) IsValidCon2For
TCQOS1Inp(th,td,ji,ec,pr,po)          = (th4 IsValidCon2For th)
    and (td4 IsValidCon2For td) and (ji4 IsValidCon2For ji)
    and (ec1 IsValidCon2For ec) and (pr1 le pr) and (po1 le
po)     ;
    not((IsBi(q2) and IsBi(q1)) or (IsInp(q2) and IsOut(q1))
    or (IsOut(q2) and IsInp(q1))) =>
q2 IsValidCon2For q1  = false                           ;
    IsBi(q2), IsBi(q1) =>
q2 IsValidRenCon2For q1  = q2 IsValidCon2For q1    ;
    IsInp(q2), IsOut(q1) =>
q2 IsValidRenCon2For q1  = q2 IsValidCon2For q1    ;
    IsOut(q2), IsInp(q1) =>
q2 IsValidRenCon2For q1  = q2 IsValidCon2For q1    ;
TCQOS2Inp(th4,td4,ji4,ec1,pr1,po1) IsValidRenCon2For
TCQOS1Bi(th,ti,td,ji,ec,th2,td2,ji2,ec2,pr,po)     =
    (th4 IsValidCon2For th) and (td4 IsValidCon2For td)
    and (ji4 IsValidCon2For ji) and (ec1 IsValidCon2For ec)
    and (pr1 le pr) and (po1 le po)                      ;
TCQOS2Out(th4,ti1,td4,ji4,ec1,pr1,po1) IsValidRenCon2For
TCQOS1Bi(th,ti,td,ji,ec,th2,td2,ji2,ec2,pr,po)     =
    (th4 IsValidCon2For th2) and (td4 IsValidCon2For td2)
    and (ji4 IsValidCon2For ji2) and (ec1 IsValidCon2For ec2)
    and (pr1 le pr) and (po1 le po)                      ;
    (IsBi(q2) and not(IsBi(q1))) or (IsInp(q2) and IsInp(q1))
    or (IsOut(q2) and IsOut(q1)) =>
q2 IsValidRenCon2For q1  = false                        ;
endtype (* TCQuality *)
(*
```

3.6.2 QoS parameter for the PP-EACLTS

The parameters defined for this mode are: a TransitDelay, and a TServiceCompletionDelay parameter, in addition to the usual TCPriority and TCProtection. The TransitDelay specifies a maximal and a threshold value. The maximal value is not compulsory, it is to be understood as a best effort request. But, as for the connection mode, if the threshold value is exceeded, this should be signalled in the TDR parameter of the TACKNOWLEDGEDDATAconfirm. The TServiceCompletionDelay is used as an indication to the service provider, that must emit a reject indication (a TOUTOFBANDREJECTACKDATAindication) if after the TACKNOWLEDGEDDATArequest this delay is reached without having received the TACKNOWLEDGEDDATAconfirm.

Due to the lack of a timed LOTOS the influence of these parameters was not traduced in a constraint.   *)

```
type TACKQuality
```

```
is TTransDelay1, TServiceCompletionDelay, TCPriority,
TCProtection
sorts TACKQOS
opns
    TACKQOS             : TransDelay1, SCD, TCPriority, TCProtection
                                  -> TACKQOS
    TACKQOSTrDel    : TACKQOS          -> TransDelay1
    TACKQOSSCD      : TACKQOS          -> SCD
    TACKQOSPri      : TACKQOS          -> TCPriority
    TACKQOSPro      : TACKQOS          -> TCProtection
    IsTACKQOS       : TACKQOS          -> Bool
    _eq_,_ne_       : TACKQOS,TACKQOS -> Bool
eqns
forall td,td1:TransDelay1, sc,sc1:SCD, pri,pri1:TCPriority,
pro,pro1:TCProtection, q,q1:TACKQOS
ofsort TransDelay1
TACKQOSTrDel(TACKQOS(td, sc, pri, pro))     = td          ;
ofsort SCD
TACKQOSSCD(TACKQOS(td, sc, pri, pro))    = sc        ;
ofsort TCPriority
TACKQOSPri(TACKQOS(td, sc, pri, pro))    = pri       ;
ofsort TCProtection
TACKQOSPro(TACKQOS(td, sc, pri, pro))    = pro       ;
ofsort Bool
IsTACKQOS(TACKQOS(td, sc, pri, pro))         = IsTrD(td)    ;
TACKQOS(td, sc, pri, pro) eq TACKQOS(td1, sc1, pri1, pro1)
    =
    (td eq td1) and (sc eq sc1) and (pri eq pri1)
    and (pro eq pro1)                                   ;
q ne q1                               = not(q eq q1)     ;
endtype (* TACKQuality *)


type TTransDelay1
is NatRepresentations
sorts   TransDelay1
opns
    MakeTD      : Nat, Nat                   -> TransDelay1
    Thld        : TransDelay1              -> Nat
    Max         : TransDelay1                -> Nat
    IsTrD       : TransDelay1              -> Bool
    _eq_,_ne_   : TransDelay1,TransDelay1 -> Bool
eqns
forall n,n1,n2,n3:Nat, q,q1:TransDelay1
ofsort Nat
Thld(MakeTD(n,n1))      = n          ;
Max(MakeTD(n,n1))      = n1    ;
ofsort Bool
IsTrD(MakeTD(n,n1))              = n le n1                  ;
MakeTD(n,n1) eq MakeTD(n2,n3)     = (n eq n2) and (n1 eq n3)
    ;
q ne q1                          = not(q eq q1)               ;
endtype (* TTransDelay1 *)


type TServiceCompletionDelay
is NatRepresentations
sorts   SCD
```

```
opns
    MakeSCD      : Nat         -> SCD
    _eq_,_ne_    : SCD,SCD     -> Bool
eqns
forall n1,n2:Nat, q,q1:SCD
ofsort Bool
    MakeSCD(n1) eq MakeSCD(n2)    = n1 eq n2      ;
    q ne q1                       = not(q eq q1) ;
endtype (* TServiceCompletionDelay *)
(*
```

3.6.3 QoS parameter for the PP-ERCLTS

The QoS parameter for the PP-ERCLTS just requires the definition of a
TServiceCompletionDelay, of a TCPriority and of a TCProtection. In comparison
with the latter QoS, no TransitDelay is defined here, but the
TServiceCompletionDelay has a similar meaning for the delay between the
TREQUESTRESPONSErequest and the TREQUESTRESPONSEconfirm.

Again, the influence of these parameters was not traduced in a constraint.   *)

```
type TRRQuality
is TServiceCompletionDelay, TCPriority, TCProtection
sorts TRRQOS
opns
    TRRQOS        : SCD, TCPriority, TCProtection        -> TRRQOS
    TRRQOSSCD     : TRRQOS                               -> SCD
    TRRQOSPri     : TRRQOS                               -> TCPriority
    TRRQOSPro     : TRRQOS                               -> TCProtection
    _eq_,_ne_     : TRRQOS,TRRQOS                        -> Bool
eqns
forall sc,sc1:SCD, pri,pri1:TCPriority, pro,pro1:TCProtection,
q,q1:TRRQOS
ofsort SCD
TRRQOSSCD(TRRQOS(sc, pri, pro)) = sc   ;
ofsort TCPriority
TRRQOSPri(TRRQOS(sc, pri, pro)) = pri  ;
ofsort TCProtection
TRRQOSPro(TRRQOS(sc, pri, pro)) = pro  ;
ofsort Bool
TRRQOS(sc, pri, pro) eq TRRQOS(sc1, pri1, pro1)    = (sc eq
sc1)
    and (pri eq pri1) and (pro eq pro1)  ;
q ne q1 = not(q eq q1) ;
endtype (* TRRQuality *)
(*
```

3.6.4 QoS parameter for the PP-EUCLTS

The QoS parameter for the PP-EUCLTS just requires the definition of a TransitDelay,
of a TCPriority and of a TCProtection. As no confirmation is returned, the TService-
CompletionDelay had no sense, and the TransitDelay specifies no threshold value.

Again, the influence of these parameters was not traduced in a constraint.   *)

```
type TCLQuality
is TTransDelay2, TCPriority, TCProtection
sorts TCLQOS
opns
    TCLQOS : TransDelay2, TCPriority, TCProtection       ->
TCLQOS
    TCLQOSTransDelay   : TCLQOS                 -> TransDelay2
    TCLQOSPri          : TCLQOS                 -> TCPriority
    TCLQOSPro          : TCLQOS                 -> TCProtection
    _eq_,_ne_          : TCLQOS,TCLQOS          -> Bool
eqns
forall td,td1:TransDelay2, pri,pri1:TCPriority,
pro,pro1:TCProtection, q,q1:TCLQOS
ofsort TransDelay2
TCLQOSTransDelay(TCLQOS(td, pri, pro))  = td   ;
ofsort TCPriority
TCLQOSPri(TCLQOS(td, pri, pro))         = pri  ;
ofsort TCProtection
TCLQOSPro(TCLQOS(td, pri, pro))         = pro  ;
ofsort Bool
TCLQOS(td, pri, pro) eq TCLQOS(td1, pri1, pro1)    = (td eq
td1)
    and (pri eq pri1) and (pro eq pro1)  ;
q ne q1 = not(q eq q1)      ;
endtype (* TCLQuality *)


type TTransDelay2
is NatRepresentations
sorts   TransDelay2
opns
    MakeTD      : Nat                 -> TransDelay2
    Max            : TransDelay2          -> Nat
    _eq_,_ne_   : TransDelay2,TransDelay2 -> Bool
eqns
forall n,n1:Nat, q,q1:TransDelay2
ofsort Nat
Max(MakeTD(n))          = n                  ;
ofsort Bool
MakeTD(n) eq MakeTD(n1)   = n eq n1            ;
q ne q1                   = not(q eq q1)       ;
endtype (* TTransDelay2 *)
(*
```

3.7 TCTSDU size parameter

The TSDU size parameter, though not incorporated in the QoS is very similar to them. It is negotiated the same way, what also causes the need for two sorts, and it must correspond to the opened directions of transmission, what again requires three different forms.

Types TSDUSize1 and TSDUSize2 respectively define values of sort TSDUS1 and TSDUS2 that are used in the different phases of the negotiation to express the proposed or the fixed minimal and maximal values.

Types TCTSDUSize1 and TCTSDUSize2 thus combine these elements into the three usual structures (Bi, Inp, Out). Some more selectors and tests and operators specifying the conditions during the negotiation are defined, with the usual meaning.
*)

```
type TSDUSize1 is NatRepresentations
sorts   TSDUS1
opns
    TSDUSize1                       : Nat, Nat, Nat, Nat  -> TSDUS1
    Min, MaxMin, MinMax, Max : TSDUS1                 -> Nat
    _eq_, _ne_                       : TSDUS1, TSDUS1      -> Bool
    IsTSDUS1                        : TSDUS1              -> Bool
    _IsValidIndFor_                 : TSDUS1, TSDUS1      -> Bool
eqns
forall n1,n2,n3,n4:Nat, ts1,ts2:TSDUS1

ofsort Nat
Min(TSDUSize1(n1,n2,n3,n4))       = n1     ;
MaxMin(TSDUSize1(n1,n2,n3,n4))    = n2     ;
MinMax(TSDUSize1(n1,n2,n3,n4))    = n3     ;
Max(TSDUSize1(n1,n2,n3,n4))       = n4     ;

ofsort Bool
ts1 eq ts2      = (Min(ts1) eq Min(ts2))
    and (MaxMin(ts1) eq MaxMin(ts2))
    and (MinMax(ts1) eq MinMax(ts2))
    and (Max(ts1) eq Max(ts2))              ;
ts1 ne ts2      = not(ts1 eq ts2)           ;
IsTSDUS1(ts1)        = (Min(ts1) le MaxMin(ts1))
    and (MaxMin(ts1) le MinMax(ts1))
    and (MinMax(ts1) le Max(ts1))              ;
    not(IsTSDUS1(ts1) and IsTSDUS1(ts2)) =>
ts2 IsValidIndFor ts1     = false            ;
    IsTSDUS1(ts1), IsTSDUS1(ts2) =>
ts2 IsValidIndFor ts1        = (MinMax(ts2) eq MinMax(ts1))
    and (Max(ts2) le Max(ts1)) and (MaxMin(ts2) eq MaxMin(ts1))
    and (Min(ts2) ge Min(ts1))              ;
endtype (* TSDUSize1 *)


type TSDUSize2 is NatRepresentations
sorts   TSDUS2
opns
    TSDUSize2        : Nat, Nat         -> TSDUS2
    Min, Max         : TSDUS2           -> Nat
    _eq_, _ne_        : TSDUS2, TSDUS2   -> Bool
    IsTSDUS2         : TSDUS2           -> Bool
eqns
forall n1,n2:Nat, ts1,ts2:TSDUS2

ofsort Nat
Min(TSDUSize2(n1,n2)) = n1    ;
Max(TSDUSize2(n1,n2)) = n2    ;

ofsort Bool
ts1 eq ts2              = (Min(ts1) eq Min(ts2))
                          and (Max(ts1) eq Max(ts2))     ;
ts1 ne ts2              = not(ts1 eq ts2)                 ;
IsTSDUS2(ts1)           = Max(ts1) ge Min(ts1)              ;
endtype (* TSDUSize2 *)
```

```
type TCTSDUSize1
is TSDUSize1
sorts  TSS1
opns
    TCTSS1Bi            : TSDUS1, TSDUS1   -> TSS1
    TCTSS1Inp           : TSDUS1            -> TSS1
    TCTSS1Out           : TSDUS1            -> TSS1
    FirMin              : TSS1          -> Nat
    FirMax              : TSS1          -> Nat
    SecMin              : TSS1          -> Nat
    SecMax              : TSS1          -> Nat
    h                   : TSS1          -> Nat
    IsBi, IsInp, IsOut     : TSS1                -> Bool
    IsTSS1                 : TSS1                -> Bool
    _IsValidIndFor_        : TSS1, TSS1          -> Bool
    _IsValidRenIndFor_     : TSS1, TSS1          -> Bool
    _eq_, _ne_             : TSS1, TSS1          -> Bool
eqns
forall
ts11, ts12, ts13, ts14 : TSDUS1, ts1, ts2 : TSS1
ofsort Nat
FirMin(TCTSS1Bi(ts11,ts12))   = Min(ts11)       ;
FirMin(TCTSS1Inp(ts11))       = Min(ts11)       ;
FirMin(TCTSS1Out(ts11))       = Min(ts11)       ;
FirMax(TCTSS1Bi(ts11,ts12))   = Max(ts11)       ;
FirMax(TCTSS1Inp(ts11))       = Max(ts11)       ;
FirMax(TCTSS1Out(ts11))       = Max(ts11)       ;
SecMin(TCTSS1Bi(ts11,ts12))   = Min(ts12)       ;
SecMax(TCTSS1Bi(ts11,ts12))   = Max(ts12)       ;
h(TCTSS1Bi(ts11,ts12))        = 0                   ;
h(TCTSS1Inp(ts11))            = succ(0)         ;
h(TCTSS1Out(ts11))            = succ(succ(0))     ;
ofsort Bool
IsBi(ts1)                 = h(ts1) eq 0                  ;
IsInp(ts1)                = h(ts1) eq succ(0)            ;
IsOut(ts1)                = h(ts1) eq succ(succ(0))      ;
IsTSS1(TCTSS1Bi(ts11,ts12))   = IsTSDUS1(ts11)
                              and IsTSDUS1(ts12)         ;
IsTSS1(TCTSS1Inp(ts11))       = IsTSDUS1(ts11)           ;
IsTSS1(TCTSS1Out(ts11))       = IsTSDUS1(ts11)           ;
TCTSS1Bi(ts11,ts12) IsValidIndFor
TCTSS1Bi(ts13,ts14)       = (ts11 IsValidIndFor ts13)
                          and (ts12 IsValidIndFor ts14) ;
TCTSS1Inp(ts11) IsValidIndFor
TCTSS1Inp(ts12)           = (ts11 IsValidIndFor ts12)   ;
TCTSS1Out(ts11) IsValidIndFor
TCTSS1Out(ts12)           = (ts11 IsValidIndFor ts12)   ;
    not(h(ts1) eq h(ts2)) =>
ts1 IsValidIndFor ts2     = false                       ;
    h(ts1) eq h(ts2) =>
ts1 IsValidRenIndFor ts2      = ts1 IsValidIndFor ts2   ;
TCTSS1Inp(ts11) IsValidRenIndFor
TCTSS1Bi(ts12,ts13)       = ts11 IsValidIndFor ts12     ;
TCTSS1Out(ts11) IsValidRenIndFor
TCTSS1Bi(ts12,ts13)       = ts11 IsValidIndFor ts13     ;
    not(h(ts1) eq h(ts2)), not(IsBi(ts2)) =>
ts1 IsValidRenIndFor ts2      = false                   ;
```

```
TCTSS1Bi(ts11,ts12) eq TCTSS1Bi(ts13,ts14)  = (ts11 eq ts13)
                                      and (ts12 eq ts14)
    ;
TCTSS1Inp(ts11) eq TCTSS1Inp(ts12)           = (ts11 eq ts12)
    ;
TCTSS1Out(ts11) eq TCTSS1Out(ts12)           = (ts11 eq ts12)
    ;
    not(h(ts1) eq h(ts2))  =>
ts1 eq ts2                      = false                    ;
ts1 ne ts2                      = not(ts1 eq ts2)          ;
endtype (* TCTSDUSize1 *)

type TCTSDUSize2
is TSDUSize2
sorts  TSS2
opns
    TCTSS2Bi          : TSDUS2, TSDUS2   -> TSS2
    TCTSS2Inp         : TSDUS2           -> TSS2
    TCTSS2Out         : TSDUS2           -> TSS2
    FirMin            : TSS2          -> Nat
    FirMax            : TSS2          -> Nat
    SecMin            : TSS2          -> Nat
    SecMax            : TSS2          -> Nat
    h                 : TSS2          -> Nat
    IsBi, IsInp, IsOut     : TSS2 -> Bool
    IsTSS2            : TSS2          -> Bool
    _eq_, _ne_        : TSS2, TSS2   -> Bool
eqns
forall
ts21, ts22, ts23, ts24 : TSDUS2, ts1, ts2 : TSS2

ofsort Nat
FirMin(TCTSS2Bi(ts21,ts22))   = Min(ts21)    ;
FirMin(TCTSS2Inp(ts21))       = Min(ts21)    ;
FirMin(TCTSS2Out(ts21))       = Min(ts21)    ;
FirMax(TCTSS2Bi(ts21,ts22))   = Max(ts21)    ;
FirMax(TCTSS2Inp(ts21))       = Max(ts21)    ;
FirMax(TCTSS2Out(ts21))       = Max(ts21)    ;
SecMin(TCTSS2Bi(ts21,ts22))   = Min(ts22)    ;
SecMax(TCTSS2Bi(ts21,ts22))   = Max(ts22)    ;
h(TCTSS2Bi(ts21,ts22))        = 0               ;
h(TCTSS2Out(ts21))            = succ(0)         ;
h(TCTSS2Inp(ts21))            = succ(succ(0))   ;

ofsort Bool
IsBi(ts1)                     = h(ts1) eq 0              ;
IsInp(ts1)                    = h(ts1) eq succ(0)        ;
IsOut(ts1)                    = h(ts1) eq succ(succ(0))    ;
IsTSS2(TCTSS2Bi(ts21,ts22))   = IsTSDUS2(ts21)
                                and IsTSDUS2(ts22)       ;
IsTSS2(TCTSS2Inp(ts21))       = IsTSDUS2(ts21)          ;
IsTSS2(TCTSS2Out(ts21))       = IsTSDUS2(ts21)          ;
TCTSS2Bi(ts21,ts22) eq TCTSS2Bi(ts23,ts24)  = (ts21 eq ts23)
                                      and (ts22 eq ts24)
    ;
TCTSS2Inp(ts21) eq TCTSS2Inp(ts22)           = (ts21 eq ts22)
    ;
TCTSS2Out(ts21) eq TCTSS2Out(ts22)           = (ts21 eq ts22)
    ;
    not(h(ts1) eq h(ts2))  =>
```

```
ts1 eq ts2                               = false
    ;
ts1 ne ts2                               = not(ts1 eq ts2)
    ;
endtype (* TCTSDUSize2 *)

type TCTSDUSize
is TCTSDUSize1, TCTSDUSize2
opns
    _IsValidCon2For_         : TSDUS2, TSDUS1    -> Bool
    _IsValidCon2For_         : TSS2, TSS1        -> Bool
    _IsValidRenCon2For_      : TSS2, TSS1        -> Bool
eqns
forall
sd11,sd12:TSDUS1, sd21,sd22:TSDUS2, ts1:TSS1, ts2:TSS2
ofsort Bool
sd21 IsValidCon2For sd11      = IsTSDUS1(sd11) and
IsTSDUS2(sd21)
    and (Min(sd21) ge Min(sd11)) and (Min(sd21) le
MaxMin(sd11))
    and (Max(sd21) le Max(sd11))
    and (Max(sd21) ge MinMax(sd11))  ;
TCTSS2Bi(sd21,sd22) IsValidCon2For
TCTSS1Bi(sd11,sd12)    = (sd21 IsValidCon2For sd12)
                       and (sd22 IsValidCon2For sd11)   ;
TCTSS2Out(sd21) IsValidCon2For
TCTSS1Inp(sd11)        = sd21 IsValidCon2For sd11        ;
TCTSS2Inp(sd21) IsValidCon2For
TCTSS1Out(sd11)        = sd21 IsValidCon2For sd11        ;
    not(h(ts1) eq h(ts2)) =>
ts2 IsValidCon2For ts1   = false                        ;
    h(ts1) eq h(ts2) =>
ts2 IsValidRenCon2For ts1      = ts2 IsValidCon2For ts1     ;
TCTSS2Inp(sd21) IsValidRenCon2For
TCTSS1Bi(sd11,sd12)        = sd21 IsValidCon2For sd12    ;
TCTSS2Out(sd21) IsValidRenCon2For
TCTSS1Bi(sd11,sd12)        = sd21 IsValidCon2For sd11    ;
    not(h(ts1) eq h(ts2)), not(IsBi(ts2)) =>
ts2 IsValidRenCon2For ts1    = false                       ;
endtype (* TCTSDUSize *)
(*
```

3.8 Auxiliary definitions

GeneralIdentifier defines an infinite set of identifiers together with equality on them.

Pair defines a generic pair of values of possibly different sorts, with equality and projections.

Element is defined in the LOTOS library of data types, whereas OtherElement and ThirdElement are isomorphic, distinct copies of it.

Doublet defines a set consisting of two elements, endowed with Boolean equality.
*)

```
type GeneralIdentifier
is Boolean
sorts Identifier
```

```
opns
    SomeIdentifier         :                        -> Identifier
    AnotherIdentifier      : Identifier            -> Identifier
    _eq_,_ne_         : Identifier, Identifier         -> Bool
eqns
forall a,a1 : Identifier

ofsort Bool
SomeIdentifier eq SomeIdentifier            = true         ;
AnotherIdentifier(a) eq SomeIdentifier      = false        ;
SomeIdentifier eq AnotherIdentifier(a)      = false        ;
AnotherIdentifier(a) eq AnotherIdentifier(a1)  = a eq a1   ;
a ne a1                            = not(a eq a1) ;
endtype (* GeneralIdentifier *)

type OtherElement is Element renamedby
sortnames OtherElement for Element

endtype (* OtherElement *)

type ThirdElement is Element renamedby
sortnames ThirdElement for Element

endtype (* ThirdElement *)

type Pair is Boolean, Element, OtherElement
sorts Pair
opns
    Pair              : Element, OtherElement   -> Pair
    First             : Pair                    -> Element
    Second            : Pair                    -> OtherElement
    _eq_, _ne_        : Pair, Pair              -> Bool
eqns
forall e1:Element, e2:OtherElement, p,p1:Pair

ofsort Element
First(Pair(e1,e2))     = e1    ;

ofsort OtherElement
Second(Pair(e1,e2))    = e2    ;

ofsort Bool
p = p1 =>  p eq p1 = true          ;
First(p) ne First(p1) => p eq p1 = false        ;
Second(p) ne Second(p1) =>   p eq p1 = false    ;
p ne p1 = not(p eq p1)                          ;
endtype (* Pair *)

type Doublet is Boolean
sorts Doublet
opns
    TheOne, TheOther  :                         -> Doublet
    _eq_, _ne_        : Doublet, Doublet    -> Bool
eqns
forall u,v:Doublet

ofsort Bool
u eq u                 = true      ;
TheOne eq TheOther     = false     ;
TheOther eq TheOne     = false     ;
u ne v             = not(u eq v)   ;
endtype (* Doublet *)
(*
```

4 Global constraints

Here begins the part concerned with the definition of the behaviour of the service.
Basically, it is an interleaving of the processes for each mode.

The connection mode is described as the conjunction of separate constraints.
TConnections is designed as being able to support a potentially infinite number of
independent connections. The process TCIdentif takes care that each TCEP be used
by just one connection. TCEIVerif expresses that the primitives that require it, but
only these ones, pass through an out-of-band TCEI.

Similarly to TConnections, TReqResps, TAckCLs and TSConLess describe a potentially infinite number of independent occurrences of the mode of transmission they
relate to. The role of TAEIControl is to avoid that a given TAddress-TAEI pair be
used by more than one transmission at a time.   *)

```
behaviour
(    TConnections [t]
     ||
     TCIdentif [t]
     ||
     TCEIVerif [t]
)
|||
(    ((TReqResps [t]
      |||
      TAckCLs [t])
      ||
      TAEIControl [t]) ||| TSConLess [t]
)

where
(*
```

5 The PP-ECOTS

Point 5 is concerned by the description of all the aspects of the behaviour of the PP-
ECOTS. As said in the previous point this service provides a potentially infinite
number of independent connections.

NOTE - Each instance of TConnection can terminate. The composite TConnections
can never terminate because the possibility always exists that a new instance of its
component TConnection is invoked.   *)

```
process TConnections [t] : noexit :=
(TConnection [t] >> stop)
|||
TConnections [t]
endproc (* TConnections *)
(*
```

5.1 Provision of a TC

The requirements on the service provider behaviour that relate to provision of a single TC are decomposed into two classes, viz. constraints that are local to a TCEP and end-to-end constraints. Constraints of the first class relate the behaviour at a TCEP to the history of TSPs executed at that TCEP. Constraints of the second class relate the behaviour at a TCEP to the history of TSPs at the other TCEP.

A suitable notion of "history" is to be defined formally. In particular, it is convenient to take account of only those events that may affect the future behaviour of the service provider, thus forgetting the events that are not relevant anymore: such a notion is referred to as "relevant history". For local constraints the relevant history is represented by the process state. For end-to-end constraints the relevant history is represented by process parameters, whose structure and operations are formulated by means of an ad-hoc designed data type definition (see 5.3.2.2.2).

Local constraints are represented by two independent parallel processes, which are different instances of process TCEP (see 5.2.1) that respectively constrain the interactions with the Calling and Called TS Users.

End-to-end constraints are represented by process TCEPAssociation (see 5.3.1), which relates the Indications that occur at each end of the TC to the corresponding primitives occurred at the other end. The provider's non-determinacy which affects this relationship is specified by this process. TCEPAssociation specifies also the non-determinacy that relates to the execution of provider-generated TSPs.

TSUserRole defines the Calling and Called TS user roles.

TSPDirection defines the Request and Indication TSP directions.

NOTE - A technical remark about termination: termination of both of the processes that represent the ends of the connection is a clearly sufficient representation of the end of the connection lifetime. This is why the parallel component that contains TCEPAssociation is described (using the [> exit construction) as able to terminate at any time, whilst TCEPAssociation is a non-terminating process.   *)

```
type TSUserRole is Doublet renamedby
sortnames       TSUserRole      for Doublet
opnnames        CallingRole     for TheOne
                CalledRole      for TheOther
endtype (* TSUserRole *)

type TSPDirection is Doublet renamedby
sortnames    TSPDirection   for Doublet
opnnames     Request        for TheOne
             Indication     for TheOther
endtype (* TSPDirection *)

process TConnection [t] : exit :=
( TCEP [t] (CallingRole) ||| TCEP [t] (CalledRole) )
||
( TCEPAssociation [t] [> exit )
endproc (* TConnection *)
(*
```

5.2 Local constraints for a TCEP

5.2.1 General

The first three processes in the following decomposition of the local constraints on the interactions of a single TC at each end point mirror the event structure: the intent is pursued of separating the constraints that apply to each component of the event structure from those that apply to the other components. The fourth process specifies constraints on the length of the UserData in some primitives.   *)

```
process TCEP [t] (role: TSUserRole) : exit :=
TCEPAddress [t]
||
TCEPIdentification [t]
||
TCEPConnect1 [t] (role)
||
TCEPUserData [t]
endproc (* TCEP *)
(*
```

5.2.2 Address and identification of a TCEP

The local constraints on the address and TCEI parts of the events at a TCEP exhibit a similar form: the value is determined on the first event, in co-operation with the TS User, and thereafter is constant. However, let us recall that for the out-of-band transfers, the TCEI is enlarged with an association identifier, which differs from a communication to another. Thus, one just imposes the in-band part of the TCEI to remain unchanged.

   NOTE - For both processes termination is allowed at any time: the end of the local (i.e. at a TCEP) lifetime of the TC is actually determined by the local ordering of service primitives (see 5.2.3.3).   *)

```
process TCEPAddress [t] : exit :=
t ?ta:TAddress ?tcei:TCEI ?tsp:TSP ; ConstantTA [t] (ta)
[> exit
endproc (* TCEPAddress *)

process ConstantTA [t] (ta:TAddress) : noexit :=
t !ta ?tcei:TCEI ?tsp:TSP ; ConstantTA [t] (ta)
endproc (* ConstantTA *)

process TCEPIdentification [t] : exit:=
t ?ta:TAddress ?tcei:TCEI ?tsp:TSP ; ConstantTCEI [t] (tcei)
[> exit
endproc (* TCEPIdentification *)

process ConstantTCEI [t] (tcei:TCEI) : noexit :=
t ?ta:TAddress ?tcei1:TCEI ?tsp:TSP [InBPart(tcei) eq
InBPart(tcei1)] ; ConstantTCEI [t] (tcei)
endproc (* ConstantTCEI *)
(*
```

5.2.3 Local ordering of ETS primitives at a TCEP

5.2.3.1 General

TCEPConnect1 specifies the constraints on the possible sequences of TSPs at one TC endpoint, applied to a single TC.

As far as the specification of local constraints is concerned, at any time the TS primitives that may be executed only depend on the history of the TS primitives executed at that TCEP. The relevant aspects of this history are mainly represented below in the names of the processes that succeed to TCEPConnect1, together with a few parameters.

The opportunity to exit let to the called TCEP in TCEPConnect1 caters for the possibility that TC release terminates the TC establishment locally at the Calling TCEP, with no TSP execution at the other (potential, but actually never observed) end of the TC.

The TCONNECT primitive executed in TCEPConnect1 is relevant history for TCEPConnect2, as negotiation constraints apply to the TCONNECT response/confirm that depend on the TCONNECT indication/request. See 3.4.3.4, for the definition of the Boolean function IsValidTCON2For, that represents TS negotiation requirements. The definition of TSUserRole is in 5.1.

Successful TC establishment enables entering the data transfer phase. Except in what concerns the re-negotiation, the behaviour in this phase is independent of the role of the TCEP, but depends on the directions of communication that have been opened. The processes that follow TCEPConnect2 describe the various ways offered by the service to close, partially or definitively, the connection: the bilateral disconnection, the unilateral disconnection, the graceful release. Unilateral disconnection is only allowed if both directions of communication were initially opened, this explains the need to export the value of the plex parameter through some of the processes. At the same time, these processes provide the local constraints on the occurrence of re-negotiation procedures. More precisely, they traduce the following rules:

- Only one re-negotiation procedure can be running at a time
- It is always initiated by the calling entity
- The re-negotiation procedure is similar to the connection opening one: it is a four primitives sketch. However, a fifth NewQOSindication primitive, induced by the occurrence of the TRENEGOTIATIONconfirm at the Calling entity, is sent back to the Called entity, to signal the moment when the data with the new QoS arrive.
- The occurrence of a re-negotiation procedure does not interfere with the behaviour of the service in what concerns the other primitives
- The structure of the QoS in the re-negotiation primitive that occur at an endpoint must reflect what the entity believes about the opening of the transmission directions. For example, if the called entity has emitted a TRELEASErequest and then receives a TRENEGOTIATIONindication, the QoS in this last primitive will have a 'Inp' structure (see 3.6.1, 3.6.1.6 for the notion of QoS structure)
- No NewQOSindication primitive occurs at the called entity if its input direction of transmission is closed
- Some rules, defined by IsValidTCON2For (see 3.4.3.4) must apply between a request or indication primitive and their confirmation

- A re-negotiation procedure can be rejected, either by the called entity, by emitting a TREJECTRENEGOTIATIONrequest instead of a TRENEGOTIATIONresponse, or by the transmission medium, just after the occurrence of the TRENEGOTIA-TIONrequest. In both cases, this is signalled to the calling entity by a TREJECTRENEGOTIATIONindication.

Let us remark that, as long as the connection is not definitely closed and whatever the value of the plex parameter is, the out-of-band communications remain possible in both directions.

NOTE - In the names of the processes that follow TCEPDTPhase, ...AfterR signifies: after the occurrence of a TRELEASErequest, similarly ...I corresponds to a TRELEASEindication (or to an unidirectional TDISCONNECT, either request or indication, that closes the direction of reception and has locally, the same effect), and ...C corresponds to a TRELEASE confirm. An unidirectional TDISCONNECT closing the direction of emission has locally the effect of a TRELEASErequest followed by a TRELEASEconfirm.

5.2.3.2 TC establishment phase at a TCEP   *)

```
process TCEPConnect1 [t] (role:TSUserRole) : exit :=
[role = CallingRole] -> t ?ta:TAddress ?tcei:TCEI ?tcr:TSP
[IsValidTCONreq(tcr) andthen (ta eq CallingOf(tcr))] ;
    TCEPConnect2 [t] (tcr,role)
[]
[role = CalledRole] -> exit
[]
[role = CalledRole] -> t ?ta:TAddress ?tcei:TCEI ?tci:TSP
[IsTCONind(tci) andthen (ta eq CalledOf(tci))] ;
    TCEPConnect2 [t] (tci,role)
endproc (* TCEPConnect1 *)

process TCEPConnect2 [t] (tc:TSP, role:TSUserRole) : exit :=
t ?ta:TAddress ?tcei:TCEI ?tc2:TSP [tc2 IsValidTCON2For tc];
    ([both eq PlexOf(tc2)] -> TCEPDTPhase [t] (role,0 of
                                                Nat,tc2)
    []
    [him eq PlexOf(tc2)] -> TCEPDTPhaseAfterRC [t] (him,role,0
                                                of Nat,tc2)
    []
    [me eq PlexOf(tc2)] -> TCEPDTPhaseAfterI [t] (me,role,0 of
                                                Nat,tc2))
[]
t ?ta:TAddress ?tcei:TCEI ?tc2:TSP [IsTDISInOut(tc2)] ; exit
endproc (* TCEPConnect2 *)
(*
```

5.2.3.3 Data transfer phase at a TCEP   *)

```
process TCEPDTPhase [t] (role:TSUserRole, renph:Nat, trn:TSP)
:
exit :=
t ?ta:TAddress ?tcei:TCEI ?tsp:TSP [IsTDISInOut(tsp)] ; exit
[]
```

```
t ?ta:TAddress ?tcei:TCEI ?tsp:TSP [IsTDISIn(tsp)] ;
TCEPDTPhaseAfterI [t] (both,role,renph,trn)
[]
t ?ta:TAddress ?tcei:TCEI ?tsp:TSP [IsTDISOut(tsp)] ;
TCEPDTPhaseAfterRC [t] (both,role,renph,trn)
[]
t ?ta:TAddress ?tcei:TCEI ?tsp:TSP [IsTRELreq(tsp)] ;
TCEPDTPhaseAfterR [t] (role,renph,trn)
[]
t ?ta:TAddress ?tcei:TCEI ?tsp:TSP [IsTRELind(tsp)] ;
TCEPDTPhaseAfterI [t] (both,role,renph,trn)
[]
t ?ta:TAddress ?tcei:TCEI ?tsp:TSP [IsTDT(tsp)] ; TCEPDTPhase
[t] (role,renph,trn)
[]
t ?ta:TAddress ?tcei:TCEI ?tsp:TSP [IsTOOBACK(tsp)] ;
TCEPDTPhase [t] (role,renph,trn)
[]
t ?ta:TAddress ?tcei:TCEI ?tsp:TSP [IsTREPind(tsp)] ;
TCEPDTPhase [t] (role,renph,trn)
[]
[role = CallingRole] ->
    ( [renph = 0 of Nat] ->
      t ?ta:TAddress ?tcei:TCEI ?tsp:TSP [IsTRENreq(tsp)
                                andthen (IsBi(TQOS1Of(tsp))
                                and IsBi(TTSDU1SizeOf(tsp)))]
;
                     TCEPDTPhase [t] (role,succ(0),tsp)
      []
      [renph = succ(0)] ->
      t ?ta:TAddress ?tcei:TCEI ?tsp:TSP [((tsp IsValidTCON2For
                                trn) andthen (IsBi(TQOS2Of(tsp))
                                and IsBi(TTSDU2SizeOf(tsp))))
                                or IsTREJRENind(tsp)] ;
                     TCEPDTPhase [t] (role,0 of Nat,trn) )
[]
[role = CalledRole] ->
    ( [renph = 0 of Nat] ->
      t ?ta:TAddress ?tcei:TCEI ?tsp:TSP [IsTRENind(tsp)
                                andthen (IsBi(TQOS1Of(tsp))
                                and IsBi(TTSDU1SizeOf(tsp)))]
;
                     TCEPDTPhase [t] (role,succ(0),tsp)
      []
      [renph = succ(0)] ->
      t ?ta:TAddress ?tcei:TCEI ?tsp:TSP [(tsp IsValidTCON2For
                                trn) andthen (IsBi(TQOS2Of(tsp))
                                and IsBi(TTSDU2SizeOf(tsp)))] ;
                     TCEPDTPhase [t](role,succ(succ(0)),trn)
      []
      [renph = succ(0)] ->
      t ?ta:TAddress ?tcei:TCEI ?tsp:TSP [IsTREJRENreq(tsp)] ;
                     TCEPDTPhase [t] (role,0 of Nat,trn)
      []
      [renph = succ(succ(0))] ->
      t ?ta:TAddress ?tcei:TCEI ?tsp:TSP [IsTNEWQOSind(tsp)] ;
                     TCEPDTPhase [t] (role,0 of Nat,trn) )

 endproc (* TCEPDTPhase *)
```

```
process TCEPDTPhaseAfterR [t] (role:TSUserRole, renph:Nat,
trn:TSP): exit :=
t ?ta:TAddress ?tcei:TCEI ?tsp:TSP [IsTDISInOut(tsp)] ; exit
[]
t ?ta:TAddress ?tcei:TCEI ?tsp:TSP [IsTDISIn(tsp)] ;
    TCEPDTPhaseAfterRI [t] (both)
[]
t ?ta:TAddress ?tcei:TCEI ?tsp:TSP [IsTDISOut(tsp)] ;
    TCEPDTPhaseAfterRC [t] (both,role,renph,trn)
[]
t ?ta:TAddress ?tcei:TCEI ?tsp:TSP [IsTRELconf(tsp)] ;
    TCEPDTPhaseAfterRC [t] (both,role,renph,trn)
[]
t ?ta:TAddress ?tcei:TCEI ?tsp:TSP [IsTRELind(tsp)] ;
    TCEPDTPhaseAfterRI [t] (both)
[]
t ?ta:TAddress ?tcei:TCEI ?tsp:TSP [IsTDTind(tsp)] ;
    TCEPDTPhaseAfterR [t] (role,renph,trn)
[]
t ?ta:TAddress ?tcei:TCEI ?tsp:TSP [IsTOOBACK(tsp)] ;
    TCEPDTPhaseAfterR [t] (role,renph,trn)
[]
t ?ta:TAddress ?tcei:TCEI ?tsp:TSP [IsTREPindInp(tsp)] ;
    TCEPDTPhaseAfterR [t] (role,renph,trn)
[]
[role = CallingRole] ->
    ( [renph = 0 of Nat] ->
     t ?ta:TAddress ?tcei:TCEI ?tsp:TSP [IsTRENreq(tsp)
                                  andthen (IsInp(TQOS1Of(tsp))
                                  and IsInp(TTSDU1SizeOf(tsp)))]
     ;
             TCEPDTPhaseAfterR [t] (role,succ(0),tsp)
      []
      [renph = succ(0)] ->
      t ?ta:TAddress ?tcei:TCEI ?tsp:TSP [((tsp IsValidTCON2For
                                  trn) andthen (IsOut(TQOS2Of(tsp))
                                  and IsOut(TTSDU2SizeOf(tsp))))
                                  or IsTREJRENind(tsp)] ;
             TCEPDTPhaseAfterR [t] (role,0 of Nat,trn) )
[]
[role = CalledRole] ->
    ( [renph = 0 of Nat] ->
     t ?ta:TAddress ?tcei:TCEI ?tsp:TSP [IsTRENind(tsp)
                                  andthen IsOut(TQOS1Of(tsp))
                                  and IsOut(TTSDU1SizeOf(tsp))]
     ;
             TCEPDTPhaseAfterR [t] (role,succ(0),tsp)
      []
      [renph = succ(0)] ->
      t ?ta:TAddress ?tcei:TCEI ?tsp:TSP [(tsp IsValidTCON2For
                                  trn) andthen (IsInp(TQOS2Of(tsp))
                                  and IsInp(TTSDU2SizeOf(tsp)))] ;
         TCEPDTPhaseAfterR [t](role,succ(succ(0)),trn)
      []
      [renph = succ(0)] ->
      t ?ta:TAddress ?tcei:TCEI ?tsp:TSP [IsTREJRENreq(tsp)] ;
         TCEPDTPhaseAfterR [t] (role,0 of Nat,trn)
      []
      [renph = succ(succ(0))] ->
```

```
        t ?ta:TAddress ?tcei:TCEI ?tsp:TSP [IsTNEWQOSind(tsp)] ;
            TCEPDTPhaseAfterR [t] (role,0 of Nat,trn) )
endproc (* TCEPDTPhaseAfterR *)

process TCEPDTPhaseAfterI [t] (pl:TPlex, role:TSUserRole,
renph:Nat, trn:TSP) : exit :=
t ?ta:TAddress ?tcei:TCEI ?tsp:TSP [IsTDISInOut(tsp)] ; exit
[]
[pl eq both] -> t ?ta:TAddress ?tcei:TCEI ?tsp:TSP
[IsTDISOut(tsp)] ; exit
[]
t ?ta:TAddress ?tcei:TCEI ?tsp:TSP [IsTRELreq(tsp)] ;
    TCEPDTPhaseAfterRI [t] (pl)
[]
t ?ta:TAddress ?tcei:TCEI ?tsp:TSP [IsTDTreq(tsp)] ;
    TCEPDTPhaseAfterI [t] (pl,role,renph,trn)
[]
t ?ta:TAddress ?tcei:TCEI ?tsp:TSP [IsTOOBACK(tsp)] ;
    TCEPDTPhaseAfterI [t] (pl,role,renph,trn)
[]
t ?ta:TAddress ?tcei:TCEI ?tsp:TSP [IsTREPindOut(tsp)] ;
    TCEPDTPhaseAfterI [t] (pl,role,renph,trn)
[]
[role = CallingRole] ->
    ( [renph = 0 of Nat] ->
      t ?ta:TAddress ?tcei:TCEI ?tsp:TSP [IsTRENreq(tsp)
          andthen (IsOut(TQOS1Of(tsp))
          and IsOut(TTSDU1SizeOf(tsp)))] ;
              TCEPDTPhaseAfterI [t] (pl,role,succ(0),tsp)
      []
      [renph = succ(0)] ->
      t ?ta:TAddress ?tcei:TCEI ?tsp:TSP [((tsp IsValidTCON2For
          trn) andthen (IsInp(TQOS2Of(tsp))
          and IsInp(TTSDU2SizeOf(tsp)))) or IsTREJRENind(tsp)] ;
              TCEPDTPhaseAfterI [t] (pl,role,0 of Nat,trn)
    )
[]
[role = CalledRole] ->
    ( [renph = 0 of Nat] ->
      t ?ta:TAddress ?tcei:TCEI ?tsp:TSP [IsTRENind(tsp)
          andthen (IsInp(TQOS1Of(tsp))
          and IsInp(TTSDU1SizeOf(tsp)))] ;
              TCEPDTPhaseAfterI [t] (pl,role,succ(0),tsp)
      []
      [renph = succ(0)] ->
      t ?ta:TAddress ?tcei:TCEI ?tsp:TSP [((tsp IsValidTCON2For
          trn) andthen (IsOut(TQOS2Of(tsp))
          and IsOut(TTSDU2SizeOf(tsp)))) or IsTREJRENreq(tsp)] ;
              TCEPDTPhaseAfterI [t] (pl,role,0 of Nat,trn)
    )
endproc (* TCEPDTPhaseAfterI *)

process TCEPDTPhaseAfterRC [t] (pl:TPlex, role: TSUserRole,
renph:Nat, trn:TSP) : exit :=
t ?ta:TAddress ?tcei:TCEI ?tsp:TSP [IsTDISInOut(tsp)] ; exit
[]
[pl eq both] -> t ?ta:TAddress ?tcei:TCEI ?tsp:TSP
[IsTDISIn(tsp)] ; exit
[]
```

```
t ?ta:TAddress ?tcei:TCEI ?tsp:TSP [IsTRELind(tsp)] ; exit
[]
t ?ta:TAddress ?tcei:TCEI ?tsp:TSP [IsTDTind(tsp)] ;
    TCEPDTPhaseAfterRC [t] (pl,role,renph,trn)
[]
t ?ta:TAddress ?tcei:TCEI ?tsp:TSP [IsTOOBACK(tsp)] ;
    TCEPDTPhaseAfterRC [t] (pl,role,renph,trn)
[]
t ?ta:TAddress ?tcei:TCEI ?tsp:TSP [IsTREPindInp(tsp)] ;
    TCEPDTPhaseAfterRC [t] (pl,role,renph,trn)
[]
[role = CallingRole] ->
    ( [renph = 0 of Nat] ->
        t ?ta:TAddress ?tcei:TCEI ?tsp:TSP [IsTRENreq(tsp)
          andthen (IsInp(TQOS1Of(tsp))
          and IsInp(TTSDU1SizeOf(tsp)))] ;
            TCEPDTPhaseAfterRC [t] (pl,role,succ(0),tsp)
        []
        [renph = succ(0)] ->
        t ?ta:TAddress ?tcei:TCEI ?tsp:TSP [((tsp IsValidTCON2For
          trn) andthen (IsOut(TQOS2Of(tsp))
          and IsOut(TTSDU2SizeOf(tsp)))) or IsTREJRENind(tsp)] ;
            TCEPDTPhaseAfterRC [t] (pl,role,0 of Nat,tsp)
    )
[]
[role = CalledRole] ->
    ( [renph = 0 of Nat] ->
        t ?ta:TAddress ?tcei:TCEI ?tsp:TSP [IsTRENind(tsp)
          andthen (IsOut(TQOS1Of(tsp))
          and IsOut(TTSDU1SizeOf(tsp)))] ;
            TCEPDTPhaseAfterRC [t] (pl,role,succ(0),tsp)
        []
        [renph = succ(0)] ->
        t ?ta:TAddress ?tcei:TCEI ?tsp:TSP [(tsp IsValidTCON2For
            trn) andthen (IsInp(TQOS2Of(tsp))
            and IsInp(TTSDU2SizeOf(tsp)))] ;
            TCEPDTPhaseAfterRC [t] (pl,role,succ(succ(0)),trn)
        []
        [renph = succ(0)] ->
        t ?ta:TAddress ?tcei:TCEI ?tsp:TSP [IsTREJRENreq(tsp)] ;
            TCEPDTPhaseAfterRC [t] (pl,role,0 of Nat,trn)
        []
        [renph = succ(succ(0))] ->
        t ?ta:TAddress ?tcei:TCEI ?tsp:TSP [IsTNEWQOSind(tsp)] ;
            TCEPDTPhaseAfterRC [t] (pl,role,0 of Nat,trn)
    )
endproc (* TCEPDTPhaseAfterRC *)

process TCEPDTPhaseAfterRI [t] (pl:TPlex) : exit :=
t ?ta:TAddress ?tcei:TCEI ?tsp:TSP [IsTDISInOut(tsp)] ; exit
[]
[pl eq both] -> t ?ta:TAddress ?tcei:TCEI ?tsp:TSP
    [IsTDISOut(tsp)] ; exit
[]
t ?ta:TAddress ?tcei:TCEI ?tsp:TSP [IsTRELconf(tsp)] ; exit
[]
t ?ta:TAddress ?tcei:TCEI ?tsp:TSP [IsTOOBACK(tsp)] ;
    TCEPDTPhaseAfterRI [t] (pl)
endproc (* TCEPDTPhaseAfterRI *)
```

```
(*
```

5.2.4 User data constraints

The length of user data parameter in TS primitives is constrained by TCEPUserData which is mainly based on what is defined in ISO 8072. For TDATAindication one must allow the possibility of a void UserData parameter, in case of loss of the TSDU (see 3.6.1.4).   *)

```
process TCEPUserData [t] : exit :=
t ?ta:TAddress ?tcei:TCEI ? tsp:TSP [IsValidUserData(tsp)] ;
TCEPUserData [t]
[]
exit
endproc (* TCEPUserData *)

type ValidUserData
is TransportServicePrimitive
opns
    IsValidUserData : TSP -> Bool
eqns
forall t : TSP
ofsort Bool
    IsTCON(t) =>
IsValidUserData(t) = Length(UserData(t)) le NatNum(3+Dec(2))
    ;
    IsTDTreq(t) =>
IsValidUserData(t) = Length(UserData(t)) gt 0           ;
    IsTDTind(t) =>
IsValidUserData(t) = ((Length(UserData(t)) eq 0) implies
                          (LSigOf(ErSigOf(t)) eq Sig)) ;
    IsTRELreq(t) or IsTRELind(t) =>
IsValidUserData(t) = true                                   ;
    IsTOOBACK1(t) =>
IsValidUserData(t) = Length(UserData(t)) gt 0          ;
    IsTRELconf(t) or IsTOOBACK2(t) or IsTREPind(t)
    or IsTREN(t) =>
IsValidUserData(t) = Length(UserData(t)) eq 0          ;
    IsTDIS(t) =>
IsValidUserData(t) = Length(UserData(t)) le NatNum(6+Dec(4))
    ;
endtype (* ValidUserData *)
(*
```

5.3 End-to-end constraints for a single TC

5.3.1 General

The role of TCEPAssociation is to express the end-to-end constraints, i.e. the link between what happens at one peer and what happens at the other and the non determinacy related to it. This mainly includes the emission of Indications at each TCEP caused by Requests at the other one. But not only Requests induce Indications, TRELEASEindication and TRENEGOTIATEconfirm also provoke the emission of an indication (resp. TRELEASEconfirm and TNEWQOSindication). In the sequel we will thus call 'Outputs' the Indication primitives (according to the interpretation in clause 2) and 'Inputs' the Request primitives plus TRELEASEindication and TRENEGOTIATEconfirm, which are thus members of the two groups. TCEPAssociation also describes the non determinacy on the execution of provider-generated TSPs.

After a first Request, TCEPAssociation is split into two independent processes, one that cares for the in-band transfers and the other for the out-of-band transfers.

NOTE - Because of the conjunction with the local constraints, there won't be out-of-band transfers until the connection is established.   *)

```
process TCEPAssociation [t] : noexit :=
( t ?ta:TAddress ?tcei:TCEI ?tcr:TSP [IsTReq(tcr)] ;
TAssocInB [t] (ta, tcei, tcr) )
|||
TOOBACK [t]
endproc (* TCEPAssociation *)
(*
```

5.3.2 In-band constraints

TAssocInB is the result of many constraints: two instances of TAssoc1, one per direction of transfer, are in independent parallel composition with TRENReject, a process that represents a possible reject by the service provider of the new qualities of service expressed in a TRENEGOTIATIONrequest. Theses processes interact with seven other constraints.

TInB restricts the access to the 'in-band' primitives. TCCorruptionLossesManager traduces the resolutions negotiated for the type of error control on the TSDU transmission. TCTransDelController, TCJitterController, TCCorruptionResErrorRateController, TCLossesResErrorRateController, TCThroughputController all describe the influence of the qualities of service they are related to on the behaviour of the service provider. Particularly, they induce, according to the threshold and compulsory values, the emission of TREPORTindication or TDISCONNECTindication primitives. Except for error rate controls, due to the lack of a timed LOTOS no complete description of these mechanisms was possible. TCTSDUSize describes the constraints on the size of the TSDUs, according to what was indicated by each entity.

The two instances of TAssoc1 interact, as some primitives must be simultaneously observed by both, i.e. TRELEASEindication and TRENEGOTIATEconfirm: which are simultaneously Outputs for one and Inputs for the other and, for technical reasons made clear in the sequel, TRELEASErequest and unidirectional TDISCON-NECTrequest (see 5.3.2.1, 5.3.2.2.1).

NOTE - The initial interaction is only constrained to be a Request. It will be a TCONNECT request because of the conjunction with the local constraints (see 5.2.3.2).   *)

```
process TAssocInB [t] (ta:TAddress, tcei:TCEI, tcr:TSP)
: noexit :=
( ( ( TAssoc1 [t] (ta, tcei, Request, Indication,
                Append(tcr,NoTPrim))
```

```
            ||
        TAssoc1 [t] (ta, tcei, Indication, Request, NoTPrim) )
    |||
   TRENReject [t]
  )
  ||
  TInB [t]
  ||
  TCCorruptionLossesManager [t] (ta, tcei, tcr)
  ||
  TCTransDelController [t] (ta, tcei)
  ||
  TCJitterController [t] (ta, tcei)
  ||
  TCCorruptionResErrorRateController [t] (ta, tcei)
  ||
  TCLossesResErrorRateController [t] (ta, tcei)
  ||
  TCThroughputController [t] (ta, tcei)
  ||
  TCTSDUSizeController [t] (ta, tcei)
)
endproc (* TAssocInB *)
(*
```

5.3.2.1 Unidirectionality

Basically, TAssoc1 observes the Inputs at one given peer and induces the corresponding Outputs at the opposite side. It is made of two processes disjointedly composed in
parallel: TAssoc2 and TRunAssoc. TAssoc2 is the process that ensures the transfer of
the primitives. The need for TRunAssoc comes from the interaction between the two
instances of TAssoc1: this situation imposes that, at both endpoints, both processes
must take part in the occurrence of all the primitives. But one must free the instances
of TAssoc2 from this constraint. Indeed, one must take care that their respective roles
be well defined, i.e. that the endpoint where one of them takes its Inputs is the
endpoint where the other emits its Outputs, and vice versa, Inputs and Outputs being
usually separate primitives. The part of TRunAssoc is thus to accept the primitives
which TAssoc2 is not involved in.

TAssoc2 relates the Outputs that may be executed at one end of the TC to the history of the Inputs executed at the other end, but catering for the provider's non-
determinacy affecting this relationship. The most complex sub process of TAssoc1 is
TCInpToOut (see 5.3.2.2.1), which represents the end-to-end constraints with reference only to the TSP part of the interactions at t. The fourth parameter of
TCInpToOut represents the history of the Inputs that are relevant w.r.t. possible future
Outputs. Let us remark that this set of possible future Outputs also depends on the
local state at the TCEP they reach, as this state determines the primitives allowed to
occur there (see 5.2.3.3). If this state happens to change, some Outputs thus stop being
allowed, and the corresponding Inputs in the history become obsolete (for instance, if
an entity emits an unidirectional TDISCONNECTrequest that closes its input
direction, this signifies it is no more interested in receiving TDATAindications, so
that the TDATArequests in the corresponding history become obsolete). The fact that
knowing the local state at the TCEP where it emits its Outputs is important for

TCInpToOut explains the reason why it also needs to observe the TRELEASErequest and the unidirectional TDISCONNECTrequest primitives at this TCEP (see 5.3.2).

   The other sub processes of TAssoc2 express the constraints on the TCEPs where the primitives must occur, i.e. : for the direction of transfer from the Calling entity to the Called one: at the Calling TCEP, only Inputs may happen, and at the Called TCEP, Outputs plus the TRELEASErequest and unidirectional TDISCONNECTrequest primitives; and vice-versa for the reverse direction of transfer.

   NOTE - The Called TCEP is not specified, it is just imposed that it is different from the Calling one. This is enough, because of the interaction with the local constraints, which already impose that each primitive occur either through the Calling or through the Called TCEP.   *)

```
process TAssoc1 [t] (ta:TAddress, tcei:TCEI,
d1,d2:TSPDirection, rh:THistory) : noexit :=
( TAssoc2 [t] (ta, tcei, d1, d2, rh)
   |||
   TRunAssoc [t] (ta, tcei, d2, d1)
)
endproc (* TAssoc1 *)

process TAssoc2 [t] (ta:TAddress, tcei:TCEI,
d1,d2:TSPDirection, rh:THistory) : noexit :=
(    (    TCEPHalf [t] (d1)
          ||
          ConstantTA [t] (ta)
          ||
          ConstantTCEI [t] (tcei)
     )
     |||
     (    TCEPHalf [t] (d2)
          ||
          DifferentTId [t] (ta, tcei)
     )
)
|| TCInpToOut [t] (ta, tcei, d1, rh, Init)
endproc (* TAssoc2 *)

process TCEPHalf [t] (d1:TSPDirection) : noexit :=
(    [d1 eq Request] ->
     t ?ta:TAddress ?tcei:TCEI ?tsp:TSP [IsTReq(tsp)
     or IsTRELind(tsp) or IsTRENconf(tsp)] ; TCEPHalf [t] (d1)
     []
     [d1 eq Indication] ->
     t ?ta:TAddress ?tcei:TCEI ?tsp:TSP [IsTInd(tsp)
     or IsTRELreq(tsp) or IsTDISIn(tsp) or IsTDISOut(tsp)] ;
        TCEPHalf [t] (d1)
)
endproc (* TCEPHalf *)

process DifferentTId [t] (ta:TAddress, tcei:TCEI) : noexit :=
t ?t1:TAddress ?tc1:TCEI ?tsp:TSP
[(TId(t1,tc1) ne TId(ta,tcei)) ] ;
     (    ConstantTA [t] (t1)
          ||
          ConstantTCEI [t] (tc1)
     )
```

```
endproc (* DifferentTA *)

process TRunAssoc [t] (ta:TAddress, tcei:TCEI,
d1,d2:TSPDirection) : noexit :=
(    TCEPRunHalf [t] (d1)
     ||
     ConstantTA [t] (ta)
     ||
     ConstantTCEI [t] (tcei)
)
|||
(    TCEPRunHalf [t] (d2)
     ||
     DifferentTId [t] (ta, tcei)
)
endproc (* TRunAssoc *)

process TCEPRunHalf [t] (d1:TSPDirection) : noexit :=
(    [d1 eq Request] ->
     t ?ta:TAddress ?tcei:TCEI ?tsp:TSP [not(IsTInd(tsp)
     or IsTRELreq(tsp) or IsTDISIn(tsp) or IsTDISOut(tsp))] ;
     TCEPRunHalf [t] (d1)
     []
     [d1 eq Indication] ->
     t ?ta:TAddress ?tcei:TCEI ?tsp:TSP [not(IsTReq(tsp)
     or IsTRELind(tsp) or IsTRENconf(tsp))] ;
     TCEPRunHalf [t] (d1)
)
endproc (* TCEPRunHalf *)
(*
```

5.3.2.2 End-to-end non-determinacy

5.3.2.2.1 General

As anticipated above, TCInpToOut is parameterized with the history ih of the Inputs
executed at one end that may play a role in respect of the possible future Outputs at
the other end. Values of sort THistory are basically sequences, on which a few
operations are defined that enable consideration of only the relevant elements of the
history (see 5.3.2.2.2 for details).

But TCInpToOut is also parameterised with a value of sort LoStInd. The need for
this value has been evoked in 5.3.2.1. The idea is that ih should always be free of
obsolete Inputs. This supposes two tasks:

- when the local state at the 'output' TCEP changes, ih must be emptied of the Inputs
  that became obsolete
- after this refreshing of ih, one must also avoid that new obsolete Inputs enter it.

In order to realise this the knowledge of the local state at the TCEP is required. This
is what the last parameter (lsi) indicates. Values of sort LoStInd are defined in type
LocalStateIndications, the names following the conventions explained in 5.2.3.1.

NOTE - The need for the refreshing of the history ih comes from the way this
history is described in the specification, and from the way precedence between
consecutive Inputs is realised (see 5.3.2.2.2). Refreshing ih prevents obsolete Inputs
from blocking valid ones of lower precedence. It is just an internal mechanism of the

464   **Appendix**

specification, in order to provide the desired observable behaviour, but let us recall it has nothing to do with a possible implementation.

The immediate form of the definition of TCInpToOut is that of a tail-recursion. At any time, the process TSPEvent specifies the constraints on the next observable event; the fourth parameter of TSPEvent is the keystone of the representation of non-determinacy (the Tops operation is defined in 5.3.2.2.2). After execution of that event, TCInpToOut is invoked with updated parameter values. The determination of these values is made according to two parameters: lsi and cl1. cl1 is of sort ClosInd. It is an indication furnished by TSPEvent about how the occurrence of the last primitive influenced the local state at the TCEP receiving the Outputs. lsi and cl1 determine whether (and how) ih should be refreshed. The value of lsi is also updated. Refreshing ih is made with the operator Clean (see 5.3.2.2.2). The second task, avoiding that new obsolete Inputs enter ih, is realised by process TSPEvent (see 5.3.2.2.3), which thus also receives lsi as last parameter.   *)

```
type LocalStateIndications
is NaturalNumber, Boolean
sorts LoStInd
opns
    Init, AfterR, AfterI, AfterRC, AfterRI  : -> LoStInd
    h                              : LoStInd -> Nat
    _eq_, _ne_                     : LoStInd, LoStInd -> Bool
eqns
forall he,he1 : LoStInd

ofsort Nat
h(Init)        = 0                         ;
h(AfterR)      = Succ(0)          ;
h(AfterI)      = Succ(h(AfterR))      ;
h(AfterRC)     = Succ(h(AfterI))      ;
h(AfterRI)     = Succ(h(AfterRC))     ;

ofsort Bool
he eq he1      = h(he) eq h(he1) ;
he ne he1      = h(he) ne h(he1) ;

endtype (* LocalStateIndications *)


process TCInpToOut [t] (t1:TAddress, tc1:TCEI,
d1:TSPDirection,
ih:THistory, lsi:LoStInd) : noexit :=
TSPEvent [t] (t1, tc1, d1, Tops(ih), ih, lsi) >>
accept ih1:THistory, cl1:ClosInd in
( [lsi eq Init] ->
    ( [cl1 eq Noth] -> TCInpToOut [t] (t1, tc1, d1, ih1, lsi)
      []
      [cl1 eq RELr] -> TCInpToOut [t] (t1, tc1, d1, ih1,
AfterR)
      []
      [cl1 eq RELi] -> TCInpToOut [t] (t1, tc1, d1,
                                  Clean(ih1,AfterI), AfterI)
      []
      [cl1 eq RELc] -> TCInpToOut [t] (t1, tc1, d1,
                                  Clean(ih1,AfterRC), AfterRC) )
  []
  [lsi eq AfterR] ->
    ( [cl1 eq Noth] -> TCInpToOut [t] (t1, tc1, d1, ih1, lsi)
      []
```

```
      [cl1 eq RELi] -> TCInpToOut [t] (t1, tc1, d1,
                                  Clean(ih1,AfterRI), AfterRI)
      []
      [cl1 eq RELc] -> TCInpToOut [t] (t1, tc1, d1,
                                  Clean(ih1,AfterRC), AfterRC) )
  []
  [lsi eq AfterI] ->
    ( [cl1 eq Noth] -> TCInpToOut [t] (t1, tc1, d1, ih1, lsi)
      []
      [cl1 eq RELr] -> TCInpToOut [t] (t1, tc1, d1,
                                  Clean(ih1,AfterR), AfterRI)
      []
      [cl1 eq RELc] -> TCInpToOut [t] (t1, tc1, d1, ih1, lsi)
    )
  []
  [(lsi eq AfterRC) or (lsi eq AfterRI)] -> TCInpToOut [t]
(t1,
                                  tc1, d1, ih1, lsi)
)
endproc (* TCInpToOut *)
(*
```

5.3.2.2.2 Data definitions

The first definition provides a construction for history elements, by means of the operations NoTPrim (the empty history) and OnTopOf (to extend a history with an earlier Input).

A few Boolean functions are also introduced, with the usual interpretation and two other operators are described:

a) Reduce: deprives a history of its top element (i.e. the Input that was executed earliest in time).

b) Append: adds a Input to a history as latest element. Note that Append acts as OnTopOf, but on the opposite end of the history.

The second definition enriches this basic data type with two operations needed to formulate the end-to-end constraints:

a) Remove: removes a given Input from a given history; if the former has multiple occurrences in the latter, then the earliest occurrence is removed.

b) Tops: yields a history that consists of those TSPs in the argument history that may lead to the next Indication.

Other functions are also introduced, viz. TDISTops1 and TDISTops2, for the conciseness of the definition. TDISTops1 and TDISTops2 yield intermediate results for the computation of Tops.
The following precedence order on TSPs is defined:

- T-DISCONNECTInOut precedes any other primitive,
- T-DISCONNECTIn and T-DISCONNECTOut precede any primitive, except T-DISCONNECTInOut and the connection establishment primitives,
- no other precedence between the TSPs.

Tops is specified as follows: for any given value ih of sort THistory, Tops(ih) is the sequence of Inputs that

a) contains only Inputs of different services, and

b) contains a Input t if, and only if, t is in ih and Inputs that are before it in ih (i.e. were executed earlier in time) are all of lower precedence than t, and

c) preserves the order of Inputs in ih.

The second definition also introduces the operator Clean. This operator takes two parameter: an history and a local state indication. According to this last indication, Clean returns an history emptied of all its obsolete Inputs.   *)

```
type TransportServiceBasicTSPHistory is
TransportServicePrimitive
sorts   THistory
opns
    NoTPrim              :                        -> THistory
    _OnTopOf_            : TSP, THistory    -> THistory
    Reduce               : THistory         -> THistory
    Append               : TSP, THistory    -> THistory
    _IsTopOf_            : TSP , THistory   -> Bool
    IsEmpty              : THistory          -> Bool
    _eq_,_ne_            : THistory, THistory   -> Bool
eqns
forall t,t1 : TSP, h,h1 : THistory

ofsort THistory
Reduce(NoTPrim)             = NoTPrim                    ;
Reduce(t OnTopOf h)         = h                                ;
Append(t,NoTPrim)          = t OnTopOf NoTPrim        ;
Append(t, t1 OnTopOf h1)       = t1 OnTopOf Append(t,h1)       ;

ofsort Bool
IsEmpty(NoTPrim)           = true     ;
IsEmpty(t OnTopOf h)       = false    ;
NoTPrim eq NoTPrim         = true     ;
NoTPrim eq (t OnTopOf h)    = false    ;
t OnTopOf h eq NoTPrim     = false    ;
    t eq t1, h eq h1 =>
t OnTopOf h eq (t1 OnTopOf h1)    = true     ;
    not(t eq t1) =>
t OnTopOf h eq (t1 OnTopOf h1)    = false       ;
    not(h eq h1) =>
t OnTopOf h eq (t1 OnTopOf h1)    = false       ;
h ne h1                    = not(h eq h1)     ;
t IsTopOf NoTPrim          = false       ;
    t eq t1 =>
t IsTopOf (t1 OnTopOf h1)        = true     ;
    not(t eq t1) =>
t IsTopOf (t1 OnTopOf h1)        = false     ;
endtype (* TransportServiceBasicTSPHistory *)

type TransportServiceTSPHistory
is TransportServiceBasicTSPHistory, LocalStateIndications
opns
    Remove : TSP, THistory         -> THistory
    Clean  : THistory, LoStInd    -> THistory
    Tops, TDISTopsUni, TDISTops1, TDISTops2
            : THistory            -> THistory
eqns
forall t,t1:TSP, h,h1:THistory, lsi:LoStInd
ofsort THistory
```

```
Remove(t, NoTPrim)                         = NoTPrim
    ;
t eq t1 => Remove(t, t1 OnTopOf h1)        = h1
    ;
t ne t1 => Remove(t, t1 OnTopOf h1)        = t OnTopOf
Remove(t,h1)    ;
Clean(NoTPrim, lsi)                = NoTPrim
    ;
    IsTREN(t) =>
Clean(t OnTopOf h1, AfterR)        = Clean(h1,AfterR)
    ;
    not(IsTREN(t)) =>
Clean(t OnTopOf h1, AfterR)        = t OnTopOf Clean(h1,AfterR)
    ;
    IsTDISOut(t) or IsTDTreq(t) or IsTRENconf(t) =>
Clean(t OnTopOf h1, AfterI)    = Clean(h1, AfterI)
    ;
    not(IsTDISOut(t) or IsTDTreq(t) or IsTRENconf(t)) =>
Clean(t OnTopOf h1, AfterI)        = t OnTopOf Clean(h1,AfterI)
    ;
    IsTDISIn(t) or IsTRELind(t) =>
Clean(t OnTopOf h1, AfterRC)       = Clean(h1, AfterRC)
    ;
    not(IsTDISIn(t) or IsTRELind(t)) =>
Clean(t OnTopOf h1, AfterRC)       = t OnTopOf
Clean(h1,AfterRC) ;
    IsTDISOut(t) or IsTDTreq(t) or IsTREN(t) =>
Clean(t OnTopOf h1, AfterRI)       = Clean(h1, AfterRI)
    ;
    not(IsTDISOut(t) or IsTDTreq(t) or IsTREN(t)) =>
Clean(t OnTopOf h1, AfterRI)       = t OnTopOf
Clean(h1,AfterRI);
Tops(NoTPrim)                      = NoTPrim
    ;
    IsTDISInOut(t) and IsTDISreq(t) =>
Tops(t OnTopOf h)                  = t OnTopOf NoTPrim
    ;
    (IsTDISIn(t) or IsTDISOut(t)) and IsTDISreq(t) =>
Tops(t OnTopOf h)                  = t OnTopOf TDISTopsUni(h)
    ;
    IsTCONreq(t) or IsTCONresp(t) =>
Tops(t OnTopOf h)                  = t OnTopOf TDISTops1(h)
    ;
    not(IsTDIS(t) or IsTCONreq(t) or IsTCONresp(t)) =>
Tops(t OnTopOf h)                  = t OnTopOf TDISTops2(h)
    ;
TDISTopsUni(NoTPrim)               = NoTPrim
    ;
    IsTDISInOut(t) and IsTDISreq(t) =>
TDISTopsUni(t OnTopOf h)               = t OnTopOf NoTPrim
    ;
    (IsTDISIn(t) or IsTDISOut(t)) and IsTDISreq(t) =>
TDISTopsUni(t OnTopOf h)               =
TDISreq(inout,UserData(t))
                                   OnTopOf (t OnTopOf NoTPrim)
    ;
    not(IsTDISreq(t)) =>
TDISTopsUni(t OnTopOf h)           = TDISTopsUni(h)
        ;
```

```
TDISTops1(NoTPrim)                      = NoTPrim
    ;
    IsTDISInOut(t) and IsTDISreq(t) =>
TDISTops1(t OnTopOf h)                  = t OnTopOf NoTPrim
    ;
    not(IsTDISInOut(t) and IsTDISreq(t)) =>
TDISTops1(t OnTopOf h)                  = TDISTops1(h)
    ;
TDISTops2(NoTPrim)                      = NoTPrim
    ;
    IsTDISInOut(t) and IsTDISreq(t) =>
TDISTops2(t OnTopOf h)                  = t OnTopOf NoTPrim
    ;
    (IsTDISIn(t) or IsTDISOut(t)) and IsTDISreq(t) =>
TDISTops2(t OnTopOf h)                  = t OnTopOf TDISTops1(h)
    ;
    not(IsTDIS(t)) =>
TDISTops2(t OnTopOf h)                  = TDISTops2(h)
    ;
endtype (* TransportServiceTSPHistory *)
(*
```

5.3.2.2.3 *Process definitions*

For a given Input history ih, TSPEvent specifies the constraints on the possible next event. The parameter eih is the sequence of executed Inputs that suffices to specify the next possible Indication. The following four requirements are formulated to cater for the provider's non-determinacy in its participation to the next event.

a) The provider never refuses to engage in a Input event (see also the definition of TSPInpEvent below).

b) An Indication may be executed that is to be related to the top Input of a non-empty eih according to IsOutputOf (see 3.4.3.4). If such an Indication occurs the corresponding Input is removed from ih.

NOTE - For a given Input, the provider is entitled to autonomously determine the values of the parameters of the Indication, provided they meet that requirement.

c) For any given top Input of a non-empty eih, the provider may offer no related Indication only if it offers to engage

1) in an Indication relating to the Input of next higher precedence in eih, if it exists, or

2) in a provider-generated Indication (see 3.4.3.4).

In case 1), requirements b, c and d apply to eih deprived of its top Input.

d) if the user refuses an Indication, an Indication that complies with case 1) or 2) of requirement c) shall be offered.

One must notice that, however the provider never refuses to engage in a Input event (as said in a)), the fact that this event be added to ih or not is determined according to the value of parameter lsi. Also, let us remark, in the definition of TSPEvent, the process TSPNonInpEvent. This process observes, at the side where the Outputs are emitted, the other primitives that cause a change of the local state (i.e. TRELEASErequest, and both unidirectional TDISCONNECTrequest).

TSPEvent exits two parameters: the updated history, ih, according to the last event and a value of sort ClosInd. These values are defined in type ClosingIndications. They indicate the effect of the last event on the local state. Noth corresponds to "no change". The three other values correspond to the relative TRELEASE primitives. As already explained (see 5.2.3.1), a TDISCONNECTInput has (locally) the effect of a TRELEASEindication and a TDISCONNECTOutput, of a TRELEASEconfirm. *)

```
type ClosingIndications
is NaturalNumber, Boolean
sorts ClosInd
opns
    Noth, RELr, RELi, RELc    : -> ClosInd
    h                    : ClosInd            -> Nat
    _eq_, _ne_           : ClosInd, ClosInd   -> Bool
eqns
forall he,he1 : ClosInd

ofsort Nat
h(Noth)     = 0                      ;
h(RELr)     = Succ(0)                ;
h(RELi)     = Succ(h(RELr))          ;
h(RELc)     = Succ(h(RELi))          ;

ofsort Bool
he eq he1 = h(he) eq h(he1)    ;
he ne he1 = h(he) ne h(he1)    ;

endtype (* ClosingIndications *)

process TSPInpEvent [t] (t1:TAddress, tc1:TCEI,
d:TSPDirection,
ih:THistory, lsi:LoStInd) : exit (THistory, ClosInd) :=
[(lsi eq Init) or (lsi eq AfterR)] ->
    ( [d eq Request] ->
        ( t ?ta:TAddress ?tcei:TCEI ?tsp:TSP [(IsTRELind(tsp)
          and (TId(t1,tc1) eq TId(ta,tcei))) or
IsTRENconf(tsp)]
              ; exit (Append(tsp,ih), Noth)
              []
              t ?ta:TAddress ?tcei:TCEI ?tsp:TSP [IsTReq(tsp)
              and (TId(t1,tc1) eq TId(ta,tcei))] ;
              exit (Append(tsp,ih), Noth) )
        []
        [d eq Indication] ->
          ( t ?ta:TAddress ?tcei:TCEI ?tsp:TSP [IsTRELind(tsp)
            and (TId(t1,tc1) ne TId(ta,tcei))] ;
            exit (Append(tsp,ih), Noth)
            []
            t ?ta:TAddress ?tcei:TCEI ?tsp:TSP [IsTReq(tsp)
            and (TId(t1,tc1) ne TId(ta,tcei))] ;
            exit (Append(tsp,ih), Noth) ) )
  []
  [lsi eq AfterI] ->
     ( [d eq Request] ->
         ( t ?ta:TAddress ?tcei:TCEI ?tsp:TSP [IsTRELind(tsp)
           and (TId(t1,tc1) eq TId(ta,tcei))] ;
           exit (Append(tsp,ih), Noth)
           []
           t ?ta:TAddress ?tcei:TCEI ?tsp:TSP [IsTReq(tsp)
           and not(IsTDT(tsp) or IsTDISOut(tsp))
           and (TId(t1,tc1) eq TId(ta,tcei))] ;
           exit (Append(tsp,ih), Noth)
           []
```

```
                t ?ta:TAddress ?tcei:TCEI ?tsp:TSP [((IsTReq(tsp)
                and (IsTDT(tsp) or IsTDISOut(tsp)))
                or IsTRENconf(tsp)) and (TId(t1,tc1) eq
TId(ta,tcei))]
                ; exit (ih, Noth) )
        []
        [d eq Indication] ->
          ( t ?ta:TAddress ?tcei:TCEI ?tsp:TSP [IsTRELind(tsp)
            and (TId(t1,tc1) ne TId(ta,tcei))] ;
            exit (Append(tsp,ih), Noth)
            []
            t ?ta:TAddress ?tcei:TCEI ?tsp:TSP [IsTReq(tsp)
            and not(IsTDT(tsp) or IsTDISOut(tsp))
            and (TId(t1,tc1) ne TId(ta,tcei))] ;
            exit (Append(tsp,ih), Noth)
            []
            t ?ta:TAddress ?tcei:TCEI ?tsp:TSP [IsTReq(tsp)
            and (IsTDT(tsp) or IsTDISOut(tsp))
            and (TId(t1,tc1) ne TId(ta,tcei))] ;
            exit (ih, Noth) ) )
[]
[lsi eq AfterRC] ->
    ( [d eq Request] ->
        ( t ?ta:TAddress ?tcei:TCEI ?tsp:TSP [IsTRENconf(tsp)]
;
            exit (Append(tsp,ih), Noth)
            []
            t ?ta:TAddress ?tcei:TCEI ?tsp:TSP [IsTReq(tsp)
            and not(IsTDISIn(tsp))
            and (TId(t1,tc1) eq TId(ta,tcei))] ;
            exit (Append(tsp,ih), Noth)
            []
            t ?ta:TAddress ?tcei:TCEI ?tsp:TSP [((IsTReq(tsp)
            and IsTDISIn(tsp)) or IsTRELind(tsp))
            and (TId(t1,tc1) eq TId(ta,tcei))] ;
            exit (ih, Noth) )
        []
        [d eq Indication] ->
          ( t ?ta:TAddress ?tcei:TCEI ?tsp:TSP [IsTReq(tsp)
            and not(IsTDISIn(tsp))
            and (TId(t1,tc1) ne TId(ta,tcei))] ;
            exit (Append(tsp,ih), Noth)
            []
            t ?ta:TAddress ?tcei:TCEI ?tsp:TSP [((IsTReq(tsp)
            and IsTDISIn(tsp)) or IsTRELind(tsp))
            and (TId(t1,tc1) ne TId(ta,tcei))] ;
            exit (ih, Noth) ) )
[]
[lsi eq AfterRI] ->
    ( [d eq Request] ->
        ( t ?ta:TAddress ?tcei:TCEI ?tsp:TSP [IsTRELind(tsp)
          and (TId(t1,tc1) eq TId(ta,tcei))] ;
            exit (Append(tsp,ih), Noth)
          []
          t ?ta:TAddress ?tcei:TCEI ?tsp:TSP [IsTReq(tsp)
          and not(IsTDT(tsp) or IsTDISOut(tsp) or IsTREN(tsp))
          and (TId(t1,tc1) eq TId(ta,tcei))] ;
            exit (Append(tsp,ih), Noth)
          []
```

```
          t ?ta:TAddress ?tcei:TCEI ?tsp:TSP [((IsTReq(tsp)
          and (IsTDT(tsp) or IsTDISOut(tsp))) or IsTREN(tsp))
          and (TId(t1,tc1) eq TId(ta,tcei))] ;
            exit (ih, Noth) )
      []
      [d eq Indication] ->
        ( t ?ta:TAddress ?tcei:TCEI ?tsp:TSP [IsTRELind(tsp)
          and (TId(t1,tc1) ne TId(ta,tcei))] ;
            exit (Append(tsp,ih), Noth)
          []
          t ?ta:TAddress ?tcei:TCEI ?tsp:TSP [IsTReq(tsp)
          and not(IsTDT(tsp) or IsTDISOut(tsp) or IsTREN(tsp))
          and (TId(t1,tc1) ne TId(ta,tcei))];
            exit (Append(tsp,ih), Noth)
          []
          t ?ta:TAddress ?tcei:TCEI ?tsp:TSP
          [IsTReq(tsp) and (IsTDT(tsp) or IsTDISOut(tsp)
          or IsTREN(tsp)) and (TId(t1,tc1) ne TId(ta,tcei))] ;
            exit (ih, Noth) ) )
endproc (* TSPInpEvent *)

process TSPNonInpEvent [t] (t1:TAddress, tc1:TCEI,
d:TSPDirection,ih:THistory) : exit (THistory, ClosInd) :=
[d eq Request] ->
    ( t ?ta:TAddress ?tcei:TCEI ?tsp:TSP  [IsTRELreq(tsp)
      and (TId(t1,tc1) ne TId(ta,tcei))] ; exit (ih, RELr)
      []
      t ?ta:TAddress ?tcei:TCEI ?tsp:TSP [IsTReq(tsp)
      and IsTDISIn(tsp) and (TId(t1,tc1) ne TId(ta,tcei))] ;
        exit (ih, RELi)
      []
      t ?ta:TAddress ?tcei:TCEI ?tsp:TSP [IsTReq(tsp)
      and IsTDISOut(tsp) and (TId(t1,tc1) ne TId(ta,tcei))] ;
        exit (ih, RELc) )
[]
[d eq Indication] ->
    ( t ?ta:TAddress ?tcei:TCEI ?tsp:TSP [IsTRELreq(tsp)
      and (TId(t1,tc1) eq TId(ta,tcei))] ;
        exit (ih, RELr)
      []
      t ?ta:TAddress ?tcei:TCEI ?tsp:TSP [IsTReq(tsp)
      and IsTDISIn(tsp) and (TId(t1,tc1) eq TId(ta,tcei))] ;
        exit (ih, RELi)
      []
      t ?ta:TAddress ?tcei:TCEI ?tsp:TSP [IsTReq(tsp)
      and IsTDISOut(tsp) and (TId(t1,tc1) eq TId(ta,tcei))] ;
        exit (ih, RELc) )
endproc (* TSPNonInpEvent *)

process TSPEvent [t] (t1:TAddress, tc1:TCEI, d1:TSPDirection,
eih,ih:THistory, lsi:LoStInd) : exit (THistory, ClosInd) :=
TSPInpEvent [t] (t1, tc1, d1, ih, lsi)
[]
TSPNonInpEvent [t] (t1, tc1, d1, ih)
[]
[not(IsEmpty(eih))] ->
    ( choice tspr,tspi:TSP []
            [(tspr IsTopOf eih) and (tspi IsOutputOf tspr)]
->
```

```
                    i ; ( t ?ta:TAddress ?tcei:TCEI !tspi
                        [((d1 eq Request) and (TId(t1,tc1) ne
                        TId(ta,tcei))) or ((d1 eq Indication)
                        and (TId(t1,tc1) eq TId(ta,tcei)))] ;
                ( [IsTRELind(tspi)] -> exit (Remove(tspr,ih), RELi)
                []
                [IsTRELconf(tspi)] -> exit (Remove(tspr,ih),
RELc)

                []
                [IsTDISIn(tspi)] -> exit (Remove(tspr,ih), RELi)
                []
                [IsTDISOut(tspi)] -> exit (Remove(tspr,ih), RELc)
                []
                [not(IsTRELind(tspi) or IsTRELconf(tspi)
                or IsTDISIn(tspi) or IsTDISOut(tspi))] ->
                        exit (Remove(tspr,ih), Noth)
            )
                    []
                    i ; TSPEvent [t] (t1, tc1, d1, Reduce(eih),
                    ih, lsi) )
        )
[]
[IsEmpty(eih)] ->
        ( choice tdi:TSP [] [ProviderGeneratedInd(tdi)] ->
          i ; ( t ?ta:TAddress ?tcei:TCEI !tdi ; exit (ih, Noth)
                []
              TSPInpEvent [t] (t1, tc1, d1, ih, lsi) )
        )
endproc (* TSPEvent *)
(*
```

5.3.2.3 Possibility of the rejection of a re-negotiation by the PP-ECOTS provider

TRENReject is in independent parallel composition with the two occurrences of TAssoc1 (see 5.3.2). It may happen that it catches an occurrence of TRENEGOTI-ATErequest instead of them, and then rejects it. This represents the possibility of an immediate rejection of a renegotiation procedure by the service provider.   *)

```
process TRENReject [t] : noexit :=
t ?ta:TAddress ?tcei:TCEI ?tsp:TSP [IsTRENreq(tsp)] ;
t !ta !tcei !TREJRENind(Provider) ; TRENReject [t]
endproc (* TRENReject *)
(*
```

5.3.2.4 Restriction to the in-band ETS primitives

TInB, cares for the fact that TAssocInB (see 5.3.2) does not interfere with the primitives used by TOOBAck (see 5.3.1), that describes the out-of-band transmissions. (See 3.4.2.1 for the definition of IsInB)   *)

```
process TInB [t] : noexit :=
t ?ta:TAddress ?tcei:TCEI ?tcr:TSP [IsInB(tcr)] ; TInB [t]
endproc (* TInB *)
(*
```

5.3.2.5 Error control

The process TCCorruptionLossesManager specifies the constraints relative to the error control mechanism as it has been negotiated by the two connected entities and described in the ErrorControl element of the QoS parameter (see 3.6.1.4). This means that, depending on the case, corruption and losses can be admitted or not, and that some information about possible damages must appear or not in the errorsignal parameter carried by the TDATAindication and TRELEASEindication primitives (see 3.5).

The knowledge of the constraints is required. They are fixed in the TCON-NECTresponse primitive, except for one parameter: the dummy octet that the calling entity requires or not to replace possibly lost ones, as this information only appears in the TCONNECTrequest and indication primitives (and similarly, the same indication from the called entity only appears in the TCONNECT response and confirm primitives) This information can be found in the TCONNECTrequest primitive, that has to be kept in memory (parameter tcr). Once the TCONNECT response has been received, the treatment of the data by process TCCorLos can begin.   *)

```
process TCCorruptionLossesManager [t] (ta:TAddress, tcei:TCEI,
tcr:TSP) : noexit :=
t ?ta1:TAddress ?tcei1:TCEI ?tsp:TSP [IsTCONresp(tsp)] ;
TCCorLos [t] (ta, ta1, tcei, tcei1, tcr, tsp)
[]
t ?ta1:TAddress ?tcei1:TCEI ?tsp:TSP [not(IsTCONresp(tsp))] ;
TCCorruptionLossesManager [t] (ta, tcei, tcr)
endproc (* TCCorruptionLossesManager *)
(*
```

TCCorLos is made of two occurrences of process TDataTransf, one for each direction of transfer. These processes only care for the TDATArequest, TDATAindication, TRELEASErequest, TRELEASEindication and for the renegotiation primitives, as these last ones could induce modifications in the constraints to be applied. Thus, TRunNonDataOrRen accepts all the other primitives.   *)

```
process TCCorLos [t] (ta,ta1:TAddress, tcei,tcei1:TCEI,
tcr,tsp:TSP) : noexit :=
TDataTransf [t] (ta, ta1, tcei, tcei1, tcr, tsp, Request,
Indication)
|||
TDataTransf [t] (ta, ta1, tcei, tcei1, tcr, tsp, Indication,
Request)
|||
TRunNonDataOrRen [t]
endproc (* TCCorLos *)

process TRunNonDataOrREN [t] : noexit :=
t ?ta:TAddress ?tcei:TCEI ?tsp:TSP [not(IsTDT(tsp)
or IsTREL(tsp) or IsTRENreq(tsp) or IsTREN2(tsp))
or IsTRELconf(tsp)] ;
TRunNonDataOrREN [t]
endproc (* TRunNonData *)
(*
```

The restrictions on the primitives that TDataTransf may accept are described by the conjunctions of the processes ConstantTA, ConstantTCEI, TDataHalf.

The rest of the process is mainly concerned by the determination of where the required constraints can be found in the QoS of the primitives received as arguments, i.e. tcr, the TCONNECTrequest and tsp, the TCONNECTresponse. This location depends on the structure of the QoS parameters, that depends on the directions of transmission opened, what is fixed by the Plex parameter carried on by the TCON-NECTrequest primitive.

If the direction that TDataTransf cares for is closed, the process does not express any constraint: it is the process TRun that unconditionally accepts every primitive. The process TDataReqToInd cares for the opened directions.   *)

```
process TDataTransf [t] (ta,ta1:TAddress, tcei,tcei1:TCEI,
tcr,tsp:TSP,di,di1:TSPDirection) : noexit :=
( ( ConstantTA [t] (ta) || ConstantTCEI [t] (tcei)
    || TDataHalf [t] (di) )
  |||
  ( ConstantTA [t] (ta1) || ConstantTCEI [t] (tcei1)
    || TDataHalf [t] (di1) ) )
||
( [di eq Request] ->
    ( [him eq PlexOf(tcr)] -> TRun [t]
      []
      ( let qos:TCQOS2 = TQOS2Of(tsp) in
        ( [me eq PlexOf(tcr)] -> TDataReqToInd [t]

    (LosD(FirErC(qos)),FirErC(qos),NoTPrim,NoLos)
        []
        [both eq PlexOf(tcr)] -> TDataReqToInd [t]

    (LosD(SecErC(qos)),SecErC(qos),NoTPrim,NoLos)
      ) ) )
    []
    [di eq Indication] ->
      ( [me eq PlexOf(tcr)] -> TRun [t]
        []
        ( let qos:TCQOS1 = TQOS1Of(tcr), qos1:TCQOS2 =
          TQOS2Of(tsp) in
          ( [him eq PlexOf(tcr)] -> TDataReqToInd [t]
                  (LosD(FirErC(qos)),FirErC(qos1),NoTPrim,NoLos)
            []
            [both eq PlexOf(tcr)] -> TDataReqToInd [t]
                  (LosD(SecErC(qos)),FirErC(qos1),NoTPrim,NoLos) )
        ) )
)
endproc (* TDataTransf *)

process TDataHalf [t] (di:TSPDirection) : noexit :=
[di eq Request] ->
    t ?ta:TAddress ?tcei:TCEI ?tsp:TSP [IsTDTreq(tsp) or
    IsTRELreq(tsp) or IsTREN2(tsp)] ; TDataHalf [t] (di)
[]
[di eq Indication] ->
    t ?ta:TAddress ?tcei:TCEI ?tsp:TSP [IsTDTind(tsp) or
    IsTRELind(tsp) or IsTRENreq(tsp)] ; TDataHalf [t] (di)
endproc (* TDataHalf *)

process TRun [t] : noexit :=
t ?ta:TAddress ?tcei:TCEI ?tsp:TSP; TRun [t]
```

```
endproc (* TRun *)
(*
```

When a TDATArequest or a TRELEASErequest occurs, provided the UserData is not void, TDataReqToInd passes it to process Modify that returns a corresponding indication. The behaviour of Modify is non determinist according to the possibilities of corruption or losses, but respects the constraints expressed by the parameters lod (useful for the Called to Calling direction as it indicates the option chosen by the calling entity for the dummy octet) and erc (that contains the other information: the definitive choice for the Cor and LosD parameters (see 3.6.1.4) and, useful in the Calling to Called direction, the option chosen by the calling entity for the dummy octet). The indication primitive returned by Modify is placed in a FIFO queue, and the sole DATA(RELEASE)indication TDataReqToInd allows to occur is the one at the top of this queue. As no reordering is allowed, it will correspond to the TDATA indication or TRELEASEindication delivered by process TAssociation (see 5.3.1). Let us recall that, according to the definition of operator IsOutputOf (see 3.4.3.4) no constraint is imposed by TAss on the parameters of ⸱the TDATA(RELEASE)indication it delivers: that is the job of TDataReqToInd.

TDataReqToInd also cares for the possibility of a renegotiation, that could modify the constraints. For the calling to called direction, it is enough to observe the TRENEGOTIATE confirm primitive, that carries all the necessary information: the definitive constraints, the dummy option specified by the called entity, and the moment when the new constraints begin to be effective. For the called to calling direction all these information are furnished by the TRENEGOTIATIONresponse primitive, except, as usual, the dummy octet required by the calling entity. Thus, the TRENEGOTIATErequest must be observed, and the LosD parameter memorised. That is the use of the fourth argument, nlod.   *)

```
process TDataReqToInd [t] (lod:LosD, erc:ErCtrl, dh:THistory,
nlod:LosD) : noexit :=
t ?ta:TAddress ?tcei:TCEI ?tsp:TSP [(IsTDTind(tsp)
    or IsTRELind(tsp)) and (tsp IsTopOf dh)] ;
TDataReqToInd [t] (lod, erc, Reduce(dh), nlod)
[]
t ?ta:TAddress ?tcei:TCEI ?tsp:TSP [IsTDTreq(tsp)
    or (IsTRELreq(tsp) and not(UserData(tsp) eq <>))] ;
(Modify (tsp, lod, erc) >> accept tsm:TSP in TDataReqToInd [t]
(lod, erc, Append(tsm,dh), nlod))
[]
t ?ta:TAddress ?tcei:TCEI ?tsp:TSP [IsTRELreq(tsp) and
                                (UserData(tsp) eq <>)] ;
TDataReqToInd [t] (lod, erc,
Append(TRELind(<>,ErSig(0,0,NoSig,
                                        NoSig)),dh), nlod)
[]
t ?ta:TAddress ?tcei:TCEI ?tsp:TSP [IsTRENreq(tsp)] ;
( let qos:TCQOS1 = TQOS1Of(tsp) in
    ( [IsOut(qos)] -> TDataReqToInd [t] (lod, erc, dh, nlod)
      []
      [IsInp(qos)] -> TDataReqToInd [t] (lod, erc, dh,
                                    LosD(FirErC(qos)))
      []
      [IsBi(qos)] -> TDataReqToInd [t] (lod, erc, dh,
                                    LosD(SecErC(qos))) )
```

```
)
[]
t ?ta:TAddress ?tcei:TCEI ?tsp:TSP [IsTRENresp(tsp)] ;
( let qos:TCQOS2 = TQOS2Of(tsp), alo:LosD =
LosD(FirErC(TQOS2Of(tsp))) in
    ( [IsInp(qos)] -> TRun [t]
      []
      [not(IsInp(qos))] -> TDataReqToInd [t] (alo, FirErC(qos),
                                                  dh, nlod) )
)
[]
t ?ta:TAddress ?tcei:TCEI ?tsp:TSP [IsTRENconf(tsp)] ;
( let qos:TCQOS2 = TQOS2Of(tsp) in
    ( [IsOut(qos)] -> TRun [t]
      []
      [IsInp(qos)] -> TDataReqToInd [t] (LosD(FirErC(qos)),
                                            FirErC(qos), dh, nlod)
      []
      [IsBi(qos)] -> TDataReqToInd [t] (LosD(SecErC(qos)),
                                            SecErC(qos), dh, nlod)
      )
)
endproc (* TDataReqToInd *)
(*
```

Modify initialises the process Modif1, that recursively transforms, octet by octet, the UserData it receives as first argument. Step by step it constructs in its second argument a new TSDU with the possibly modified octets. The third, fourth and fifth arguments respectively contain the indications relative to the corruption, the losses and our old fellow: the dummy parameter. The last seven arguments are natural numbers used as counters: n1 for the rank of the first modified (corrupted or lost) octet, n2 does the same for the last of these octets, n3 counts the number of corrupted octets (in fact it is just necessary to know whether there are corrupted octets or not, in order to signal it when required, but making n3 count the number of corrupted octets was the easiest solution), n4 does the same for the losses (what is necessary here, as an empty string is delivered when all the octets are lost), n5 and n6 respectively retain the rank of the first and of the last octet lost (they are used when non signalled corruption is allowed, one just worries for the losses), finally n7 indicates, at each step, the rank of the treated octet.   *)

```
process Modify (tsp:TSP, lod:LosD, erc:ErCtrl) :
exit (TSP) :=
Modif1 (UserData(tsp),<>,Cor(erc),LosD(erc),lod,0 of Nat,0 of
Nat,0 of Nat,0 of Nat,0 of Nat,0 of Nat,succ(0))
>> accept os:OctetString, ers:TErSig in
    ( [IsTDT(tsp)] -> exit(TDTind(os,ers))
      []
      [IsTREL(tsp)] -> exit(TRELind(os,ers)) )
endproc (* Modify *)

process Modif1 (os1,os2:OctetString, co:Cor, lo,lod:LosD,
n1,n2,n3,n4,n5,n6,n7:Nat) : exit (OctetString,TErSig) :=
[os1 eq <>] ->
    ( [succ(n4) eq n7] -> exit(<>,ErSig(0 of Nat,0 of
                                      Nat,NoSig,Sig))
      []
```

```
      [succ(n4) ne n7] ->
        ( [IsCorSig(co)] ->
            ( [(n3 eq 0) and (n4 eq 0)] ->
                exit(os2,ErSig(n1,n2,NoSig,NoSig))
              []
              [(n3 ne 0) and (n4 eq 0)] ->
                exit(os2,ErSig(n1,n2,Sig,NoSig))
              []
              [(n3 eq 0) and (n4 ne 0)] ->
                exit(os2,ErSig(n1,n2,NoSig,Sig))
              []
              [(n3 ne 0) and (n4 ne 0)] ->
                exit(os2,ErSig(n1,n2,Sig,Sig))
            )
          []
          [not(IsCorSig(co))] ->
            ( [n4 eq 0] ->
                exit(os2,ErSig(0,0 of Nat,NoSig,NoSig))
              []
              [n4 ne 0] ->
                exit(os2,ErSig(n5,n6,NoSig,Sig))
            )
        )
    )
[]
[os1 ne <>] ->       ( choice oc,oc1:Octet, os:OctetString []
        [(os1 eq (oc + os)) and (oc1 ne oc)] ->
          ( Modif1 (os,os2 + oc,co,lo,lod,n1,n2,n3,n4,
                                          n5,n6,succ(n7))
            []
            [IsCor(co)] ->
              i; ([n1 eq 0] -> Modif1 (os,os2 + oc1,co,lo,lod,
                             n7,n7,succ(n3),n4,n5,n6,succ(n7))
                  []
                  [n1 ne 0] -> Modif1 (os,os2 + oc1,co,lo,lod,
                             n1,n7,succ(n3),n4,n5,n6,succ(n7))
                 )
            []
            [IsLSig(lo)] ->
              i; ([(n1 eq 0) and IsD(lod)] ->
                  Modif1 (os,os2 + DumOf(lod),co,lo,lod,n7,n7,
n3,succ(n4),n7,n7,succ(n7))
                  []
                  [(n1 eq 0) and IsNoD(lod)] ->
                  Modif1 (os,os2 + oc1,co,lo,lod, n7,n7,n3,
                                  succ(n4),n7,n7,succ(n7))
                  []
                  [(n1 ne 0) and (n5 eq 0) and IsD(lod)] ->
                  Modif1 (os,os2 + DumOf(lod),co,lo,lod,n1,n7,
n3,succ(n4),n7,n7,succ(n7))
                  []
                  [(n1 ne 0) and (n5 eq 0) and IsNoD(lod)] ->
                  Modif1 (os,os2 + oc1,co,lo,lod,n1,n7,
n3,succ(n4),n7,n7,succ(n7))
                  []
                  [(n1 ne 0) and (n5 ne 0) and IsD(lod)] ->
```

```
                         Modif1 (os,os2 + DumOf(lod),co,lo,lod,n1,n7,
         n3,succ(n4),n5,n7,succ(n7))
                         []
                         [(n1 ne 0) and (n5 ne 0) and IsNoD(lod)] ->
                         Modif1 (os,os2 + oc1,co,lo,lod,n1,n7,n3,
                                     succ(n4),n5,n7,succ(n7))
                     )
             )
         )
endproc (* Modif1 *)
(*
```

5.3.2.6 Control of the throughput

TCThroughputController expresses the constraints indicated by the Throughput quality of service (see 3.6.1.1). Because of the lack of a timed LOTOS, it was not possible to describe this mechanism. The fact that at a given instant, the actual throughput offered by the service provider is observed under the threshold or the compulsory value is represented by an internal event i. Nothing describes the role of the Max value. Anyway, some other constraints were translatable in LOTOS:

- When the service provider notices that its ability to accept data is under the threshold value, it sends a TREPindication primitive to the emitting entity. It does it before accepting any new data, i.e. any new TDATArequest or TRELEASErequest from this entity. A TREPindication has also to be sent to the peer entity, but with no constraint on the moment when (the order of successive TREPindication at the receiving entity is not necessarily the one of the corresponding primitives at the emitting entity).
- When the service provider notices that its ability to accept data is under the compulsory value, it knows it will close the connection by a TDISCONNECTindication. No more primitives are thus exchanged with the emitting entity, except TREPindication concerning the other direction of transmission or primitives that could also provoke a closure: other TDISCONNECTrequest or indication and TRELEASEconfirm. (TRELEASErequest and TRELEASEindication are not allowed as they are concerned by the transmission of data).
- Only one TREPindication can be emitted between two successive TDATArequest except if a re-negotiation procedure occurred in the mean time. In this case, a second TREPindication can occur after it.
- When a re-negotiation occurs that changes the threshold value for a direction, all the TREPindication relative to this direction and not yet delivered are forgotten.
- TCThroughputController stops controlling the throughput capabilities of the service provider when it notices its direction of transmission is closed

The structure of the first processes composing TCTroughputController is quite similar to the one of TCCorruptionLossesManager. The main differences come from the fact that no problems such as the ones caused by the dummy parameter are encountered here, but that new ones arise concerning the management of the TREPind and TDISind primitives during a re-negotiation. From the fact also that the two occurrences of TThrControl must interact.

At the 'emitting' TCEP of the direction it controls, TThrControl 'sees' all the primitives it needs. This does not includes primitives that it must not block when it intends to close the connection (see above). At the 'receiving' TCEP, it is only concerned by the TDISCONNECTindication and TREPORTindication it emits. The distinction between disconnect or report primitives relative to a given direction and a given QoS is made thanks to the parameters they carry (see 3.5 for the definition of the parameters TQualityReport and TDISReason). These constraints are expressed by the processes TRunThrControl, that accepts all the primitives TThrControl neglects, and TThrCHalf. The rest of the process TThrControl has the same role as in TDataReqToInd (see 5.3.2.5).   *)

```
process TCThroughputController [t] (ta:TAddress, tcei:TCEI) :
noexit :=
t ?ta1:TAddress ?tcei1:TCEI ?tsp:TSP [IsTCONresp(tsp)] ;
TCThroughput [t] (ta, ta1, tcei, tcei1, tsp)
[]
t ?ta1:TAddress ?tcei1:TCEI ?tsp:TSP [not(IsTCONresp(tsp))] ;
TCThroughputcontroller [t] (ta, tcei)
endproc (* TCThroughputController *)

process TCThroughput [t] (ta,ta1:TAddress, tcei,tcei1:TCEI,
tsp:TSP) : noexit :=
( TThrControl [t] (ta, ta1, tcei, tcei1, tsp, Request,
                                          Indication)
    |||
   TRunThrControl [t] (ta, ta1, tcei, tcei1, Request,
Indication) )
||
( TThrControl [t] (ta, ta1, tcei, tcei1, tsp, Indication,
                                          Request)
    |||
   TRunThrControl [t] (ta, ta1, tcei, tcei1, Indication,
Request) )
endproc (* TCThroughput *)

process TThrControl [t] (ta,ta1:TAddress, tcei,tcei1:TCEI,
                         tsp:TSP, di,di1:TSPDirection) : noexit :=
( ( ConstantTA [t] (ta) || ConstantTCEI [t] (tcei)
    || TThrCHalf [t] (di) )
  |||
   ( ConstantTA [t] (ta1) || ConstantTCEI [t] (tcei1)
    || TThrCHalf [t] (di1) ) )
||
( let qos:TCQOS2 = TQOS2Of(tsp) in
( [di eq Request] ->
    ( [me eq PlexOf(tsp)] -> TRun [t]
      []
      ( let j1:Thr2 = FirThr(qos), j2:Thr2 = SecThr(qos) in
        ( [him eq PlexOf(tsp)] ->
            TThrCont [t] (Max(j1),Thld(j1),Min(j1),false,false,
            false,false,NoTPGPrim,TId(ta,tcei),di)
          []
          [both eq PlexOf(tsp)] ->
            TThrCont [t] (Max(j2),Thld(j2),Min(j1),false, false,
            false,false,NoTPGPrim,TId(ta,tcei),di) )
      ) )
```

```
    []
    [di eq Indication] ->
      ( [him eq PlexOf(tsp)] -> TRun [t]
        []
        ( let j1:Thr2 = FirThr(qos) in
              TThrCont [t] (Max(j1),Thld(j1),Min(j1),false,false,
              false,false,NoTPGPrim,TId(ta1,tcei1),di) )
      )
) )

endproc (* TThrControl *)

process TThrCHalf [t] (di:TSPDirection) : noexit :=
[di eq Request] ->
t ?ta1:TAddress ?tcei1:TCEI ?tsp:TSP [not(IsTRELconf(tsp)
or IsTDIS(tsp) or IsTREPind(tsp)) or IsTDISOut(tsp)
or (IsTDISind(tsp) andthen IsQualCause(DISReasonOf(tsp))
andthen (IsCauThr(DCau(DISReasonOf(tsp)))
and (DDir(DISReasonOf(tsp)) eq Output)))
or (IsTREPindOut(tsp) andthen
IsCauThr(RCau(QualReportOf(tsp))))] ; TThrCHalf [t] (di)
[]
[di eq Indication] ->
t ?ta1:TAddress ?tcei1:TCEI ?tsp:TSP [IsTNEWQOSind(tsp)
or IsTRENconf(tsp) or (IsTDISind(tsp)
andthen IsQualCause(DISReasonOf(tsp))
andthen (IsCauThr(DCau(DISReasonOf(tsp)))
and (DDir(DISReasonOf(tsp)) eq Input))) or (IsTREPindInp(tsp)
andthen IsCauThr(RCau(QualReportOf(tsp))))] ; TThrCHalf [t]
(di)

endproc (* TThrCHalf *)

process TRunThrControl [t] (ta,ta1:TAddress, tcei,tcei1:TCEI,
di,di1:TSPDirection) : noexit :=
( ConstantTA [t] (ta) || ConstantTCEI [t] (tcei)
    || TRunThrCHalf [t] (di) )
|||
( ConstantTA [t] (ta1) || ConstantTCEI [t] (tcei1)
    || TRunThrCHalf [t] (di1) )
endproc (* TRunThrControl *)

process TRunThrCHalf [t] (di:TSPDirection) : noexit :=
[di eq Request] ->
t ?ta1:TAddress ?tcei1:TCEI ?tsp:TSP [(IsTRELconf(tsp)
or IsTDIS(tsp) or IsTREPind(tsp)) and not(IsTDISOut(tsp))
and not(IsTDISind(tsp) andthen IsQualCause(DISReasonOf(tsp))
andthen (IsCauThr(DCau(DISReasonOf(tsp)))
and (DDir(DISReasonOf(tsp)) eq Output)))
and not(IsTREPindOut(tsp) andthen
IsCauThr(RCau(QualReportOf(tsp))))] ; TRunThrCHalf [t] (di)
[]
[di eq Indication] ->
t ?ta1:TAddress ?tcei1:TCEI ?tsp:TSP [not(IsTNEWQOSind(tsp)
or IsTRENconf(tsp)) and not(IsTDISind(tsp)
andthen IsQualCause(DISReasonOf(tsp))
andthen (IsCauThr(DCau(DISReasonOf(tsp)))
and (DDir(DISReasonOf(tsp)) eq Input)))
and not(IsTREPindInp(tsp) andthen
IsCauThr(RCau(QualReportOf(tsp))))] ; TRunThrCHalf [t] (di)
```

```
endproc (* TRunThrCHalf *)
(*
```
The core of the mechanism of control is TThrCont, that describes the constraints explained above on the occurrence of the various primitives concerned. The first three arguments are the maximal, the threshold and the minimal values defined for the throughput quality of service (see 3.6.1.1). The next four Booleans indicate whether some events occurred or not: tdd is set to true after each occurrence of a TDATArequest (but not after the occurrence of a TRELEASErequest, as this signifies the closure of the direction of transmission). tdd is a condition for the detection of a throughput under the threshold value. When this occurs, what is represented by event i, tdd is set to false and tri to true. This signifies that a TREPindication has to be emitted, and blocks the acceptance of any new data. According to the way the threshold is evaluated, a problem regarding the threshold value will always be noticed before a problem with the compulsory value. When tri is true, a problem with the compulsory value can thus be detected. If this happens, dii is set to true, in order to authorise the emission of a TDISCONNECTindication and to block some primitives, as explained above. The role of trd is to prevent the occurrence of successive TREPindication. When a problem relative to the threshold value is detected, a TREPindication is put in the set pgih, to the intention of the peer entity. The fact that pgih is a set (and not a FIFO queue for instance) represents the absence of constraints on the moment when the primitives must occur at the peer entity.

TThrCont also cares for the influence of a re-negotiation. For the calling to called direction, all the necessary indications (in particular the moment when the new conditions become effective) are given by the TRENEGOTIATEconfirm. For the opposite direction, it is in the TRENEGOTIATEresponse. But one must also care for the fact that old TREPind should not be delivered to the peer entity when it ceases to be interested in it, i.e. at the occurrence of the TRENEGOTIATEconfirm for the calling entity, and of the TNEWQOSindication for the called entity. At this moment, the concerned pgih is 'purged' of all the old TREPindication (see the data type definitions at the end of the clause).   *)

```
process TThrCont [t] (max,thld,min:Nat, tdd,tri,trd, dii:Bool,
pgih:TPGIndHist, ti:TCId, di:TSPDirection) :
noexit :=
[not(tri or dii)] ->
t ?ta:TAddress ?tcei:TCEI ?tsp:TSP [IsTDTreq(tsp)] ;
TThrCont [t] (max, thld, min, true, tri, false, dii, pgih, ti,
di)
[]
[not(tri or dii)] ->
t ?ta:TAddress ?tcei:TCEI ?tsp:TSP [IsTRELreq(tsp)] ;
TThrCont [t] (max, thld, min, false, tri, trd, dii, pgih, ti,
di)
[]
[tdd] ->
i; TThrCont [t] (max, thld, min, false, true, trd, dii,
Insert(TREPind(TQR(Throughput(QVal1(thld)),Input)),pgih), ti,
di)
[]
[tri and not(trd)] ->
t ?ta:TAddress ?tcei:TCEI
!TREPind(TQR(Throughput(QVal1(thld)),
```

```
Output)) ; TThrCont [t] (max, thld, min, tdd, false, true,
dii,
pgih, ti, di)
[]
[tri] ->
i; TThrCont [t] (max, thld, min, tdd, tri, trd, true,
Insert(TDISind(inout,ProvQual(Throughput(QVal1(max)),Input),<>
),
pgih),ti, di)
[]
[dii] ->
t ?ta:TAddress ?tcei:TCEI !TDISind(inout,ProvQual(Throughput
(QVal1(max)),Output),<>) ; TThrCont [t] (max, thld, min, tdd,
tri, trd, dii, pgih, ti, di)
[]
t ?ta:TAddress ?tcei:TCEI ?tsp:TSP [tsp IsIn pgih] ; TThrCont
[t] (max, thld, min, tdd, tri, trd, dii, Remove(tsp,pgih), ti,
di)
[]
t ?ta:TAddress ?tcei:TCEI ?tsp:TSP [IsTDISOut(tsp)] ; TThrCont
[t] (max, thld, min, false, false, trd, dii, pgih, ti, di)
[]
[not(dii)] ->
t ?ta:TAddress ?tcei:TCEI ?tsp:TSP [(TId(ta,tcei) eq ti) and
IsTInd(tsp) and not(IsTRENconf(tsp))] ; TThrCont [t] (max,
thld,
min, tdd, tri, trd, dii, pgih, ti, di)
[]
[not(dii)] ->
t ?ta:TAddress ?tcei:TCEI ?tsp:TSP [IsTRENreq(tsp)] ; TThrCont
[t] (max, thld, min, tdd, tri, trd, dii, pgih, ti, di)
[]
[not(dii)] ->
t ?ta:TAddress ?tcei:TCEI ?tsp:TSP [IsTRENresp(tsp)] ;
( let qos:TCQOS2 = TQOS2Of(tsp) in
    ( [IsInp(qos)] -> TThrCont [t] (max, thld, min, tdd, tri,
                                    trd, dii, pgih, ti, di)
      []
      ( let j1:Thr2 = FirThr(qos) in
            [not(IsInp(qos))] -> TThrCont [t] (Max(j1),
Thld(j1),
            Min(j1), true, false, false, false, pgih, ti, di) )
    ) )
[]
t ?ta:TAddress ?tcei:TCEI ?tsp:TSP [IsTRENconf(tsp)
and (di eq Indication)] ; TThrCont [t] (max, thld, min, tdd,
tri, trd, dii, Purge(QVal1(thld),pgih), ti, di)
[]
[not(dii)] ->
t ?ta:TAddress ?tcei:TCEI ?tsp:TSP [IsTRENconf(tsp)
and (di eq Request)] ;
( let qos:TCQOS2 = TQOS2Of(tsp) in
    ( [IsOut(qos)] -> TThrCont [t] (max, thld, min, tdd, tri,
                                    trd, dii, pgih, ti, di)
      []
      ( let j1:Thr2 = FirThr(qos), j2:Thr2 = SecThr(qos) in
        ( [him eq PlexOf(tsp)] -> TThrCont [t] (Max(j1),
                          Thld(j1), Min(j1), true, false,
                          false, false, pgih, ti, di)
          []
```

```
        [both eq PlexOf(tsp)] -> TThrCont [t] (Max(j2),
                            Thld(j2), Min(j1), true, false,
                            false, false, pgih, ti, di) )
    ) ) )
[]
t ?ta:TAddress ?tcei:TCEI ?tsp:TSP [IsTNEWQOSind(tsp)
and (di eq Request)] ; TThrCont [t] (max, thld, min, tdd, tri,
trd, dii, Purge(QVal1(thld),pgih), ti, di)
endproc (* TThrCont *)


type BasicTransportServiceProviderGeneratedIndHistory
is Set actualizedby TransportServicePrimitive using
sortnames  TSP              for Element
           Bool        for FBool
           TPGIndHist for Set
opnnames   NoTPGPrim  for {}
endtype (* BasicTransportServiceProviderGeneratedIndHistory *)


type TransportServiceProviderGeneratedIndHistory
is BasicTransportServiceProviderGeneratedIndHistory
opns
    Purge : QVal, TPGIndHist -> TPGIndHist
eqns
forall qv:QVal, t:TSP, pgh:TPGIndHist

ofsort TPGIndHist
Purge(qv,NoTPGPrim)     = NoTPGPrim               ;
    IsTDIS(t) =>
Purge(qv,Insert(t,pgh))   = Insert(t,Purge(qv,pgh))    ;
    IsTREPind(t), ((ThldVal(RCau(QualReportOf(t))) eq qv)
    or (qv eq All)) =>
Purge(qv,Insert(t,pgh))   = Insert(t,Purge(qv,pgh))    ;
    IsTREPind(t), ((ThldVal(RCau(QualReportOf(t))) ne qv)
    and (qv ne All)) =>
Purge(qv,Insert(t,pgh))   = Purge(qv,pgh)              ;
endtype (* TransportServiceProviderGeneratedIndHistory *)
(*
```

5.3.2.7 Control of the transit delay

This process has a structure very similar to the one of TCThroughputController. Again, due to the lack of a timed LOTOS, the way the parameter is controlled is not described. The described constraints are almost the same as for the throughput. The main difference is that, for a given direction of transmission, the parameter is no more controlled at the 'emitting' point, but at the 'receiving' point. The two situations are thus symmetrical with respect to the entities. This causes some problems for the calling to called direction, when treating a re-negotiation. In this case, the new qualities are indicated by the TRENEGOTIATEresponse, but the moment when they become effective is given by the occurrence of TNEWQOS indication. This requires to memorise the new qualities without applying they between the occurrence of the two primitives. This explains the role of the last two arguments of TTrDCont. Also, during the period between the occurrence of the two primitives, as the data that arrive are still submitted to the former qualities, they must not cause the sending of a TREPind to the peer entity, as it would arrive after the TRENEGOTIATEconfirm,

and thus be neglected. This is the role of the Boolean ren, that is true only during this
period.   *)

```
process TCTransDelController [t] (ta:TAddress, tcei:TCEI) :
noexit :=
t ?ta1:TAddress ?tcei1:TCEI ?tsp:TSP [IsTCONresp(tsp)] ;
TCTransDel [t] (ta, ta1, tcei, tcei1, tsp)
[]
t ?ta1:TAddress ?tcei1:TCEI ?tsp:TSP [not(IsTCONresp(tsp))] ;
TCTransDelController [t] (ta, tcei)
endproc (* TCTransDelController *)

process TCTransDel [t] (ta,ta1:TAddress, tcei,tcei1:TCEI,
tsp:TSP) : noexit :=
( TTrDControl [t] (ta, ta1, tcei, tcei1, tsp, Request,
Indication)
   |||
  TRunTrDControl [t] (ta, ta1, tcei, tcei1, Request,
Indication) )
||
( TTrDControl [t] (ta, ta1, tcei, tcei1, tsp, Indication,
Request)
   |||
  TRunTrDControl [t] (ta, ta1, tcei, tcei1, Indication,
Request) )
endproc (* TCTransDel *)

process TTrDControl [t] (ta,ta1:TAddress, tcei,tcei1:TCEI,
tsp:TSP, di,di1:TSPDirection) : noexit :=
( ( ConstantTA [t] (ta) || ConstantTCEI [t] (tcei)
    || TTrDCHalf [t] (di) )
   |||
  ( ConstantTA [t] (ta1) || ConstantTCEI [t] (tcei1)
    || TTrDCHalf [t] (di1) ) )
||
( let qos:TCQOS2 = TQOS2Of(tsp) in
( [di eq Request] ->
    ( [me eq PlexOf(tsp)] -> TRun [t]
      []
      ( let j1:TrD2 = FirTrD(qos), j2:TrD2 = SecTrD(qos) in
        ( [him eq PlexOf(tsp)] ->
            TTrDCont [t] (Max(j1),Thld(j1),0 of Nat,false,false,
            false,NoTPGPrim,di,TId(ta,tcei),0 of Nat,0 of Nat)
          []
          [both eq PlexOf(tsp)] ->
            TTrDCont [t] (Max(j2),Thld(j2),0 of Nat,false,false,
            false,NoTPGPrim,di,TId(ta,tcei),0 of Nat,0 of Nat) )
    ) )
  []
  [di eq Indication] ->
    ( [him eq PlexOf(tsp)] -> TRun [t]
      []
      ( let j1:TrD2 = FirTrD(qos) in
            TTrDCont [t] (Max(j1),Thld(j1),0 of
Nat,false,false,
        false,NoTPGPrim,di,TId(ta1,tcei1),0 of Nat,0 of Nat) )
    )
) )
```

```
endproc (* TTrDControl *)

process TTrDCHalf [t] (di:TSPDirection) : noexit :=
[di eq Request] ->
t ?ta1:TAddress ?tcei1:TCEI ?tsp:TSP [IsTRENconf(tsp)
or IsTNEWQOSind(tsp) or (IsTDISind(tsp)
andthen IsQualCause(DISReasonOf(tsp))
andthen (IsCauTrD(DCau(DISReasonOf(tsp)))
and (DDir(DISReasonOf(tsp)) eq Output)))
or (IsTREPindOut(tsp) andthen
IsCauTrD(RCau(QualReportOf(tsp)))))] ; TTrDCHalf [t] (di)
[]
[di eq Indication] ->
t ?ta1:TAddress ?tcei1:TCEI ?tsp:TSP [not(IsTRELconf(tsp)
or IsTDIS(tsp) or IsTREPind(tsp))
or (IsTDISind(tsp) andthen IsQualCause(DISReasonOf(tsp))
andthen (IsCauTrD(DCau(DISReasonOf(tsp)))
and (DDir(DISReasonOf(tsp)) eq Input))) or (IsTREPindInp(tsp)
andthen IsCauTrD(RCau(QualReportOf(tsp)))))] ; TTrDCHalf [t]
(di)

endproc (* TTrDCHalf *)

process TRunTrDControl [t] (ta,ta1:TAddress, tcei,tcei1:TCEI,
di,di1:TSPDirection) : noexit :=
( ConstantTA [t] (ta) || ConstantTCEI [t] (tcei)
   || TRunTrDCHalf [t] (di) )
|||
( ConstantTA [t] (ta1) || ConstantTCEI [t] (tcei1)
   || TRunTrDCHalf [t] (di1) )
endproc (* TRunTrDControl *)

process TRunTrDCHalf [t] (di:TSPDirection) : noexit :=
[di eq Request] ->
t ?ta1:TAddress ?tcei1:TCEI ?tsp:TSP [not(IsTRENconf(tsp)
or IsTNEWQOSind(tsp) or (IsTDISind(tsp)
andthen IsQualCause(DISReasonOf(tsp))
andthen (IsCauTrD(DCau(DISReasonOf(tsp)))
and (DDir(DISReasonOf(tsp)) eq Output))) or (IsTREPindOut(tsp)
andthen IsCauTrD(RCau(QualReportOf(tsp)))))] ;
TTrDCHalf [t] (di)
[]
[di eq Indication] ->
t ?ta1:TAddress ?tcei1:TCEI ?tsp:TSP [(IsTRELconf(tsp)
or IsTDIS(tsp) or IsTREPind(tsp))
and not(IsTDISind(tsp) andthen IsQualCause(DISReasonOf(tsp))
andthen (IsCauTrD(DCau(DISReasonOf(tsp)))
and (DDir(DISReasonOf(tsp)) eq Input)))
and not(IsTREPindInp(tsp)
and IsCauTrD(RCau(QualReportOf(tsp)))))] ; TTrDCHalf [t] (di)

endproc (* TRunTrDCHalf *)

process TTrDCont [t] (max,thld,acttrd:Nat, tri,dii,ren:Bool,
pgih:TPGIndHist, di:TSPDirection, ti:TCId, nmax,nthld:Nat) :
noexit :=
[not(tri or dii)] ->
t ?ta:TAddress ?tcei:TCEI ?tsp:TSP [IsTDTind(tsp)
                                    or IsTRELind(tsp)] ;
( choice acttrd:Nat []
```

```
      ( [acttrd le thld] ->
        TTrDCont [t] (max, thld, acttrd, tri, dii, ren, pgih, di,
          ti, nmax, nthld)
        []
        [not(max eq thld) and (acttrd gt thld)
          and (acttrd le max)] ->
        ( [ren] ->
          TTrDCont [t] (max, thld, acttrd, true, dii, ren, pgih,
          di, ti, nmax, nthld)
          []
          [not(ren)] ->
          TTrDCont [t] (max, thld, acttrd, true, dii, ren,
          Insert(TREPind(TQR(TransitDelay(QVal1(thld),
          QVal1(acttrd)),Output)),pgih), di, ti, nmax, nthld) )
        []
        [acttrd gt max] ->
        TTrDCont [t] (max, thld, acttrd, tri, true, ren,
          Insert(TDISind(inout,ProvQual(TransitDelay(QVal1(max),
          QVal1(acttrd)),Output),<>),pgih), di, ti, nmax, nthld)
)
)
[]
[tri] ->
t ?ta:TAddress ?tcei:TCEI
!TREPind(TQR(TransitDelay(QVal1(thld),
QVal1(acttrd)),Input)) ; TTrDCont [t] (max, thld, acttrd,
false,
dii, ren, pgih, di, ti, nmax, nthld)
[]
[dii] ->
t ?ta:TAddress ?tcei:TCEI
!TDISind(inout,ProvQual(TransitDelay(
QVal1(max),QVal1(acttrd)),Input),<>) ; TTrDCont [t] (max,
thld,
acttrd, tri, dii, ren, pgih, di, ti, nmax, nthld)
[]
t ?ta:TAddress ?tcei:TCEI ?tsp:TSP [tsp IsIn pgih] ;
TTrDCont [t] (max, thld, acttrd, tri, dii, ren,
Remove(tsp,pgih), di, ti, nmax, nthld)
[]
[not(dii)] ->
t ?ta:TAddress ?tcei:TCEI ?tsp:TSP [(TId(ta,tcei) ne ti)
and IsTReq(tsp) and not(IsTRENresp(tsp))] ; TTrDCont [t] (max,
thld, acttrd, tri, dii, ren, pgih, di, ti, nmax, nthld)
[]
[di eq Request] ->
( [not(tri or dii)] ->
  t ?ta:TAddress ?tcei:TCEI ?tsp:TSP [IsTRENind(tsp)] ;
  TTrDCont [t] (max, thld, acttrd, tri, dii, ren, pgih, di,
ti,
  nmax, nthld)
  []
  [not(tri or dii)] ->
  t ?ta:TAddress ?tcei:TCEI ?tsp:TSP [IsTNEWQOSind(tsp)] ;
  TTrDCont [t] (nmax, nthld, acttrd, false, dii, false, pgih,
  di, ti, nmax, nthld)
  []
  t ?ta:TAddress ?tcei:TCEI ?tsp:TSP [IsTRENconf(tsp)] ;
  TTrDCont [t] (max, thld, acttrd, false, dii, ren,
  Purge(QVal1(nthld),pgih), di, ti, nmax, nthld)
```

```
[]
[not(tri or dii)] ->
t ?ta:TAddress ?tcei:TCEI ?tsp:TSP [IsTRENresp(tsp)] ;
( let qos:TCQOS2 = TQOS2Of(tsp) in
  ( [IsOut(qos)] -> TRun [t]
    []
    ( let r1:TrD2 = FirTrD(qos), r2:TrD2 = SecTrD(qos) in
  ( [IsInp(qos)] ->
      TTrDCont [t] (max, thld, acttrd, tri, dii, true, pgih,
      di, ti, Max(r1), Thld(r1))
    []
    [IsBi(qos)] ->
      TTrDCont [t] (max, thld, acttrd, tri, dii, true, pgih,
      di, ti, Max(r2), Thld(r2))
  ) ) ) )
)
[]
[di eq Indication] ->
(  t ?ta:TAddress ?tcei:TCEI ?tsp:TSP [IsTNEWQOSind(tsp)] ;
  TTrDCont [t] (max, thld, acttrd, tri, dii, ren,
  Purge(QVal1(thld),pgih), di, ti, nmax, nthld)
  []
  [not(tri or dii)] ->
  t ?ta:TAddress ?tcei:TCEI ?tsp:TSP [IsTRENconf(tsp)] ;
  ( let qos:TCQOS2 = TQOS2Of(tsp) in
    ( [IsInp(qos)] -> TRun [t]
      []
      ( let r1:TrD2 = FirTrD(qos) in
      [not(IsInp(qos))] ->
        TTrDCont [t] (Max(r1), Thld(r1), acttrd, tri, dii,
ren,
                        pgih, di, ti, nmax, nthld) ) ) )
)
endproc (* TTrDCont *)
(*
```

5.3.2.8 Control of the transit delay jitter

This process is the same as the previous one (see 5.3.2.7), just the names change. In
practice of course, differences exist, but they come from they way the parameter is
determined, what is not represented here.   *)

```
process TCJitterController [t] (ta:TAddress, tcei:TCEI) :
exit :=
t ?ta1:TAddress ?tcei1:TCEI ?tsp:TSP [IsTCONresp(tsp)] ;
TCJitter [t] (ta, ta1, tcei, tcei1, tsp)
[]
t ?ta1:TAddress ?tcei1:TCEI ?tsp:TSP [not(IsTCONresp(tsp))] ;
TCJitterController [t] (ta, tcei)
endproc (* TCJitterController *)

process TCJitter [t] (ta,ta1:TAddress, tcei,tcei1:TCEI,
tsp:TSP): noexit :=
( TJitControl [t] (ta, ta1, tcei, tcei1, tsp, Request,
                                        Indication)
  |||
  TRunJitControl [t] (ta, ta1, tcei, tcei1, Request,
Indication) )
```

```
||
( TJitControl [t] (ta, ta1, tcei, tcei1, tsp, Indication,
                                                    Request)
   |||
   TRunJitControl [t] (ta, ta1, tcei, tcei1, Indication,
Request) )
endproc (* TCJitter *)

process TJitControl [t] (ta,ta1:TAddress, tcei,tcei1:TCEI,
tsp:TSP, di,di1:TSPDirection) : noexit :=
( ( ConstantTA [t] (ta) || ConstantTCEI [t] (tcei)
    || TJitCHalf [t] (di) )
  |||
  ( ConstantTA [t] (ta1) || ConstantTCEI [t] (tcei1)
    || TJitCHalf [t] (di1) ) )
||
( let qos:TCQOS2 = TQOS2Of(tsp) in
( [di eq Request] ->
    ( [me eq PlexOf(tsp)] -> TRun [t]
      []
      ( let j1:Jit2 = FirJit(qos), j2:Jit2 = SecJit(qos) in
        ( [him eq PlexOf(tsp)] ->
            TJitCont [t] (Max(j1),Thld(j1),0 of Nat,false,false,
            false,NoTPGPrim,di,TId(ta,tcei),0 of Nat,0 of Nat)
          []
          [both eq PlexOf(tsp)] ->
            TJitCont [t] (Max(j2),Thld(j2),0 of Nat,false,false,
            false,NoTPGPrim,di,TId(ta,tcei),0 of Nat,0 of Nat) )
      ) )
  []
  [di eq Indication] ->
    ( [him eq PlexOf(tsp)] -> TRun [t]
      []
      ( let j1:Jit2 = FirJit(qos) in
          TJitCont [t] (Max(j1),Thld(j1),0 of
Nat,false,false,
      false,NoTPGPrim,di,TId(ta1,tcei1),0 of Nat,0 of Nat) )
    )
) )
endproc (* TJitControl *)

process TJitCHalf [t] (di:TSPDirection) : noexit :=
[di eq Request] ->
t ?ta1:TAddress ?tcei1:TCEI ?tsp:TSP [IsTRENconf(tsp)
or IsTNEWQOSind(tsp) or (IsTDISind(tsp)
andthen IsQualCause(DISReasonOf(tsp))
andthen (IsCauJit(DCau(DISReasonOf(tsp)))
and (DDir(DISReasonOf(tsp)) eq Output)))
or (IsTREPindOut(tsp) andthen
IsCauJit(RCau(QualReportOf(tsp))))] ; TJitCHalf [t] (di)
[]
[di eq Indication] ->
t ?ta1:TAddress ?tcei1:TCEI ?tsp:TSP [not(IsTRELconf(tsp)
or IsTDIS(tsp) or IsTREPind(tsp)) or (IsTDISind(tsp)
andthen IsQualCause(DISReasonOf(tsp))
andthen (IsCauJit(DCau(DISReasonOf(tsp)))
and (DDir(DISReasonOf(tsp)) eq Input)))
or (IsTREPindInp(tsp) andthen
IsCauJit(RCau(QualReportOf(tsp))))] ; TJitCHalf [t] (di)
```

```
endproc (* TJitCHalf *)

process TRunJitControl [t] (ta,ta1:TAddress,
tcei,tcei1:TCEI, di,di1:TSPDirection) : noexit :=
( ConstantTA [t] (ta) || ConstantTCEI [t] (tcei)
    || TRunJitCHalf [t] (di) )
|||
( ConstantTA [t] (ta1) || ConstantTCEI [t] (tcei1)
    || TRunJitCHalf [t] (di1) )
endproc (* TRunJitControl *)

process TRunJitCHalf [t] (di:TSPDirection) : noexit :=
[di eq Request] ->
t ?ta1:TAddress ?tcei1:TCEI ?tsp:TSP [not(IsTRENconf(tsp)
or IsTNEWQOSind(tsp) or (IsTDISind(tsp)
andthen IsQualCause(DISReasonOf(tsp))
andthen (IsCauJit(DCau(DISReasonOf(tsp)))
and (DDir(DISReasonOf(tsp)) eq Output)))
or (IsTREPindOut(tsp) andthen
IsCauJit(RCau(QualReportOf(tsp)))))] ; TJitCHalf [t] (di)
[]
[di eq Indication] ->
t ?ta1:TAddress ?tcei1:TCEI ?tsp:TSP [(IsTRELconf(tsp)
or IsTDIS(tsp) or IsTREPind(tsp)) and not(IsTDISind(tsp)
andthen IsQualCause(DISReasonOf(tsp))
andthen (IsCauJit(DCau(DISReasonOf(tsp)))
and (DDir(DISReasonOf(tsp)) eq Input)))
and not(IsTREPindInp(tsp)
andthen IsCauJit(RCau(QualReportOf(tsp))))] ;
TJitCHalf [t] (di)

endproc (* TRunJitCHalf *)

process TJitCont [t] (max,thld,actjit:Nat, tri,dii,ren:Bool,
pgih:TPGIndHist, di:TSPDirection, ti:TCId, nmax,nthld:Nat):
noexit :=
[not(tri or dii)] ->
t ?ta:TAddress ?tcei:TCEI ?tsp:TSP [IsTDTind(tsp)
                                    or IsTRELind(tsp)] ;
( choice actjit:Nat []
    ( [actjit le thld] ->
      TJitCont [t] (max, thld, actjit, tri, dii, ren, pgih, di,
       ti, nmax, nthld)
      []
      [not(max eq thld) and (actjit gt thld)
      and (actjit le max)] ->
      ( [ren] ->
            TJitCont [t] (max, thld, actjit, true, dii, ren,
                          pgih, di, ti, nmax, nthld)
        []
        [not(ren)] ->
            TJitCont [t] (max, thld, actjit, true, dii, ren,
                    Insert(TREPind(TQR(Jitter(QVal1(thld),
                    QVal1(actjit)),Output)),pgih), di, ti,
                    nmax, nthld) )
      []
      [actjit gt max] ->
      TJitCont [t] (max, thld, actjit, tri, true, ren,
```

```
            Insert(TDISind(inout,ProvQual(Jitter(QVal1(max),
            QVal1(actjit)),Output),<>), pgih), di, ti, nmax, nthld)
)
)
[]
[tri] ->
t ?ta:TAddress ?tcei:TCEI !TREPind(TQR(Jitter(QVal1(thld),
QVal1(actjit)),Input)) ; TJitCont [t] (max, thld, actjit,
false,
dii, ren, pgih, di, ti, nmax, nthld)
[]
[dii] ->
t ?ta:TAddress ?tcei:TCEI
!TDISind(inout,ProvQual(Jitter(QVal1(max),QVal1(actjit)),Input
),
<>) ; TJitCont [t] (max, thld, actjit, tri, dii, ren, pgih,
di,
ti, nmax, nthld)
[]
t ?ta:TAddress ?tcei:TCEI ?tsp:TSP [tsp IsIn pgih] ;
TJitCont [t] (max, thld, actjit, tri, dii, ren,
Remove(tsp,pgih), di, ti, nmax, nthld)
[]
[not(dii)] ->
t ?ta:TAddress ?tcei:TCEI ?tsp:TSP [(TId(ta,tcei) ne ti)
and IsTReq(tsp) and not(IsTRENresp(tsp))] ; TJitCont [t] (max,
thld, actjit, tri, dii, ren, pgih, di, ti, nmax, nthld)
[]
[di eq Request] ->
( [not(tri or dii)] ->
  t ?ta:TAddress ?tcei:TCEI ?tsp:TSP [IsTRENind(tsp)] ;
  TJitCont [t] (max, thld, actjit, tri, dii, ren, pgih, di,
ti,
    nmax, nthld)
  []
  [not(tri or dii)] ->
  t ?ta:TAddress ?tcei:TCEI ?tsp:TSP [IsTNEWQOSind(tsp)] ;
  TJitCont [t] (nmax, nthld, actjit, false, dii, false, pgih,
    di, ti, nmax, nthld)
  []
  t ?ta:TAddress ?tcei:TCEI ?tsp:TSP [IsTRENconf(tsp)] ;
  TJitCont [t] (max, thld, actjit, false, dii, ren,
    Purge(QVal1(nthld),pgih), di, ti, nmax, nthld)
  []
  t ?ta:TAddress ?tcei:TCEI ?tsp:TSP [IsTRENresp(tsp)] ;
  ( let qos:TCQOS2 = TQOS2Of(tsp) in
    ( [IsOut(qos)] -> TRun [t]
      []
      ( let r1:Jit2 = FirJit(qos), r2:Jit2 = SecJit(qos) in
    ( [IsInp(qos)] ->
        TJitCont [t] (max, thld, actjit, tri, dii, true, pgih,
                      di, ti, Max(r1), Thld(r1))
      []
      [IsBi(qos)] ->
        TJitCont [t] (max, thld, actjit, tri, dii, true, pgih,
                      di, ti, Max(r2), Thld(r2))
    ) ) ) )
)
[]
[di eq Indication] ->
```

```
( t ?ta:TAddress ?tcei:TCEI ?tsp:TSP [IsTNEWQOSind(tsp)] ;
  TJitCont [t] (max, thld, actjit, tri, dii, ren,
                Purge(QVal1(thld),pgih), di, ti, nmax, nthld)
  []
  [not(tri or dii)] ->
  t ?ta:TAddress ?tcei:TCEI ?tsp:TSP [IsTRENconf(tsp)] ;
  ( let qos:TCQOS2 = TQOS2Of(tsp) in
    ( [IsInp(qos)] -> TRun [t]
      []
      ( let r1:Jit2 = FirJit(qos) in
      [not(IsInp(qos))] ->
        TJitCont [t] (Max(r1), Thld(r1), actjit, tri, dii,
ren,
            pgih, di, ti, nmax, nthld) ) ) )
)
endproc (* TJitCont *)
(*
```

5.3.2.9 Control processes for the residual error rate

The structure of these processes is again very similar to the previous ones. They have to face the same problems for the re-negotiation procedure. However it is possible here to describe the whole mechanism. Though many different cases are to be envisaged, the logic behind it is quite simple. For instance for the corruption (the situation is similar for the losses): the first three parameters of TCRERCont respectively indicate the maximal number of errors tolerated before closing the connection, the maximal number of errors tolerated before emitting a TREPORTindication and the size of the 'sequence' in which these values are calculated. Successive sequences never overlap: once the number of controlled data (given by argument nbrdat) reaches the value ws, it is set back to 0, and the same happens to nbrer that counts the number of erroneous transmitted data.

The only thing really peculiar to these processes is the parameter 'act', which expresses whether the error rate control is activated or not. This comes from the fact (see 3.6.1.4) that residual error rates are not always specified, depending on the negotiated type of error control. The value of act for each direction of transmission is thus initially determined according to the error control required for this direction, what is found in TCONNECT response, and must be updated in the sequel, following possible re-negotiations.

5.3.2.9.1 Corruptions error rate   *)

```
process TCCorruptionResErrorRateController [t] (ta:TAddress,
tcei:TCEI) : noexit :=
t ?ta1:TAddress ?tcei1:TCEI ?tsp:TSP [IsTCONresp(tsp)] ;
TCCorResErRate [t] (ta, ta1, tcei, tcei1, tsp)
[]
t ?ta1:TAddress ?tcei1:TCEI ?tsp:TSP [not(IsTCONresp(tsp))];
TCCorruptionResErrorRateController [t] (ta, tcei)

endproc (* TCCorruptionResErrorRateController *)

process TCCorResErRate [t] (ta,ta1:TAddress,tcei,tcei1:TCEI,
tsp:TSP): noexit :=
( TCRERControl [t] (ta, ta1, tcei, tcei1, tsp, Request,
                                    Indication)
```

```
   | | |
   TRunCRERControl [t] (ta, ta1, tcei, tcei1, Request,
                                         Indication) )
 | |
( TCRERControl [t] (ta, ta1, tcei, tcei1, tsp, Indication,
                                         Request)
   | | |
   TRunCRERControl [t] (ta, ta1, tcei, tcei1, Indication,
                                         Request) )
endproc (* TCCorResErRate *)

process TCRERControl [t] (ta,ta1:TAddress, tcei,tcei1:TCEI,
tsp:TSP, di,di1:TSPDirection) : noexit :=
( ( ConstantTA [t] (ta) || ConstantTCEI [t] (tcei)
    || TCRERCHalf [t] (di) )
  | | |
  ( ConstantTA [t] (ta1) || ConstantTCEI [t] (tcei1)
    || TCRERCHalf [t] (di1) ) )
| |
( let qos:TCQOS2 = TQOS2Of(tsp) in
( [di eq Request] ->
    ( [me eq PlexOf(tsp)] -> TRun [t]
      []
      [him eq PlexOf(tsp)] ->
        ( let co1:Cor = Cor(FirErC(qos)) in
          ( [IsCorSig(co1)] ->
            ( let r1:RER2 = RER2Of(co1) in TCRERCont [t]
                (MaxErNumber(Max(r1)),MaxErNumber(Thld(r1)),
                SequenceSize(Max(r1)), 0 of Nat,0 of Nat,

    true,false,false,false,NoTPGPrim,di,TId(ta,tcei),
              0 of Nat,0 of Nat,0 of Nat,false) )
            []
            [not(IsCorSig(co1))] ->
            TCRERCont [t] (0 of Nat,0 of Nat,0 of Nat,0 of Nat,
            0 of Nat,false,false,false,false, NoTPGPrim,di,
            TId(ta,tcei),0 of Nat,0 of Nat,0 of Nat,false)
          ) )
      []
      [both eq PlexOf(tsp)] ->
        ( let co1:Cor = Cor(SecErC(qos)) in
          ( [IsCorSig(co1)] ->
            ( let r1:RER2 = RER2Of(co1) in
            TCRERCont [t] (MaxErNumber(Max(r1)),
            MaxErNumber(Thld(r1)),SequenceSize(Max(r1)),
            0 of Nat,0 of Nat,true,false,false,false,NoTPGPrim,
            di,TId(ta,tcei),0 of Nat,0 of Nat,0 of Nat,false) )
            []
            [not(IsCorSig(co1))] ->
            TCRERCont [t] (0 of Nat,0 of Nat,0 of Nat,0 of Nat,
            0 of Nat,false,false,false,false,NoTPGPrim,di,
            TId(ta,tcei),0 of Nat,0 of Nat,0 of Nat,false)
          ) )
      )
  []
  [di eq Indication] ->
    ( [him eq PlexOf(tsp)] -> TRun [t]
      []
      [not(him eq PlexOf(tsp))] ->
```

```
        ( let col:Cor = Cor(FirErC(qos)) in
          ( [IsCorSig(col)] ->
            ( let r1:RER2 = RER2Of(col) in
            TCRERCont [t] (MaxErNumber(Max(r1)),
            MaxErNumber(Thld(r1)),SequenceSize(Max(r1)),
            0 of Nat,0 of Nat,true,false,false,false,NoTPGPrim,
            di,TId(ta1,tcei1),0 of Nat,0 of Nat,0 of Nat,false)
)
            []
            [not(IsCorSig(col))] ->
            TCRERCont [t] (0 of Nat,0 of Nat,0 of Nat,0 of Nat,
            0 of Nat,false,false,false,false,NoTPGPrim,di,
            TId(ta1,tcei1),0 of Nat,0 of Nat,0 of Nat,false)
        ) ) )
) )
endproc (* TCRERControl *)

process TCRERCHalf [t] (di:TSPDirection) : noexit :=
[di eq Request] ->
t ?ta1:TAddress ?tcei1:TCEI ?tsp:TSP [IsTRENconf(tsp)
or IsTNEWQOSind(tsp) or (IsTDISind(tsp)
andthen IsQualCause(DISReasonOf(tsp))
andthen (IsCauCER(DCau(DISReasonOf(tsp)))
and (DDir(DISReasonOf(tsp)) eq Output)))
or (IsTREPindOut(tsp) andthen
IsCauCER(RCau(QualReportOf(tsp)))))] ; TCRERCHalf [t] (di)
[]
[di eq Indication] ->
t ?ta1:TAddress ?tcei1:TCEI ?tsp:TSP [not(IsTRELconf(tsp)
or IsTDIS(tsp) or IsTREPind(tsp)) or (IsTDISind(tsp)
andthen IsQualCause(DISReasonOf(tsp))
andthen (IsCauCER(DCau(DISReasonOf(tsp)))
and (DDir(DISReasonOf(tsp)) eq Input)))
or (IsTREPindInp(tsp) andthen
IsCauCER(RCau(QualReportOf(tsp)))))] ; TCRERCHalf [t] (di)
endproc (* TCRERCHalf *)

process TRunCRERControl [t] (ta,ta1:TAddress, tcei,tcei1:TCEI,
di,di1:TSPDirection) : noexit :=
( ConstantTA [t] (ta) || ConstantTCEI [t] (tcei)
    || TRunCRERCHalf [t] (di) )
|||
( ConstantTA [t] (ta1) || ConstantTCEI [t] (tcei1)
    || TRunCRERCHalf [t] (di1) )
endproc (* TRunCRERControl *)

process TRunCRERCHalf [t] (di:TSPDirection) : noexit :=
[di eq Request] ->
t ?ta1:TAddress ?tcei1:TCEI ?tsp:TSP [not(IsTRENconf(tsp)
or IsTNEWQOSind(tsp) or (IsTDISind(tsp)
andthen IsQualCause(DISReasonOf(tsp))
andthen (IsCauCER(DCau(DISReasonOf(tsp)))
and (DDir(DISReasonOf(tsp)) eq Output)))
or (IsTREPindOut(tsp) andthen
IsCauCER(RCau(QualReportOf(tsp))))))] ; TRunCRERCHalf [t] (di)
[]
[di eq Indication] ->
```

```
t ?ta1:TAddress ?tcei1:TCEI ?tsp:TSP [(IsTRELconf(tsp)
or IsTDIS(tsp) or IsTREPind(tsp)) and not(IsTDISind(tsp)
andthen IsQualCause(DISReasonOf(tsp))
andthen (IsCauCER(DCau(DISReasonOf(tsp)))
and (DDir(DISReasonOf(tsp)) eq Input)))
and not(IsTREPindInp(tsp) andthen
IsCauCER(RCau(QualReportOf(tsp))))] ; TRunCRERCHalf [t] (di)
endproc (* TRunCRERCHalf *)

process TCRERCont [t] (max,thld,ws,nbrer,nbrdat:Nat,
act,tri,dii,ren:Bool, pgih:TPGIndHist, di:TSPDirection,
ti:TCId, nmax,nthld,nws:Nat, nact:Bool) : noexit :=
[not(act)] ->
t ?ta:TAddress ?tcei:TCEI ?tsp:TSP [IsTDTind(tsp)
or IsTRELind(tsp)] ; TCRERCont [t] (max, thld, ws, nbrer,
nbrdat, act, true, dii, ren, pgih, di, ti, nmax, nthld, nws,
nact)
[]
[act and not(tri or dii)] ->
t ?ta:TAddress ?tcei:TCEI ?tsp:TSP [IsTDTind(tsp)
or IsTRELind(tsp)] ;
    ( [CSigOf(ErSigOf(tsp)) eq NoSig] ->
        ( [nbrdat ne ws] -> TCRERCont [t] (max, thld, ws,
                              nbrer,succ(nbrdat), act, tri,
dii,
                              ren, pgih, di, ti, nmax, nthld,
                              nws, nact)
            []
            [nbrdat eq ws] -> TCRERCont [t] (max, thld, ws,
                              0 of Nat, succ(0), act, tri, dii,
                              ren, pgih, di, ti, nmax, nthld,
                              nws, nact) )
        []
        [CSigOf(ErSigOf(tsp)) eq Sig] ->
            ( [nbrdat ne ws] ->
                ( [nbrer ne max] ->
                    ( [nbrer ne thld] ->
                    TCRERCont [t] (max, thld, ws, succ(nbrer),
                        succ(nbrdat), act, tri, dii, ren,
                        pgih, di, ti, nmax, nthld, nws, nact)
                      []
                      [nbrer eq thld] ->
                        ( [ren] ->
                    TCRERCont [t] (max, thld, ws, succ(nbrer),
                        succ(nbrdat), act, true, dii, ren,
                        pgih, di, ti, nmax, nthld, nws, nact)
                        []
                        [not(ren)] ->
                    TCRERCont [t] (max, thld, ws, succ(nbrer),
                        succ(nbrdat), act, true, dii, ren,
                        Insert(TREPind(TQR(CorruptErrorRate(
                        QVal2(thld,ws)), Output)), pgih),
                        di, ti, nmax, nthld, nws, nact) ) )
                  []
                  [nbrer eq max] ->
                    TCRERCont [t] (max, thld, ws, succ(nbrer),
                        succ(nbrdat), act, tri, true, ren,

                        Insert(TDISind(inout,ProvQual(
```

```
                                CorruptErrorRate(QVal2(max,ws)),
                                Output),<>),pgih), di,
                                ti, nmax, nthld, nws, nact) )
                    []
                    [nbrdat eq ws] ->
                       ( [0 ne max] ->
                            ( [0 ne thld] ->
                              TCRERCont [t] (max, thld, ws, succ(0),
                                  succ(0), act, tri, dii, ren, pgih, di,
ti,
                                  nmax, nthld, nws, nact)
                                []
                                [0 eq thld] ->
                                ( [ren] ->
                              TCRERCont [t] (max, thld, ws, succ(0),
                                  succ(0), act, true, dii, ren, pgih, di,
                                  ti, nmax, nthld, nws, nact)
                                   []
                                   [not(ren)] ->
                              TCRERCont [t] (max, thld, ws, succ(0),
                                  succ(0), act, true, dii, ren,
                                  Insert(TREPind(TQR(CorruptErrorRate(
                                  QVal2(thld,ws)),Output)),pgih), di, ti,
                                  nmax, nthld, nws, nact) ) )
                       []
                       [0 eq max] ->
                         TCRERCont [t] (max, thld, ws, succ(0),
                             succ(0), act, tri, true, ren,
                             Insert(TDISind(inout,ProvQual(
                             CorruptErrorRate(QVal2(max,ws)),Output),
                             <>),pgih), di, ti, nmax,nthld,
                             nws, nact) ) )
          )
[]
[tri] ->
t ?ta:TAddress ?tcei:TCEI !TREPind(TQR(CorruptErrorRate(
QVal2(thld,ws)),Input)) ; TCRERCont [t] (max, thld, ws, nbrer,
nbrdat, act, false, dii, ren, pgih, di, ti, nmax, nthld, nws,
nact)
[]
[dii] ->
t ?ta:TAddress ?tcei:TCEI !TDISind(inout,ProvQual(
CorruptErrorRate(QVal2(max,ws)),Input),<>) ;
TCRERCont [t] (max, thld, ws, nbrer, nbrdat, act, tri, dii,
ren,
pgih, di, ti, nmax, nthld, nws, nact)
[]
t ?ta:TAddress ?tcei:TCEI ?tsp:TSP [tsp IsIn pgih] ;
TCRERCont [t] (max, thld, ws, nbrer, nbrdat, act, tri, dii,
ren,
Remove(tsp,pgih), di, ti, nmax, nthld, nws, nact)
[]
[not(dii)] ->
t ?ta:TAddress ?tcei:TCEI ?tsp:TSP [(TId(ta,tcei) ne ti) and
IsTReq(tsp) and not(IsTRENresp(tsp))] ; TCRERCont [t] (max,
thld, ws, nbrer, nbrdat, act, tri, dii, ren, pgih, di, ti,
nmax,
nthld, nws, nact)
[]
[di eq Request] ->
```

```
( [not(tri or dii)] ->
  t ?ta:TAddress ?tcei:TCEI ?tsp:TSP [IsTRENind(tsp)] ;
  TCRERCont [t] (max, thld, ws, nbrer, nbrdat, act, tri, dii,
  ren, pgih, di, ti, nmax, nthld, nws, nact)
  []
  [not(tri or dii)] ->
  t ?ta:TAddress ?tcei:TCEI ?tsp:TSP [IsTNEWQOSind(tsp)] ;
  TCRERCont [t] (nmax, nthld, nws, 0 of Nat, 0 of Nat, nact,
  false, dii, false,pgih, di, ti, nmax, nthld, nws, nact)
  []
  [not(dii)] ->
  t ?ta:TAddress ?tcei:TCEI ?tsp:TSP [IsTRENresp(tsp)] ;
  ( let qos:TCQOS2 = TQOS2Of(tsp) in
    ( [IsOut(qos)] -> TRun [t]
      []
      [IsInp(qos)] ->
    ( let co1:Cor = Cor(FirErC(qos)) in
        ( [IsCorSig(co1)] ->
              ( let r1:RER2 = RER2Of(co1) in
                TCRERCont [t] (max, thld, ws, nbrer, nbrdat, act,
                    tri, dii, true,  pgih, di, ti,
                    MaxErNumber(Max(r1)), MaxErNumber(Thld(r1)),
                    SequenceSize(Max(r1)),true) )
          []
          [not(IsCorSig(co1))] ->
            TCRERCont [t] (max, thld, ws, nbrer, nbrdat, act,
                tri, dii, true,  pgih, di, ti, nmax, nthld, nws,
                false)
        ) )
      []
      [IsBi(qos)] ->
    ( let ec2:Cor = Cor(SecErC(qos)) in
        ( [IsCorSig(ec2)] ->
              ( let r2:RER2 = RER2Of(ec2) in
                TCRERCont [t] (max, thld, ws, nbrer, nbrdat, act,
                    tri, dii, true,  pgih, di, ti,
                    MaxErNumber(Max(r2)), MaxErNumber(Thld(r2)),
                    SequenceSize(Max(r2)),true) )
          []
          [not(IsCorSig(ec2))] ->
            TCRERCont [t] (max, thld, ws, nbrer, nbrdat, act,
tri, dii, true,  pgih, di, ti, nmax, nthld, nws, false)
        ) ) ) )

  []
  t ?ta:TAddress ?tcei:TCEI ?tsp:TSP [IsTRENconf(tsp)] ;
  ( let qos:TCQOS2 = TQOS2Of(tsp) in
    ( [IsInp(qos)] ->
    ( let co1:Cor = Cor(FirErC(qos)) in
        ( [IsCorSig(co1)] ->
            TCRERCont [t] (max, thld, ws, nbrer, nbrdat, nact,
                false, dii, ren, Purge(QVal2(nthld,nws),pgih), £
                di, ti, nmax, nthld, nws, nact)
          []
          [not(IsCorSig(co1))] ->
            TCRERCont [t] (max, thld, ws, nbrer, nbrdat, nact,
                false, dii, ren, Purge(All,pgih), di, ti, nmax,
                nthld, nws, nact) ) )
      []
      [IsBi(qos)] ->
```

```
      ( let ec2:Cor = Cor(SecErC(qos)) in
          ( [IsCorSig(ec2)] ->
             TCRERCont [t] (max, thld, ws, nbrer, nbrdat, nact,
                 false, dii, ren, Purge(QVal2(nthld,nws),pgih),
                 di, ti, nmax, nthld, nws, nact)
            []
            [not(IsCorSig(ec2))] ->
             TCRERCont [t] (max, thld, ws, nbrer, nbrdat, nact,
             false, dii, ren, Purge(All,pgih), di, ti, nmax,
             nthld, nws, nact) ) )
   ) )
)
[]
[di eq Indication] ->
( [not(tri or dii)] ->
   t ?ta:TAddress ?tcei:TCEI ?tsp:TSP [IsTRENconf(tsp)] ;
   ( let qos:TCQOS2 = TQOS2Of(tsp) in
     ( [IsInp(qos)] -> TRun [t]
       []
       [not(IsInp(qos))] ->
     ( let co1:Cor = Cor(FirErC(qos))  in
         ( [IsCorSig(co1)] ->
               ( let r1:RER2 = RER2Of(co1) in
                 TCRERCont [t] (MaxErNumber(Max(r1)),
                    MaxErNumber(Thld(r1)),
SequenceSize(Max(r1)),
                    0 of Nat, 0 of Nat, true, tri, dii, ren,
                    pgih, di, ti, max, thld, ws, nact))
           []
           [not(IsCorSig(co1))] ->
            TCRERCont [t] (max, thld, ws, nbrer, nbrdat, false,
            tri, dii, true, pgih, di, ti, nmax, nthld, nws,
nact)
   ) ) )
   []
   t ?ta:TAddress ?tcei:TCEI ?tsp:TSP [IsTNEWQOSind(tsp)] ;
     ( let co1:Cor = Cor(FirErC(qos))  in
         ( [IsCorSig(co1)] ->
            TCRERCont [t] (max, thld, ws, nbrer, nbrdat, act,
            tri, dii, ren, Purge(QVal2(thld,nws),pgih), di, ti,
            nmax, nthld, nws, nact)
           []
           [not(IsCorSig(co1))] ->
            TCRERCont [t] (max, thld, ws, nbrer, nbrdat, act,
            tri, dii, ren, Purge(All,pgih), di, ti, nmax, nthld,
            nws, nact) ) )
) )
endproc (* TCRERCont *)
(*
```

5.3.2.9.2 *Losses error rate*   *)

```
process TCLossesResErrorRateController [t] (ta:TAddress,
tcei:TCEI) : noexit :=
t ?ta1:TAddress ?tcei1:TCEI ?tsp:TSP [IsTCONresp(tsp)] ;
TCLosResErRate [t] (ta, ta1, tcei, tcei1, tsp)
[]
t ?ta1:TAddress ?tcei1:TCEI ?tsp:TSP [not(IsTCONresp(tsp))];
TCLossesResErrorRateController [t] (ta, tcei)
```

```
endproc (* TCLossesResErrorRateController *)

process TCLosResErRate [t] (ta,ta1:TAddress, tcei,tcei1:TCEI,
tsp:TSP): noexit :=
( TLRERControl [t] (ta, ta1, tcei, tcei1, tsp, Request,
                                              Indication)
   |||
   TRunLRERControl [t] (ta, ta1, tcei, tcei1, Request,
                                              Indication) )
||
( TLRERControl [t] (ta, ta1, tcei, tcei1, tsp, Indication,
                                              Request)
   |||
   TRunLRERControl [t] (ta, ta1, tcei, tcei1, Indication,
                                              Request) )
endproc (* TCLosResErRate *)


process TLRERControl [t] (ta,ta1:TAddress, tcei,tcei1:TCEI,
tsp:TSP, di,di1:TSPDirection) : noexit :=
( ( ConstantTA [t] (ta) || ConstantTCEI [t] (tcei) ||
    TLRERCHalf [t] (di) )
   |||
   ( ConstantTA [t] (ta1) || ConstantTCEI [t] (tcei1) ||
    TLRERCHalf [t] (di1) ) )
||
( let qos:TCQOS2 = TQOS2Of(tsp) in
( [di eq Request] ->
   ( [me eq PlexOf(tsp)] -> TRun [t]
    []
    [him eq PlexOf(tsp)] ->
       ( let lo1:LosD = LosD(FirErC(qos)) in
          ( [IsLSig(lo1)] ->
           ( let r1:RER2 = RER2Of(lo1) in
            TLRERCont [t] (MaxErNumber(Max(r1)),
               MaxErNumber(Thld(r1)), SequenceSize(Max(r1)),
               0 of Nat, 0 of Nat,true,false,false,false,
               NoTPGPrim,di,TId(ta,tcei),0 of Nat,0 of Nat,
               0 of Nat,false) )
           []
           [not(IsLSig(lo1))] ->
            TLRERCont [t] (0 of Nat,0 of Nat,0 of Nat,0 of Nat,
               0 of Nat,false,false,false,false,NoTPGPrim,di,
               TId(ta,tcei),0 of Nat,0 of Nat,0 of Nat,false)
          ) )
       []
       [both eq PlexOf(tsp)] ->
         ( let lo1:LosD = LosD(SecErC(qos)) in
           ( [IsLSig(lo1)] ->
             ( let r1:RER2 = RER2Of(lo1) in
             TLRERCont [t] (MaxErNumber(Max(r1)),
                MaxErNumber(Thld(r1)),SequenceSize(Max(r1)),
                0 of Nat,0 of Nat,true,false,false,false,
                NoTPGPrim,di,TId(ta,tcei),0 of Nat,0 of Nat,
                0 of Nat,false) )
             []
             [not(IsLSig(lo1))] ->
```

```
                 TLRERCont [t] (0 of Nat,0 of Nat,0 of Nat,0 of Nat,
                      0 of Nat,false,false,false,false,NoTPGPrim,di,
                      TId(ta,tcei),0 of Nat,0 of Nat,0 of Nat,false)
              ) )
         )
  []
  [di eq Indication] ->
     ( [him eq PlexOf(tsp)] -> TRun [t]
       []
       [not(him eq PlexOf(tsp))] ->
          ( let lol:LosD = LosD(FirErC(qos)) in
            ( [IsLSig(lol)] ->
               ( let rl:RER2 = RER2Of(lol) in
               TLRERCont [t] (MaxErNumber(Max(rl)),
                   MaxErNumber(Thld(rl)),SequenceSize(Max(rl)),
                   0 of Nat,0 of Nat,true,false,false,false,
                   NoTPGPrim,di,TId(tal,tceil),0 of Nat,0 of Nat,
                   0 of Nat,false) )
               []
               [not(IsLSig(lol))] ->
               TLRERCont [t] (0 of Nat,0 of Nat,0 of Nat,0 of Nat,
                   0 of Nat,false,false,false,false,NoTPGPrim,di,
                   TId(tal,tceil),0 of Nat,0 of Nat,0 of Nat,false)
           ) ) )
) )
endproc (* TLRERControl *)

process TLRERCHalf [t] (di:TSPDirection) : noexit :=
[di eq Request] ->
t ?tal:TAddress ?tceil:TCEI ?tsp:TSP [IsTRENconf(tsp)
or IsTNEWQOSind(tsp) or (IsTDISind(tsp)
andthen IsQualCause(DISReasonOf(tsp))
andthen (IsCauLER(DCau(DISReasonOf(tsp)))
and (DDir(DISReasonOf(tsp)) eq Output)))
or (IsTREPindOut(tsp) andthen
IsCauLER(RCau(QualReportOf(tsp))))] ; TLRERCHalf [t] (di)
[]
[di eq Indication] ->
t ?tal:TAddress ?tceil:TCEI ?tsp:TSP [not(IsTRELconf(tsp)
or IsTDIS(tsp) or IsTREPind(tsp)) or (IsTDISind(tsp)
andthen IsQualCause(DISReasonOf(tsp))
andthen (IsCauLER(DCau(DISReasonOf(tsp)))
and (DDir(DISReasonOf(tsp)) eq Input)))
or (IsTREPindInp(tsp) andthen
IsCauLER(RCau(QualReportOf(tsp))))] ; TLRERCHalf [t] (di)
endproc (* TLRERCHalf *)

process TRunLRERControl [t] (ta,tal:TAddress, tcei,tceil:TCEI,
di,dil:TSPDirection) : noexit :=
( ConstantTA [t] (ta) || ConstantTCEI [t] (tcei)
    || TRunLRERCHalf [t] (di) )
|||
( ConstantTA [t] (tal) || ConstantTCEI [t] (tceil)
    || TRunLRERCHalf [t] (dil) )
endproc (* TRunLRERControl *)

process TRunLRERCHalf [t] (di:TSPDirection) : noexit :=
[di eq Request] ->
```

```
t ?ta1:TAddress ?tcei1:TCEI ?tsp:TSP [not(IsTRENconf(tsp)
or IsTNEWQOSind(tsp) or (IsTDISind(tsp)
andthen IsQualCause(DISReasonOf(tsp))
andthen (IsCauLER(DCau(DISReasonOf(tsp)))
and (DDir(DISReasonOf(tsp)) eq Output)))
or (IsTREPindOut(tsp) andthen
IsCauLER(RCau(QualReportOf(tsp)))))] ; TRunLRERCHalf [t] (di)
[]
[di eq Indication] ->
t ?ta1:TAddress ?tcei1:TCEI ?tsp:TSP [(IsTRELconf(tsp)
or IsTDIS(tsp) or IsTREPind(tsp)) and not(IsTDISind(tsp)
andthen IsQualCause(DISReasonOf(tsp))
andthen (IsCauLER(DCau(DISReasonOf(tsp)))
and (DDir(DISReasonOf(tsp)) eq Input)))
and not(IsTREPindInp(tsp) andthen
IsCauLER(RCau(QualReportOf(tsp))))] ; TRunLRERCHalf [t] (di)
endproc (* TRunLRERCHalf *)


process TLRERCont [t] (max,thld,ws,nbrer,nbrdat:Nat,
act,tri,dii,ren:Bool, pgih:TPGIndHist, di:TSPDirection,
ti:TCId,
nmax,nthld,nws:Nat, nact:Bool) : noexit :=
[not(act)] ->
t ?ta:TAddress ?tcei:TCEI ?tsp:TSP [IsTDTind(tsp)
or IsTRELind(tsp)] ; TLRERCont [t] (max, thld, ws, nbrer,
nbrdat, act, true, dii, ren, pgih, di, ti, nmax, nthld, nws,
nact)
[]
[act and not(tri or dii)] ->
t ?ta:TAddress ?tcei:TCEI ?tsp:TSP [IsTDTind(tsp)
or IsTRELind(tsp)] ;
    ( [LSigOf(ErSigOf(tsp)) eq NoSig] ->
            ( [nbrdat ne ws] -> TLRERCont [t] (max, thld, ws,
                         nbrer, succ(nbrdat), act, tri,
                         dii, ren, pgih, di, ti, nmax,
                         nthld, nws, nact)
               []
               [nbrdat eq ws] -> TLRERCont [t] (max, thld, ws,
                         0 of Nat, succ(0), act, tri,
                         dii, ren, pgih, di, ti, nmax,
                         nthld, nws, nact) )
      []
      [LSigOf(ErSigOf(tsp)) eq Sig] ->
            ( [nbrdat ne ws] ->
                ( [nbrer ne max] ->
                    ( [nbrer ne thld] ->
                      TLRERCont [t] (max, thld, ws,
succ(nbrer),
                        succ(nbrdat), act, tri, dii, ren, pgih,
                        di, ti, nmax, nthld, nws, nact)
                     []
                     [nbrer eq thld] ->
                       ( [ren] ->
                       TLRERCont [t] (max, thld, ws,
succ(nbrer),
                         succ(nbrdat), act, true, dii, ren, pgih,
                         di, ti, nmax, nthld, nws, nact)
                           []
                           [not(ren)] ->
```

```
                              TLRERCont [t] (max, thld, ws,
succ(nbrer),
                              succ(nbrdat), act, true, dii, ren,
                              Insert(TREPind(TQR(LossesErrorRate(
                              QVal2(thld,ws)),Output)), pgih),
                              di, ti, nmax, nthld, nws, nact) ) )
                         []
                      [nbrer eq max] ->
                        TLRERCont [t] (max, thld, ws, succ(nbrer),
                        succ(nbrdat), act, tri, true, ren,
                        Insert(TDISind(inout,ProvQual(
                        LossesErrorRate(QVal2(max,ws)),Output),<>),
                        pgih), di, ti, nmax, nthld, nws, nact) )
                    []
                    [nbrdat eq ws] ->
                      ( [0 ne max] ->
                         ( [0 ne thld] ->
                         TLRERCont [t] (max, thld, ws, succ(0),
                         succ(0), act, tri, dii, ren, pgih, di, ti,
                         nmax, nthld, nws, nact)
                           []
                          [0 eq thld] ->
                            ( [ren] ->
                         TLRERCont [t] (max, thld, ws, succ(0),
                         succ(0), act, true, dii, ren, pgih, di, ti,
                         nmax, nthld, nws, nact)
                                []
                              [not(ren)] ->
                         TLRERCont [t] (max, thld, ws, succ(0),
                         succ(0), act, true, dii, ren,
Insert(TREPind(

    TQR(LossesErrorRate(QVal2(thld,ws)),Output)),
                         pgih), di, ti, nmax, nthld, nws, nact) ) )
                       []
                       [0 eq max] ->
                         TLRERCont [t] (max, thld, ws, succ(0),
                         succ(0), act, tri, true, ren,
Insert(TDISind(
                         inout,ProvQual(LossesErrorRate(
                         QVal2(max,ws)),Output),<>),pgih),
                         di, ti, nmax, nthld, nws, nact) ) )
     )
[]
[tri] ->
t ?ta:TAddress ?tcei:TCEI
!TREPind(TQR(LossesErrorRate(QVal2(thld,ws)),Input)) ;
TLRERCont [t] (max, thld, ws, nbrer, nbrdat, act, false, dii,
ren, pgih, di, ti, nmax, nthld, nws, nact)
[]
[dii] ->
t ?ta:TAddress ?tcei:TCEI !TDISind(inout,ProvQual(
LossesErrorRate(QVal2(max,ws)),Input),<>) ;
TLRERCont [t] (max, thld, ws, nbrer, nbrdat, act, tri, dii,
ren,
pgih, di, ti, nmax, nthld, nws, nact)
[]
t ?ta:TAddress ?tcei:TCEI ?tsp:TSP [tsp IsIn pgih] ;
```

```
TLRERCont [t] (max, thld, ws, nbrer, nbrdat, act, tri, dii,
ren,
Remove(tsp,pgih), di, ti, nmax, nthld, nws, nact)
[]
[not(dii)] ->
t ?ta:TAddress ?tcei:TCEI ?tsp:TSP [(TId(ta,tcei) ne ti) and
IsTReq(tsp) and not(IsTRENresp(tsp))] ;
TLRERCont [t] (max, thld, ws, nbrer, nbrdat, act, tri, dii,
ren,
pgih, di, ti, nmax, nthld, nws, nact)
[]
[di eq Request] ->
( [not(tri or dii)] ->
  t ?ta:TAddress ?tcei:TCEI ?tsp:TSP [IsTRENind(tsp)] ;
  TLRERCont [t] (max, thld, ws, nbrer, nbrdat, act, tri, dii,
    ren, pgih, di, ti, nmax, nthld, nws, nact)
  []
  [not(tri or dii)] ->
  t ?ta:TAddress ?tcei:TCEI ?tsp:TSP [IsTNEWQOSind(tsp)] ;
  TLRERCont [t] (nmax, nthld, nws, 0 of Nat, 0 of Nat, nact,
    false, dii, false, pgih, di, ti, nmax, nthld, nws, nact)
  []
  [not(dii)] ->
  t ?ta:TAddress ?tcei:TCEI ?tsp:TSP [IsTRENresp(tsp)] ;
  ( let qos:TCQOS2 = TQOS2Of(tsp) in
    ( [IsOut(qos)] -> TRun [t]
      []
      [IsInp(qos)] ->
    ( let ec1:LosD = LosD(FirErC(qos)) in
        ( [IsLSig(ec1)] ->
              ( let r1:RER2 = RER2Of(ec1) in
               TLRERCont [t] (max, thld, ws, nbrer, nbrdat, act,
               tri, dii, true,  pgih, di, ti, MaxErNumber(Max(r1)),
               MaxErNumber(Thld(r1)), SequenceSize(Max(r1)),true) )
          []
          [not(IsLSig(ec1))] ->
              TLRERCont [t] (max, thld, ws, nbrer, nbrdat, act,
              tri, dii, true,  pgih, di, ti, nmax, nthld,
              nws, false)
        ) )
      []
      [IsBi(qos)] ->
    ( let ec2:LosD = LosD(SecErC(qos)) in
        ( [IsLSig(ec2)] ->
              ( let r2:RER2 = RER2Of(ec2) in
               TLRERCont [t] (max, thld, ws, nbrer, nbrdat, act,
               tri, dii, true, pgih, di, ti, MaxErNumber(Max(r2)),
               MaxErNumber(Thld(r2)), SequenceSize(Max(r2)),true) )
          []
          [not(IsLSig(ec2))] ->
              TLRERCont [t] (max, thld, ws, nbrer, nbrdat, act,
              tri, dii, true, pgih, di, ti, nmax, nthld, nws,
              false)
        ) ) ) )

  []
  t ?ta:TAddress ?tcei:TCEI ?tsp:TSP [IsTRENconf(tsp)] ;
  ( let qos:TCQOS2 = TQOS2Of(tsp) in
    ( [IsInp(qos)] ->
    ( let ec1:LosD = LosD(FirErC(qos)) in
```

```
      ( [IsLSig(ec1)] ->
          TLRERCont [t] (max, thld, ws, nbrer, nbrdat, nact,
          false, dii, ren, Purge(QVal2(nthld,nws),pgih), di,
          ti, nmax, nthld, nws, nact)
        []
        [not(IsLSig(ec1))] ->
          TLRERCont [t] (max, thld, ws, nbrer, nbrdat, nact,
          false, dii, ren, Purge(All,pgih), di, ti, nmax,
          nthld, nws, nact) ) )
     []
     [IsBi(qos)] ->
   ( let ec2:LosD = LosD(SecErC(qos)) in
       ( [IsLSig(ec2)] ->
          TLRERCont [t] (max, thld, ws, nbrer, nbrdat, nact,
          false, dii, ren, Purge(QVal2(nthld,nws),pgih), di,
          ti, nmax, nthld, nws, nact)
        []
        [not(IsLSig(ec2))] ->
          TLRERCont [t] (max, thld, ws, nbrer, nbrdat, nact,
          false, dii, ren, Purge(All,pgih), di, ti, nmax,
          nthld, nws, nact) ) )
  ) )
)
[]
[di eq Indication] ->
( [not(tri or dii)] ->
  t ?ta:TAddress ?tcei:TCEI ?tsp:TSP [IsTRENconf(tsp)] ;
  ( let qos:TCQOS2 = TQOS2Of(tsp) in
    ( [IsInp(qos)] -> TRun [t]
      []
      [not(IsInp(qos))] ->
    ( let ec1:LosD = LosD(FirErC(qos))  in
       ( [IsLSig(ec1)] ->
             ( let r1:RER2 = RER2Of(ec1) in
               TLRERCont [t] (MaxErNumber(Max(r1)),
               MaxErNumber(Thld(r1)), SequenceSize(Max(r1)),
               0 of Nat, 0 of Nat, true, tri, dii, ren, pgih,
               di, ti, max, thld, ws, nact))
         []
         [not(IsLSig(ec1))] ->
           TLRERCont [t] (max, thld, ws, nbrer, nbrdat, false,
           tri, dii, true, pgih, di, ti, nmax, nthld, nws,
nact)
  ) ) )
  []
  t ?ta:TAddress ?tcei:TCEI ?tsp:TSP [IsTNEWQOSind(tsp)] ;
    ( let ec1:LosD = LosD(FirErC(qos))  in
       ( [IsLSig(ec1)] ->
           TLRERCont [t] (max, thld, ws, nbrer, nbrdat, act,
           tri, dii, ren, Purge(QVal2(thld,nws),pgih), di, ti,
           nmax, nthld, nws, nact)
         []
         [not(IsLSig(ec1))] ->
           TLRERCont [t] (max, thld, ws, nbrer, nbrdat, act,
           tri, dii, ren, Purge(All,pgih), di, ti, nmax, nthld,
           nws, nact) ) )
) )
endproc (* TLRERCont *)
(*
```

5.3.2.10 **Control of the TCTSDU size**

This process controls whether the size of the TSDU an entity wants to transmit really
fits in the bounds this entity indicated to the service provider. It must first get these
indications that are found in the TCONNECTconfirm for the calling entity and in the
TCONNECTresponse for the called entity. It must also care for the re-negotiation
procedures, in order to update the TSDU sizes. SizeContR corresponds to the Calling
entity, and SizeContI to the Called one.   *)

```
process TCTSDUSizeController [t] (ta:TAddress, tcei:TCEI) :
noexit :=
( ConstantTA [t] (ta) || ConstantTCEI [t] (tcei)
  ||
  WaitConf [t] )
|||
( DifferentTId [t] (ta,tcei)
  ||
  WaitResp [t] )
endproc (* TCTSDUSizeController *)

process WaitResp [t] : noexit :=
t ?ta:TAddress ?tcei:TCEI ?tsp:TSP [IsTCONresp(tsp)] ;
( let ttss:TSS2 = TTSDU2SizeOf(tsp) in
    ( [IsBi(ttss)] ->
      SizeContI [t] (SecMin(ttss),SecMax(ttss))
      []
      [IsOut(ttss)] ->
      SizeContI [t] (FirMin(ttss),FirMax(ttss))
      []
      [IsInp(ttss)] -> TRun [t] ) )
[]
t ?ta:TAddress ?tcei:TCEI ?tsp:TSP [not(IsTCONresp(tsp))] ;
WaitResp [t]

endproc (* WaitResp *)

process WaitConf [t] : noexit :=
t ?ta:TAddress ?tcei:TCEI ?tsp:TSP [IsTCONconf(tsp)] ;
( let ttss:TSS2 = TTSDU2SizeOf(tsp) in
    ( [not(IsOut(ttss))] ->
      SizeContR [t] (FirMin(ttss),FirMax(ttss))
      []
      [IsOut(ttss)] -> TRun [t] ) )
[]
t ?ta:TAddress ?tcei:TCEI ?tsp:TSP [not(IsTCONconf(tsp))] ;
WaitConf [t]

endproc (* WaitConf *)

process SizeContR [t] (min,max:Nat) : noexit :=
t ?ta:TAddress ?tcei:TCEI ?tsp:TSP [(IsTDTreq(tsp)
or IsTRELreq(tsp)) and (Length(UserData(tsp)) ge min)
and (Length(UserData(tsp)) le max)] ; SizeContR [t] (min,max)
[]
t ?ta:TAddress ?tcei:TCEI ?tsp:TSP [IsTRENconf(tsp)] ;
( let ttss:TSS2 = TTSDU2SizeOf(tsp) in
    ( [not(IsOut(ttss))] ->
      SizeContR [t] (FirMin(ttss),FirMax(ttss))
      []
      [IsOut(ttss)] -> TRun [t] ) )
```

```
[]
t ?ta:TAddress ?tcei:TCEI ?tsp:TSP [not(IsTDTreq(tsp)
or IsTRELreq(tsp) or IsTRENconf(tsp))] ; SizeContR [t]
(min,max)
endproc (* SizeContR *)

process SizeContI [t] (min,max:Nat) : noexit :=
t ?ta:TAddress ?tcei:TCEI ?tsp:TSP [(IsTDTreq(tsp)
or IsTRELreq(tsp)) and (Length(UserData(tsp)) ge min)
and (Length(UserData(tsp)) le max)] ; SizeContI [t] (min,max)
[]
t ?ta:TAddress ?tcei:TCEI ?tsp:TSP [IsTRENresp(tsp)] ;
( let ttss:TSS2 = TTSDU2SizeOf(tsp) in
    ( [IsBi(ttss)] ->
      SizeContI [t] (SecMin(ttss),SecMax(ttss))
      []
      [IsOut(ttss)] ->
      SizeContI [t] (FirMin(ttss),FirMax(ttss))
      []
      [IsInp(ttss)] -> TRun [t] ) )
[]
t ?ta:TAddress ?tcei:TCEI ?tsp:TSP [not(IsTDTreq(tsp)
or IsTRELreq(tsp) or IsTRENresp(tsp))] ; SizeContI [t]
(min,max)
endproc (* SizeContI *)
(*
```

5.3.3 Out-of-band constraints

TOOBACK describes the constraints on the Out-of-Band transmissions. It is made of three sub processes.

TAckOOBs is the conjunction of a potentially infinite number of instances of TAckOOB, each one describing the possible behaviours of a single Out-of-Band communication at its two endpoints.

OOBTCEIVerif imposes that the primitives that require it, and only these ones, pass through an OOBTCEI (see 3.3).

OOBTCEIControl verifies that no OOBTCEI is used by more than one Out-of-Band communication at the same time.

TAckOOB is made of two sub processes, TAckOOBAssoc that takes off the description of the behaviour at both endpoints, and TAckOOBTA that cares for the primitives to be issued at the right TCEP.   *)

```
process TOOBACK [t] : noexit :=
TAckOOBs [t]
||
OOBTCEIVerif [t]
||
OOBTCEIControl [t] ({} of OOBTCEIRec)
endproc (* TOOBACK *)

process TAckOOBs [t] : noexit :=
TAckOOBs [t] ||| TAckOOB [t]
endproc (* TAckOOBs *)
```

```
process TAckOOB [t] : noexit :=
(TAckOOBAssoc [t] || TAckOOBTA [t]) >> stop
endproc (* TAckOOB *)

process TAckOOBAssoc [t] : exit :=
t ?ta:TAddress ?tcei:TCEI ?tsp:TSP [IsValidTOOBACKreq(tsp)];
(   t ?ta:TAddress ?tcei:TCEI ?tspi:TSP [tspi IsOutputOf tsp];
    (   t ?ta:TAddress ?tcei:TCEI ?tspc:TSP [tspc IsOutputOf
                                                tspi] ; exit
        []
        i ; t ?ta:TAddress ?tcei:TCEI ?tsprj:TSP
            [IsTOOBREJACKind(tsprj)] ; exit
    )
    []
    i ; t ?ta:TAddress ?tcei:TCEI ?tsprj:TSP
            [IsTOOBREJACKind(tsprj)] ;
    (   t ?ta:TAddress ?tcei:TCEI ?tspi:TSP [tspi IsOutputOf
tsp]
        ; exit
        []
        i ; exit
    )
)
endproc (* TAckOOBAssoc *)

process TAckOOBTA [t] : exit :=
t ?ta:TAddress ?tcei:TCEI ?tsp:TSP ; TAckOOBTA2 [t] (ta,tcei)
endproc (* TAckOOBTA *)

process TAckOOBTA2 [t] (ta:TAddress, tcei:TCEI) : exit :=
t ?ta1:TAddress ?tcei1:TCEI ?tsp:TSP [not(TId(ta1,tcei1) eq
TId(ta,tcei)) iff IsTOOBACKind(tsp)] ; TAckOOBTA2 [t]
(ta,tcei)
[]
exit
endproc (* TAckOOBTA2 *)

process OOBTCEIVerif [t] : noexit :=
t ?ta:TAddress ?tcei:TCEI ?tsp:TSP [not(IsTOOBACKind(tsp)) iff
IsOOBTCEI(tcei)] ; OOBTCEIVerif [t]
endproc (* OOBTCEIVerif *)

process OOBTCEIControl [t] (oc:OOBTCEIRec) : noexit :=
t ?ta:TAddress ?tcei:TCEI ?tsp:TSP
    [IsTOOBACKreq(tsp) implies not(TId(ta,tcei) IsIn oc)] ;
    ([IsTOOBACKreq(tsp)] -> OOBTCEIControl [t]
                                    (Insert(TId(ta,tcei),oc))
    []
    [IsTOOBACKconf(tsp)] -> OOBTCEIControl [t]
                                    (Remove(TId(ta,tcei),oc))
    []
    [IsTOOBREJACKind(tsp)] -> OOBTCEIControl [t]
                                    (Remove(TId(ta,tcei),oc))
    []
    [IsTOOBACKind(tsp)] -> OOBTCEIControl [t] (oc))
endproc (* OOBTCEIControl *)
```

```
type OOBTCEIRec is Set actualizedby TCEIdentification using
sortnames  TCId        for Element
           Bool        for FBool
           OOBTCEIRec  for Set
endtype (* OOBTCEIRec *)
(*
```

5.4 Identification of the TCs

The only reason for TCIdentif is to initialise TCIdent with an empty set of sort TIdHist, what can not be done directly in the behaviour expression where TCIdentif appears.

The use of TCIdent is to ensure that no TCEP is used by more than one connection at the same moment. This requires, for TCIdent, the ability to determine when an attempt to establish a connection through a TCEP begins, what is easy, and when a busy TCEP is released by the failure of the attempt to connect or by the closure of the connection, what is much more difficult to determine because of the various scenari, which also depend on the type of the connection (simplex or duplex), that can lead to a closure. TCIdent thus needs a way to recall, for every busy TCEP, all the events that occurred through it and that could be relevant to its liberation. This is done by the use of the next coming data types.

HistoryElement defines different values of sort HistElem, that will serve to record the relevant events occuring on a TCEP:

- Cb indicates a duplex connection: both directions of transmission must be closed,
- Ch indicates a simplex connection with just the ability to receive data: only this direction needs to be closed,
- Cm indicates a simplex connection with just the ability to send data: only this direction needs to be closed,
- Ri indicates that the direction of reception is closed,
- Rcf indicates that the direction of emission is closed.   *)

```
type HistoryElement is NaturalNumber, Boolean
sorts HistElem
opns
    Nul, Cb, Ch,
    Cm, Ri, Rcf    : -> HistElem
    h              : HistElem              -> Nat
    _eq_, _ne_     : HistElem, HistElem  -> Bool
eqns
forall he,he1 : HistElem

ofsort Nat
h(Nul) = 0                  ;
h(Cb)  = Succ(0)       ;
h(Ch)  = Succ(h(Cb))   ;
h(Cm)  = Succ(h(Ch))   ;
h(Ri)  = Succ(h(Cm))   ;
h(Rcf) = Succ(h(Ri))   ;

ofsort Bool
he eq he1 = h(he) eq h(he1)  ;
he ne he1 = h(he) ne h(he1)  ;
```

```
endtype (* HistoryElement *)
(*
```
BasicHistoryRecord defines a set of sort HistRec where all the history elements
relative to a same TCEP are recorded, and HistoryRecord improves it with the operator IsFinished that determines, according to the content of an HistoryRecord, whether
the related TCEP is free again.   *)

```
type BasicHistoryRecord is Set actualizedby HistoryElement
using
sortnames  HistElem        for Element
           Bool            for FBool
           HistRec         for Set
endtype (* BasicHistoryRecord *)

type HistoryRecord
is BasicHistoryRecord
opns    IsFinished : HistRec -> Bool
eqns
forall hr : HistRec

ofsort Bool
IsFinished(hr)    = ((Cb IsIn hr) and (Ri IsIn hr)
    and (Rcf IsIn hr)) or ((Ch IsIn hr) and (Ri IsIn hr))
    or ((Cm IsIn hr) and (Rcf IsIn hr))  ;
endtype (* HistoryRecord *)
(*
```
TIdHistory creates a pair that links the identifier of a TCEP to its HistoryRecord. *)

```
type TIdHistory is Pair actualizedby TCEIdentification,
HistoryRecord
using
sortnames  TCId        for Element
           HistRec     for OtherElement
           Bool        for FBool
           TIdHist     for Pair
opnnames   TIdHist     for Pair
           TId             for First
           HistRec     for Second
endtype (* TIdHistory *)
(*
```
TIdHistoryBasicRecord defines the set, of sort TIdHistRec, that TCIdent uses to
record all the TidHistories of all the 'active' TCEPs and TIdHistoryRecord improves it
with some operators TCIdent uses to manage this set:

- Add determines whether a TIdHistory for the TCEP whose identifier it has received
  as first argument, already exists in the TIdHistoryRecord it has received as third
  argument. If yes, it inserts in the HistoryRecord of this TIdHistory the
  HistoryElement it has received as second argument. If not, it inserts in the
  TIdHistoryRecord a new TIdHistory made of its first two arguments.
- Rem removes from a TIdHistory it receives as second argument, the TIdHistory
  related to the TCEP whose identifier it receives as first argument.
- IsTIdIn determines whether a TIdHistory exists in the TIdHistoryRecord it receives
  as second argument, that is relative to the TCEP whose identifier it receives as first
  argument.

- IsFinishedIn determines, according to what is recorded in the TIdHistoryRecord it receives as second argument, whether the TCEP whose identifier it receives as first argument is free again.
- HistOf returns the HistoryRecord from the TIdHistory in the TIdHistoryRecord it receives as second argument, that corresponds to the TCEP whose identifier it receives as first argument.   *)

```
type TIdHistoryBasicRecord is Set actualizedby TIdHistory
using
sortnames  TIdHist     for Element
           Bool        for FBool
           TIdHistRec  for Set
opnnames   {{}}        for {}
endtype (* TIdHistoryBasicRecord *)

type TIdHistoryRecord is TIdHistoryBasicRecord
opns
    Add              : TCId, HistElem, TIdHistRec  -> TIdHistRec
    Rem              : TCId, TIdHistRec            -> TIdHistRec
    _IsTIdIn_         : TCId, TIdHistRec       -> Bool
    _IsFinishedIn_    : TCId, TIdHistRec        -> Bool
    HistOf       : TCId, TIdHistRec            -> HistRec
eqns
forall he : HistElem, ti : TCId, ih : TIdHist, hr : TIdHistRec
ofsort TIdHistRec
    not(ti IsTIdIn hr) =>
Add(ti,he,hr)       = Insert(TIdHist(ti,Insert(he,{})),hr)   ;
    ti IsTIdIn hr =>
Add(ti,he,hr)       = Insert(TIdHist(ti,Insert(he,
                          HistOf(ti,hr))),Rem(ti,hr))         ;
Rem(ti,{{}})        = {{}}                                    ;
    ti eq TId(ih) =>
Rem(ti,Insert(ih,hr))    = Rem(ti,hr)                         ;
    ti ne TId(ih) =>
Rem(ti,Insert(ih,hr))    = Insert(ih,Rem(ti,hr))             ;
ofsort Bool
ti IsTIdIn hr            = HistOf(ti,hr) ne {}               ;
ti IsFinishedIn hr      = IsFinished(HistOf(ti,hr))          ;
ofsort HistRec
HistOf(ti,{{}})         = {}                                  ;
    ti eq TId(ih) =>
HistOf(ti,Insert(ih,hr))       = HistRec(ih)
    ;
    ti ne TId(ih) =>
HistOf(ti,Insert(ih,hr))       = HistOf(ti,hr)
    ;
endtype (* TIdHistoryRecord *)

process TCIdentif [t] : noexit :=
TCIdent [t] ({{}} of TIdHistRec)
endproc (* TCIdentif *)
(*
```

Using these data types, TCIdent, whose sole argument is the TIdHistoryRecord, watches all the primitives, and according to their nature and to the TCEP where they happen, completes the TIdHistoryRecord, by adding new HistoryElements to TIdHistories that already exist, or by creating new TIdHistories, or by removing TIdHistories when it notices the TCEP is free again.

The constraint imposed to the occurrence of event t is that TCONNECTrequest and TCONNECTindication primitives can only happen at a free TCEP, i.e. a TCEP that has no TIdHistory.   *)

```
process TCIdent [t] (Use:TIdHistRec) : noexit :=
t ?ta:TAddress ?tcei:TCEI ?tsp:TSP
[IsTCON1(tsp) implies not(TId(ta,tcei) IsTIdIn Use)] ;
( let ti:TCId = TId(ta,tcei) in
    ( [IsTDISInOut(tsp)] -> TCIdent [t] (Rem(ti,Use))
    []
    [IsTCON1(tsp)] ->
        ([both eq PlexOf(tsp)] ->
            TCIdent [t] (Add(ti,Cb,Use))
         []
         [him eq PlexOf(tsp)] ->
            TCIdent [t] (Add(ti,Ch,Use))
         []
         [me eq PlexOf(tsp)] ->
            TCIdent [t] (Add(ti,Cm,Use))
        )
    []
    [IsTRELind(tsp) or IsTDISIn(tsp)] ->
       ([ti IsFinishedIn Add(ti,Ri,Use)] ->
          TCIdent [t] (Rem(ti,Use))
        []
        [not(ti IsFinishedIn Add(ti,Ri,Use))] ->
          TCIdent [t] (Add(ti,Ri,Use))
       )
         []
         [IsTRELconf(tsp) or IsTDISOut(tsp)] ->
               ([ti IsFinishedIn Add(ti,Rcf,Use)] ->
                    TCIdent [t] (Rem(ti,Use))
                []
                [not(ti IsFinishedIn Add(ti,Rcf,Use))] ->
                    TCIdent [t] (Add(ti,Rcf,Use))
               )
    []
    [IsTDT(tsp) or IsTCON2(tsp) or IsTRELreq(tsp)
     or IsTREN(tsp) or IsTREPind(tsp) or IsTOOBACK(tsp)] ->
    TCIdent [t] (Use)
    )
)
endproc (* TCIdent *)
(*
```

5.5 Verification of the TCEI

According to what is explained in clause 2, this process verifies that the structure of a TCEI is the one required for the primitive passed through it. This is done by the use of the test NeedsOOBTCEI (see 3.4.3.4)   *)

```
process TCEIVerif [t] : noexit :=
```

```
t ?ta:TAddress ?tcei:TCEI ?tsp:TSP [IsOOBTCEI(tcei) iff
NeedsOOBTCEI(tsp)] ; TCEIVerif [t]
endproc (* TCEIVerif *)
(*
```

6 The PP-ERCLTS

As TConnections, TReqResps presents a potentially infinite number of independent occurrences of its transmission mode, each one described by process TReqResp. TReqResp has a structure similar to the one of TConnection: after a first primitive (TREQUESTRESPONSErequest) TReqResp is split in four concurrent processes, two of them describing constraints local to each entity (LocalReq and LocalResp) and the other two describing end to end constraints, one for each direction (ReqtoResp and ResptoReq).

LocalReq corresponds to the calling entity. It expresses that this entity expects one answer to its request: either a confirmation, or a TREJECTREQUESTRESPONSEindication emitted by the service provider.

LocalResp corresponds to the called entity. This entity should receive a TREQUESTRESPONSEindication and thus answer by a response, except if, in the mean time, the service provider has emitted a TREJECTRESPONSEindication. It is also possible that no indication ever arrives to the called entity. At the calling side, this is signalled by a reject primitive, but nothing happens at the called side. This explains the opportunity left to this side of an immediate exit.

ReqtoResp describes the link from the calling to the called entity. It receives as parameter the TREQUESTRESPONSErequest primitive. At the called side, it can emit the corresponding indication and/or a reject indication.

NOTE - Although the combination TREJECTREQUESTRESPONSEindication followed by TREQUESTRESPONSEindication seems to be allowed by this process, it will never occur as it is not accepted by the interacting LocalResp process.

ResptoReq describes the link from the called to the calling entity. It expects a TREQUESTRESPONSEresponse from the called entity, in order to transmit the corresponding confirmation to the calling side. However, it can also emit at this side a reject primitive.   *)

```
process TReqResps [t] : noexit :=
TReqResp [t] ||| TReqResps [t]
endproc (* ReqResps *)

process TReqResp [t] : noexit :=
t ?ta:TAddress ?taei:TAEI ?tsp:TSP [IsTREQRESPreq(tsp)
andthen (ta eq CallingOf(tsp))] ;
((( LocalReq [t] (ta,taei,tsp) ||| LocalResp [t] )
     ||
   ( ReqtoResp [t] (ta,taei,tsp) ||| ResptoReq [t] (ta,taei) ))
>> stop)
endproc (* ReqResp *)
```

```
process LocalReq [t] (ta:TAddress, taei:TAEI, tsp:TSP) : exit
:=
t !ta !taei ?tcf:TSP [tcf IsValidTCON2For tsp] ; exit
[]
t !ta !taei ?trj:TSP [IsTREJREQRESPind(trj)] ; exit
endproc (* LocalReq *)

process LocalResp [t] : exit :=
t ?ta:TAddress ?taei:TAEI ?tsp:TSP [IsTREQRESPind(tsp)
andthen (ta eq CalledOf(tsp))] ;
( t !ta !taei ?trp:TSP [trp IsValidTCON2For tsp] ; exit
  []
    t !ta !taei ?trj:TSP [IsTREJREQRESPind(trj)] ; exit
)
[]
exit

endproc (* LocalResp *)

process ReqtoResp [t] (ta:TAddress, taei:TAEI, tsp:TSP) :
exit :=
( t ?ta1:TAddress ?taei1:TAEI ?tind:TSP [(tind IsOutputOf tsp)
and (TId(ta,taei) ne TId(ta1,taei1))] ; exit
  []
  exit )
|||
( t ?ta1:TAddress ?taei1:TAEI ?trj:TSP [IsTREJREQRESPind(trj)
and (TId(ta,taei) ne TId(ta1,taei1))] ; exit
  []
  exit )
endproc (* ReqtoResp *)

process ResptoReq [t] (ta:TAddress, taei:TAEI) : exit :=
( t ?ta1:TAddress ?taei1:TAEI ?trp:TSP [IsTREQRESPresp(trp)] ;
  ( t !ta !taei ?tcf:TSP [tcf IsOutputOf trp] ; exit
    []
    exit )
[]
exit )
|||
( t !ta !taei ?trj:TSP [IsTREJREQRESPind(trj)] ; exit
  []
  exit )
endproc (* ResptoReq *)
(*
```

7 The PP-EACLTS

TAckCLs (as usual), provides a potentially infinite number of independent transmissions, represented by TAckCL.

This last process is composed of two interacting constraints TAckCLAssoc and ConLessTransDelControl. As the mechanism of an acknowledged connectionless

transmission is quite simple, one process (TAckCLAssoc) is enough to describe it at
both sides.   *)

```
process TAckCLs [t] : noexit :=
TAckCL [t] ||| TAckCLs [t]
endproc (* TAckCLs *)

process TAckCL [t] : noexit :=
TAckCLAssoc [t] >> stop
endproc (* TAckCL *)

process TAckCLAssoc [t] : exit :=
t ?ta:TAddress ?taei:TAEI ?tsp:TSP
[IsTACKDTreq(tsp) and (ta eq CallingOf(tsp))] ;
(   t ?ta1:TAddress ?tspi:TSP [(tspi IsOutputOf tsp) andthen
                                    (ta1 eq CalledOf(tsp))]   ;
    (   t !ta !taei ?tspc:TSP [tspc IsOutputOf tspi] ; exit
        []
          i; t !ta !taei ?tsprj:TSP [IsTREJACKDTind(tsprj)] ;
exit
    )
    []
    i ; t !ta !taei ?tsprj:TSP [IsTREJACKDTind(tsprj)] ;
        t ?ta1:TAddress ?tspi:TSP [(tspi IsOutputOf tsp)
        andthen (ta1 eq CalledOf(tsp))] ; exit
    []
    i ; t !ta !taei ?tsprj:TSP [IsTREJACKDTind(tsprj)] ; exit
)
endproc (* TAckCLAssoc *)
(*
```

7.1 TAEIs management

The role of TAEIControl for the PP-ERCLTS and the PP-EACLTS is similar to the
one of TCIdentif for the PP-ECOTS: it prevents the fact that a same Transport
Association End-point Identifier (TAEI see 3.3) be used by more than one
acknowledged or requestresponse transmission at a time. It must thus determine when
such transmissions begin or end, and memorise the TAEIs currently used, what is
done with the TAEIRecord.   *)

```
type TAEIRecord is Set actualizedby TAEIdentification using
sortnames  TAId           for Element
           Bool           for FBool
           TAEIRec     for Set
endtype (* TAEIRecord *)

process TAEIControl [t] : noexit :=
TAEIController [t] ({} of TAEIRec)
endproc (* TAEIControl *)

process TAEIController [t] (r:TAEIRec) : noexit :=
t ?ta:TAddress ?taei:TAEI ?tsp:TSP [(IsTREQRESPreq(tsp)
or IsTREQRESPind(tsp) or IsTACKDTreq(tsp))
```

```
and not(TId(ta,taei) IsIn r)] ;
TAEIController [t] (Insert(TId(ta,taei),r))
[]
t ?ta:TAddress ?taei:TAEI ?tsp:TSP [(IsTClREJ(tsp)
or IsTREQRESPresp(tsp) or IsTREQRESPconf(tsp) or
IsTACKDTconf(tsp)) and (TId(ta,taei) IsIn r)] ;
TAEIController [t] (Remove(TId(ta,taei),r))
[]
t ?ta:TAddress ?tsp:TSP ; TAEIController [t] (r)
endproc (* TAEIController *)
(*
```

8 The PP-EUCLTS

The specification of this mode is just a copy of the one given in [ISO 8072], as the mechanism we retained for the connectionless mode is the same.

ConnectionLess relates the indications that may be executed to requests that have occurred. A set is used to describe the history of indications.   *)

```
process TSConLess [t] : noexit :=
ConnectionLess [t] (NoTClReq)
endproc (* TSConLess *)

process ConnectionLess [t] (rh : TClReqHistory) : noexit :=
t ?ta:TAddress ?tsp:TSP [IsTUDTreq(tsp) and
(Length(UserData(tsp)) le MaxTUDTLength)] ;
ConnectionLess [t] (Insert(tsp, rh))
 []
[rh ne NoTClReq] ->
(choice tspr, tspi : TSP
 []
[tspr IsIn rh and (tspi IsOutputOf tspr)] ->
t ?ta:TAddress !tspi [ta eq CallingOf(tspi)]     ;
 (ConnectionLess [t] (Remove(tspr, rh))
  []
  i ; ConnectionLess [t] (rh))
  []
  i ; ConnectionLess [t] (Remove(tspr, rh)))
endproc (* ConnectionLess *)

type BasicTransportServiceConnectionlessReqHistory
is Set actualizedby TransportServicePrimitive using
sortnames   TSP                       for Element
            Bool                   for FBool
            TClReqHistory          for Set
opnnames    NoTClReq         for {}
endtype (* BasicTransportServiceConnectionlessReqHistory *)

type MaxTUDTLength
is DecNatRepr
opns
    MaxTUDTLength : -> Nat
eqns
```

```
ofsort Nat
MaxTUDTLength = NatNum(6 + (3 + (4 + (8 + Dec(8)))))    ;
endtype (* MaxTUDTLength *)

endspec (* EnhancedTransportService *)
```

Springer-Verlag and the Environment

MIX
Papier aus verantwortungsvollen Quellen
Paper from responsible sources
FSC® C105338

If you have any concerns about our products,
you can contact us on
ProductSafety@springernature.com

In case Publisher is established outside the EU,
the EU authorized representative is:
Springer Nature Customer Service Center GmbH
Europaplatz 3, 69115 Heidelberg, Germany

Printed by Libri Plureos GmbH
in Hamburg, Germany